T0331807

NEUTRINO COSMOLOGY

The role that neutrinos have played in the evolution of the Universe is one of the most fascinating research areas that has stemmed from the interplays between cosmology, astrophysics and particle physics. In this self-contained book, the authors bring together all aspects of the role of neutrinos in cosmology, spanning from leptogenesis to primordial nucleosynthesis, and from their role in CMB and structure formation to the problem of their direct detection.

The book starts by guiding the reader through aspects of fundamental neutrino physics, the standard cosmological model and statistical mechanics in the expanding Universe, before discussing the history of neutrinos in chronological order from the very early stages until today. This timely book will interest graduate students and researchers in astrophysics, cosmology and particle physics, who work with either a theoretical or experimental focus.

JULIEN LESGOURGUES currently works at EPFL in Lausanne and in the Theory Division at CERN, Switzerland. His research is focused on cosmology and on the theoretical interpretation of cosmic microwave background and large scale structure astrophysical data.

GIANPIERO MANGANO is a Senior Researcher in Theoretical Physics at INFN, Italy. His research interests cover several areas in cosmology, such as primordial nucleosynthesis and the physics of neutrinos in the early Universe, and cosmological aspects of noncommutative spacetimes at the Planck scale.

GENNARO MIELE is Associate Professor in Theoretical Physics at the University of Naples 'Federico II', Italy. His main research interest concerns cosmology, where he has been working on primordial nucleosynthesis and neutrino cosmology.

SERGIO PASTOR is a Researcher in the Instituto de Física Corpuscular, CSIC – Universitat de València, Spain. His main research interests include neutrino physics, in particular the study of their role in astrophysical and cosmological scenarios, and other topics in astroparticle physics.

NEUTRINO COSMOLOGY

JULIEN LESGOURGUES

École Polytechnique Fédérale de Lausanne (EPFL), Switzerland
CERN, Switzerland
CNRS – Université de Savoie, France

GIANPIERO MANGANO

Istituto Nazionale di Fisica Nucleare (INFN), Italy

GENNARO MIELE

University of Naples Federico II, Italy

SERGIO PASTOR

IFIC, CSIC – Universitat de València, Spain

CAMBRIDGE
UNIVERSITY PRESS

CAMBRIDGE
UNIVERSITY PRESS

University Printing House, Cambridge CB2 8BS, United Kingdom

Published in the United States of America by Cambridge University Press, New York

Cambridge University Press is part of the University of Cambridge.

It furthers the University's mission by disseminating knowledge in the pursuit of
education, learning and research at the highest international levels of excellence.

www.cambridge.org
Information on this title: www.cambridge.org/9781107013957

© J. Lesgourgues, G. Mangano, G. Miele, S. Pastor 2013

This publication is in copyright. Subject to statutory exception
and to the provisions of relevant collective licensing agreements,
no reproduction of any part may take place without the written
permission of Cambridge University Press.

First published 2013

A catalogue record for this publication is available from the British Library

Library of Congress Cataloguing in Publication data
Lesgourgues, Julien.
Neutrino cosmology / Julien Lesgourgues, École Polytechnique Federale de Lausanne (EPFL), Switzerland,
CERN, Switzerland, CNRS – Universite de Savoie, France, Gianpiero Mangano, Istituto Nazionale di Fisica
Nucleare (INFN), Italy, Gennaro Miele, University of Naples "Frederico II", Italy, Sergio Pastor, CSIC –
Universitat de Valencia, Spain.
pages cm
Includes bibliographical references and index.
ISBN 978-1-107-01395-7 (hardback)
1. Neutrinos. 2. Neutrino astrophysics. 3. Cosmology. I. Mangano, Gianpiero. II. Miele, Gennaro.
III. Pastor, Sergio. IV. Title.
QC793.5.N42L47 2013
539.7′215 – dc23 2012042364

ISBN 978-1-107-01395-7 Hardback

Cambridge University Press has no responsibility for the persistence or accuracy of
URLs for external or third-party internet websites referred to in this publication,
and does not guarantee that any content on such websites is, or will remain, accurate
or appropriate.

Contents

Preface   *page* ix

1   The basics of neutrino physics   1
   1.1   The electroweak Standard Model   2
   1.2   Spontaneous symmetry breaking and fermion masses   5
   1.3   The basic properties of neutrinos: interactions, masses
       and oscillations   9
       1.3.1   Neutrino interactions in the low energy limit   9
       1.3.2   Dirac and Majorana masses   16
       1.3.3   The seesaw mechanism   22
       1.3.4   Flavour oscillations in vacuum   25
       1.3.5   Flavour oscillations in matter   30
   1.4   Neutrino experiments   35
       1.4.1   Oscillation experiments and three-neutrino mixing   35
       1.4.2   Oscillation experiments and sterile neutrinos   40
       1.4.3   Neutrino mass scale experiments   42
       1.4.4   Dirac or Majorana? Neutrinoless double-β decay   46
   1.5   Nonstandard neutrino–electron interactions   50

2   Overview of the Standard Cosmological Model   53
   2.1   The homogeneous and isotropic universe   55
       2.1.1   The dynamics of expansion   55
       2.1.2   Distances in the universe   65
   2.2   Statistical mechanics in the expanding universe   70
       2.2.1   The relativistic Boltzmann equation   70
       2.2.2   When equilibrium holds   80
   2.3   The expansion stages   83
       2.3.1   Inflation   83

	2.3.2	Radiation and matter domination	87
	2.3.3	Λ (or dark energy) domination	92
2.4	A first look at photon and neutrino backgrounds		95
	2.4.1	Photon decoupling and the formation of the cosmic microwave background	95
	2.4.2	The cosmic neutrino background	98

3   Neutrinos in the early ages   **106**
3.1	The baryon number of the universe		107
3.2	Sakharov conditions		109
3.3	C, CP, B, out of equilibrium and all that		112
	3.3.1	C and CP violation	112
	3.3.2	Baryon and lepton number violation	113
	3.3.3	Relating baryon and lepton numbers	119
	3.3.4	The out-of-equilibrium decay scenario	121
3.4	Basics of leptogenesis		125
	3.4.1	Standard leptogenesis and Majorana neutrinos	126
	3.4.2	Leptogenesis and neutrino oscillation: Two right-handed neutrinos	131

4   Neutrinos in the MeV age   **134**
4.1	Neutrino decoupling		135
4.2	Neutrino oscillations in the expanding universe		143
	4.2.1	Effective matter potentials	143
	4.2.2	Density matrix formalism	145
	4.2.3	Flavour oscillations and relic neutrino distortions	152
	4.2.4	Flavour oscillations and relic neutrino asymmetries	154
	4.2.5	Active–sterile oscillations	160
4.3	Big Bang nucleosynthesis		166
	4.3.1	Neutron–proton chemical equilibrium	170
	4.3.2	The nuclear network	173
	4.3.3	Light-element observations	176
	4.3.4	Theory vs. data	180
4.4	Bounds on neutrino properties from Big Bang nucleosynthesis		181
	4.4.1	Extra relativistic degrees of freedom	183
	4.4.2	Relic neutrino asymmetries	185
	4.4.3	Nonstandard neutrino electromagnetic properties and interactions	189
	4.4.4	Sterile neutrinos and Big Bang nucleosynthesis	193

5   Neutrinos in the cosmic microwave background epoch   198
   5.1   Cosmic microwave background anisotropies   199
       5.1.1   Overview   199
       5.1.2   Perturbation equations   201
       5.1.3   Adiabatic and isocurvature modes   208
       5.1.4   Power spectra and transfer functions   211
       5.1.5   Acoustic oscillations   213
       5.1.6   Temperature anisotropies   220
       5.1.7   Polarization anisotropies   233
       5.1.8   Tensor perturbations   234
   5.2   Neutrino perturbations   236
       5.2.1   Perturbation equations   236
       5.2.2   Neutrino isocurvature modes   240
       5.2.3   Adiabatic mode in the presence of neutrinos   242
       5.2.4   Free-streaming length   244
       5.2.5   Linear evolution of neutrino perturbations   248
       5.2.6   Practical implementation and approximations   249
   5.3   Effects of neutrinos on primary cosmic microwave background anisotropies   253
       5.3.1   How can decoupled species affect the cosmic microwave background?   253
       5.3.2   Effects of massless neutrinos   255
       5.3.3   Effects of massive neutrinos   262
       5.3.4   Effects of interacting neutrinos   266
   5.4   Bounds on neutrinos from primary cosmic-microwave-background anisotropies   267
       5.4.1   Cosmic microwave background and homogeneous cosmology data sets   267
       5.4.2   Neutrino abundance   268
       5.4.3   Neutrino masses   271

6   Recent times: neutrinos and structure formation   273
   6.1   Linear matter power spectrum   274
       6.1.1   Neutrinoless universe with cold dark matter   275
       6.1.2   Neutrinoless universe with cold dark matter and baryons   284
       6.1.3   Impact of massless neutrinos   290
       6.1.4   Impact of hot dark matter   293
       6.1.5   Impact of warm dark matter   312

6.2	Nonlinear matter power spectrum	317
	6.2.1 *N*-body simulations	317
	6.2.2 Analytic approaches	323
6.3	Impact of neutrinos on secondary cosmic microwave background anisotropies	324
	6.3.1 Late integrated Sachs–Wolfe effect	324
	6.3.2 Cosmic microwave background lensing	327
6.4	Observing the large-scale structure	328
	6.4.1 Galaxy and cluster power spectrum	329
	6.4.2 Cluster mass function	331
	6.4.3 Galaxy weak lensing	332
	6.4.4 Cosmic microwave background lensing	334
	6.4.5 Lyman alpha forests	334
	6.4.6 21-cm surveys	335
6.5	Large-scale structure bounds on neutrino properties	335
	6.5.1 Active neutrino masses	335
	6.5.2 Neutrino abundance and light sterile neutrinos	339
	6.5.3 Nonstandard properties of active neutrinos	342
	6.5.4 Heavy sterile neutrinos (warm dark matter)	344
7	Cosmological neutrinos today	348
7.1	The ultimate dream: detecting cosmological neutrinos	349
	7.1.1 Scatterings: G_F^2 effects are too small	349
	7.1.2 The order G_F interactions and the Stodolsky effect	350
	7.1.3 Massive neutrinos and β-decaying nuclei	354
7.2	Beyond the ultimate dream: neutrino anisotropies in the sky	359
	7.2.1 Neutrino last scattering surface	359
	7.2.2 Massless neutrinos	359
	7.2.3 Massive neutrinos	360
	References	362
	Index	375

Preface

To Arianna, Carmen, Isabelle, and María José

When neutrinos first came on the scene in 1930, their father, Wolfgang Pauli, confessed to his colleague, the astronomer Walter Baade, that to save energy conservation in β-decays (quoted in Hoyle, 1967),

I have done a terrible thing today, something which no theoretical physicist should ever do. I have suggested something that can never be verified experimentally.

This was perhaps the only time Pauli was mistaken. Less than 30 years later, neutrinos were discovered by Reines and Cowan.

Since then, we have learned so many things about neutrinos that Pauli himself would be very surprised. More than this, understanding neutrino properties has always brought new insights into the whole field of fundamental interactions, and new theoretical paradigms.

Today we know quite accurately how to describe their feeble interactions with matter, from the very first attempts of Fermi to the succesful Standard Model of electroweak interactions. Many pieces of information have been collected in laboratory experiments, the traditional setting of particle physics. The study of neutrino interactions has been pursued at accelerators and reactors and, more recently, by sending neutrino beams produced at accelerators to underground laboratories. Accelerator experiments have also confirmed that there are only three generations of light neutrinos which are weakly interacting.

The main breakthrough in neutrino physics over the last few decades came from a different environment: astrophysics. The solar neutrino problem – an observed deficit of neutrino flux from the sun – along with the atmospheric neutrino anomaly, has led to the discovery that neutrinos are massive particles. We do not understand their mass spectrum yet, nor why they are such light particles. On the experimental side, remarkable improvements are expected in the next few years, both in measuring the neutrino mass scale using tritium β-decay, and in understanding the real *nature* of neutrinos as Dirac or Majorana particles. At the same time, intense theoretical activity is going on, addressing the neutrino mass problem, which is seen as a possible clue to unveiling the behaviour of fundamental interactions at high energy scales.

The most spectacular property of neutrinos is deeply rooted in quantum mechanics. As first suggested by Bruno Pontecorvo, neutrinos exhibit oscillations among different flavours during their time evolution, leading to an incredibly rich phenomenology. The parameters characterizing this oscillatory behaviour are currently quite well known, because of the interplay of a variety of different experimental techniques and neutrino sources of both terrestrial and astrophysical origin. Just at the time of writing this book, the last missing piece of the puzzle, one of the neutrino mixing angles, was measured with good precision. No doubt neutrino oscillation physics will represent a leading research line in the coming years.

Since the early works on the synthesis of light nuclei in the Big Bang model in the late forties, it was clear that neutrinos are not simply passive spectators during the expansion of the universe. Through their weak interactions with other particles, as well as their *gravity*, they influence a wide variety of phenomena which took place in the early stages of the life of the universe, till very recent epochs. This means that using observations of astrophysical quantities, related to specific phases of the expansion history, we have a further way to constrain neutrino properties at different energy, time and length scales, which in some cases are not accessible to laboratory experiments. Just to give a few examples, some constraints on the number of light weakly interacting neutrinos and on their mass scale were obtained using observations of primordial ^{4}He and of the total energy density of the universe well before laboratory experiments could provide comparable information.

This vast arena in which to test neutrino properties is usually referred to as *neutrino cosmology*, and that is what this book is about. By its very nature, it is a multidisciplinary reasearch field, where the different expertises and backgrounds of theoretical and experimental particle physicists, astrophysicists and cosmologists find a meeting point and a common language. It is a branch of an even broader scientific activity, commonly called *astroparticle physics*, aimed at understanding fundamental interactions by exploiting observations of very large objects, such as astrophysical sources, or the universe as a whole.

In the last two decades, we have witnessed a big boost along this research line, due to outstanding improvements in the number and quality of astrophysical observations. Large galaxy surveys, detailed maps of the cosmic microwave background, observations of primordial nuclear abundances and new ways to trace the expansion history of the universe are just a few examples of this experimental effort. Whereas only 20 years ago neutrino cosmology was in its infancy, and theoretical physicists were typically satisfied by order-of-magnitude calculations made on the back of the envelope, the situation has changed rapidly since then. Observations currently require a much more detailed analysis, and provide several new tests of theoretical models.

This book is a summary of the history of the universe from a neutrino perspective. The first two chapters introduce the three important theoretical tools that will be widely used in the following: the basics of neutrino interactions and properties in the framework of the Standard Model of particle physics, the homogeneous and isotropic cosmological model, and some concepts of kinetic theory. We have done our best to present a pedagogical and self-contained discussion of these topics, and we are not sure that we have succeeded in this respect. Indeed, the subject of this book is intrinsically multidisciplinary, and covering all topics in a detailed and self-consistent manner – while keeping the number of pages reasonable – was a major challenge. Readers who are not familiar with, say, quantum field theory, general relativity, gauge issues in cosmology or the theory of inflation will need further reading in more specialized books or reviews. In any case, we tried at least to introduce all the concepts that are necessary for understanding the remaining chapters. We hope that this part may also trigger the reader's interest in further studying the topics he or she might be unfamiliar with. To this end, we give a long list of possible references.

The remaining chapters are devoted to different aspects of the role of neutrinos in cosmology, in *chronological order*, as they intervene in the evolution of the universe, from the very early stages till today. Chapter 3 addresses the issue of baryogenesis, the dynamical production of the baryon asymmetry observed in our universe, and in particular a scenario deeply related to neutrino properties, called leptogenesis. Chapter 4 deals with the dynamics of neutrino oscillations in a cosmological setting, and also with primordial nucleosynthesis, one of the main pillars of the cosmological model, providing a lot of information about neutrino physics. Chapter 5 explains the properties of cosmic microwave background anisotropies, which contain a huge quantity of information about the whole history of the universe, and shows how they are impacted by neutrinos. Chapter 6 describes the dynamics of structure formation on very large scales – those of galaxies, clusters, etc. – which is crucially affected by the abundance, mass and properties of neutrinos. Finally, Chapter 7 presents a summary of the methods which have been proposed so far to detect the relic neutrino background in the laboratory, and a brief discussion of the anisotropies that such detectors would see if they could ever become operational. As we will see, this is a very challenging task, the ultimate dream of a *neutrino cosmologist*.

In this book we will adopt the signature $(-+++)$, except in Chapters 1 and 4, which are more particle physics oriented, where we adopt the more widely used $(+---)$. Unless otherwise mentioned, we use natural units.

If we look into some of the available Web archives for scientific papers related to both neutrinos and cosmology, the query will return a number of publications of the order of several thousands. This gives an idea of how intense the activity is in this research field. In the following pages, the reader will not find a complete analysis of all possible models and ideas proposed so far. Some alternatives to the mainstream scenarios – sometimes extremely interesting and intriguing – have not been considered in our discussion, and are cited in our (rather long) list of references. We apologize for all omissions. However, in writing this book, our guideline has been to try to present the main physical aspects of the phenomena neutrinos are involved in, rather than to go through all their possible variations. In a sense, what we have mainly considered is a *standard* neutrino cosmology, describing what is currently well established on solid theoretical and experimental bases. We hope this might be helpful for students and researchers who are interested in approaching this fascinating research field, starting from different cultural backgrounds. If this ambitious goal is achieved even partially, we will be happy with our contribution to a process that is well on the way, namely, the emergence of a homogeneous community of theoretical and experimental particle physicists, cosmologists and astrophysicists.

This book is the result of the authors' friendship over many years. However, it would not have been written were not for enlightening discussions and collaborations with many of our colleagues. Several topics that the reader hopefully will find interesting in the following pages are the outcome of their work and enthusiasm, and of their sharing with us their knowledge and experience.

We warmly thank Benjamin Audren, Steve Blanchet, Diego Blas, Alexei Boyarski, Marco Cirelli, Gaëlle Giesen, Martin Hirsch, Michal Malinský, Oleg Ruchayskiy, Pasquale Serpico, Mikhail Shaposhnikov and Mariam Tórtola for reading a draft version of this book.

We are also very much indebted to Alfredo Cocco, Alexander Dolgov, Salvatore Esposito, Giuliana Fiorillo, Jan Hamann, Steen Hannestad, Steen Hansen, Fabio Iocco, Alessandro Melchiorri, Marcello Messina, Marco Peloso, Serguey Petcov, Massimo Pietroni, Ofelia Pisanti, Georg Raffelt, Antonio Riotto, Thomas Tram, José W.F. Valle, Matteo Viel and Yvonne Wong.

Matteo Viel and his collaborators Martin G. Haehnelt and Volker Springel deserve special thanks for allowing us to use their beautiful N-body simulations on the cover of this book.

We are also pleased to acknowledge the Cambridge University Press staff for their help and continuous support.

Our families have given the strongest support in this adventure. This book was written by many hands: those of Aitana, Apolline, Arianna, Carmen, Constance, Davide, Diane, Héctor, Isabelle, María José and Matteo.

1

The basics of neutrino physics

Like the actors in ancient Greek tragedy and comedy, neutrinos play more than one role in the drama of the expanding universe. They couple to gravity and contribute to Einstein equations which rule the expansion dynamics. Furthermore, they interact in the primordial plasma with charged leptons and hadrons via electroweak interactions, until the rates for these processes become so low compared with the typical expansion rate that they *decouple* and start to propagate freely along geodesic lines. Any quantitative description of their role in cosmology thus requires several inputs from the theory of fundamental interactions, as well as a knowledge of their basic properties, such as masses and, in some cases, the features of neutrino flavour oscillations.

Neutrino interactions have been well understood since the first theory of β-decay proposed by Enrico Fermi in 1934, and now are succesfully and beautifully described by the unified picture of electroweak interactions. In the low energy limit the strength of these interactions is encoded in a single coupling, the Fermi coupling constant G_F, whose value, combined with the Newton constant, fixes the time of neutrino decoupling. From the strong experimental evidence in favour of neutrino oscillation, we also know that neutrinos are massive particles, and this, as we will see at length in the following, has a strong impact on how structures, i.e., inhomogeneities in the universe, grow on certain length scales.

As a viaticum for this journey in the land of neutrino cosmology, it seemed worth-while to the authors to provide the reader with certain minimal information on the basic properties of neutrinos, both at the level of their theoretical formulation and from the experimental point of view.

Unfortunately, to keep a self-contained summary of these topics reasonably short requires the reader to be acquainted with the basics of quantum field theory and of the gauge principle, which are treated in full detail in many excellent textbooks (e.g., Itzykson and Zuber, 1980; Halzen and Martin, 1984; Weinberg, 1995; Peskin and Schroeder, 1995). In case he or she is familiar with neutrino physics, it is then possible to skip this chapter, though it might be useful to go through it anyway to become familiar with our notation. For all other readers, the following sections can represent only a too-brief synthesis of the present understanding of neutrino properties, hopefully sufficient for them to comfortably read the rest of this book, and likewise hopefully to trigger their curiosity for a deeper understanding of neutrino physics.

Here is a summary of this introductory chapter. After a short review of the Standard Model of fundamental interactions, which covers only the details of its electroweak sector, we describe the main observable properties of neutrinos – interaction processes, Dirac and Majorana masses and flavour oscillations – including a summary of bounds on a certain class of exotic interactions which are beyond our present understanding of microscopic physics but are typically predicted by extensions of the Standard Model which represent its *ultraviolet completion*. We then give a résumé of experimental results on flavour oscillation experiments, laboratory neutrino mass bounds and neutrinoless double-β decay, the last being an *experimentum crucis* to test their Dirac or Majorana nature.

1.1 The electroweak Standard Model

Gauge symmetry has proven to be a powerful guideline to building up a satisfactory theory of fundamental interactions. Strong and electroweak interactions are described by a relativistic quantum field theory based on the gauge symmetry principle for the group $SU(3)_C \times SU(2)_L \times U(1)_Y$, where C, L and Y denote *colour*, *left-handed chirality* and *weak hypercharge* (Glashow, 1961; Weinberg, 1972; Salam, 1968; Fritzsch *et al.*, 1973; Gross and Wilczek, 1973; Politzer, 1973; Weinberg, 1973). The model is so successful that it is now usually referred to as the 'Standard Model' (SM) of elementary particles.

Whereas the strong sector $SU(3)_C$ symmetry remains unbroken, and hence is an exact symmetry at any energy level, the electroweak forces undergo spontaneous symmetry breaking via the Higgs mechanism, which reduces the symmetry of the model at low energies to $SU(2)_L \times U(1)_Y \to U(1)_Q$, with Q being the electric charge. In the following we will focus our attention on the electroweak sector only, because neutrinos, like all leptons, do not carry colour charge and are not strongly interacting.

The requirement that a field theory is gauge-invariant under a particular symmetry group strictly fixes the form of the interaction and the number of gauge bosons. Unfortunately, it leaves quite a high level of arbitrariness in the choice of the irreducible representations (IR) of the gauge group to accommodate fermions and the Higgs scalar bosons. The only constraint is provided by the cancellation of

Table 1.1 *Elementary fermions in the SM*

Generation	1st	2nd	3rd
quarks	u	c	t
	d	s	b
leptons	ν_e	ν_μ	ν_τ
	e^-	μ^-	τ^-

Table 1.2 *Electroweak quantum numbers of fermions in the SM*

Fermion IRs under $SU(2)_L \times U(1)_Y$			I	I_3	Y	Q
$L_{eL} \equiv \begin{pmatrix} \nu_{eL} \\ e_L \end{pmatrix}$	$L_{\mu L} \equiv \begin{pmatrix} \nu_{\mu L} \\ \mu_L \end{pmatrix}$	$L_{\tau L} \equiv \begin{pmatrix} \nu_{\tau L} \\ \tau_L \end{pmatrix}$	$1/2$	$1/2$ $-1/2$	-1	0 -1
$l_{eR} \equiv e_R$	$l_{\mu R} \equiv \mu_R$	$l_{\tau R} \equiv \tau_R$	0	0	-2	-1
$Q_{1L} \equiv \begin{pmatrix} u_L \\ d_L \end{pmatrix}$	$Q_{2L} \equiv \begin{pmatrix} c_L \\ s_L \end{pmatrix}$	$Q_{3L} \equiv \begin{pmatrix} t_L \\ b_L \end{pmatrix}$	$1/2$	$1/2$ $-1/2$	$1/3$	$2/3$ $-1/3$
$q_{uR}^U \equiv u_R$	$q_{cR}^U \equiv c_R$	$q_{tR}^U \equiv t_R$	0	0	$4/3$	$2/3$
$q_{dR}^D \equiv d_R$	$q_{sR}^D \equiv s_R$	$q_{bR}^D \equiv b_R$			$-2/3$	$-1/3$

the chiral anomaly, a condition which must be fulfilled if gauge symmetry should also be respected at the quantum level.

The currently known fermionic elementary particles (spin $s = 1/2$) are split into three generations of *quarks* and *leptons*; see Table 1.1. Each generation of fermions is described by the IRs of the electroweak gauge group as shown in Table 1.2, where we also report their charges.

By I we denote the weak isospin, which is $1/2$ for $SU(2)_L$ doublets and 0 for singlets, respectively, whereas I_3 is its third component. The electric charge Q is given by the Gell-Mann–Nishijima relation $Q = I_3 + Y/2$.

The three-generation electroweak Lagrangian density is

$$\mathcal{L} = i\overline{L'_{\alpha L}}\slashed{D}L'_{\alpha L} + i\,\overline{Q'_{\alpha L}}\slashed{D}Q'_{\alpha L} + i\overline{l'_{\alpha R}}\slashed{D}l'_{\alpha R}$$
$$+ i\overline{q'^{D}_{\alpha R}}\slashed{D}q'^{D}_{\alpha R} + i\overline{q'^{U}_{\alpha R}}\slashed{D}q'^{U}_{\alpha R} - \frac{1}{4}\vec{F}_{\mu\nu}\cdot\vec{F}^{\mu\nu} - \frac{1}{4}B_{\mu\nu}B^{\mu\nu}$$
$$+ \left(D_\rho\Phi\right)^\dagger\left(D^\rho\Phi\right) + \mu^2\Phi^\dagger\Phi - \lambda\left(\Phi^\dagger\Phi\right)^2$$

Table 1.3 *Electroweak quantum numbers of the Higgs doublet*

Higgs doublet	I	I_3	Y	Q
$\Phi(x) \equiv \begin{pmatrix} \phi_+(x) \\ \phi_0(x) \end{pmatrix}$	1/2	1/2 $-1/2$	$+1$	1 0

$$- \left(Y_{\alpha\beta}^{\prime l} \, \overline{L_{\alpha L}'} \, \Phi \, l_{\beta R}' + Y_{\alpha\beta}^{\prime l*} \, \overline{l_{\beta R}'} \, \Phi^\dagger L_{\alpha L}' \right)$$

$$- \left(Y_{\alpha\beta}^{\prime D} \, \overline{Q_{\alpha L}'} \, \Phi \, q_{\beta R}^{\prime D} + Y_{\alpha\beta}^{\prime D*} \, \overline{q_{\beta R}^{\prime D}} \, \Phi^\dagger Q_{\alpha L}' \right)$$

$$- \left(Y_{\alpha\beta}^{\prime U} \, \overline{Q_{\alpha L}'} \, (i\sigma_2 \Phi^*) q_{\beta R}^{\prime U} + Y_{\alpha\beta}^{\prime U*} \, \overline{q_{\beta R}^{\prime U}} \, (-i\Phi^T \sigma_2) \, Q_{\alpha L}' \right), \tag{1.1}$$

where Φ is the Higgs doublet, whose properties are reported in Table 1.3. σ_2 is a Pauli matrix, and α is the generation index. In the following, repeated indices are summed over, unless differently specified. The covariant derivative D_μ is defined as

$$D_\mu \equiv \partial_\mu + ig\vec{A}_\mu \cdot \frac{\vec{\sigma}}{2} + ig'B_\mu \frac{Y}{2}, \tag{1.2}$$

with $\vec{A}^\mu \equiv (A_1^\mu, A_2^\mu, A_3^\mu)$ and B^μ denoting the gauge boson fields of the $SU(2)_L$ and $U(1)_Y$ factors.

The canonical kinetic (and self-interacting for the $SU(2)_L$ factor) term for gauge bosons is written in terms of the electroweak tensors $\vec{F}^{\mu\nu} \equiv (F_1^{\mu\nu}, F_2^{\mu\nu}, F_3^{\mu\nu})$ and $B^{\mu\nu}$, where

$$F_a^{\mu\nu} = \partial^\mu A_a^\nu - \partial^\nu A_a^\mu - g \sum_{b,c=1}^{3} \varepsilon_{abc} A_b^\mu A_c^\nu$$

$$B^{\mu\nu} = \partial^\mu B^\nu - \partial^\nu B^\mu. \tag{1.3}$$

In the expression (1.1), fermion fields are marked by a *prime* to denote that these fields in general are not mass eigenstates, as will be discussed in detail in the next sections. Equation (1.1) contains in the first two lines the kinetic and electroweak interaction terms for leptons and quarks and the pure gauge boson term, whereas the third line accounts for the Higgs sector responsible for the symmetry breaking. Finally, the last three lines correspond to the Yukawa terms characterized by the complex couplings $Y_{\alpha\beta}^{\prime l}$, $Y_{\alpha\beta}^{\prime D}$ and $Y_{\alpha\beta}^{\prime U}$. They are responsible for charged leptons and quark masses and mixing.

We note that in its minimal version, there are no right-handed neutrino states ν_R in the SM, which would be a singlet under all symmetry group factors. This implies that *active* neutrinos ν_L remain massless, because there are no mass terms which appear as a consequence of symmetry breaking, differently from charged leptons and quarks. The extension of the model to massive neutrinos will be discussed in the following.

From the first two lines of Eq. (1.1) one can extract the charged-current and neutral-current weak interaction Lagrangian densities, denoted by $\mathcal{L}_I^{(CC)}$ and $\mathcal{L}_I^{(NC)}$. In particular, one gets

$$\mathcal{L}_I^{(CC)} = -\frac{g}{2\sqrt{2}} J_W^\mu W_\mu + \text{h.c.,} \tag{1.4}$$

where $J_W^\mu = J_{W,L}^\mu + J_{W,Q}^\mu$ and

$$J_{W,L}^\mu = 2\left(\overline{\nu'_{eL}}\gamma^\mu e'_L + \overline{\nu'_{\mu L}}\gamma^\mu \mu'_L + \overline{\nu'_{\tau L}}\gamma^\mu \tau'_L\right)$$

$$J_{W,Q}^\mu = 2\left(\overline{u'_L}\gamma^\mu d'_L + \overline{c'_L}\gamma^\mu s'_L + \overline{t'_L}\gamma^\mu b'_L\right). \tag{1.5}$$

The gauge boson field $W^\mu \equiv (A_1^\mu - iA_2^\mu)/\sqrt{2}$ by definition annihilates a W^+ boson and creates a W^- boson. The neutral current density

$$\mathcal{L}_I^{(NC)} = -\frac{g}{2\cos\theta_W} J_Z^\mu Z_\mu + \text{h.c.,} \tag{1.6}$$

where $J_Z^\mu = J_{Z,L}^\mu + J_{Z,Q}^\mu$ and

$$J_{Z,L}^\mu = 2\left(g_L^\nu \overline{\nu'_{\alpha L}}\gamma^\mu \nu'_{\alpha L} + g_L^l \overline{l'_{\alpha L}}\gamma^\mu l'_{\alpha L} + g_R^l \overline{l'_{\alpha R}}\gamma^\mu l'_{\alpha R}\right)$$

$$J_{Z,Q}^\mu = 2\left(g_L^U \overline{q'^U_{\alpha L}}\gamma^\mu q'^U_{\alpha L} + g_R^U \overline{q'^U_{\alpha R}}\gamma^\mu q'^U_{\alpha R} + g_L^D \overline{q'^D_{\alpha L}}\gamma^\mu q'^D_{\alpha L} + g_R^D \overline{q'^D_{\alpha R}}\gamma^\mu q'^D_{\alpha R}\right). \tag{1.7}$$

The gauge field Z^μ is defined *via* the rotation $Z^\mu = \cos\theta_W A_3^\mu - \sin\theta_W B^\mu$, where $\tan\theta_W = g'/g$ and $e = g\sin\theta_W$. Finally, the couplings $g_L^{\nu,l,U,D}$ and $g_R^{l,U,D}$ are given by the relations

$$g_L^f = I_3^f - Q_f \sin^2\theta_W \tag{1.8}$$

$$g_R^f = -Q_f \sin^2\theta_W, \tag{1.9}$$

which are summarized in Table 1.4.

1.2 Spontaneous symmetry breaking and fermion masses

It is easy to see that a naive mass term such as $\overline{e_L}e_R + \text{h.c.}$ is not allowed in the Lagrangian density (1.1) because it would spoil the symmetry invariance under the gauge group $SU(2)_L \times U(1)_Y$. However, when this group symmetry is broken,

Table 1.4 *Neutral-current couplings for the elementary fermions*

g_L	g_R
$g_L^\nu = \frac{1}{2}$	
$g_L^l = -\frac{1}{2} + \sin^2\theta_W$	$g_R^l = \sin^2\theta_W$
$g_L^U = \frac{1}{2} - \frac{2}{3}\sin^2\theta_W$	$g_R^U = -\frac{2}{3}\sin^2\theta_W$
$g_L^D = -\frac{1}{2} + \frac{1}{3}\sin^2\theta_W$	$g_R^D = \frac{1}{3}\sin^2\theta_W$

masses are produced via the celebrated Higgs–Englert–Brout–Guralnik–Hagen–Kibble mechanism (Englert and Brout, 1964; Guralnik *et al.*, 1964; Higgs, 1964a,b).

The dynamics of the Higgs field Φ is ruled by the term in $\mathcal{L}$

$$\mathcal{L}_H = \left(D_\rho\Phi\right)^\dagger \left(D^\rho\Phi\right) - V(\Phi) = \left(D_\rho\Phi\right)^\dagger \left(D^\rho\Phi\right) + \mu^2\Phi^\dagger\Phi - \lambda\left(\Phi^\dagger\Phi\right)^2. \tag{1.10}$$

In quantum field theory, the minimum of the potential defines the ground state around which the fields are expanded in terms of creation and annihilation operators. Quantum excitations on the ground state correspond to particle states. Note that only neutral fields with vanishing spin (scalar) may have nontrivial ground states; otherwise this would spoil the electric charge conservation and the invariance under spatial rotations. The Higgs field ground state value $\langle\Phi\rangle$, hereafter referred to as the *vacuum expectation value* (vev), can be written in the form

$$\langle\Phi\rangle = \frac{1}{\sqrt{2}}\begin{pmatrix} 0 \\ v \end{pmatrix}, \tag{1.11}$$

where v is a real positive quantity. By substituting $\langle\Phi\rangle$ in $V(\Phi)$ one gets the minimum of the energy density for $v = \sqrt{\mu^2/\lambda}$, and around the minimum, in the *unitary gauge*, the Higgs doublet reads

$$\Phi = \frac{1}{\sqrt{2}}\begin{pmatrix} 0 \\ v + H(x) \end{pmatrix}, \tag{1.12}$$

$H(x)$ being a real scalar field.

The value of v is known, because the first striking effect of symmetry breaking is that three gauge bosons become massive, the charged $W^\pm$ and Z. In particular, $m_W = gv/2 = m_Z\cos\theta_W$. From the experimental value of the Fermi constant G_F – see the next section – we get $v \sim 246$ GeV.

Substituting (1.12) in the Yukawa couplings reported in the last three lines of Eq. (1.1), we see how fermion mass terms are produced after the symmetry breaking. Let us consider, for example, the term of $\mathcal{L}$ coupling leptons with the Higgs field,

$$\mathcal{L}_{H,L} = -Y''_{\alpha\beta} \, \overline{L'_{\alpha L}} \, \Phi \, l'_{\beta R} + \text{h.c.} \tag{1.13}$$

Once the Higgs field is developed around its minimum, one gets

$$\mathcal{L}_{H,L} = -\frac{v + H(x)}{\sqrt{2}} \left(Y''_{\alpha\beta} \, \overline{l'_{\alpha L}} \, l'_{\beta R} + \text{h.c.} \right). \tag{1.14}$$

The term of Eq. (1.14) proportional to the vev provides the mass term for charged leptons, whereas the contribution proportional to $H(x)$ accounts for the trilinear coupling between charged leptons and the scalar boson H. Because the couplings Y'' are generally not diagonal in the three-generations space, one must diagonalize them before interpreting (1.14) as a genuine mass term. A generic complex matrix such as Y'' can be transformed into a diagonal form Y^l through a biunitary transformation

$$V_L^{l\dagger} \, Y'' \, V_R^l = Y^l \qquad Y^l_{\alpha\beta} = y^l_\alpha \, \delta_{\alpha\beta}. \tag{1.15}$$

With the transformed leptonic fields defined as

$$l_R = V_R^{l\dagger} l'_R \quad \text{and} \quad l_L = V_L^{l\dagger} l'_L, \tag{1.16}$$

the term $\mathcal{L}_{H,L}$ can be rewritten as

$$\mathcal{L}_{H,L} = -\frac{v + H(x)}{\sqrt{2}} \sum_\alpha y_{l\alpha} \left(\overline{l_{\alpha L}} \, l_{\alpha R} + \text{h.c.} \right). \tag{1.17}$$

From (1.17) one gets $m_\alpha = y^l_\alpha \, v/\sqrt{2}$ for $\alpha = e, \mu, \tau$. In terms of these masses one can also rewrite the interaction term between charged leptons and $H(x)$ as

$$\mathcal{L}^l_{H,L} = -H(x) \sum_\alpha \frac{m_\alpha}{v} \left(\overline{l_{\alpha L}} \, l_{\alpha R} + \text{h.c.} \right), \tag{1.18}$$

which simply states that a heavier lepton is more strongly coupled to the Higgs field than a lighter one.

When the weak charged current $J^\mu_{W,L}$ is rewritten in terms of mass eigenstates $l_{\alpha L}$,

$$J^\mu_{W,L} = 2 \, \overline{v'_{\alpha L}} \, (V_L)_{\alpha\beta} \, \gamma^\mu \, l_{\beta L}. \tag{1.19}$$

As long as neutrinos do not receive any mass term by the Higgs mechanism, because no right-handed partners ν_R have been introduced so far, we can redefine the neutrino field as $\nu_L = V_L^\dagger \nu_L'$ and get

$$J_{W,L}^\mu = 2 \overline{\nu_{\alpha L}} \, \gamma^\mu \, l_{\alpha L}, \tag{1.20}$$

where by definition $l_{\alpha L} \equiv (e_L^-, \mu_L^-, \tau_L^-)$ and $\nu_{\alpha L} \equiv (\nu_{eL}, \nu_{\mu L}, \nu_{\tau L})$.

Concerning $J_{Z,L}^\mu$ of Eq. (1.7), one can easily see that by virtue of the unitarity of the matrices V_L^l and V_R^l it remains unchanged; hence one can simply replace the *primed* fields with the *unprimed* ones in (1.7).

For the Yukawa terms for quark fields one proceeds in the very same way. In the unitary gauge

$$\mathcal{L}_{H,Q} = -\frac{v + H(x)}{\sqrt{2}} \left(Y_{\alpha\beta}'^D \, \overline{q_{\alpha L}'^D} \, q_{\beta R}'^D + Y_{\alpha\beta}'^U \, \overline{q_{\alpha L}'^U} \, q_{\beta R}'^U + \text{h.c.} \right). \tag{1.21}$$

Thus, by simultaneously diagonalizing the matrices of Yukawa couplings $(Y'^D)_{\alpha\beta}$ and $(Y'^U)_{\alpha\beta}$ via biunitary transformations V_L^D, V_R^D, V_L^U and V_R^U and defining the transformed quark fields as

$$q_R^D = V_R^{D\dagger} q_R'^D, \qquad q_L^D = V_L^{D\dagger} q_L'^D \tag{1.22}$$

$$q_R^U = V_R^{U\dagger} q_R'^U, \qquad q_L^U = V_L^{U\dagger} q_L'^U, \tag{1.23}$$

we can rewrite the mass term in $\mathcal{L}_{H,Q}$ as

$$\mathcal{L}_{H,Q} = - \sum_{\alpha=d,s,b} m_\alpha \left(1 + \frac{H(x)}{v} \right) \overline{q_{\alpha L}^D} \, q_{\alpha R}^D$$

$$- \sum_{\alpha=u,c,t} m_\alpha \left(1 + \frac{H(x)}{v} \right) \overline{q_{\alpha L}^U} \, q_{\alpha R}^U + \text{h.c.} \tag{1.24}$$

In this case, however, as all quarks are massive and have different masses, we have no freedom to arbitrarily rotate D or U quarks in the hadronic weak charged current

$$J_{W,Q}^\mu = 2 \overline{q_{\alpha L}^U} \gamma^\mu \left(V_L^{U\dagger} V_L^D \right)_{\alpha\beta} q_{\beta L}^D. \tag{1.25}$$

The unitary matrix

$$V \equiv V_L^{U\dagger} V_L^D \tag{1.26}$$

is the Cabibbo–Kobayashi–Maskawa (CKM) mixing matrix (Cabibbo, 1963; Kobayashi and Maskawa, 1973), which depends upon three angles θ_{12} (the Cabibbo angle, θ_C, up to very small corrections), θ_{23} and θ_{13} and one phase δ,

$$V = \begin{pmatrix} c_{12}c_{13} & s_{12}c_{13} & s_{13}e^{-i\delta} \\ -s_{12}c_{23} - c_{12}s_{23}s_{13}e^{i\delta} & c_{12}c_{23} - s_{12}s_{23}s_{13}e^{i\delta} & s_{23}c_{13} \\ s_{12}s_{23} - c_{12}c_{23}s_{13}e^{i\delta} & -c_{12}s_{23} - s_{12}c_{23}s_{13}e^{i\delta} & c_{23}c_{13} \end{pmatrix}, \tag{1.27}$$

where $c_{ij} = \cos\theta_{ij}$, $s_{ij} = \sin\theta_{ij}$, and $0 \leq \theta_{ij} \leq \pi/2$. To see that V can always be reduced to this form one has to recall that an arbitrary 3×3 unitary matrix has nine real parameters, but we have to subtract five free parameters connected with the single and independent rephasing of quark fields (the global rephasing still remains a symmetry of the system). In the case of quarks the three mixing angles satisfy a hierarchical structure, $s_{12} = 0.22535 \pm 0.00065$ (Beringer *et al.*, 2012), $s_{23} \sim s_{12}^2$, $s_{13} \sim s_{12}^4$.

As can be proven by studying the CP transformation (charge conjugation and parity) of the SM Lagrangian density written in terms of mass eigenstates, the only possible source of CP violation is encoded in the presence of the complex phase δ in V (1.27). Because CP violation processes have been observed in the hadron phenomenology, this has been ascribed to the presence of a nonvanishing value of δ. Indeed, the preferred value for this parameter is $\sin\delta \sim 0.93$ (Beringer *et al.*, 2012). All CP-violating effects in the quark sector can be expressed in terms of a single parameter, the Jarlskog parameter, which is invariant under the phase convention of quark fields:

$$J = -\mathrm{Im}\left[V_{us}V_{cd}V_{cs}^*V_{ud}^*\right] = c_{12}c_{23}c_{13}^2 s_{12}s_{23}s_{13}s_\delta \sim 3 \times 10^{-5}. \qquad (1.28)$$

In complete analogy to the leptonic case, one can show that unitarity of the matrices V_L^D, V_L^U, V_R^D, V_R^U implies that the expression for $J_{Z.Q}^\mu$ remains the same after the primed quark fields are replaced with the mass eigenstates.

1.3 The basic properties of neutrinos: interactions, masses and oscillations

1.3.1 Neutrino interactions in the low energy limit

The two terms in the SM Lagrangian density, $\mathcal{L}_I^{(CC)}$ and $\mathcal{L}_I^{(NC)}$ – see (1.4) and (1.6), respectively – describe a three-body process involving two fermions and $W^\pm$ and Z gauge bosons, whose mass is on the order of 100 GeV. Whenever the typical range for energies and momenta carried by the leptons (or quarks) is much smaller than this value, the gauge bosons produced in the trilinear vertex can only propagate as virtual particles. We will see that this is, for example, typically the case in almost all relevant cases in cosmology in which we will be interested in the following. The gauge propagators

$$G_{\mu\nu}^W(x-x') \equiv \langle 0|T\,W_\mu(x)W_\nu^\dagger(x')|0\rangle = \lim_{\epsilon\to 0} i\int \frac{d^4p}{(2\pi)^4} \frac{-g_{\mu\nu} + \frac{p_\mu p_\nu}{m_W^2}}{p^2 - m_W^2 + i\epsilon} e^{-ip\cdot(x-x')}$$

$$G_{\mu\nu}^Z(x-x') \equiv \langle 0|T\,Z_\mu(x)Z_\nu^\dagger(x')|0\rangle = \lim_{\epsilon\to 0} i\int \frac{d^4p}{(2\pi)^4} \frac{-g_{\mu\nu} + \frac{p_\mu p_\nu}{m_Z^2}}{p^2 - m_Z^2 + i\epsilon} e^{-ip\cdot(x-x')}$$

$$(1.29)$$

can then be considered in their short-range limit,

$$G_{\mu\nu}^W(x - x') \xrightarrow{p^\mu \ll m_W} i\frac{g_{\mu\nu}}{m_W^2}\delta^4(x - x') \tag{1.30}$$

$$G_{\mu\nu}^Z(x - x') \xrightarrow{p^\mu \ll m_Z} i\frac{g_{\mu\nu}}{m_Z^2}\delta^4(x - x'). \tag{1.31}$$

Hence, the weak charged-current and neutral-current processes at tree level in the low energy limit are described by the effective Lagrangians

$$\mathcal{L}_{\text{eff}}^{(CC)} = -\frac{g^2}{8m_W^2}J_W^{\mu\dagger}J_{\mu W} = -\frac{G_F}{\sqrt{2}}J_W^{\mu\dagger}J_{\mu W} \tag{1.32}$$

$$\mathcal{L}_{\text{eff}}^{(NC)} = -\frac{g^2}{4\cos^2\theta_W m_Z^2}J_Z^{\mu\dagger}J_{\mu Z} = -2\frac{G_F}{\sqrt{2}}\rho J_Z^{\mu\dagger}J_{\mu Z}, \tag{1.33}$$

where $G_F \equiv \sqrt{2}g^2/(8m_W^2) = 1.166 \times 10^{-5}$ GeV^{-2} is the Fermi constant and $\rho \equiv m_W^2/(m_Z^2\cos^2\theta_W)$, which is equal to unity in the SM.

The interaction terms $\mathcal{L}_{\text{eff}}^{(CC)}$ and $\mathcal{L}_{\text{eff}}^{(NC)}$ can mediate a set of purely four-lepton processes such as the ones reported in Tables 1.5 and 1.6. In the following we will treat only some of them in detail, but we will show how to use the results contained in Tables 1.5 and 1.6 for a simple generalization to all the others.

Let us consider in particular neutrino–electron elastic scattering, $v_e + e^- \to v_e + e^-$ and $v_{\mu(\tau)} + e^- \to v_{\mu(\tau)} + e^-$. From Eqs. (1.32) and (1.33) one can easily get the amplitudes

$$\mathcal{A}_{v_x e^- \to v_x e^-} = -\frac{G_F}{\sqrt{2}}[\bar{u}_{v_x}\gamma^\rho(1 - \gamma_5)u_{v_x}][\bar{u}_e\gamma_\rho(g_V^l - g_A^l\gamma_5)u_e] \tag{1.34}$$

$$\mathcal{A}_{v_e e^- \to v_e e^-} = -\frac{G_F}{\sqrt{2}}\left\{[\bar{u}_{v_e}\gamma^\rho(1 - \gamma_5)u_e]\left\{[\bar{u}_e\gamma_\rho(1 - \gamma_5)u_{v_e}]\right.\right.$$
$$+ [\bar{u}_{v_e}\gamma^\rho(1 - \gamma_5)u_{v_e}][\bar{u}_e\gamma_\rho(g_V^l - g_A^l\gamma_5)u_e]\}$$
$$= -\frac{G_F}{\sqrt{2}}[\bar{u}_{v_e}\gamma^\rho(1 - \gamma_5)u_{v_e}][\bar{u}_e\gamma_\rho((1 + g_V^l) - (1 + g_A^l)\gamma_5)u_e],$$
$$\tag{1.35}$$

where $x = \mu, \tau$, $g_V^l \equiv g_L^l + g_R^l$, and $g_A^l \equiv g_L^l - g_R^l$ (see Table 1.4). Note that to obtain the final form of $\mathcal{A}_{v_e e^- \to v_e e^-}$ we have used one of the Fierz rearrangement formulas. In Tables 1.5 and 1.6 are reported the squared amplitudes for several pure weak leptonic processes at three level. The list is not complete, but the missing processes can be obtained using crossing symmetry.

Table 1.5 *The matrix elements for various processes with electronic neutrinos, where* $\tilde{g}_L^l \equiv 1 + g_L^l$

Process $1 + 2 \longrightarrow 3 + 4$	$2^{-5} G_F^{-2} S \sum_s \lvert \mathcal{A}_{12 \to 34} \rvert^2$
$\nu_e + \bar{\nu}_e \longrightarrow \nu_e + \bar{\nu}_e$	$4\,(p_1 \cdot p_4)(p_2 \cdot p_3)$
$\nu_e + \nu_e \longrightarrow \nu_e + \nu_e$	$2\,(p_1 \cdot p_2)(p_3 \cdot p_4)$
$\nu_e + \bar{\nu}_e \longrightarrow \nu_{\mu(\tau)} + \bar{\nu}_{\mu(\tau)}$	$(p_1 \cdot p_4)(p_2 \cdot p_3)$
$\nu_e + \bar{\nu}_{\mu(\tau)} \longrightarrow \nu_e + \bar{\nu}_{\mu(\tau)}$	$(p_1 \cdot p_4)(p_2 \cdot p_3)$
$\nu_e + \nu_{\mu(\tau)} \longrightarrow \nu_e + \nu_{\mu(\tau)}$	$(p_1 \cdot p_2)(p_3 \cdot p_4)$
$\nu_e + \bar{\nu}_e \longrightarrow e^+ + e^-$	$4[\tilde{g}_L^{l2}(p_1 \cdot p_4)(p_2 \cdot p_3)$ $+ g_R^{l2}(p_1 \cdot p_3)(p_2 \cdot p_4)$ $+ \tilde{g}_L^l g_R^l m_e^2 (p_1 \cdot p_2)]$
$\nu_e + e^- \longrightarrow \nu_e + e^-$	$4[\tilde{g}_L^{l2}(p_1 \cdot p_2)(p_3 \cdot p_4)$ $+ g_R^{l2}(p_1 \cdot p_4)(p_2 \cdot p_3)$ $- \tilde{g}_L^l g_R^l m_e^2 (p_1 \cdot p_3)]$
$\bar{\nu}_e + e^- \longrightarrow \bar{\nu}_e + e^-$	$4[g_R^{l2}(p_1 \cdot p_2)(p_3 \cdot p_4)$ $+ \tilde{g}_L^{l2}(p_1 \cdot p_4)(p_2 \cdot p_3)$ $- \tilde{g}_L^l g_R^l m_e^2 (p_1 \cdot p_3)]$

Note: S is a symmetrization factor, the product of a factor 1/2 for each pair of identical particles in initial and final states and a factor 2 if there are two identical particles in the initial state.
Source: Dolgov *et al.* (1997).

Starting from Eq. (1.34), it is easy to obtain the differential cross section (see, e.g., Halzen and Martin, 1984):

$$\frac{d}{dy}\sigma_{\nu_x e^- \to \nu_x e^-} = \frac{G_F^2 s}{4\pi} \left[\left(g_V^l + g_A^l\right)^2 + \left(g_V^l - g_A^l\right)^2 (1 - y)^2 \right.$$
$$\left. - 2\frac{m_e^2}{s} \left((g_V^l)^2 - (g_A^l)^2\right) y \right], \tag{1.36}$$

where s stands for the Mandelstam variable $s \equiv (p_1 + p_2)^2$ and $y \equiv 1 - (p_2 \cdot p_3 / p_1 \cdot p_2)$ is a Bjorken variable (using the notation for the momenta reported in Tables 1.5 and 1.6). Note that $0 \le y \le 1$. In the electron rest frame $y = (E_1 - E_3)/E_1$ (inelasticity parameter). One can easily get the analogous expression (1.36) for antineutrinos by simply replacing $g_A^l \to -g_A^l$, obtaining

$$\frac{d}{dy}\sigma_{\bar{\nu}_x e^- \to \bar{\nu}_x e^-} = \frac{G_F^2 s}{4\pi} \left[\left(g_V^l - g_A^l\right)^2 + \left(g_V^l + g_A^l\right)^2 (1 - y)^2 \right.$$
$$\left. - 2\frac{m_e^2}{s} \left((g_V^l)^2 - (g_A^l)^2\right) y \right]. \tag{1.37}$$

Table 1.6 *The matrix elements for various processes with muon neutrinos*

Process $1 + 2 \longrightarrow 3 + 4$	$2^{-5} G_F^{-2} S \sum_s \vert \mathcal{A}_{12 \to 34} \vert^2$
$\nu_\mu + \overline{\nu}_\mu \longrightarrow \nu_\mu + \overline{\nu}_\mu$	$4 (p_1 \cdot p_4)(p_2 \cdot p_3)$
$\nu_\mu + \nu_\mu \longrightarrow \nu_\mu + \nu_\mu$	$2 (p_1 \cdot p_2)(p_3 \cdot p_4)$
$\nu_\mu + \overline{\nu}_\mu \longrightarrow \nu_{e(\tau)} + \overline{\nu}_{e(\tau)}$	$(p_1 \cdot p_4)(p_2 \cdot p_3)$
$\nu_\mu + \overline{\nu}_{e(\tau)} \longrightarrow \nu_\mu + \overline{\nu}_{e(\tau)}$	$(p_1 \cdot p_4)(p_2 \cdot p_3)$
$\nu_\mu + \nu_{e(\tau)} \longrightarrow \nu_\mu + \nu_{e(\tau)}$	$(p_1 \cdot p_2)(p_3 \cdot p_4)$
$\nu_\mu + \overline{\nu}_\mu \longrightarrow e^+ + e^-$	$4[g_L^{l2}(p_1 \cdot p_4)(p_2 \cdot p_3)$ $+ g_R^{l2}(p_1 \cdot p_3)(p_2 \cdot p_4)$ $+ g_L^l g_R^l m_e^2(p_1 \cdot p_2)]$
$\nu_\mu + e^- \longrightarrow \nu_\mu + e^-$	$4[g_L^{l2}(p_1 \cdot p_2)(p_3 \cdot p_4)$ $+ g_R^{l2}(p_1 \cdot p_4)(p_2 \cdot p_3)$ $- g_L^l g_R^l m_e^2(p_1 \cdot p_3)]$
$\overline{\nu}_\mu + e^- \longrightarrow \overline{\nu}_\mu + e^-$	$4[g_R^{l2}(p_1 \cdot p_2)(p_3 \cdot p_4)$ $+ g_L^{l2}(p_1 \cdot p_4)(p_2 \cdot p_3)$ $- g_L^l g_R^l m_e^2(p_1 \cdot p_3)]$

Note: The elements also apply to ν_τ.
Source: Dolgov *et al.* (1997).

The analogous results for ν_e and $\overline{\nu}_e$ can be deduced by using Eqs. (1.36) and (1.37) and replacing $g_V^l \to 1 + g_V^l$ and $g_A^l \to 1 + g_A^l$. Compare, for example, Eqs. (1.34) and (1.35). After performing the integration over the Bjorken variable y, one gets the expressions for the total cross sections, which are reported in Tables 1.7 and 1.8.

Concerning the reaction $\nu_e + \overline{\nu}_e \to e^+ + e^-$ and the reverse process $e^+ + e^- \to \nu_e + \overline{\nu}_e$ (see Table 1.7), note that the difference in the prefactors multiplying the expression $[(s - m_e^2)(\tilde{g}_L^{l2} + g_R^{l2}) + 6 m_e^2 \, \tilde{g}_L^l \, g_R^l]$ is due just to kinematic and statistical reasons (the factor S is different for the two expressions). The same holds for ν_μ.

In cosmology a relevant role is played by charged current processes involving ν_e ($\overline{\nu}_e$) and nucleons,

$$\nu_e + n \to e^- + p \qquad \overline{\nu}_e + p \to e^+ + n$$

$$e^- + p \to \nu_e + n \qquad n \to e^- + \overline{\nu}_e + p$$

$$e^+ + n \to \overline{\nu}_e + p \qquad e^- + \overline{\nu}_e + p \to n, \tag{1.38}$$

because their rates are among the key parameters which fix the amounts of light elements synthesized in the early universe, in particular that of ^{4}He.

Table 1.7 *The cross sections for different purely leptonic processes with the same notations as in Tables 1.5 and 1.6*

Process $1 + 2 \longrightarrow 3 + 4$	$3\pi G_F^{-2} \sigma_{12 \to 34}$
$\nu_e + \bar{\nu}_e \longrightarrow \nu_e + \bar{\nu}_e$	s
$\nu_e + \nu_e \longrightarrow \nu_e + \nu_e$	$\frac{3}{2} s$
$\nu_e + \bar{\nu}_e \longrightarrow \nu_{\mu(\tau)} + \bar{\nu}_{\mu(\tau)}$	$\frac{1}{4} s$
$\nu_e + \bar{\nu}_{\mu(\tau)} \longrightarrow \nu_e + \bar{\nu}_{\mu(\tau)}$	$\frac{1}{4} s$
$\nu_e + \nu_{\mu(\tau)} \longrightarrow \nu_e + \nu_{\mu(\tau)}$	$\frac{3}{4} s$
$\nu_e + \bar{\nu}_e \longrightarrow e^+ + e^-$	$\frac{4}{\sqrt{s}}\sqrt{s - 4m_e^2}[(s - m_e^2)(\tilde{g}_L^{l2} + g_R^{l2}) + 6m_e^2 \tilde{g}_L^l g_R^l]$
$e^+ + e^- \longrightarrow \nu_e + \bar{\nu}_e$	$\frac{\sqrt{s}}{2\sqrt{s-4m_e^2}}[(s - m_e^2)(\tilde{g}_L^{l2} + g_R^{l2}) + 6m_e^2 \tilde{g}_L^l g_R^l]$
$\nu_e + e^- \longrightarrow \nu_e + e^-$	$(3s\,\tilde{g}_L^{l2} + s\,g_R^{l2} - 3m_e^2\,\tilde{g}_L^l\,g_R^l)$
$\bar{\nu}_e + e^- \longrightarrow \bar{\nu}_e + e^-$	$(3s\,g_R^{l2} + s\,\tilde{g}_L^{l2} - 3m_e^2\,\tilde{g}_L^l\,g_R^l)$

Table 1.8 *The cross sections for different purely leptonic processes with the same notations as in Tables 1.5 and 1.6*

Process $1 + 2 \longrightarrow 3 + 4$	$3\pi G_F^{-2} \sigma_{12 \to 34}$
$\nu_\mu + \bar{\nu}_\mu \longrightarrow \nu_\mu + \bar{\nu}_\mu$	s
$\nu_\mu + \nu_\mu \longrightarrow \nu_\mu + \nu_\mu$	$\frac{3}{2} s$
$\nu_\mu + \bar{\nu}_\mu \longrightarrow \nu_{e(\tau)} + \bar{\nu}_{e(\tau)}$	$\frac{1}{4} s$
$\nu_\mu + \bar{\nu}_{e(\tau)} \longrightarrow \nu_\mu + \bar{\nu}_{e(\tau)}$	$\frac{1}{4} s$
$\nu_\mu + \nu_{e(\tau)} \longrightarrow \nu_\mu + \nu_{e(\tau)}$	$\frac{3}{4} s$
$\nu_\mu + \bar{\nu}_\mu \longrightarrow e^+ + e^-$	$\frac{4}{\sqrt{s}}\sqrt{s - 4m_e^2}[(s - m_e^2)(g_L^{l2} + g_R^{l2}) + 6m_e^2 g_L^l g_R^l]$
$e^+ + e^- \longrightarrow \nu_\mu + \bar{\nu}_\mu$	$\frac{\sqrt{s}}{2\sqrt{s-4m_e^2}}[(s - m_e^2)(g_L^{l2} + g_R^{l2}) + 6m_e^2 g_L^l g_R^l]$
$\nu_\mu + e^- \longrightarrow \nu_\mu + e^-$	$(3s\,g_L^{l2} + s\,g_R^{l2} - 3m_e^2\,g_L^l\,g_R^l)$
$\bar{\nu}_\mu + e^- \longrightarrow \bar{\nu}_\mu + e^-$	$(3s\,g_R^{l2} + s\,g_L^{l2} - 3m_e^2\,g_L^l\,g_R^l)$

Among them, the neutron decay process $n \to p + e^- + \bar{\nu}_e$ is particularly important because it affects the n/p ratio at the onset of primordial nucleosynthesis. From a historical point of view, it was studied by Fermi, who first described the low-energy weak interaction via a four-fermion Hamiltonian. We here describe this process in some detail. It will also be further discussed in Chapter 4. In the low energy limit, the amplitude for β-decay, $n(p_n) \to p(p_p) + e^-(p_e) + \bar{\nu}_e(p_{\bar{\nu}})$ (see

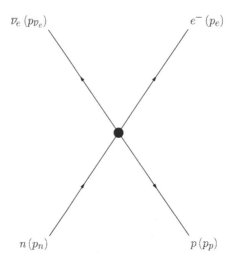

Figure 1.1 The Feynman diagram for the reaction $n \rightarrow p + e^- + \bar{\nu}_e$ at tree level.

Fig. 1.1), is obtained by $\mathcal{L}_{\text{eff}}^{(\text{CC})}$ reported in (1.32),

$$\mathcal{A}_{n \rightarrow pe^- \bar{\nu}_e} = -\frac{G_F}{\sqrt{2}} V_{ud} \, \langle p | \bar{u}_u \gamma^\rho (1 - \gamma_5) u_d | n \rangle \, [\bar{u}_e \gamma_\rho (1 - \gamma_5) u_{\nu_e}] \tag{1.39}$$

where V_{ud} is the corresponding entry of the CKM matrix. In principle, the matrix element of the hadronic current $\langle p | \bar{u}_u \gamma^\rho (1 - \gamma_5) u_d | n \rangle$ should be computed using quantum chromodynamics (QCD). Unfortunately the nonperturbative regime of QCD is still hardly treatable. Typically, to study such a matrix element, one exploits the symmetry constraints it should obey. Let us split the hadronic current into its vector and axial terms,

$$V_W^\mu(x) = \bar{u}_u(x) \gamma^\mu u_d(x) \tag{1.40}$$

$$A_W^\mu(x) = \bar{u}_u(x) \gamma^\mu \gamma_5 u_d(x). \tag{1.41}$$

Because of Lorentz covariance, the Fourier transform of the matrix element $\langle p | V_W^\mu | n \rangle$ can be written in full generality as

$$\langle p | V_W^\mu | n \rangle = \bar{u}_p(p_p) [f_1(q^2) \gamma^\mu + f_2(q^2)(p_n^\mu + p_p^\mu)$$
$$+ f_3(q^2)(p_n^\mu - p_p^\mu)] u_n(p_n), \tag{1.42}$$

where $q \equiv p_p - p_n$, and $f_1(q^2)$, $f_2(q^2)$ and $f_3(q^2)$ are independent form factors, which can be functions of the nontrivial Lorentz invariants only. By using the Dirac equation for spinors u_p and u_n, under the assumption of a negligible mass difference between the neutron and proton, $m_n \approx m_p \approx m_N$, we can recast (1.42) in the form

$$\langle p|V_W^\mu|n\rangle = \bar{u}_p(p_p)\left[\gamma^\mu F_1(Q^2) + \frac{i\sigma^{\mu\rho}q_\rho}{2m_N}F_2(Q^2) + \frac{q^\mu}{m_N}F_3(Q^2)\right]u_n(p_n),$$

(1.43)

where $Q^2 \equiv -q^2$ and F_1, F_2 and F_3 are three independent complex form factors, combinations of f_1, f_2 and f_3. In complete analogy, one can write for the axial current

$$\langle p|A_W^\mu|n\rangle = \bar{u}_p(p_p)\left[\gamma^\mu G_A(Q^2) + \frac{p_n^\mu + p_p^\mu}{m_N}G_3(Q^2) + \frac{q^\mu}{m_N}G_P(Q^2)\right]\gamma_5 u_n(p_n).$$

(1.44)

Invariance of QCD under time reversal and isospin implies that all form factors are real and in particular $F_3(Q^2) = G_3(Q^2) = 0$. Moreover, because in β-decay $m_e^2 \leq q^2 \leq (m_n - m_p)^2$, the form factors intervening in Eqs. (1.43) and (1.44) are effectively computed for $Q^2 \approx 0$.

The isospin invariance constrains even more the remaining form factors of vector current for $Q^2 = 0$. In particular, the conserved vector current (CVC) hypothesis, first considered in the 1950s and based on isospin invariance of the QCD Lagrangian, connects $F_1(0)$ and $F_2(0)$ to the electromagnetic form factors of nucleons that in the Q^2 limit are well known,

$$F_1(0) = 1$$

$$F_2(0) = \frac{\mu_p - \mu_n}{\mu_N},$$

(1.45)

where μ_n and μ_p are neutron and proton magnetic moments, respectively, and $\mu_N = e/(2m_p)$ is the nuclear magneton. Note that similar considerations can be done for the axial current, leading to the well-known partial conserved axial current hypothesis (PCAC), which connects the value of some of the axial parameters at $Q^2 = 0$ with pion mass and pion/nucleon coupling. This issue will not be discussed because it is somewhat beyond the scope of this book.

In the neutron rest frame, because $q^2 \ll m_N^2$, we can neglect terms proportional to F_2 and G_P in Eqs. (1.43) and (1.44). Hence we get

$$V_{ud}\langle p|\bar{u}_u\gamma^\rho(1 - \gamma_5)u_d|n\rangle \approx \bar{u}_p(p_p)\gamma^\mu(C_v - \gamma_5 C_A)u_n(p_n),$$

(1.46)

where $C_v = V_{ud}F_1(0)$ and $C_A = V_{ud}G_A(0)$. By using such an expression and in the limit of very large nucleon mass (keeping fixed $\Delta m \equiv m_n - m_p$), known as the Born limit, one gets the prediction for neutron lifetime

$$\tau_n^{-1} = \frac{G_F^2 \left(C_V^2 + 3C_A^2\right)}{2\pi^3} m_e^5 \int_1^{\frac{\Delta m}{m_e}} d\epsilon \, \epsilon \left(\epsilon - \frac{\Delta m}{m_e}\right)^2 \left(\epsilon^2 - 1\right)^{\frac{1}{2}}. \qquad (1.47)$$

Inserting the value of the vector coupling $C_V = V_{ud} = 0.97425 \pm 0.00022$ and for the ratio $C_A/C_V = -1.2701 \pm 0.0025$ (Beringer *et al.*, 2012) deduced from the study of the decay product angular distribution in neutron decay, we obtain $\tau_n \simeq 961$ s, which has to be compared with the experimental result $\tau_n^{\text{exp}} = 880.1 \pm 1.1$ s (Beringer *et al.*, 2012). The Born limit gives too high a value for the neutron lifetime, the difference being on the order of 10%. A similar level of approximation is also expected for the other reaction rates (1.38), because they are all mediated by the same interaction Hamiltonian. We will say more about this in Chapter 4.

1.3.2 Dirac and Majorana masses

As we will discuss in the following, a large number of experimental results since the pioneering Homestake experiment led by Ray Davis in the late 1960s have accumulated impressive evidence that the three active neutrinos ν_e, ν_μ and ν_τ undergo flavour oscillations and (at least two of them) have nonzero mass. On the other hand, the energy spectrum of the electron emitted in the β-decay of tritium puts an upper bound on neutrino mass scale which is currently on the order of 2 eV; see later.

The nature of the masses of neutrinos and their small values compared with the masses of the charged lepton partners have been puzzling theoretical physicists for decades and are still an open problem. There is, however, a class of simple and elegant mechanisms which have been proposed, the so-called *seesaw*, which is rooted in the idea that neutrino masses are the effect at low energy of some new physics related to a mass scale higher than the electroweak breaking parameter, given by the Higgs vev v. A closely related issue is to understand if neutrinos coincide with their antiparticles or not, in other words, if they are *Majorana* or *Dirac* fermions. We review here the main concepts which we think the reader should become familiar with (if not already) and which will be used in the following. Much more detailed analysis can be found in several review papers and books (see, e.g., Giunti and Kim, 2007).

Dirac mass

A *Dirac* mass term for neutrinos can be introduced in the SM as for quarks or charged leptons. To this end, it is enough to add three right-handed neutrino fields $\nu_{\alpha R}$ with $\alpha = e, \mu, \tau$ which are *sterile*, in the sense that they are singlet under the whole gauge group and are not interacting with all other particles. Yet their role is crucial in the lepton Yukawa sector of the SM Lagrangian density (1.1), which now takes the form

$$\mathcal{L}_{H,L} = -Y_{\alpha\beta}^{\prime l} \overline{L_{\alpha L}^{\prime}} \, \Phi \, l_{\beta R}^{\prime} - Y_{\alpha\beta}^{\prime v} \overline{L_{\alpha L}^{\prime}} (i\sigma_2 \Phi^*) v_{\beta R}^{\prime} + \text{h.c.} \qquad (1.48)$$

After spontaneous symmetry breaking, as we discussed for quarks, the mass terms are obtained after the matrices $Y^{\prime l}$ and $Y^{\prime v}$ are diagonalized through biunitary transformations

$$V_L^{l\dagger} Y^{\prime l} V_R^{l} = Y^{l}, \qquad Y_{\alpha\beta}^{l} = y_{\alpha}^{l} \delta_{\alpha\beta}$$

$$V_L^{v\dagger} Y^{\prime v} V_R^{v} = Y^{v}, \qquad Y_{\alpha\beta}^{v} = y_{\alpha}^{v} \delta_{\alpha\beta}. \qquad (1.49)$$

Transforming the leptonic fields as follows,

$$l_R = (V_R^{l\dagger}) l_R^{\prime}, \qquad l_L = (V_L^{l\dagger}) l_L^{\prime}$$

$$v_R = (V_R^{v\dagger}) v_R^{\prime}, \qquad v_L = (V_L^{v\dagger}) v_L^{\prime}, \qquad (1.50)$$

we finally get

$$\mathcal{L}_{H,L} = -\frac{v + H(x)}{\sqrt{2}} \left(\sum_{\alpha=e,\mu,\tau} y_{\alpha}^{l} \overline{l_{\alpha L}} \, l_{\alpha R} + \sum_{k=1,2,3} y_{k}^{v} \overline{v_{kL}} \, v_{kR} + \text{h.c.} \right). \qquad (1.51)$$

The neutrino masses are then given by $m_k = y_k^v v / \sqrt{2}$ and the neutrino fields[1] v_{kL} and v_{kR} combine to form a Dirac bispinor $v_k \equiv v_{kL} + v_{kR}$. Hereafter, neutrino fields with latin indices denote mass eigenstates. If we define the unitary matrix

$$U \equiv V_L^{l\dagger} V_L^{v}, \qquad (1.52)$$

the charged current can be written either in the neutrino *flavour* eigenbasis $v_{\alpha L}$,

$$J_{W,L}^{\mu} = 2\overline{v_{\alpha L}} \, \gamma^{\mu} \, l_{\alpha L}, \qquad (1.53)$$

or in the *mass* eigenbasis v_{kL},

$$J_{W,L}^{\mu} = 2\overline{v_{kL}} \, \gamma^{\mu} \, U_{k\beta}^{\dagger} \, l_{\beta L}. \qquad (1.54)$$

The U matrix is the analogue for leptons of the CKM, and is usually called the *Pontecorvo–Maki–Nakagawa–Sakata* matrix (Pontecorvo, 1957, 1958; Maki *et al.*, 1962), relating the two different bases, $v_{\alpha L} \equiv U_{\alpha k} v_{kL}$. Note that as for quarks, even in the case of massive neutrinos, neutral current interactions remain unchanged.

Unless the matrix U is diagonal, the lepton numbers for each flavour, L_e, L_μ and L_τ, are not conserved, whereas the total leptonic charge $L = L_e + L_\mu + L_\tau$ is

---

[1] See Schechter and Valle 1980, 1981 for an alternative two-component notation for the neutrinos.

still a global symmetry of the Lagrangian. In fact, because the electron, muon and tau are defined as negatively charged particles with definite mass, and hence there is no difference between mass and flavour eigenstates in this case, rewriting the leptonic Yukawa coupling in terms of $\nu_{\alpha L}$'s, we have

$$\mathcal{L}_{\mathrm{H},L} = -\frac{v + H(x)}{\sqrt{2}} \left(\sum_{\alpha=e,\mu,\tau} y_\alpha^l \, \overline{l_{\alpha L}} \, l_{\alpha R} + \sum_{k=1,2,3} y_k^\nu \, \overline{\nu_{\alpha L}} \, U_{k\alpha} \, \nu_{kR} \right) + \text{h.c.} \quad (1.55)$$

Although the current $J_{W,L}^\mu$ and the first term in (1.55) are invariant under the transformations $l_\alpha \to e^{i\delta_\alpha} l_\alpha$ and $\nu_{\alpha L} \to e^{i\delta_\alpha} \nu_{\alpha L}$, the neutrino term in (1.55) spoils this symmetry, because there is no transformation on ν_{kR} which leaves $\mathcal{L}_{\mathrm{H},L}$ invariant, unless y_k^ν are all equal (neutrino mass degeneracy). Because experimentally this is not the case, the only remaining symmetry corresponds to the choice $\delta_\alpha = \delta$ for all α, associated with total lepton number.

As for quarks, a convenient parameterization of the mixing matrix U is written in terms of three angles and one complex phase,

$$U = \begin{pmatrix} c_{12}c_{13} & s_{12}c_{13} & s_{13}e^{-i\delta} \\ -s_{12}c_{23} - c_{12}s_{23}s_{13}e^{i\delta} & c_{12}c_{23} - s_{12}s_{23}s_{13}e^{i\delta} & s_{23}c_{13} \\ s_{12}s_{23} - c_{12}c_{23}s_{13}e^{i\delta} & -c_{12}s_{23} - s_{12}c_{23}s_{13}e^{i\delta} & c_{23}c_{13} \end{pmatrix}, \quad (1.56)$$

where $c_{ij} = \cos\theta_{ij}$, $s_{ij} = \sin\theta_{ij}$, and $0 \le \theta_{ij} \le \pi/2$. A nonzero value for δ leads to CP violation in the leptonic sector, which is again parameterized by a Jarlskog invariant J, defined as in (1.28).

Majorana mass

We recall the expression for the Dirac equation for a free field ψ written in terms of its Weyl components ψ_L and ψ_R of definite chirality,

$$i\slashed{\partial}\,\psi_L = m\,\psi_R$$
$$i\slashed{\partial}\,\psi_R = m\,\psi_L. \quad (1.57)$$

For $m = 0$ the two equations decouple and can be treated separately. For a massless neutrino there is no need to introduce both chiral components. Indeed, in the SM, only the left-handed neutrino was considered at the beginning. For a massive particle, the two components are intertwined as they propagate in time according to (1.57). Even if one introduces a right-handed component, ψ_R might not be independent but rather related to ψ_L. As is well known, such a field was first introduced by Ettore Majorana in 1937 (Majorana, 1937), and it again contains only two independent components as a Weyl spinor.

Consider the charge conjugation C acting on a spinor ψ,

$$\psi(x) \xrightarrow{C} \psi^C(x) = \xi_C \, C \overline{\psi}^T(x) = i\xi_C \gamma^2 \gamma^0 \overline{\psi}^T(x)$$

$$\overline{\psi}(x) \xrightarrow{C} \overline{\psi^C}(x) = -\xi_C^* \psi^T(x) C^\dagger, \tag{1.58}$$

where $|\xi_C|^2 = 1$. It is easy to show that $\psi_L^C = C\overline{\psi_L}^T(x)$ behaves as a right-handed field. In fact, because $P_L C = C P_L$, with $P_L \equiv (1 - \gamma_5)/2$ the left-handed projector, we have

$$P_L \, C \overline{\psi_L}^T = C \left(\overline{\psi_L} \, P_L\right)^T = C \left((P_R \, \psi_L)^\dagger \, \gamma^0\right)^T = 0. \tag{1.59}$$

Of course an analogous relation holds for ψ_R^C. A Majorana spinor $\psi = \psi^C$ is thus represented by

$$\psi = \psi_L + C \overline{\psi_L}^T = \psi_L + \psi_L^C \tag{1.60}$$

and satisfies the equation

$$i\slashed{\partial}\,\psi_L = m \, C \overline{\psi_L}^T. \tag{1.61}$$

Of course, only neutral fermions such as neutrinos can be a Majorana field, and only for massive neutrinos can one observe the difference between the field's Dirac or Majorana nature, because otherwise, the two Eqs. (1.57) completely decouple and the right-handed component becomes completely inert.

The neutrino mass terms and their form are thus crucial. For simplicity, let us start by considering one neutrino only. Chirality and Lorentz invariance impose that a generic mass term has to mix the adjoint of a spinor of definite chirality with a spinor of opposite chirality, as $\overline{\nu_L}\nu_R$ or $\overline{\nu_R}\nu_L$. A proper Majorana mass term can then be written as

$$\mathcal{L}_{\text{mass}}^{\text{M}} = -\frac{1}{2}m \, \overline{\nu_L^C}\nu_L + \text{h.c.}, \tag{1.62}$$

where the extra factor $1/2$ takes into account the double counting of degrees of freedom. The free Majorana Lagrangian density becomes

$$\mathcal{L}^{\text{M}} = \frac{1}{2}\left[\overline{\nu_L}\, i \overleftrightarrow{\slashed{\partial}} \nu_L + \overline{\nu_L^C}\, i \overleftrightarrow{\slashed{\partial}} \nu_L^C - m \left(\overline{\nu_L^C}\nu_L + \overline{\nu_L}\,\nu_L^C\right)\right]$$

$$= \frac{1}{2}\overline{\nu}\left(i \overleftrightarrow{\slashed{\partial}} - m\right)\nu, \tag{1.63}$$

where $\nu \equiv \nu_L + \nu_L^C$. Using the anticommutation property of the fermionic field and the explicit expression for C, one finds

$$\nu_L^T \, i \overleftrightarrow{\slashed{\partial}}^T \overline{\nu_L}^T = \overline{\nu_L}\, i \overleftrightarrow{\slashed{\partial}} \nu_L, \tag{1.64}$$

and (1.63) can be expressed in terms of the only independent field ν_L,

$$\mathcal{L}^{\text{M}} = \overline{\nu_L} \, i \overleftrightarrow{\not{\partial}} \nu_L - \frac{m}{2} \left(-\nu_L^T \, \mathcal{C}^\dagger \, \nu_L + \overline{\nu_L} \, \mathcal{C} \, \overline{\nu_L}^T \right). \qquad (1.65)$$

The kinetic term is as in the Dirac case, whereas the mass term takes a completely different form.

We have already mentioned that a Dirac neutrino mass term, in general, breaks the conservation of lepton number for each flavour, whereas the total lepton number is still preserved. Such a residual symmetry, whose conserved charge counts the number of leptons minus antileptons summed over the flavour, is lost whenever Majorana mass terms are introduced. In fact, under the rephasing of fermion fields $l_\alpha \to e^{i\delta} l_\alpha$ and $\nu_{\alpha L} \to e^{i\delta} \nu_{\alpha L}$ the mass term in (1.65) becomes

$$-\frac{m}{2} \left(-\nu_L^T \, \mathcal{C}^\dagger \, \nu_L + \overline{\nu_L} \, \mathcal{C} \, \overline{\nu_L}^T \right) \longrightarrow -\frac{m}{2} \left(-e^{i 2\delta} \, \nu_L^T \, \mathcal{C}^\dagger \, \nu_L + e^{-i 2\delta} \, \overline{\nu_L} \, \mathcal{C} \, \overline{\nu_L}^T \right).$$

$$(1.66)$$

For small neutrino masses one can treat the Majorana mass term present in a theory such as the electroweak SM as a perturbation. Total lepton number can still be defined and is conserved in all processes for which Majorana mass terms are not involved. Including its effect at first order would produce observable processes where $\Delta L = \pm 2$. This is just the case of neutrinoless double-β decay, which will be discussed in the next sections.

By looking at (1.2) one can easily find that the mass term in (1.65) has the third component of weak isospin $I_3 = 1$ and correspondingly $Y = -2$. Hence it is not a weak-isospin singlet as required by gauge invariance but rather behaves as a weak-isospin triplet. For this reason, such a term cannot be obtained by spontaneous symmetry breaking from a corresponding gauge-invariant Yukawa term involving a single Higgs doublet. On the other hand, by using two Higgs doublets, whose product contains a weak-isospin triplet, one can construct the mass dimension-5 coupling term

$$\mathcal{L}_5 = \frac{g}{\mathcal{M}} (L_L^T \, \sigma_2 \, \Phi) \mathcal{C}^\dagger (\Phi^T \, \sigma_2 \, L_L) + \text{h.c.,} \qquad (1.67)$$

where $\mathcal{M}$ is a mass scale. We recall that a dimension 5 operator is not renormalizable, so that in the usual approach, in which only renormalizable theories are thought to be valid candidates for a consistent description of fundamental interactions, the SM with the addition of such a term should be regarded as a low-energy effective theory. Its ultraviolet completion will be some more fundamental and renormalizable model which contains new heavy degrees of freedom which only show up as asymptotic states (real particles) at energies on the order of $\mathcal{M}$ or higher. Yet, at low energies, they can contribute to physical processes as virtual particles and mediate new interaction terms among the SM degrees of freedom. The Majorana term would be just one possible example, the shadow of the full theory at the electroweak scale and below, exactly as neutron β decay is the effect of a virtual W boson at the MeV scale.

Turning back to the expression (1.67), spontaneous symmetry breaking provides the Majorana mass term

$$\mathcal{L}^{\mathrm{M}}_{\mathrm{mass}} = \frac{1}{2} \left(\frac{g\, v^2}{\mathcal{M}} \right) v_L^T C^\dagger v_L + \mathrm{h.c.} \tag{1.68}$$

Because v^2 gives the scale of typical Dirac masses, m_D, this implies the neutrino mass

$$m_\nu \simeq \frac{m_\mathrm{D}^2}{\mathcal{M}}, \tag{1.69}$$

which is the usual relation obtained via the *seesaw* mechanism, which will be discussed later on.

In the more realistic situation of three Majorana neutrinos $\nu_{\alpha L}$, the mass term can generally be written as

$$\mathcal{L}^{\mathrm{M}}_{\mathrm{mass}} = \frac{1}{2} v_{\alpha L}^{\prime T} C^\dagger M_{\alpha\beta}^L v_{\beta L}' + \mathrm{h.c.}, \tag{1.70}$$

where $M_{\alpha\beta}^L$ is a complex symmetric matrix. It can be diagonalized via a unitary transformation V_L^ν as

$$\left(V_L^\nu \right)^T M^L V_L^\nu = M \quad \text{with} \quad M_{ij} = m_{\nu_i}\, \delta_{ij} \quad i, j = 1, 2, 3. \tag{1.71}$$

In terms of the mass eigenstate Majorana neutrino fields $v_{iL} = \left(V_L^{\nu\dagger} \right)_{i\alpha} v_{\alpha L}'$, the Lagrangian density $\mathcal{L}^{\mathrm{M}}_{\mathrm{mass}}$ and the weak charged current then become

$$\mathcal{L}^{\mathrm{M}}_{\mathrm{mass}} = \frac{1}{2} \sum_{i=1}^3 m_{\nu_i}\, v_{iL}^T C^\dagger v_{iL} + \mathrm{h.c.} = -\frac{1}{2} \sum_{i=1}^3 m_{\nu_i}\, \overline{v_{iL}^C} v_{iL} + \mathrm{h.c.} \tag{1.72}$$

$$J^\mu_{W,L} = 2 \overline{v_{iL}}\, U_{i\alpha}^\dagger\, \gamma^\mu\, l_{\alpha L}, \tag{1.73}$$

where the mixing matrix U is defined as $U \equiv V_L^{l\dagger} V_L^\nu$. Differently from the Dirac case, the matrix U contains three phases which cannot be removed, as can be done for two of them in the Dirac case. In general, one can write U as

$$U = U^\mathrm{D}\, D^\mathrm{M}, \tag{1.74}$$

where U^D is defined in (1.56) and

$$D^\mathrm{M} = \mathrm{diag}(1, e^{i\lambda_2}, e^{i\lambda_3}). \tag{1.75}$$

1.3.3 The seesaw mechanism

To complete this brief description of Dirac/Majorana neutrinos we now consider the well-known *seesaw mechanism* (Minkowski, 1977; Gell-Mann *et al.*, 1979; Yanagida, 1979; Mohapatra and Senjanovic, 1980; Schechter and Valle, 1980), which is often considered the most natural way to explain the extremely small masses of neutrinos compared with those of charged fermions.

There are three types of seesaw models, which all share the feature that the effect is due to the exchange of some heavy particle, but differ in their specific properties.

Type I: $SU(3)_C \times SU(2)_L \times U(1)_Y$-singlet fermions

In the electroweak SM the only active neutrino components are $\nu'_{\alpha L}$ and in fact they are the only neutrinos we directly detect. If N_s right-handed singlets ν_{sR} are introduced, one can have a Dirac mass term

$$\mathcal{L}^{\mathrm{D}}_{\mathrm{mass}} = -\overline{\nu_{sR}}\, M^{\mathrm{D}}_{s\alpha}\, \nu_{\alpha L} + \mathrm{h.c.} \tag{1.76}$$

with $M^{\mathrm{D}}_{s\alpha}$ an $N_s \times 3$ complex matrix. We can also introduce a Majorana mass term for the ν_{sR} in the total Lagrangian density,

$$\mathcal{L}^{R}_{\mathrm{mass}} = \frac{1}{2}\nu^{T}_{sR}\, \mathcal{C}^{\dagger}\, M^{R}_{ss'}{}^{*}\, \nu_{s'R} + \mathrm{h.c.}, \tag{1.77}$$

with $M^{R}_{ss'}$ an $N_s \times N_s$ complex matrix. Because ν_R is a gauge singlet, this term is gauge-invariant. In general, when both Majorana and Dirac mass terms are present, one can construct a $(3 + N_s) \times (3 + N_s)$ complex symmetric mass matrix

$$M^{\mathrm{D+M}} \equiv \begin{pmatrix} M^{L} & M^{\mathrm{D}^{T}} \\ M^{\mathrm{D}} & M^{R} \end{pmatrix}. \tag{1.78}$$

Defining the row vector $\mathbf{N}'^{T}_{L} \equiv (\nu'_{eL}, \nu'_{\mu L}, \nu'_{\tau L}, \nu^{C}_{s_1 R}, \dots, \nu^{C}_{s_{N_s} R})$, we can sum up all three mass terms in the single expression

$$\mathcal{L}^{\mathrm{D+M}}_{\mathrm{mass}} = \frac{1}{2}\mathbf{N}'^{T}_{L}\, \mathcal{C}^{\dagger}\, M^{\mathrm{D+M}}\mathbf{N}'_{L} + \mathrm{h.c.} \tag{1.79}$$

Because the matrix $M^{\mathrm{D+M}}$ has the same properties as the Majorana matrix M^{L}, it can be diagonalized by means of a proper unitary transformation of neutrino fields $\mathbf{N}'_{L} = W\, \mathbf{n}_{L}$ such that

$$W^{T} M^{\mathrm{D+M}} W = M, \qquad M_{ij} = m_{\nu_i}\delta_{ij}, \qquad i, j = 1, \dots, N_s + 3, \tag{1.80}$$

$\mathbf{n}_{L} \equiv (\nu_{1L}, \dots, \nu_{(N_s+3)L})$ being mass eigenstate fields. Because the leptonic weak charged current is written in terms of three active left-handed neutrinos $\nu'_{\alpha L}$ only, once rewritten in terms of $\mathbf{n}_L$, it reads

$$J_{W,L}^{\mu} = 2\overline{\nu'_{\alpha L}} \gamma^{\mu} l'_{\alpha L} = 2\overline{\nu_{iL}} (U^{\dagger})_{i\alpha} \gamma^{\mu} l_{\alpha L}, \tag{1.81}$$

where $U_{\alpha i} = (V_L^{l\dagger})_{\alpha\beta}(W)_{\beta i}$. As can be seen from the last term of Eq. (1.81), the three combinations $\nu_{\alpha L} = U_{\alpha i} \nu_{iL}$ represent the flavour eigenstate neutrinos, those which are coupled to distinct charged leptons. Similarly, the combination of mass eigenstates providing sterile degrees of freedom reads $\nu_{sR}^C = (W)_{si} \nu_{iL}$. Such a combination contains the mixing of active and sterile neutrinos. It is interesting to observe that the rectangular matrix U is not unitary, because $UU^{\dagger} = 1$, whereas $U^{\dagger}U$ is different from the identity. This implies that in the Dirac–Majorana scheme the weak neutral current can mediate transitions among different massive neutrinos.

Let us consider the case where in the matrix $M^{\mathrm{D+M}}$ we have a vanishing M^L, and M^R has entries much larger than those of M^{D}. This is physically reasonable, because the Majorana mass term for left-handed neutrinos is forbidden from SM gauge invariance. Moreover, one expects M^R to be related to physics beyond the SM and hence connected to a mass scale higher than M^{D}. The neutrino mass matrix in this case is

$$M^{\mathrm{D+M}} = \begin{pmatrix} 0 & M^{\mathrm{D}^T} \\ M^{\mathrm{D}} & M^R \end{pmatrix}. \tag{1.82}$$

If all eigenvalues of M^R are much larger than the elements of M^{D}, the full mass matrix $M^{\mathrm{D+M}}$ can be diagonalized in the $\nu_L - \nu_R^C$ space as

$$W^T M^{\mathrm{D+M}} W \approx \begin{pmatrix} M'_{\mathrm{light}} & 0 \\ 0 & M'_{\mathrm{heavy}} \end{pmatrix} \tag{1.83}$$

with

$$W \approx \begin{pmatrix} 1 - \frac{1}{2}M^{\mathrm{D}\dagger}(M^R M^{R\dagger})^{-1}M^{\mathrm{D}} & [(M^R)^{-1}M^{\mathrm{D}}]^{\dagger} \\ -(M^R)^{-1}M^{\mathrm{D}} & 1 - \frac{1}{2}(M^R)^{-1}M^{\mathrm{D}}M^{\mathrm{D}\dagger}(M^{R\dagger})^{-1} \end{pmatrix}, \tag{1.84}$$

where up to the first order in the $(M^R)^{-1}M^{\mathrm{D}}$ expansion, the matrices on the principal diagonal of the l.h.s. of Eq. (1.83) are $M'_{\mathrm{light}} \approx -M^{\mathrm{D}^T}(M^R)^{-1}M^{\mathrm{D}}$ and $M'_{\mathrm{heavy}} \approx M^R$ and the corresponding eigenvectors are $\nu'_L + \nu'^C_L$ and $\nu'_R + \nu'^C_R$, respectively. One can further diagonalize the symmetric matrices M'_{light} and M'_{heavy} in their internal spaces via the unitary transformations V_L^{ν} and V_R^{ν} acting on active and sterile degrees of freedom, respectively,

$$M_{\mathrm{light}} = -V_L^{\nu T} M^{\mathrm{D}^T} \frac{1}{M^R} M^{\mathrm{D}} V_L^{\nu} \qquad M_{\mathrm{heavy}} = V_R^{\nu T} M^R V_R^{\nu}, \tag{1.85}$$

with Majorana mass eigenvectors given by $\nu = V_L^{\nu\dagger} \nu'_L + V_L^{\nu*} \nu'^C_L$ and $N = V_R^{\nu\dagger} \nu'_R + V_R^{\nu*} \nu'^C_R$. From these relations it is clear that in the simplest case of one left-handed and one right-handed neutrino only, the lightest degree-of-freedom mass would be given by $m_{\text{light}} \approx (m^D)^2/m^R$, whereas $m_{\text{heavy}} \approx m^R$.

Type II: $SU(2)_L$-triplet scalars

One can generate neutrino masses via the exchange of a $SU(2)_L$-triplet scalar Δ (Magg and Wetterich, 1980; Schechter and Valle, 1980; Lazarides *et al.*, 1981; Mohapatra and Senjanovic, 1981; Wetterich, 1981). The triplet should be colour-singlet and carry hypercharge $Y = +2$. In the minimal model, there is only one such scalar, which can be represented in the $SU(2)_L$ space as a 2×2 matrix,

$$\Delta = \begin{pmatrix} \Delta^+/\sqrt{2} & \Delta^{++} \\ \Delta^0 & -\Delta^+/\sqrt{2} \end{pmatrix}. \tag{1.86}$$

The new terms in the Lagrangian density which involve Δ are (Davidson *et al.*, 2008)

$$\mathcal{L}_\Delta = \text{Tr}\left[\left(D_\mu \Delta\right)^\dagger \left(D_\mu \Delta\right) \right] - M_\Delta^2 \, \text{tr}\left(\Delta^\dagger \Delta\right)$$

$$+ \frac{1}{2} \left(\sum_{\alpha,\beta=e,\mu,\tau} \lambda_{L\alpha\beta} \, L_{\alpha L}^T \, C^\dagger i\sigma_2 \, \Delta \, L_{\beta L} + M_\Delta \lambda_\Phi \, \Phi^T i\sigma_2 \Delta^\dagger \Phi + \text{h.c.} \right).$$

$$\tag{1.87}$$

The quantity M_Δ is a real mass parameter, λ_L a symmetric 3×3 matrix of complex Yukawa couplings, and λ_Φ a dimensionless complex coupling. It is worth stressing that in the previous expression we have not explicitly written the Δ self-interaction potential and interaction terms of the form $\Phi\Phi^\dagger\Delta\Delta^\dagger$ which in principle can be also included (see Schmidt, 2007 for a detailed analysis), but they are not crucial for the generation of a neutrino mass term. The exchange of scalar triplets generates an effective dimension-5 operator $LL\Phi\Phi$ which yields a Majorana mass

$$(m_{\text{II}})_{\alpha\beta} = \lambda_{L\alpha\beta} \frac{\lambda_\Phi \, v^2}{M_\Delta}. \tag{1.88}$$

Such a simple model adds 11 new parameters to those of SM, 8 real and 3 imaginary. Lepton number violation is due to the simultaneous presence of λ_L and λ_Φ, which forbids consistent assignment of a lepton charge to Δ.

Type III: $SU(2)_L$-triplet fermions

A further class of models providing effective dimension-5 operators $LL\Phi\Phi$ can be generated by the tree-level exchange of $SU(2)_L$-triplet right-handed fermions Σ_{Ri}^a (Foot *et al.*, 1989; Ma, 1998; Ma and Roy, 2002), where the index i runs over possible heavy mass eigenstates, whereas a is an $SU(2)_L$ index. The fields Σ_{Ri}^a are colour singlets and carry a vanishing hypercharge

$$\Sigma_{Ri} = \begin{pmatrix} \Sigma_{Ri}^0/\sqrt{2} & \Sigma_{Ri}^+ \\ \Sigma_{Ri}^- & -\Sigma_{Ri}^0/\sqrt{2} \end{pmatrix}. \tag{1.89}$$

The relevant Lagrangian terms have a form that is similar to the singlet-fermion case, but with different contractions of the $SU(2)$ indices. Following the notation of (Ma and Roy, 2002; Davidson *et al.*, 2008), we can write the Yukawa term involving the triplet as

$$\mathcal{L}_\Sigma = \overline{L_{\alpha L}} \lambda_{\Sigma\alpha i} \Sigma_{Ri} \, i\sigma_2 \Phi^* - \frac{1}{2} \sum_i M_i \, \mathrm{Tr}\left(\Sigma_{Ri}^T C^\dagger \Sigma_{Ri}\right) + \text{h.c.} \tag{1.90}$$

In the previous expression M_i are real mass parameters, whereas λ_Σ is a 3×3 matrix of dimensionless, complex Yukawa couplings. Again, the exchange of fermion triplets generates an effective dimension-5 operator which provides the neutrino mass term

$$(m_{\mathrm{III}})_{\alpha\beta} = \lambda_{\Sigma\alpha k} \frac{v^2}{M_k} \lambda_{\Sigma\beta k}. \tag{1.91}$$

As in the standard seesaw model, in this scenario one also has to add 18 new parameters to those of the Standard Model, 12 real and 6 imaginary, whereas lepton number is violated by the simultaneous presence of both λ_Σ and M_k.

We see that in all cases, the seesaw mechanism can explain very small neutrino mass scales, provided there is a strong hierarchy between Majorana masses and Dirac masses. It was originally noticed that with Dirac masses on the order of the electroweak scale, and Majorana masses of the order of the grand unification scale, one obtains the correct order of magnitude for active neutrino masses (roughly, a fraction of eV). This relation between the mass of the lightest known particles and two very different symmetry-breaking scales is usually considered as the beauty of the seesaw model. However, the same mechanism can also be invoked in the context of the Neutrino Minimal Standard Model (νMSM) (Asaka *et al.*, 2005; Asaka and Shaposhnikov, 2005; Boyarsky *et al.*, 2009a) with the different motivation of explaining neutrino masses (as well as baryogenesis and dark matter) without introducing any new physics above the electroweak scale. In this model, Majorana masses are chosen to be on the order of the standard model Higgs mass, whereas Dirac masses are substantially smaller than the electroweak scale.

1.3.4 Flavour oscillations in vacuum

Neutrino oscillations were first proposed by Bruno Pontecorvo in the late 1950s, inspired by the $K^0 - \overline{K^0}$ system (Pontecorvo, 1957, 1958). Because at that time the only known neutrino was ν_e, Pontecorvo assumed the existence of an extra sterile neutrino (Pontecorvo, 1968). When in the early 1960s the muon neutrino was discovered (Danby *et al.*, 1962), Pontecorvo also suggested that $\nu_e \rightarrow \nu_\mu$ (or $\nu_e \rightarrow \nu_s$) oscillations were possible in the sun (Pontecorvo, 1968). This intuition was later confirmed by the Homestake experiment (Cleveland *et al.*, 1998), and at present it is at the basis of the solution of the solar neutrino problem.

Neutrino oscillations are a quantum mechanical phenomenon that is due to neutrinos being produced via charged current interaction as flavour states, which, as has been argued before, are linear superpositions of mass eigenstates. If the energy resolution of energy and momentum measurements of all particles involved in the process were so high as to allow the inference of the mass of the produced neutrino, no oscillations would be observed. On the other hand, if differences among neutrino masses are exceedingly small, energy and momentum resolutions are typically not good enough to distinguish among neutrino mass eigenstates. In this case the outgoing wave packet is a linear superposition of different mass states and the effect of interference (oscillations) can be observed. For the sake of simplicity, in the following we will consider neutrinos as plane waves and will restrict our analysis to the relativistic regime. This is not a significant limitation, because only relativistic neutrinos can usually be detected,[2] either via nuclei conversions or using elastic scattering on electrons. This experimental issue will be treated in some detail in a following section.

We denote by $\mathcal{U}$ the $(3 + N_s) \times (3 + N_s)$ *square* mixing matrix, built in terms of the two matrices W and V_L^l defined in the previous section. A particular flavour *ket* state (including sterile degrees of freedom) can be written as

$$|\nu_\alpha\rangle = \sum_{k=1}^{3+N_s} \mathcal{U}_{\alpha k}^* |\nu_k\rangle \quad \text{with} \quad \alpha = e, \mu, \tau, s_1, \ldots, s_{N_s}. \tag{1.92}$$

Notice that the complex conjugate matrix $\mathcal{U}^*$ appears, because one should pick up in the expression of neutrino fields the term containing creation operators to construct ket states out of the vacuum. We assume the normalization conditions $\langle \nu_\alpha | \nu_\beta \rangle = \delta_{\alpha\beta}$ and $\langle \nu_i | \nu_j \rangle = \delta_{ij}$. Because massive neutrinos are Hamiltonian eigenstates, a generic $|\nu_\alpha\rangle$ produced at $t = 0$ evolves in time as

$$|\nu_\alpha(t)\rangle = \sum_{k=1}^{3+N_s} \mathcal{U}_{\alpha k}^* e^{-iE_k t} |\nu_k\rangle. \tag{1.93}$$

---

[2] Note that we consider the detection of the relic cosmological sea of neutrinos, possibly nonrelativistic particles, in Chapter 7.

Because $\mathcal{U}^\dagger \mathcal{U} = 1$, we get $|\nu_k\rangle = \sum_\beta \mathcal{U}_{\beta k} |\nu_\beta\rangle$, and we can rewrite Eq. (1.93) in the following way:

$$|\nu_\alpha(t)\rangle = \sum_\beta \left(\sum_{k=1}^{3+N_s} \mathcal{U}_{\alpha k}^* \, e^{-iE_k t} \, \mathcal{U}_{\beta k} \right) |\nu_\beta\rangle. \tag{1.94}$$

We are interested in the evolution of active neutrino flavour states, so from this point on the Greek indices label one of the three active neutrinos. In the combination (1.94) the coefficient proportional to $|\nu_\beta\rangle$ is given by

$$\langle \nu_\beta | \nu_\alpha(t) \rangle = \sum_{k=1}^{3+N_s} \mathcal{U}_{\alpha k}^* \, e^{-iE_k t} \, \mathcal{U}_{\beta k} \tag{1.95}$$

because the restriction of $\mathcal{U}$ to active states is the matrix U introduced in the previous section. From this expression one gets the probability that on performing a measurement of neutrino flavour at time t one detects a flavour β:

$$P_{\nu_\alpha \to \nu_\beta}(t) = \left| \langle \nu_\beta | \nu_\alpha(t) \rangle \right|^2 = \sum_{k,j=1}^{3+N_s} U_{\alpha k}^* \, U_{\beta k} U_{\alpha j} \, U_{\beta j}^* \, e^{-i(E_k - E_j)t}. \tag{1.96}$$

For relativistic neutrinos

$$E_i = \sqrt{|\vec{p}|^2 + m_i^2} \approx |\vec{p}| + \frac{m_i^2}{2|\vec{p}|} \tag{1.97}$$

and by substituting this expression into (1.96) one gets eventually

$$P_{\nu_\alpha \to \nu_\beta}(t) = \left| \langle \nu_\beta | \nu_\alpha(t) \rangle \right|^2 = \sum_{k,j=1}^{3+N_s} U_{\alpha k}^* \, U_{\beta k} U_{\alpha j} \, U_{\beta j}^* \, \exp\left\{ -i \frac{\Delta m_{kj}^2}{2|\vec{p}|} t \right\}, \tag{1.98}$$

where $\Delta m_{kj}^2 \equiv m_k^2 - m_j^2$.

In a neutrino experiment the time dependence of the flavour transition probability cannot be followed, but rather one can measure how the probabilities depend on L, the distance of the detection point from the origin of the neutrino beam. It is therefore customary to replace the time dependence with the variable L/c. Equation (1.98) in natural units reads

$$P_{\nu_\alpha \to \nu_\beta}(L) = \sum_{k,j=1}^{3+N_s} U_{\alpha k}^* \, U_{\beta k} U_{\alpha j} \, U_{\beta j}^* \, \exp\left\{ -i \frac{\Delta m_{kj}^2}{2|\vec{p}|} L \right\}. \tag{1.99}$$

It is interesting to note that the quartic expression $U_{\alpha k}^* \, U_{\beta k} U_{\alpha j} \, U_{\beta j}^*$ is independent of the chosen parameterization and rephasing, and this is also the case for Majorana mixing matrices such as (1.74). Of course, for $L = 0$, (1.99) becomes $P_{\nu_\alpha \to \nu_\beta}(0) = \sum_{k,j=1}^{3+N_s} U_{\alpha k}^* U_{\beta k} U_{\alpha j} U_{\beta j}^* = \delta_{\alpha\beta}$.

By observing that $\sum_{k,j=1}^{3+N_s} = \sum_{k>j} + \sum_{j>k} + \sum_{k=j}$ we can split the expression (1.99) into two parts:

$$P_{\nu_\alpha \to \nu_\beta}(L) = \sum_{k=1}^{3+N_s} |U_{\alpha k}|^2 |U_{\beta k}|^2 + 2\text{Re} \left[\sum_{k>j} U_{\alpha k}^* U_{\beta k} U_{\alpha j} U_{\beta j}^* \exp\left\{ -i \frac{\Delta m_{kj}^2}{2|\vec{p}|} L \right\} \right]$$

(1.100)

or analogously

$$P_{\nu_\alpha \to \nu_\beta}(L) = \delta_{\alpha\beta} - 4 \sum_{k>j} \text{Re} \left[U_{\alpha k}^* U_{\beta k} U_{\alpha j} U_{\beta j}^* \right] \sin^2 \left(\frac{\Delta m_{kj}^2}{4|\vec{p}|} L \right)$$

$$+ 2 \sum_{k>j} \text{Im} \left[U_{\alpha k}^* U_{\beta k} U_{\alpha j} U_{\beta j}^* \right] \sin \left(\frac{\Delta m_{kj}^2}{2|\vec{p}|} L \right). \quad (1.101)$$

The probability that starting with a ν_α one still detects the same flavour at a distance L is called the *survival probability* and it is obtained from the previous expression for $\alpha = \beta$:

$$P_{\nu_\alpha \to \nu_\alpha}(L) = 1 - 4 \sum_{k>j} |U_{\alpha k}|^2 |U_{\alpha j}|^2 \sin^2 \left[\frac{\Delta m_{kj}^2}{2|\vec{p}|} L \right]. \quad (1.102)$$

The total transition probability, namely the probability of observing a change in flavour, is given by $1 - P_{\nu_\alpha \to \nu_\alpha}$, from unitarity.

If the neutrino production or detection occurs on a distance whose uncertainty is much larger than the oscillation lengths $L_{kj}^{\text{osc}} = 4\pi |\vec{p}|/\Delta m_{kj}^2$, the expression (1.100) should be averaged over distances larger than L_{kj}^{osc} and thus the second term of the r.h.s. of Eq. (1.100) cancels out. In this case the probability of flavour oscillation becomes

$$\langle P_{\nu_\alpha \to \nu_\beta} \rangle = \sum_{k=1}^{3+N_s} |U_{\alpha k}|^2 |U_{\beta k}|^2. \quad (1.103)$$

The same expression is valid in case of incoherent detection or production as well.

A fully analogous treatment can be applied to antineutrinos, which are produced by weak charged current via positive charged lepton transitions, $l_\alpha^+ \to \bar{\nu}_\alpha$, pair creation, $l_\alpha^- \bar{\nu}_\alpha$, and neutrino–antineutrino pair creation $\nu_\alpha \bar{\nu}_\alpha$ mediated by weak neutral current.

Flavour antineutrinos $|\bar{\nu}_\alpha\rangle$, like neutrinos, can be expressed in terms of mass eigenstates as $|\bar{\nu}_\alpha\rangle = \sum_{k=1}^{3+N_s} U_{\alpha k} |\bar{\nu}_k\rangle$. Strictly speaking, in the case of Majorana neutrinos there are no independent antiparticles. However, it is customary to call the states with positive helicity antineutrinos. By using the same steps we went through till now for neutrinos, one can write the antineutrino oscillation probability

$$P_{\bar{\nu}_\alpha \to \bar{\nu}_\beta}(L) = \sum_{k,j=1}^{3+N_s} U_{\alpha k} U_{\beta k}^* U_{\alpha j}^* U_{\beta j} \exp\left\{-i\frac{\Delta m_{kj}^2}{2|\vec{p}|}L\right\}. \tag{1.104}$$

The CPT, CP and T transformations relate transition amplitudes between neutrinos to those between antineutrinos. In particular, the action of the CP operator transforms neutrinos into antineutrinos with reversed helicities:

$$|\nu_\alpha\rangle \xleftrightarrow{\text{CP}} |\bar{\nu}_\alpha\rangle. \tag{1.105}$$

On the other hand, the time reversal transformation T interchanges initial and final states:

$$|\nu_\alpha\rangle \to |\nu_\beta\rangle \xleftrightarrow{\text{T}} |\nu_\beta\rangle \to |\nu_\alpha\rangle \tag{1.106}$$

or

$$|\bar{\nu}_\alpha\rangle \to |\bar{\nu}_\beta\rangle \xleftrightarrow{\text{T}} |\bar{\nu}_\beta\rangle \to |\bar{\nu}_\alpha\rangle. \tag{1.107}$$

Even though CP and T are not separately exact symmetries of the electroweak SM, CPT symmetry is always obeyed. This is the well-known CPT theorem, which is confirmed by all present data (the reader might have a look at a summary of CPT violation bounds in Beringer *et al.*, 2012). This symmetry makes it possible to connect the amplitudes of the processes

$$|\nu_\alpha\rangle \to |\nu_\beta\rangle \xleftrightarrow{\text{CPT}} |\bar{\nu}_\beta\rangle \to |\bar{\nu}_\alpha\rangle \tag{1.108}$$

and applied to survival probabilities guarantees the equality

$$P_{\nu_\alpha \to \nu_\alpha} = P_{\bar{\nu}_\alpha \to \bar{\nu}_\alpha}. \tag{1.109}$$

A not-vanishing value for the difference $P_{\nu_\alpha \to \nu_\beta} - P_{\bar{\nu}_\beta \to \bar{\nu}_\alpha}$ would represent a clear signal of CPT violation in elementary particle physics, whereas to test the amount of CP violation in the leptonic sector one has to measure the difference $P_{\nu_\alpha \to \nu_\beta} - P_{\bar{\nu}_\alpha \to \bar{\nu}_\beta}$. Using the expression (1.101) and the analogous expression for antineutrinos, this difference reads

$$P_{\nu_\alpha \to \nu_\beta}(L) - P_{\bar{\nu}_\alpha \to \bar{\nu}_\beta}(L) = 4\sum_{k>j} \text{Im}\left[U_{\alpha k}^* U_{\beta k} U_{\alpha j} U_{\beta j}^*\right] \sin\left(\frac{\Delta m_{kj}^2}{2|\vec{p}|}L\right), \tag{1.110}$$

which shows that CP violation in the neutrino sector can be found only in flavour transition processes, because $\text{Im}\left[U_{\alpha k}^* U_{\alpha k} U_{\alpha j} U_{\alpha j}^*\right] = 0$ for any k, j.

1.3.5 Flavour oscillations in matter

As shown in the previous section, flavour oscillation is a consequence of the different velocities of propagation of neutrino mass eigenstates and of the neutrino mixing matrix. Because the propagation is a crucial ingredient, a medium different from vacuum would sensibly change the oscillation mechanism. Let us compute such an effect by determining the potential experienced by an electron neutrino propagating in a medium consisting of a gas of unpolarized electrons.[3]

The four-leptons weak Hamiltonian (1.32), after a Fierz rearrangement, reads

$$\mathcal{H}_{\text{eff}}^{(CC)}(x) = \frac{G_F}{\sqrt{2}} \left[\overline{\nu_e}(x)\gamma^\mu(1 - \gamma_5)\nu_e(x) \right] \left[\overline{e}(x)\gamma_\mu(1 - \gamma_5)e(x) \right]. \quad (1.111)$$

A gas of unpolarized electrons with a statistical distribution function $f(E_e)$ is quantum mechanically described by the density matrix

$$\rho_e = \int \frac{d^3 p_e}{(2\pi)^3 2 E_e} f(E_e) \frac{1}{2} \sum_{h_e = \pm 1} |e^-(p_e, h_e)\rangle \langle e^-(p_e, h_e)|, \quad (1.112)$$

with h_e denoting helicity and where we are assuming the normalization condition $\||e^-(p_e, h_e)\rangle\|^2 = 2E_e$. Note that

$$\text{Tr}[\rho_e] = \int \frac{d^3 p_e}{(2\pi)^3} f(E_e) = n_e, \quad (1.113)$$

with n_e the electron number density. The effective potential due to charged current interaction, V_{CC}, is then obtained by averaging (1.111) over the background particle distribution and multiplying by their number density; i.e.,

$$\overline{\mathcal{H}_{\text{eff}}^{(CC)}}(x) \equiv \text{Tr}\left[\rho_e \, \mathcal{H}_{\text{eff}}^{(CC)}(x) \right] = \frac{G_F}{\sqrt{2}} \overline{\nu_e}(x)\gamma^\mu(1 - \gamma_5)\nu_e(x) \int \frac{d^3 p_e}{(2\pi)^3 2 E_e}$$

$$\times f(E_e) \frac{1}{2} \sum_{h_e = \pm 1} \langle e^-(p_e, h_e)|\overline{e}(x)\gamma_\mu(1 - \gamma_5)e(x)|e^-(p_e, h_e)\rangle. \quad (1.114)$$

Using standard results about Dirac spinors and gamma matrix traces,

$$\sum_{h_e = \pm 1} \langle e^-(p_e, h_e)|\overline{e}(x)\gamma_\mu(1 - \gamma_5)e(x)|e^-(p_e, h_e)\rangle$$

$$= \text{Tr}\left[(\not{p}_e + m_e)\gamma_\mu(1 - \gamma_5) \right] = 4p_{e\mu}, \quad (1.115)$$

---

[3] In the early universe, neutrinos propagate in a background also containing other charged particles, for example, positrons and baryons. When applying the results of this section to this case, we should also include their contribution to understand matter effects in neutrino oscillations.

and because of the isotropy of the electron background,

$$\int \frac{d^3 p_e}{(2\pi)^3} f(E_e) \frac{p_{e\mu}}{E_e} = n_e \, \delta_{0\mu}, \tag{1.116}$$

the effective Hamiltonian (1.114) becomes

$$\overline{\mathcal{H}_{\text{eff}}^{(\text{CC})}}(x) = V_{\text{CC}} \, \overline{\nu_{eL}}(x) \gamma^0 \nu_{eL}(x), \tag{1.117}$$

where

$$V_{\text{CC}} = \sqrt{2} \, G_F \, n_e. \tag{1.118}$$

is the charged-current effective potential.

The same approach can be applied to the weak neutral-current interaction term (1.33)

$$\mathcal{H}_{\text{eff}}^{(\text{NC})}(x) = \frac{G_F}{\sqrt{2}} \sum_{e,\mu,\tau} \left[\overline{\nu_\alpha}(x) \gamma^\mu (1 - \gamma_5) \nu_\alpha(x) \right] \sum_f \left[\overline{f}(x) \gamma_\mu (g_V^f - g_A^f \gamma_5) f(x) \right],$$

$$\tag{1.119}$$

and in this case the effective potential experienced by a neutrino of any flavour α in a background of unpolarized f fermions is

$$V_{\text{NC}}^f = \sqrt{2} \, G_F \, n_f \, g_V^f, \tag{1.120}$$

where n_f is the number density of fermions f. In a typical astrophysical environment, such as the sun, the only fermions available are electrons, protons and neutrons. Hence the total neutral current effective potential, $V_{\text{NC}} = \sum_f V_{\text{NC}}^f$, is

$$V_{\text{NC}} = \sqrt{2} \, G_F \left[n_e \, g_V^e + n_p \, g_V^p + n_n \, g_V^n \right]. \tag{1.121}$$

From Table 1.4 and recalling that $g_V^f \equiv g_L^f + g_R^f$, we get

$$g_V^e = -\frac{1}{2} + 2 \sin^2 \theta_W$$

$$g_V^p = 2 g_V^U + g_V^D = \frac{1}{2} - 2 \sin^2 \theta_W = -g_V^e$$

$$g_V^n = g_V^U + 2 g_V^D = -\frac{1}{2}, \tag{1.122}$$

and because of the neutrality of the background, which ensures that $n_e = n_p$, we finally obtain

$$V_{\text{NC}} = -\frac{1}{2} \sqrt{2} \, G_F \, n_n. \tag{1.123}$$

As long as the typical neutrino wavelength is large compared to the typical background interparticle spacing, we can describe neutrino propagation in a medium by introducing a neutrino index of refraction n_ν as

$$n_\nu = \frac{p_{\text{matter}}}{p}, \tag{1.124}$$

with p_{matter} and p the linear momentum in matter and in vacuum, respectively, for a given energy $E = \sqrt{p^2 + m_\nu^2}$. Because for relativistic neutrinos $p_{\text{matter}} = p - V_{\text{CC}} - V_{\text{NC}}$, we obtain

$$n_\nu - 1 = -\frac{1}{p}(V_{\text{CC}} + V_{\text{NC}}). \tag{1.125}$$

From the expression of V_{CC} for electron neutrinos, and using V_{NC} experienced by any kind of neutrinos, one can write the effective potential for a generic ν_α with $\alpha = e, \mu, \tau$:

$$V_\alpha = V_{\text{CC}}\, \delta_{e\alpha} + V_{\text{NC}} = \sqrt{2}\, G_{\text{F}} \left(n_e\, \delta_{e\alpha} - \frac{1}{2} n_n \right). \tag{1.126}$$

Consider a generic neutrino flavour eigenstate $|\nu_\alpha\rangle = \sum_{k=1}^{3+N_s} U_{\alpha k}^* |\nu_k\rangle$. The $|\nu_k\rangle$ are momentum $\vec{p}$ and mass eigenstates; hence one can define the free Hamiltonian $\mathcal{H}_0$ as the operator such that

$$\mathcal{H}_0 |\nu_k\rangle = E_k |\nu_k\rangle \quad \text{with } E_k = \sqrt{|\vec{p}|^2 + m_k^2} \tag{1.127}$$

and the interaction Hamiltonian $\mathcal{H}_I$, which is instead diagonal in the flavour basis:

$$\mathcal{H}_I |\nu_\alpha\rangle = V_\alpha |\nu_\alpha\rangle. \tag{1.128}$$

In the Schrödinger picture a neutrino, initially in the α-flavour state, evolves according to the equation

$$i \frac{d}{dt} |\nu_\alpha(t)\rangle = (\mathcal{H}_0 + \mathcal{H}_I) |\nu_\alpha(t)\rangle \tag{1.129}$$

with the initial condition $|\nu_\alpha(0)\rangle = |\nu_\alpha\rangle$.

The probability of $\alpha \to \beta$ flavour transition at time t is $P_{\nu_\alpha \to \nu_\beta}(t) = |\mathcal{A}_{\nu_\alpha \to \nu_\beta}(t)|^2 \equiv |\langle \nu_\beta | \nu_\alpha(t)\rangle|^2$. Using (1.129) we get an evolution equation for $\mathcal{A}_{\nu_\alpha \to \nu_\beta}(t)$,

$$i \frac{d}{dL} \mathcal{A}_{\nu_\alpha \to \nu_\beta}(L) = \left(U_\beta^k E_k U_k^{\dagger \gamma} + \delta_\beta^\gamma V_\gamma \right) \mathcal{A}_{\nu_\alpha \to \nu_\gamma}(L), \tag{1.130}$$

where we have used the path length travelled by neutrinos rather than time as the evolution parameter. Because the typical energies of neutrinos are much larger than their masses, it is safe to consider for the moment the relativistic limit (1.97)

$$i\frac{d}{dL}\mathcal{A}_{\nu_\alpha\to\nu_\beta}(L) = \left(|\vec{p}| + \frac{m_1^2}{2|\vec{p}|} + V_{NC}\right)\mathcal{A}_{\nu_\alpha\to\nu_\beta}(L)$$

$$+ \left(U_\beta^k \frac{\Delta m_{k1}^2}{2|\vec{p}|} U_k^{\dagger\gamma} + \delta_\beta^\gamma \delta_{\beta e} V_{CC}\right)\mathcal{A}_{\nu_\alpha\to\nu_\gamma}(L), \quad (1.131)$$

where $\Delta m_{k1}^2 \equiv m_k^2 - m_1^2$, and we have denoted as m_1 the smallest mass eigenvalue. Note that the first line of the r.h.s. of Eq. (1.131) can be absorbed by a local rephasing,

$$\mathcal{A}_{\nu_\alpha\to\nu_\beta} \Longrightarrow \mathcal{A}_{\nu_\alpha\to\nu_\beta} \exp\left\{-i\left(|\vec{p}| + \frac{m_1^2}{2|\vec{p}|}\right)L - i\int_0^L dx' \, V_{NC}(x')\right\}, \quad (1.132)$$

and the evolution equation then becomes

$$i\frac{d}{dL}\mathcal{A}_{\nu_\alpha\to\nu_\beta}(L) = \left(U_\beta^k \frac{\Delta m_{k1}^2}{2|\vec{p}|} U_k^{\dagger\gamma} + \delta_\beta^\gamma \delta_{\beta e} V_{CC}\right)\mathcal{A}_{\nu_\alpha\to\nu_\gamma}(L)$$

$$\equiv (\mathcal{H}_F)_\beta^\gamma \, \mathcal{A}_{\nu_\alpha\to\nu_\gamma}(L). \quad (1.133)$$

It can easily be proved that Majorana phases do not affect the effective Hamiltonian $\mathcal{H}_F$; hence the possible Majorana nature of neutrinos cannot be revealed either in vacuum or in matter.

To better elucidate the properties of Eq. (1.133), consider the electron and muon neutrino bidimensional space, and two corresponding mass eigenstates ν_1 and ν_2. In this case $\mathcal{H}_F$ reduces to the following 2×2 hermitian matrix:

$$\mathcal{H}_F = \frac{1}{4|\vec{p}|}(\Delta m^2 + 2|\vec{p}| V_{CC})\mathbb{I}$$

$$+ \frac{1}{4|\vec{p}|}\begin{pmatrix} -\Delta m^2 \cos(2\theta) + 2|\vec{p}| V_{CC} & \Delta m^2 \sin(2\theta) \\ \Delta m^2 \sin(2\theta) & \Delta m^2 \cos(2\theta) - 2|\vec{p}| V_{CC} \end{pmatrix},$$

$$(1.134)$$

where θ is the mixing angle

$$|\nu_e\rangle = +\cos\theta \, |\nu_1\rangle + \sin\theta \, |\nu_2\rangle$$

$$|\nu_\mu\rangle = -\sin\theta \, |\nu_1\rangle + \cos\theta \, |\nu_2\rangle \quad (1.135)$$

and

$$\Delta m^2 \equiv m_2^2 - m_1^2. \quad (1.136)$$

The first term on the r.h.s. of Eq. (1.135) is proportional to the identity operator and produces a common phase factor in $\mathcal{A}_{\nu_\alpha\to\nu_\beta}$. The second term can be diagonalized via an orthogonal transformation

$$O(\theta_{\mathrm{M}}) = \begin{pmatrix} \cos\theta_{\mathrm{M}} & \sin\theta_{\mathrm{M}} \\ -\sin\theta_{\mathrm{M}} & \cos\theta_{\mathrm{M}} \end{pmatrix}, \tag{1.137}$$

obtaining

$$\mathcal{H}_{\mathrm{M}} \equiv \begin{pmatrix} -\Delta m_{\mathrm{M}}^2 & 0 \\ 0 & \Delta m_{\mathrm{M}}^2 \end{pmatrix}$$

$$= O^T(\theta_{\mathrm{M}}) \begin{pmatrix} -\Delta m^2 \cos(2\theta) + 2\,|\vec{p}|\,V_{\mathrm{CC}} & \Delta m^2 \sin(2\theta) \\ \Delta m^2 \sin(2\theta) & \Delta m^2 \cos(2\theta) - 2\,|\vec{p}|\,V_{\mathrm{CC}} \end{pmatrix} O(\theta_{\mathrm{M}}). \tag{1.138}$$

The mixing angle in matter is thus

$$\tan 2\,\theta_{\mathrm{M}} = \tan 2\,\theta \left[1 - \frac{2\,|\vec{p}|\,V_{\mathrm{CC}}}{\Delta m^2 \cos 2\theta} \right]^{-1} \tag{1.139}$$

and

$$\Delta m_{\mathrm{M}}^2 = \sqrt{\left(\Delta m^2 \cos 2\theta - 2\,|\vec{p}|\,V_{\mathrm{CC}}\right)^2 + \left(\Delta m^2 \sin 2\theta\right)^2}. \tag{1.140}$$

The evolution equation ruled by $\mathcal{H}_{\mathrm{F}}$ (1.134) shows a resonance for

$$\Delta m^2 \cos(2\theta) = 2\,|\vec{p}|\,V_{\mathrm{CC}}. \tag{1.141}$$

By substituting the expression (1.118) for V_{CC}, one finds that the resonance occurs for an electron number density given by

$$n_e^{\mathrm{res}} = \frac{\Delta m^2 \cos(2\theta)}{2\sqrt{2}\,G_{\mathrm{F}}\,|\vec{p}|}. \tag{1.142}$$

In this case, $\theta_{\mathrm{M}} = \pi/4$ and one can have the largest transition from one flavour to the other. This effect was first pointed out by Wolfenstein (1978) and by Mikheev and Smirnov (1985, 1986), and it is now commonly known as the MSW effect. It has been proved to play a relevant role in astrophysical environments such as the sun and type II supernovae, where the matter density radial profile can provide the correct condition for a resonant conversion. In particular, it describes the flavour conversion mechanism at the basis of the solution to the solar neutrino problem.

An identical approach can be applied in the case of ν_e–ν_s neutrino mixing. In this case, the only change to be done is the replacement in (1.134) of $V_{\mathrm{CC}} \Longrightarrow V_{\mathrm{CC}} + V_{\mathrm{NC}}$ because sterile states are not interacting via weak currents, and the resonant condition will involve both electron and neutron number densities.

1.4 Neutrino experiments

1.4.1 Oscillation experiments and three-neutrino mixing

The flavour oscillation in a neutrino beam can be detected in two different ways: through the observation of a particular neutrino flavour initially absent in the beam (*appearance experiment*), or by the depletion of neutrinos of a particular flavour (*disappearance experiment*).

In appearance experiments, because of the smallness of the background, one can reach a high sensitivity level even for very small mixing angles. This is not the case for a disappearance experiment. In this case, from the nominal intensity of the initial neutrino beam for a fixed flavour, and by measuring the number of interactions occurring in the detector, one determines a possible depletion factor. The statistical nature of the study and the level uncertainty on the expected number of events even in the case of no oscillation limit the sensitivity reachable by such a method for small mixing angles.

To classify the different types of neutrino oscillation experiments it is useful to recall the expression (1.101) for the probability that a flavour transition $\nu_\alpha \to \nu_\beta$ occurs at a certain distance L from a ν_α production point, assuming a monochromatic neutrino of energy E. In particular, let us consider the simplified toy model of two flavour and two massive neutrinos only. In this case Eq. (1.101) simplifies to

$$P_{\nu_\alpha \to \nu_\beta}(L) = \sin^2(2\theta) \sin^2 \left(1.27 \frac{\Delta m^2 [\mathrm{eV}^2]}{4E[\mathrm{MeV}]} L[\mathrm{m}] \right)$$

$$= \sin^2(2\theta) \sin^2 \left(1.27 \frac{\Delta m^2 [\mathrm{eV}^2]}{4E[\mathrm{GeV}]} L[\mathrm{km}] \right), \qquad (1.143)$$

where Δm^2 and the mixing angle θ have been defined in Eqs. (1.136) and (1.135), and $E \approx |\vec{p}|$. There are two main neutrino (antineutrino) sources: nuclear reactors (NR), where electron antineutrinos with energy of order MeV are produced in the β-decay of heavy nuclei, such as ^{235}U, ^{238}U, ^{239}Pu and ^{241}Pu, and high-energy neutrino beams produced by accelerators in the energy range 1–100 GeV, coming from the decay of pions, kaons and muons, and hence composed of both muon and electron neutrinos and antineutrinos.

In the case of NR neutrinos, the sensitivities to Δm^2 can be very different according to the distance (baseline) from the neutrino source to the detector. In order to measure an oscillation effect sensitive to Δm^2, one has to satisfy the condition $\Delta m^2 L/(2E) \sim 1$. By virtue of this relation, short ($\sim$10 m) (SBL), long ($\sim$1 km) (LBL) and very long ($\sim$10^2 km) (VLBL) baseline reactor experiments can be sensitive to $\Delta m^2 \gtrsim 0.1$, 10^{-3} and 10^{-5} eV^2, respectively. In particular,

CHOOZ (Apollonio *et al.*, 2003) and, more recently, Double CHOOZ (Ardellier *et al.*, 2006), Daya Bay (An *et al.*, 2012) and RENO (Ahn *et al.*, 2012) provide good examples of LBL reactor experiments which have significantly contributed to bounding the allowed region for neutrino oscillation parameters.

For the accelerator neutrinos, one can use the same classification concerning the length of the baseline as for NR neutrinos. In this case we denote detection lengths on the order of 1 km as SBL. Experiments of this kind use neutrinos with $E \sim$ GeV, produced by pions and kaons decaying in a tunnel of 100 m. Due to chirality properties of weak interactions, the beam is mainly made of ν_μ's or $\bar{\nu}_\mu$'s, depending on the particular procedure for π and K focalization, with the additional contribution of 1% of residual $\bar{\nu}_\mu$'s or ν_μ's, respectively. A similar level of contamination due to ν_e is expected as well. The sensitivity of this kind of SBL accelerator experiments on Δm^2, such as CHARM (Allaby *et al.*, 1987), BNL-E776 (Seto, 1988), CHORUS (Eskut *et al.*, 1997), NOMAD (Astier *et al.*, 2001), LSND (Athanassopoulos *et al.*, 1996) and NuTeV (Zeller *et al.*, 2002), is on the order of 1 eV^2, and they can scrutinize different flavour transition channels depending on the setup of each experiment.

Accelerator neutrinos produced by muon decay at rest can provide similar sensitivity on Δm^2 but with energies of tens of MeV and baseline on the order of a few meters. This is the case for the $\bar{\nu}_\mu \rightarrow \bar{\nu}_e$ experiments LSND (Athanassopoulos *et al.*, 1996) and KARMEN (Armbruster *et al.*, 2002).

More energetic neutrinos can be produced by intense proton beam accelerators. Prompt neutrinos of a few hundred GeV emerge as decay products of charmed particles. This is the case in experiments such as BEBC (Grassler *et al.*, 1986), CHARM (Allaby *et al.*, 1987) and CDHSW (Dydak *et al.*, 1984). The sensitivity of such SBL ($\sim$1 km) experiments is on the order of $\Delta m^2 \gtrsim 10^2$ eV^2.

With a source-detector distance on the order of 10^3 km, LBL accelerator experiments represent a relevant option to perform appearance searches as well in some cases. LBL accelerator experiments using neutrinos with energy higher than 1 GeV are for example MINOS (Michael *et al.*, 2006), OPERA (Acquafredda *et al.*, 2006) and Tokai-to-Kamioka (T2K) (Itow *et al.*, 2001). Their sensitivity is $\Delta m^2 \gtrsim 10^{-3}$ eV^2.

The VLBL, which should use intense neutrino sources placed at distances greater than in the LBL cases, are currently under study. Possible candidates are Super-Beam (Bandyopadhyay *et al.*, 2009), Beta-Beam (Burguet-Castell *et al.*, 2004) and Neutrino Factory (Cervera *et al.*, 2000). In this case the sensitivity reachable is on the order of $\Delta m^2 \gtrsim 10^{-4}$ eV^2.

Finally, particular attention has to be paid to the so-called solar (SOL) and atmospheric (ATM) neutrino experiments, which have represented till recently the only evidence of neutrino oscillation. In particular, SOL neutrinos have been for a long time an interesting arena in which to test possible exotic particle physics models.

The sun is a powerful source of MeV electron neutrinos, produced in the thermonuclear fusion reactions occurring in the core. The solar neutrino flux at the Earth is as large as about 6×10^{10} cm^{-2} s^{-1}, but in spite of this huge number, solar neutrino detection has always been a big challenge for particle physicists. The detection technique first adopted was the *radiochemical* method, based on a nuclide weak transmutation due to ν_e capture. This was the case for the famous Homestake experiment (Cleveland *et al.*, 1998) that in 1970 detected the solar neutrino flux for the first time, putting in evidence a large discrepancy between the measured flux and the one expected on the basis of the predictions of the solar model. In particular, the experimental data indicated a deficit of ν_e. This anomaly was for a long time known as the Solar Neutrino Problem, and it triggered an enormous amount of theoretical work to study the real meaning of the measurement in terms of possible modifications of the solar model or using alternative particle physics scenarios.

The radiochemical technique has been applied since 1990 in GALLEX/GNO (Altmann *et al.*, 2000) and SAGE (Abdurashitov *et al.*, 1999) to measure also the low-energy part of solar neutrino flux, which is also the aim of current liquid scintillator detectors such as BOREXINO (Alimonti *et al.*, 2002). In the meantime, since the late 1980s the Kamiokande (Fukuda *et al.*, 1994) experiment first, and later the Super-Kamiokande (Fukuda *et al.*, 1998), have obtained a detailed neutrino image of the sun by detecting solar neutrinos via the elastic scattering reaction $\nu_\alpha + e^- \rightarrow \nu_\alpha + e^-$, which is enhanced for electron neutrinos. In recent years, the SNO (Ahmad *et al.*, 2002) collaboration used a real-time heavy water Cherenkov detector to provide definite proof that the explanation of the solar neutrino deficit was indeed neutrino oscillations. This result is due to the different characteristics of SNO than of the other water Cherenkov detectors. In fact, SNO probed solar neutrino flux via three interaction channels: charged current processes $\nu_e + d \rightarrow p + p + e^-$, neutral current processes $\nu_\alpha + d \rightarrow p + n + \nu_\alpha$, and the elastic scattering reactions $\nu_\alpha + e^- \rightarrow \nu_\alpha + e^-$.

ATM neutrinos are the byproduct of the interaction between cosmic rays, consisting mainly of protons or heavier nuclei, and the atoms of the atmosphere. From these hadronic interactions, one has a copious production of π's, which mainly decay as $\pi^+ \rightarrow \mu^+ + \nu_\mu$ or $\pi^- \rightarrow \mu^- + \bar{\nu}_\mu$. The produced muons (antimuons) can decay leptonically before reaching the ground as $\mu^- \rightarrow e^- + \bar{\nu}_e + \nu_\mu$ or $\mu^+ \rightarrow e^+ + \nu_e + \bar{\nu}_\mu$. From the multiplicities of neutrinos in the different processes one can easily see that for low-energy neutrinos $E \leq 1$ GeV, the ratio of fluxes of muon neutrinos or antineutrinos over electron neutrinos should be almost equal to two. To measure ATM neutrinos, starting from the 1960s, several observation campaigns were carried on using deep underground sites to place scintillator detectors to track muons. In such locations, the residual cosmic muon radiation is strongly peaked around the downgoing direction, whereas ATM neutrinos tend to produce muons isotropically, hence also in upgoing and horizontal directions. From the end of the 1980s ATM started to be observed from large underground Cherenkov detectors such as Kamiokande. The real breakthrough was represented by the up–down asymmetry of high-energy reactions produced by ATM neutrinos and first observed with significant statistics from Super-Kamiokande. The results were in good agreement with similar data obtained by SOUDAN2 (Allison *et al.*, 1999) and MACRO (Ambrosio *et al.*, 1998).

Several comprehensive three-flavour/three-massive-neutrino analyses have been performed in the literature to take into account all neutrino oscillation phenomenology. In the following we consider the results of the studies Fogli, 2011, 2012; Schwetz *et al.*, 2011. According to the notations in Fogli, 2011, we define

$$\delta m^2 = m_2^2 - m_1^2 > 0 \tag{1.144}$$

and

$$\Delta m^2 = m_3^2 - \frac{m_2^2 + m_1^2}{2}, \tag{1.145}$$

where $\Delta m^2 > 0 \; (< 0)$ corresponds to the normal (inverted) mass spectrum hierarchy. Solar and long-baseline reactor neutrino experiments have measured the mass-mixing parameters $(\delta m^2, \theta_{12})$ in the $\nu_e \to \nu_e$ channel, whereas atmospheric and LBL experiments have measured $(\Delta m^2, \theta_{23})$ in the $\nu_\mu \to \nu_\mu$ channel. Finally, SBL reactor experiments, which are mainly sensitive to $(\Delta m^2, \theta_{13})$, have provided information on the allowed range for the mixing angle θ_{13}.

In δm^2-sensitive oscillation searches, the relevant 3ν variables are $(\delta m^2, \sin^2 \theta_{12}, \sin^2 \theta_{13})$, with a very minor dependence of solar neutrinos on $\pm \Delta m^2$. On the other side, in Δm^2-sensitive oscillation searches, one can take as free parameters $(\pm \Delta m^2, \sin^2 \theta_{23}, \sin^2 \theta_{13})$, whereas $(\delta m^2, \theta_{12})$ are fixed at their best-fit values from the analysis of δm^2-sensitive data. In the global fit the common parameter $\sin^2 \theta_{13}$ is constrained by both classes of oscillation searches, and the results are presented as bounds on the whole set of parameters

$$(\delta m^2, \; \sin^2 \theta_{12}, \; \sin^2 \theta_{13}, \; \sin^2 \theta_{23}, \; |\Delta m^2|, \delta/\pi). \tag{1.146}$$

The results of the global analysis (Fogli *et al.*, 2012) are reported in Table 1.9 in terms of allowed ranges for each of the oscillation parameters in Eq. (1.146). A graphical summary of the allowed regions for the neutrino mixing parameters is given in Figs. 1.2 and 1.3, based on an update of Schwetz *et al.*, 2011.

Table 1.9 *Results of the global 3ν oscillation analysis*

Parameter	Best fit	1σ range	2σ range	3σ range
$\delta m^2/10^{-5}$ eV^2 (NH or IH)	7.54	7.32–7.80	7.15–8.00	6.99–8.18
$\sin^2 \theta_{12}/10^{-1}$ (NH or IH)	3.07	2.91–3.25	2.75–3.42	2.59–3.59
$\Delta m^2/10^{-3}$eV^2 (NH)	2.43	2.33–2.49	2.27–2.55	2.19–2.62
$\Delta m^2/10^{-3}$eV^2 (IH)	2.42	2.31–2.49	2.26–2.53	2.17–2.61
$\sin^2 \theta_{13}/10^{-2}$ (NH)	2.41	2.16–2.66	1.93–2.90	1.69–3.13
$\sin^2 \theta_{13}/10^{-2}$ (IH)	2.44	2.19–2.67	1.94–2.91	1.71–3.15
$\sin^2 \theta_{23}/10^{-1}$ (NH)	3.86	3.65–4.10	3.48–4.48	3.31–6.37
$\sin^2 \theta_{23}/10^{-1}$ (IH)	3.92	3.70–4.31	3.53–4.84 ⊕ 5.43–6.41	3.35–6.63
δ/π (NH)	1.08	0.77–1.36	–	–
δ/π (IH)	1.09	0.83–1.47	–	–

Source: (Fogli *et al.*, 2012).

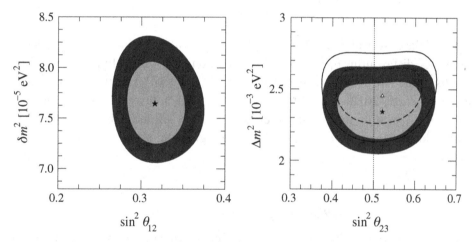

Figure 1.2 Determination of the solar and atmospheric neutrino oscillation parameters from a global three-flavour analysis at 90% C.L. and 3σ. The solar region (left panel) is mainly determined by solar + KamLAND neutrino data, whereas the atmospheric parameters (right panel) are basically fixed by atmospheric and MINOS LBL data. Lines (shaded regions) correspond to normal (inverted) neutrino mass hierarchy. Courtesy of M. A. Tórtola.

Very recently, new relevant results have been announced by two LBL accelerator experiments probing the $\nu_\mu \to \nu_e$ appearance channel, which is governed by the (Δm^2, θ_{13}) parameters (although with an additional dependence on θ_{23} and δ, absent in SBL reactor experiments). In particular, the T2K experiment has observed 6 electron-like events with an estimated background of 1.5 events, rejecting $\theta_{13} = 0$ at the level of 2.5σ (Abe *et al.*, 2011). Shortly after, the MINOS experiment

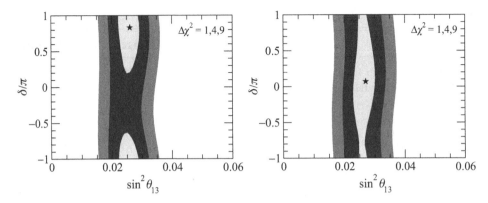

Figure 1.3 Contours of $\Delta\chi^2 = 1, 4, 9$ in the $\sin^2\theta_{13} - \delta$ plane from the analysis of global neutrino oscillation data (including T2K and MINOS, as well as Daya Bay and RENO rate data). The undisplayed oscillation parameters have been minimized over. Left (right) panel is for normal (inverted) neutrino mass hierarchy. Courtesy of M. A. Tórtola.

reported the observation of 62 electronlike events with an estimated background of 49 events, disfavouring $\theta_{13} = 0$ at 1.5σ (Adamson *et al.*, 2011). Finally, in 2012 the results of the NR neutrino experiments Double CHOOZ (Abe *et al.*, 2012), Daya Bay (An *et al.*, 2012) and RENO (Ahn *et al.*, 2012) have confirmed a nonzero value of θ_{13} with up to a 5σ statistical significance. All these data, are consistent with values of the third mixing angle in the region $\sin^2\theta_{13} \simeq 0.015 - 0.035$ (see Fig. 1.3).

1.4.2 Oscillation experiments and sterile neutrinos

The three-flavour neutrino scheme with nonzero mixing that we have just discussed has been very successful in accounting for most of the results from oscillation experiments, measuring solar, atmospheric, accelerator and reactor neutrinos. However, there exist a few experimental results (anomalies) that cannot be explained in this framework. If neutrino oscillations are responsible for all the experimental data, a solution might require additional, sterile neutrino species. These particles, which are predicted by many theoretical models beyond the SM, are neutral leptons insensitive to weak interactions and their only interaction is gravitational. Their masses are usually large (see, e.g., Chapter 3), although lighter sterile neutrinos are rarer but possible. Here we will briefly summarize these anomalies observed in neutrino experiments and the extended schemes, of four or more light neutrinos, that could explain them. An updated review of this subject is given in (Abazajian *et al.*, 2012).

Long-standing evidence, at more than 3σ, of $\bar{\nu}_\mu \to \bar{\nu}_e$ oscillations comes from the LSND experiment (Aguilar-Arevalo *et al.*, 2001). Because LSND is a SBL experiment, its results pointed out a $\Delta m^2 \sim 1\ \mathrm{eV}^2$, much larger than those required from a three-neutrino analysis, δm^2 and Δm^2, as shown in Table 1.9. This result revitalized interest in the possible existence of one or more sterile neutrinos with masses which could generate the additional squared-mass difference, but which are not contributing to the number of active neutrinos determined by LEP experiments through the measurement of the invisible width of the Z boson, $N_a = 2.9840 \pm 0.0082$ (LEP, 2006). At the same time, the KARMEN experiment (Armbruster *et al.*, 2002), very similar but not identical to LSND, has provided no support for such evidence, although a joint analysis of the two experiments (Church *et al.*, 2002) shows that their data sets are compatible with oscillations at Δm^2 either in a band from 0.2 to $1\ \mathrm{eV}^2$ or in a region around $7\ \mathrm{eV}^2$.

The initial results of the MiniBooNE experiment, designed to check the LSND results with larger distance L and energy E but similar ratio L/E, did not support such oscillation evidence in the neutrino mode $\nu_\mu \to \nu_e$ (Aguilar-Arevalo *et al.*, 2009). More recently, the collaboration presented new results for the antineutrino oscillation channel supporting the LSND evidence (Aguilar-Arevalo *et al.*, 2010) and again requiring $\Delta m^2 \sim 1\ \mathrm{eV}^2$. In a new preliminary data release by the MiniBooNE collaboration, this significance has decreased, but it is still consistent with the LSND signal. In addition, an unexplained excess of electron-like events is observed in MiniBooNE at low energies. The simultaneous interpretation of LSND (antineutrino) and MiniBooNE (neutrino and antineutrino) results in terms of sterile neutrino oscillations is only possible assuming CP violation or some other exotic scenarios, as reviewed in (Abazajian *et al.*, 2012).

A new anomaly supporting oscillations with sterile neutrinos has recently appeared from a revaluation of reactor antineutrino fluxes, which found a 3% increase relative to previous calculations of the mean flux (Mention *et al.*, 2011). As a result, data from reactor neutrino experiments at very short distances can be interpreted as an apparent 6% deficit of $\bar{\nu}_e$. This is known as the reactor antineutrino anomaly and is again compatible with sterile neutrinos having a $\Delta m^2 > 1$ eV^2. Finally, independent experimental evidence for ν_e disappearance at very short baselines exists from the gallium radioactive source experiments GALLEX and SAGE.

The existence of all these experimental hints at sterile neutrinos and a mass scale at the eV is intriguing, but so far a fully consistent picture has not emerged. Many analyses have been performed trying to explain all experimental data with 1 or 2 additional sterile neutrinos, known as the $3+1$ or $3+2$ schemes, with the corresponding additional mixing parameters. Here we will briefly describe a particular calculation, but it should be noted that none of these schemes describes all data well, as explained in detail in Abazajian *et al.*, 2012. However, for the topics of this book the potential existence of sterile neutrino oscillations would lead to important cosmological consequences, such as extra radiation from fully or partly thermalized sterile neutrinos or a larger hot or warm dark matter component. We consider these aspects in Chapters 4 and 6, but we anticipate that these $3 + 1$ or $3 + 2$ scenarios are in tension with cosmological constraints, such as the bound on the sum of neutrino masses.

Let us consider the simplest scheme, with a $3 + 1$ neutrino mixing (Okada and Yasuda, 1997; Bilenky *et al.*, 1998, 1999; Maltoni *et al.*, 2004). The effective flavour transition and survival probabilities in SBL experiments are given by

$$P^{\text{SBL}}_{\nu_\alpha \to \nu_\beta} = \sin^2 2\vartheta_{\alpha\beta} \sin^2 \left(\frac{\Delta m^2_{41} L}{4E} \right)$$

$$P^{\text{SBL}}_{\nu_\alpha \to \nu_\alpha} = 1 - \sin^2 2\vartheta_{\alpha\alpha} \sin^2 \left(\frac{\Delta m^2_{41} L}{4E} \right) \tag{1.147}$$

for $\alpha, \beta = e, \mu, \tau, s$ and $\alpha \neq \beta$, with $\sin^2 2\vartheta_{\alpha\beta} = 4|U_{\alpha 4}|^2 |U_{\beta 4}|^2$ and $\sin^2 2\vartheta_{\alpha\alpha} = 4|U_{\alpha 4}|^2 \left(1 - |U_{\alpha 4}|^2 \right)$. The same expressions are valid for antineutrinos. From Eqs. (1.147) one gets that all effective SBL oscillation probabilities depend on the absolute value of the largest squared-mass difference $\Delta m^2_{41} = m^2_4 - m^2_1$ only. By performing a global fit of all experimental data in a $3 + 1$ scheme, one obtains the best-fit values of the oscillation parameters listed in Table 1.10. In Figure 1.4 are reported the allowed regions in the $\sin^2 2\vartheta_{e\mu} - \Delta m^2_{41}$, $\sin^2 2\vartheta_{ee} - \Delta m^2_{41}$ and $\sin^2 2\vartheta_{\mu\mu} - \Delta m^2_{41}$ planes and the marginal $\Delta\chi^2$'s for Δm^2_{41}, $\sin^2 2\vartheta_{e\mu}$, $\sin^2 2\vartheta_{ee}$ and $\sin^2 2\vartheta_{\mu\mu}$ (Giunti, 2011). One can see that the preferred values are around 1 eV^2 for Δm^2_{41} and 0.1–0.001 for the $\sin^2 2\vartheta_{\alpha\beta}$. A similar analysis can be performed in a $3 + 2$ model, leading to an improved goodness of fit. The interested reader can find more details on the proposed mixing scenarios, either standard or non-standard (with CP violation, etc.), including light sterile neutrinos in Giunti, 2011; Abazajian *et al.*, 2012.

1.4.3 Neutrino mass scale experiments

Oscillation experiments are only sensitive to neutrino mass differences. In particular, the results of Table 1.9 imply only that at least one neutrino should have a mass greater than or equal to $\sqrt{\Delta m^2} \sim 0.05$ eV. Up to now the best strategy proposed to measure neutrino masses has exploited the kinematic effects that a finite neutrino mass induces on the β-decay of nuclei or on the decay features of mesons and leptons, like π and τ.

Table 1.10 *Values of* χ^2, *number of degrees of freedom (NDF), goodness of fit (GoF) and best-fit values of the mixing parameters obtained in* $3 + 1$ *and* $3 + 2$ *fits of SBL oscillation data*

	$3 + 1$	$3 + 2$
χ^2_{min}	100.2	91.6
NDF	104	100
GoF	59%	71%
Δm^2_{41} [eV^2]	0.89	0.90
$|U_{e4}|^2$	0.025	0.017
$|U_{\mu 4}|^2$	0.023	0.019
Δm^2_{51} [eV^2]		1.61
$|U_{e5}|^2$		0.017
$|U_{\mu 5}|^2$		0.0061
η		1.51π
$\Delta\chi^2_{\mathrm{PG}}$	24.1	22.2
NDF$_{\mathrm{PG}}$	2	5
PGoF	6×10^{-6}	5×10^{-4}

Note: The last three lines give the results of the parameter goodness-of-fit test (Maltoni and Schwetz, 2003): $\Delta\chi^2_{\mathrm{PG}}$, number of degrees of freedom (NDF$_{\mathrm{PG}}$) and parameter goodness of fit (PGoF) (Giunti and Laveder, 2011a).

Among such methods, the measure of the spectrum of the emitted electron in the β-decay of nuclides,

$$N(A, Z) \rightarrow N(A, Z + 1) + e^- + \bar{\nu}_e, \tag{1.148}$$

certainly represents the most sensitive technique. The differential decay rate for Cabibbo-allowed β-decay of a nucleus $N(A, Z)$ can be written as

$$\frac{d\Gamma}{dE_e} = \frac{G_F^2}{2\pi^3} \cos^2\theta_C |\mathcal{M}|^2 F(Z, E_e) E_e \, p_e \, E_\nu \, p_\nu, \tag{1.149}$$

where $\mathcal{M}$ is the nuclear matrix element, and $F(Z, E_e)$ is the Fermi function describing the electromagnetic interaction between the electron and the final-state nucleus. Because the final nuclide is much heavier than the electron, one can neglect its kinetic energy; hence $E_\nu = Q_\beta - T$, where $T = E_e - m_e$ is the kinetic energy of the electron, and the Q-value of the process is given by $Q_\beta = M_i - M_f - m_e$, where M_i and M_f are the masses of the initial and final nuclide, respectively. The maximum value for the electron kinetic energy is then $T_{\mathrm{max}} = Q_\beta - m_{\nu_e}$, where

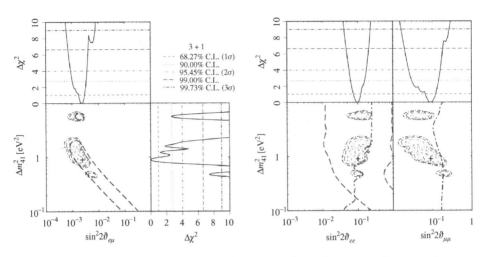

Figure 1.4 The allowed regions in the $\sin^2 2\vartheta_{e\mu}$–Δm^2_{41}, $\sin^2 2\vartheta_{ee}$–Δm^2_{41} and $\sin^2 2\vartheta_{\mu\mu}$–$\Delta m^2_{41}$ planes and marginal $\Delta \chi^2$'s obtained from a global fit in a 3 + 1 neutrino scheme. The best-fit point is shown as a cross. Left panel: the isolated long-dashed contours enclose the regions allowed at 3σ by the analysis of appearance data (LSND (Aguilar-Arevalo *et al.*, 2001), KARMEN (Armbruster *et al.*, 2002), NOMAD (Astier *et al.*, 2003), MiniBooNE (Aguilar-Arevalo *et al.*, 2009, 2010)). Right panel: the isolated dash–dotted lines are the 3σ exclusion curves obtained from reactor neutrino data and from CDHSW and atmospheric neutrino data. The isolated long-dashed lines delimit the region allowed at 99% C.L. by the gallium anomaly (Giunti and Laveder, 2011b). The detailed analysis is reported in Giunti and Laveder, 2011a. Reprinted with permission from Giunti and Laveder, 2011a. Copyright 2011 by the American Physical Society.

we are temporarily assuming that ν_e is a mass eigenstate. This approximation will be dropped later. In this case $p_\nu = \sqrt{(Q_\beta - T)^2 - m^2_{\nu_e}}$ and we have

$$\frac{d\Gamma}{dT} = \frac{G^2_F}{2\pi^3} \cos^2 \theta_C |\mathcal{M}|^2 \, F(Z, E_e) \, E_e \, p_e \, (Q_\beta - T) \sqrt{(Q_\beta - T)^2 - m^2_{\nu_e}}. \tag{1.150}$$

To determine the T range in which the presence of a finite neutrino mass affects the electron spectrum more sensibly, it is convenient to study the ratio between the differential decay rates in the case of finite m_{ν_e} and for a massless ν_e,

$$R(T) = \frac{\frac{d\Gamma}{dT}\big|_{m_{\nu_e}}}{\frac{d\Gamma}{dT}\big|_0} = \frac{\sqrt{(Q_\beta - T)^2 - m^2_{\nu_e}}}{(Q_\beta - T)}. \tag{1.151}$$

Because $dR/dT \leq 0$ for $T \in [0, T_{\max}]$, the maximum difference in the electron spectrum between the massive and massless neutrino cases occurs for values of T close to the end point. Moreover, it is easy to show that the ratio between the fraction of electrons emitted in the interval of energy $[Q_\beta - \Delta T, Q_\beta]$, denoted by $n(\Delta T)$, and the total number of events of electron emission, $n(Q_\beta)$, is

$$\frac{n(\Delta T)}{n(Q_\beta)} \propto \left(\frac{\Delta T}{Q_\beta}\right)^3. \tag{1.152}$$

Hence, to maximize the effect, one has to look for nuclides whose β-decay is characterized by a small Q_β-value. This is the case, for example, of tritium β-decay,

$$^3\text{H} \rightarrow {}^3\text{He} + e^- + \bar{\nu}_e, \tag{1.153}$$

which has $Q_\beta = 18.591$ keV. In principle, one could study the effect of a finite neutrino mass by observing that in this case the end point is shifted from Q_β (valid for the neutrino massless case) to $Q_\beta - m_{\nu_e}$. Unfortunately, such a rapid decrease of the electron spectrum cannot be probed in a statistically significant way because of the almost vanishing number of events expected near the end point. For this reason, one prefers to use the Kurie function,

$$K(T) \equiv \sqrt{\frac{d\Gamma/dT}{\frac{G_F^2}{2\pi^3}\cos^2\theta_C |\mathcal{M}|^2 F(Z, E_e) E_e\, p_e}}$$

$$= \left[(Q_\beta - T)\sqrt{(Q_\beta - T)^2 - m_{\nu_e}^2}\right]^{1/2}. \tag{1.154}$$

A departure from linear behavior in T of $K(T)$ represents the effect of a finite neutrino mass.

Because of neutrino mixing, the ν_e produced in tritium decay cannot be considered as a propagating state. This means that the real decay process is rather an incoherent sum of reactions

$$^3\text{H} \rightarrow {}^3\text{He} + e^- + \bar{\nu}_k, \tag{1.155}$$

where $\bar{\nu}_k$ is a mass eigenstate. The amplitude of this process can be obtained by using (1.32) and (1.92) (valid for both Dirac and Majorana neutrinos):

$$\mathcal{A}_{^3\text{H}\rightarrow{}^3\text{He}+e^-+\bar{\nu}_k} = -\frac{G_F}{\sqrt{2}} V_{ud}\, U_{\alpha k}\, \langle {}^3\text{He}|\bar{u}_u\gamma^\rho(1-\gamma_5)u_d|{}^3\text{H}\rangle$$

$$\times \langle e^-\, \bar{\nu}_k|\bar{u}_e\gamma_\rho(1-\gamma_5)u_{\nu_k}|0\rangle$$

$$= -\frac{G_F}{\sqrt{2}} V_{ud}\, U_{\alpha k}\, \langle {}^3\text{He}|\bar{u}_u\gamma^\rho(1-\gamma_5)u_d|{}^3\text{H}\rangle\, \bar{u}_e\gamma_\rho(1-\gamma_5)u_{\nu_k}.$$

$$\tag{1.156}$$

The mixing matrix U differs in the Dirac and Majorana cases, because of the presence of extra phases in the latter. However, in the amplitude squared modulus such a difference cannot be appreciated. As for the case of negligible mixing, we can define a Kurie function

$$K(T) = \left[(Q_\beta - T) \sum_{k=1}^{3+N_s} |U_{ek}|^2 \sqrt{(Q_\beta - T)^2 - m_k^2} \right]^{1/2}. \tag{1.157}$$

In the region where $Q_\beta - T \gg m_k$ one can approximate (1.157) as

$$K(y) \approx \left[(y + m_\beta) \sqrt{y(y + 2m_\beta)} \right]^{1/2}, \tag{1.158}$$

where

$$m_\beta^2 = \sum_{k=1}^{3+N_s} |U_{ek}|^2 m_k^2 \tag{1.159}$$

and $y \equiv Q_\beta - T - m_\beta = T_{\max} - T$. In the case of no extra sterile states ($N_s = 0$) and of a standard parameterization of the mixing matrix (1.56), the expression (1.159) becomes

$$m_\beta^2 = c_{12}^2 \, c_{13}^2 \, m_1^2 + s_{12}^2 \, c_{13}^2 \, m_2^2 + s_{13}^2 \, m_3^2. \tag{1.160}$$

The present experimental bound is $m_\beta \lesssim 2$ eV (Beringer *et al.*, 2012). The next future KArlsruhe TRItium Neutrino (KATRIN) β-experiment will be able to reduce such a limit by one order of magnitude (Osipowicz *et al.*, 2001). Besides tritium, it also seems promising to study the β-decay of the ^{187}Re isotope (Sangiorgio, 2006), which has the lowest known Q_β value, only 2.47 keV. The future sensitivity in this case will also reach values on the order of a fraction of eV.

1.4.4 Dirac or Majorana? Neutrinoless double-β decay

As we saw already, oscillation experiments cannot tell us about the real nature of neutrinos, if they are Dirac or Majorana particles. The most promising method for answering this question is probably the study of double-β decay of nuclei, as recently reviewed, e.g., in Rodejohann, 2011; Gómez-Cadenas *et al.*, 2012.

There are several nuclides able to decay via the process $N(A, Z) \rightarrow N(A, Z + 2) + 2\,e^- + 2\,\bar\nu_e$ (hereafter denoted as $2\beta_{2\nu}^-$) or $N(A, Z) \rightarrow N(A, Z - 2) + 2\,e^+ + 2\,\nu_e$ ($2\beta_{2\nu}^+$). These processes are second-order in the Fermi constant G_F. The neutrinoless double-β decays of nuclei are similar channels where there are no neutrinos emitted and are usually denoted as $2\beta_{0\nu}^-$ or $2\beta_{0\nu}^+$ if together with the daughter nuclide, two electrons or two positrons are emitted, respectively. They violate the total lepton number and are forbidden if neutrinos are massive Dirac particles, whereas they are allowed if neutrinos are massive Majorana particles.

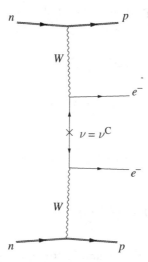

Figure 1.5 Schematic description of a neutrinoless double-β process $n + n \rightarrow p + p + 2\,e^-$.

From a kinematic point of view, because $2\beta_{2\nu}^{\pm}$ is effectively a four-body decay (the daughter nuclide is almost at rest), the two charged leptons have a continuous spectrum, whereas for $2\beta_{0\nu}^{\pm}$, the charged leptons are emitted monochromatically. This characteristic makes it possible to distinguish between the two classes of decay channel.

We note that if a nuclide can decay via a 2β process then it is crucial that its β decay should be forbidden; otherwise the typical time for observing a 2β decay will be much longer than its lifetime and the 2β transition will never be observed. This requirement represents a strong selection rule among the β-decaying nuclei. The nuclei decaying through $2\beta^+$ have typical Q-values smaller than those of nuclei decaying via $2\beta^-$. For this reason it is common to consider $2\beta^-$ processes only.

In neutrinoless double-β decay the total lepton number is violated by two units, $\Delta L = \pm 2$ for $2\beta_{0\nu}^{\mp}$. Figure 1.5 helps us to understand why such a process is forbidden in the electroweak SM in the absence of neutrinos with Majorana mass. In SM, in fact, if the upper weak vertex emits an antineutrino, the lower one ought to absorb a neutrino. Hence, if $\nu \neq \bar{\nu}$, such a process is forbidden. Such a mismatch between particle and antiparticle is worsened by a similar mismatch in the helicities. If a neutrino with positive helicity is emitted in the upper vertex, a negative one has to be absorbed in the lower one. It follows that the process is allowed only if the neutrino coincides with its antiparticle and it is massive, to solve the helicity mismatch. For a Majorana neutrino the amplitude of $2\beta_{0\nu}^{\pm}$ will thus be proportional to the Majorana mass.

This result can easily be proven by computing the neutrino propagator corresponding to the neutrino internal line of Fig. 1.5:

$$G(x_1 - x_2) = \langle 0|T \left[v_{eL}(x_1) v_{eL}^T(x_2) \right] |0\rangle$$

$$= \frac{1 - \gamma_5}{2} \sum_{k=1}^{3} U_{ek}^2 \langle 0|T \left[v_{kL}(x_1) v_{kL}^T(x_2) \right] |0\rangle \frac{1 - \gamma_5^T}{2}$$

$$= -\frac{1 - \gamma_5}{2} \sum_{k=1}^{3} U_{ek}^2 \langle 0|T \left[v_{kL}(x_1) \overline{v_{kL}}(x_2) \right] |0\rangle \, C \, \frac{1 - \gamma_5^T}{2}$$

$$= -i \sum_{k=1}^{3} U_{ek}^2 \int \frac{d^4 p}{(2\pi)^4} \frac{m_k}{p^2 - m_k^2 + i\varepsilon} e^{-ip\cdot(x_1-x_2)} \frac{1 - \gamma_5}{2} C. \quad (1.161)$$

Because in the $2\beta_{0\nu}^{\pm}$ processes of nuclides the energy available for neutrinos is much larger than m_k, $G(x_1 - x_2)$ can be approximated as follows:

$$G(x_1 - x_2) = -i \langle m_{ee} \rangle \int \frac{d^4 p}{(2\pi)^4} \frac{1}{p^2 + i\varepsilon} e^{-ip\cdot(x_1-x_2)} \frac{1 - \gamma_5}{2} C \quad (1.162)$$

with

$$\langle m_{ee} \rangle = |U_{e1}|^2 m_1 + e^{2i\lambda_2} |U_{e2}|^2 m_2 + e^{i(\lambda_3 - \delta)} |U_{e3}|^2 m_3$$

$$= c_{12}^2 c_{13}^2 m_1 + e^{2i\lambda_2} s_{12}^2 c_{13}^2 m_2 + e^{i(\lambda_3 - \delta)} s_{13}^2 m_3, \quad (1.163)$$

where the definition of the Majorana mixing parameters is given in (1.74) and (1.75). A typical analysis of the effective mass is represented in terms of a plot of its value versus the smallest neutrino mass, while the Majorana phases and/or the oscillation parameters are varied. These results are reported in Fig. 1.6, for which the 3σ ranges and the best-fit values of the oscillation parameters have been used, respectively. The shaded area is of interest because it can be covered only if the CP phases are nontrivial, i.e., if $\alpha \equiv \lambda_2$ and $\beta \equiv (\lambda_3 - \delta)/2$ are $\neq 0, \pi/2$. The values $\alpha, \beta = 0, \pi/2$ correspond to CP-conserving situations, associated with positive or negative signs of the neutrino masses, and the resulting span of $\langle m_{ee} \rangle$ is also indicated in the figure.

The neutrinoless double-β decay channel that we have discussed so far corresponds to the so-called neutrino mass mechanism, but this does not imply that this will be the dominant one. In general, other processes that violate total lepton number and exist in particle physics models beyond the SM can lead to $2\beta_{0\nu}$, i.e., to the appearance of two electrons in the final state with no missing energy (as reviewed in Hirsch, 2011; Rodejohann, 2011). However, one can show that independent of which contribution to neutrinoless double-β decay dominates, neutrinos

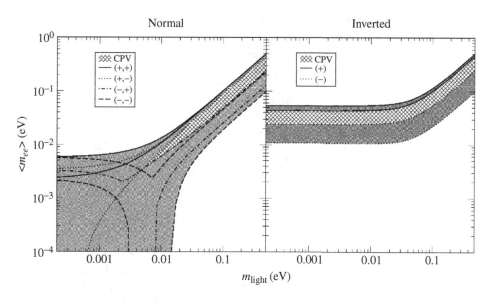

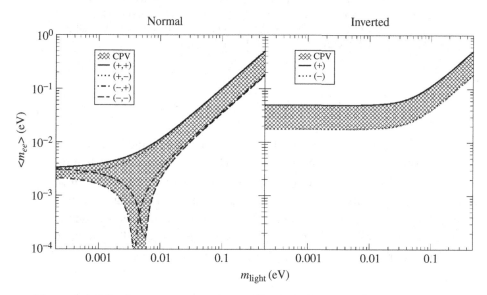

Figure 1.6 Effective mass versus the smallest neutrino mass for the 3σ ranges (top panel) and best-fit values (bottom panel) of the oscillation parameters. CP conserving and violating areas are indicated. Figure taken from (Rodejohann, 2011). Copyright 2011 World Scientific.

are guaranteed to have a nonzero Majorana mass if $2\beta_{0\nu}$ decay is observed. This is the "black box" theorem from Schechter and Valle, 1982a.

If the mass mechanism dominates, the inverse of the half-life for the $2\beta_{0\nu}$ decay process, which is the observable that experiments could measure (or limit), can be

Table 1.11 *Experimental limits at 90% C.L. on the most interesting isotopes for* $2\beta_{0\nu}$

Isotope	$T_{1/2}^{0\nu}$ [yrs]	Experiment	$\langle m_{ee}\rangle_{\min}^{\lim}$ [eV]	$\langle m_{ee}\rangle_{\max}^{\lim}$ [eV]
^{48}Ca	5.8×10^{22}	CANDLES	3.55	9.91
^{76}Ge	1.9×10^{25}	HDM	0.21	0.53
	1.6×10^{25}	IGEX	0.25	0.63
^{82}Se	3.2×10^{23}	NEMO-3	0.85	2.08
^{96}Zr	9.2×10^{21}	NEMO-3	3.97	14.39
^{100}Mo	1.0×10^{24}	NEMO-3	0.31	0.79
^{116}Cd	1.7×10^{23}	SOLOTVINO	1.22	2.30
^{130}Te	2.8×10^{24}	CUORICINO	0.27	0.57
^{136}Xe	5.0×10^{23}	DAMA	0.83	2.04
^{150}Nd	1.8×10^{22}	NEMO-3	2.35	5.08

Sources: CANDLES (Umehara *et al.*, 2008); HDM (Klapdor-Kleingrothaus *et al.*, 2001); IGEX (Aalseth *et al.*, 2002); NEMO-3 ^{82}Se (Arnold *et al.*, 2005), ^{96}Zr (Argyriades *et al.*, 2010), ^{100}Mo (Arnold *et al.*, 2005) and ^{150}Nd (Argyriades *et al.*, 2009); SOLOTVINO (Danevich *et al.*, 2003); CUORICINO (Arnaboldi *et al.*, 2005); DAMA (Bernabei *et al.*, 2002). Table taken from Rodejohann, 2011.

written as

$$\left[T_{1/2}^{2\beta_{0\nu}}\right]^{-1} = \langle m_{ee}\rangle^2 \, |M^{0\nu}|^2 \, F^{2\beta_{0\nu}}, \qquad (1.164)$$

where $F^{2\beta_{0\nu}}$ is a phase space factor that can be calculated quite accurately and $M^{0\nu}$ stands for a nuclear structure matrix element (NME). Theoretical evaluations of the NMEs for the different isotopes exist from nuclear models, but still with considerable uncertainties. This must be taken into account in converting the experimental limits on $T_{1/2}^{2\beta_{0\nu}}$ into bounds on the corresponding neutrino mass parameter $\langle m_{ee}\rangle$. A list of current limits (C.L.) from $2\beta_{0\nu}$ searches is given in Table 1.11, whereas upcoming experimental efforts are reviewed in Gómez-Cadenas *et al.*, 2012.

1.5 Nonstandard neutrino–electron interactions

Most particle physics models beyond the SM that account for neutrino masses naturally lead to nonstandard interactions for neutrinos (NSI), whose value strongly depends upon the model. For instance, NSI may arise from the structure of the charged and neutral current weak interactions in seesaw-type extended models (Schechter and Valle, 1980). Here we will focus on NSI arising only through four-fermion operators $(\bar{\nu}\nu)(\bar{f}f)$, where f is a charged lepton or a quark, but does not lead to new charged lepton interactions at tree level.

For cosmological neutrinos the most relevant period is their decoupling process, described in detail in Section 4.1, which takes place in a phase of the evolution of the universe where $e^{\pm}$ pairs are the only abundant charged leptons in the plasma. Thus, we consider only the NSI related to electrons. In the low energy limit, along with the standard neutral and charged current interactions of Eqs. (1.32) and (1.33), we consider the additional effective Lagrangian density

$$\mathcal{L}_{\text{NSI}} = -2\sqrt{2}G_{\text{F}} \left(\epsilon_{\alpha\beta}^{L} \, \overline{\nu_{\alpha L}} \, \gamma^{\mu} \, \nu_{\beta L} \, \overline{e_{L}} \, \gamma_{\mu} \, e_{L} + \epsilon_{\alpha\beta}^{R} \, \overline{\nu_{\alpha L}} \, \gamma^{\mu} \, \nu_{\beta L} \, \overline{e_{R}} \, \gamma_{\mu} \, e_{R}\right). \quad (1.165)$$

The NSI parameters $\epsilon_{\alpha\beta}^{L,R}$ can induce a breaking of lepton universality ($\alpha = \beta$) or rather a flavour-changing contribution ($\alpha \neq \beta$). Their values can be constrained by a variety of laboratory experiments (as discussed in Berezhiani and Rossi, 2002; Davidson *et al.*, 2003; Barranco *et al.*, 2006, 2008; Forero and Guzzo, 2011) or solar neutrino data (Bolaños *et al.*, 2009). In what follows we summarize the present bounds and refer the reader to these analyses for further details. In considering the bounds on the $\epsilon_{\alpha\beta}^{L,R}$ parameters, it is important to notice that they are usually obtained taking only one at a time, or at most combining two of them (such as the pair ϵ_{ee}^{L} and ϵ_{ee}^{R}). This implies that the derived constraints are expected to be weaker but more robust than in a more general analysis where many NSI parameters are simultaneously included, because cancellations may occur.

The magnitude of the NSI parameters can be constrained from the analysis of data from neutrino–electron scattering experiments, which probe the SM electroweak predictions with good precision (de Gouvea and Jenkins, 2006). The cross section for the $\nu_e\,e$ elastic scattering interaction was measured by the LSND experiment (Auerbach *et al.*, 2001), whereas $\bar{\nu}_e\,e$ scattering data were obtained from the Irvine (Reines *et al.*, 1976) and MUNU (Daraktchieva *et al.*, 2003) reactor experiments. The presence of NSI parameters leads to a change in the corresponding cross section that can be used to constrain the values of six parameters, $\epsilon_{e\alpha}^{L,R}$, by comparing with the SM prediction (including electroweak corrections), as described for instance in (Davidson *et al.*, 2003). We also have data on $\nu_{\mu}\,e$ and $\bar{\nu}_{\mu}\,e$ scattering from the CHARM II collaboration (Vilain *et al.*, 1994), which are are sensitive to six other NSI parameters $\epsilon_{\mu\alpha}^{L,R}$. Finally, as pointed out in (Berezhiani and Rossi, 2002), neutrino NSI can be also constrained by measuring the $e^{+}e^{-} \rightarrow \nu\bar{\nu}\gamma$ cross section, as done in LEP experiments. This is actually the only way to get bounds on the $\epsilon_{\tau\tau}^{L,R}$ parameters from laboratory data.

To obtain constraints on the relevant NSI parameters from these experimental data, we first follow the approach adopted in most papers. It consists of varying only one parameter at a time and fixing the remaining parameters to zero. However, we should keep in mind that this kind of analysis is fragile and might miss potential cancellations in the determination of the restrictions upon the NSI parameters.

According to Barranco *et al.*, 2008, one obtains the following 90% C.L. ranges:

$$-0.03 < \epsilon_{ee}^{L} < 0.08 \qquad 0.004 < \epsilon_{ee}^{R} < 0.151$$

$$|\epsilon_{\mu\mu}^{L}| < 0.03 \qquad |\epsilon_{\mu\mu}^{R}| < 0.03$$

$$-0.46 < \epsilon_{\tau\tau}^{L} < 0.24 \qquad -0.25 < \epsilon_{\tau\tau}^{R} < 0.43$$

$$|\epsilon_{e\mu}^{L}| < 0.13 \qquad |\epsilon_{e\mu}^{R}| < 0.13$$

$$|\epsilon_{e\tau}^{L}| < 0.33 \qquad 0.05 < |\epsilon_{e\tau}^{R}| < 0.28$$

$$|\epsilon_{\mu\tau}^{L}| < 0.1 \qquad |\epsilon_{\mu\tau}^{R}| < 0.1. \tag{1.166}$$

When the corresponding flavour-changing NSI parameters are left free in the analysis of $\nu_e \, e$ and $\bar{\nu}_e \, e$ scattering data, the bounds on $\epsilon_{ee}^{L,R}$ and $\epsilon_{e\tau}^{L,R}$ are relaxed, as shown in Barranco *et al.*, 2006. In a different approach, Barranco *et al.*, 2008 performed an analysis with free flavour-conserving NSI (six parameters). Using the same data as for the bounds in Eq. (1.166), they found the following 90% C.L. ranges:

$$-0.14 < \epsilon_{ee}^{L} < 0.09 \qquad -0.03 < \epsilon_{ee}^{R} < 0.18$$

$$-0.033 < \epsilon_{\mu\mu}^{L} < 0.055 \qquad -0.04 < \epsilon_{\mu\mu}^{R} < 0.053$$

$$-0.6 < \epsilon_{\tau\tau}^{L} < 0.4 \qquad -0.6 < \epsilon_{\tau\tau}^{R} < 0.6. \tag{1.167}$$

Slightly better bounds for the $\epsilon_{ee}^{L,R}$ were obtained in Bolaños *et al.*, 2009, including also solar neutrino data, and in Forero and Guzzo, 2011 with the addition of new reactor data from TEXONO (Deniz *et al.*, 2010).

In addition, NSI also lead to corrections at the one-loop level to processes such as the decays of the electroweak gauge bosons or lepton flavour-violating decays of charged leptons. Using conservative assumptions, Davidson *et al.*, 2003 found from the decay rates of the electroweak gauge bosons that

$$|\epsilon_{\tau\tau}^{L,R}| \lesssim 0.5, \tag{1.168}$$

again at 90% C.L. Instead, from the strong experimental limit on the branching ratio $\text{Br}(\mu^- \to e^- e^+ e^-) < 10^{-12}$ one obtains a severe bound on the flavour-changing parameter (90% C.L.)

$$|\epsilon_{e\mu}^{L,R}| \lesssim 5 \times 10^{-4}. \tag{1.169}$$

2

Overview of the Standard Cosmological Model

Cosmology is the quantitative study of the properties and evolution of the universe as a whole. Since the discovery of the redshift–distance relationship by Hubble in 1929, observations have supported the idea of an expanding universe, which can be beautifully described in terms of the Friedmann and Lemaître solution of the Einstein equations. The basis of this solution is the empirical observation that on sufficiently large scales, and at earlier times, the universe is remarkably homogeneous and isotropic. This experimental fact has been promoted to the role of a guiding assumption, the Cosmological Principle. Assuming that our observation point is not privileged, in the spirit of the Copernican revolution, one is naturally led to the conclusion that all observations made at different places in the universe should look pretty much the same independent of direction. Homogeneity and isotropy single out a unique form for the spacetime metric, the basic ingredient of Einstein theory. Cosmological models can then be quantitatively worked out after specification of the matter content, which acts as the source for curvature. Results can be then compared with astrophysical data, which in the last decades have reached a remarkable precision.

Actually, the Cosmological Principle works only on scales larger than 100 Mpc, yet it is a powerful assumption. In fact, several observables, such as the distribution in the sky of the cosmic microwave background (CMB), show inhomogeneities which are quite small, so that they can be treated as perturbations of a reference model (i.e., a reference metric) which is homogeneous and isotropic.

The idea of an expanding universe implies that all matter was characterized by a higher density in the past, and was also hotter than today, back to an initial singularity where quantum aspects of gravity are expected to have been important, and one is forced to abandon the classical Einstein theory. Apart from this initial stage, the *hot Big Bang* model (in which, as said, as we proceed toward earlier times all particle species are denser and with larger mean energy) provides quantitative predictions for many observational features which can be tested experimentally, such as the CMB formation, the primordial production of light nuclei, the matter–antimatter asymmetry, the evolution of density perturbations and the formation of large-scale structures, just to mention some key topics.

This model is self-consistent, provided that at early times, the expansion was accelerated during a phase known as *inflation*. This stage is now one of the cornerstones of the standard cosmological model, because it predicts very robust signatures that influence the later evolution, such as the fact that inhomogeneities develop starting from a nearly scale-invariant initial spectrum, or that CMB photons coming from very different directions share the same distribution of energy, up to very small fluctuations of order 10^{-5}, as first detected by the COBE satellite (Smooth *et al.*, 1992). The success of inflation is only hampered by the fact that a full and satisfactory description of the details of the model is still missing. As has been said, inflation is a paradigm searching for a theory.

The validity of the hot Big Bang model is deeply related to the observation that CMB photons are characterized by a remarkably accurate Bose–Einstein distribution in wavelength, with a present temperature $T_0 = 2.725 \pm 0.002$ K at 95% C.L. (Mather *et al.*, 1999). This is a clear signal that photons were once in thermodynamic equilibrium with charged particles (electrons, nuclei, etc.), and thus that the universe was filled by a hot plasma of ionized nuclei and charged leptons. The natural tool to describe the properties of this plasma is relativistic statistical mechanics within an expanding background. As long as interactions among particles are strong enough to guarantee equilibrium conditions, it is possible to introduce the concept of temperature and describe the time evolution of all species using equilibrium statistical mechanics, or its macroscopic counterpart, equilibrium thermodynamics.

The empirical fact that systems spontaneously evolve toward equilibrium configurations holds only if the system is unperturbed. In the universe, if the rate of expansion is too high, particles may fail to reach equilibrium at certain epochs. This observation is crucial to explain the production of light nuclei during primordial nucleosynthesis, the relic abundance of baryons, the expected relic dark matter density today, and some (as yet unobserved) features of the cosmological neutrino background. To describe all these phenomena, it is necessary to abandon equilibrium thermodynamics and exploit the tools of kinetic theory. The evolution of particle distributions in phase space is then given by a quite complicated nonlinear integro-differential equation, the Boltzmann equation. Fortunately, in several cases, this equation can be greatly simplified and reduced to an ordinary differential equation.

This chapter is an overview of the standard homogeneous and isotropic cosmological model (perturbations will first appear on the stage in Chapter 5) and of the tools of kinetic theory and thermodynamics in an expanding universe. We then give a first look at the properties and evolution of the cosmological neutrino background (in the following CNB). The discussion will be quite brief and mainly aimed at introducing the most relevant concepts and theoretical tools which will be widely used in the following chapters. The reader may wish to deepen several aspects of our treatment and refer to excellent monographs on cosmology and general relativity (such as Weinberg, 1972; Wald, 1984; Kolb and Turner, 1994; Dodelson, 2003; Mukhanov, 2005; Weinberg, 2008) and on kinetic theory and statistical mechanics (such as Huang, 1987; Bernstein, 1988).

2.1 The homogeneous and isotropic universe

2.1.1 The dynamics of expansion

The assumption that our universe is spatially homogeneous (physical conditions are the same at every point of a fixed-time hypersurface) and isotropic (they are independent of direction at any given point) implies that it is possible to choose a suitable set of coordinates in which the spacetime metric $g_{\mu\nu}$ is remarkably simple and the whole symmetries of the system are clearly manifest. With this choice the metric takes the standard Friedmann–Robertson–Walker (FRW) expression (Friedmann, 1922, 1924; Robertson, 1935; Walker, 1936),

$$g_{\mu\nu}dx^{\mu}dx^{\nu} = -dt^2 + a^2(t)\left(\frac{dr^2}{1-kr^2} + r^2\left(d\theta^2 + \sin^2\theta d\phi^2\right)\right), \quad (2.1)$$

where t is the "physical time", whereas r, θ and ϕ are the spatial comoving coordinates, which label the points of the 3-dimensional constant-time slice. They are not sensitive to the expansion dynamics and define comoving observers, namely those which have constant values of r, θ and ϕ as time flows. The time variable t represents the proper time as measured by this family of observers. The *physical* distances are weighted by the scale factor $a(t)$, and are thus increasing with time in an expanding universe. The scale factor evolution is eventually fixed by the dynamics encoded in Einstein equations. It is defined up to an arbitrary constant rescaling, which corresponds to a rescaling of comoving variables. In particular, this constant can be chosen in such a way that $a = 1$ today; i.e., comoving distances are chosen as the present physical distances.

The parameter k specifies the spatial curvature of the model. By a suitable redefinition of the radial variable r it can be reduced to the canonical values $k = +1, 0, -1$ corresponding to a constant-positive-curvature 3-space (a 3-sphere), a flat 3-dimensional plane and a negative-curvature 3-dimensional space, respectively. Notice that in general, it is not possible to set both $a = 1$ today and $k = \pm 1$. Their relationship will be given by the Friedmann equation in the following.

For the flat case, which will be frequently used in the following as a reference model, and is currently favoured by experimental observations, we can use cartesian coordinates so that the FRW metric becomes

$$g_{\mu\nu}dx^\mu dx^\nu = -dt^2 + a^2(t)\left(dx^2 + dy^2 + dz^2\right). \tag{2.2}$$

In several applications it is convenient to use the conformal time η defined in terms of the physical time t by

$$\eta \equiv \int \frac{dt}{a(t)}. \tag{2.3}$$

With this definition the metric then becomes

$$g_{\mu\nu}dx^\mu dx^\nu = a^2(\eta)\left(-d\eta^2 + \frac{dr^2}{1 - kr^2} + r^2\left(d\theta^2 + \sin^2\theta d\phi^2\right)\right). \tag{2.4}$$

The variable η has a simple interpretation. As in special relativity, the world line of a (freely falling) photon satisfies the condition

$$g_{\mu\nu}dx^\mu dx^\nu = 0. \tag{2.5}$$

This relation holds if one chooses a local inertial coordinate frame in a sufficiently small neighbourhood of a given point, and being an invariant statement should be valid in any curved spacetime. Thus, the comoving distance $\chi(t)$ travelled by the photon in the time interval $t - t_i$ is

$$\chi(t) = \int_{t_i}^{t} \frac{dt'}{a(t')} = \eta - \eta_i. \tag{2.6}$$

For geodesics corresponding to propagation along directions with fixed θ and ϕ, we have $d\chi = dr/\sqrt{1 - kr^2}$, and it is useful to rewrite the FRW metric using χ as a coordinate,

$$g_{\mu\nu}dx^\mu dx^\nu = -dt^2 + a^2(t)\left(d\chi^2 + r^2(\chi)(d\theta^2 + \sin^2\theta d\phi^2)\right), \tag{2.7}$$

where the function $r^2(\chi)$ is

$$r^2(\chi) = \begin{cases} \sinh^2\chi, & k = -1 \\ \chi^2, & k = 0 \\ \sin^2\chi, & k = +1. \end{cases} \tag{2.8}$$

The motion of a massless particle such as a photon is a particular case of geodesic motion, corresponding to a null geodesic. In general, the freely falling motion of a particle in a given spacetime metric is given by the condition that the four-velocity vector

$$u^\mu = \frac{dx^\mu}{d\lambda} \tag{2.9}$$

is parallel transported along the particle trajectory. The parameter λ is some monotonically increasing variable along the particle path. This gives the geodesic equation

$$\frac{du^\mu}{d\lambda} + \Gamma^\mu_{\nu\rho} u^\nu u^\rho = 0, \tag{2.10}$$

where the Christoffel symbols are related to first derivatives of the metric

$$\Gamma^\mu_{\nu\rho} = \frac{g^{\mu\sigma}}{2} \left(\frac{\partial g_{\sigma\rho}}{\partial x^\nu} + \frac{\partial g_{\nu\sigma}}{\partial x^\rho} - \frac{\partial g_{\nu\rho}}{\partial x^\sigma} \right), \tag{2.11}$$

with $g^{\mu\nu} g_{\nu\rho} = \delta^\mu_\rho$, δ^μ_ρ being the Kronecker delta. A straightforward computation shows that in the FRW metric the only nonvanishing components are the following, for $k = 0$:

$$\Gamma^j_{0i} = \frac{\dot{a}}{a} \delta^j_i, \quad \Gamma^0_{ij} = \frac{\dot{a}}{a} g_{ij}, \tag{2.12}$$

with the dot denoting the derivative with respect to physical time, whereas if one adopts conformal time

$$\Gamma^0_{00} = \frac{a'}{a}, \quad \Gamma^j_{0i} = \frac{a'}{a} \delta^j_i, \quad \Gamma^0_{ij} = \frac{a'}{a} g_{ij}, \tag{2.13}$$

where the prime is the derivative with respect to conformal time. For $k \neq 0$ the fully spatial components Γ^k_{ij} are also nonvanishing, and their computation is left to the reader as an exercise.

For a particle of mass m one can also introduce the energy–momentum 4-vector as $P^\mu = m u^\mu$, whose components satisfy the mass shell condition

$$g_{\mu\nu} P^\mu P^\nu = -m^2 \tag{2.14}$$

and also obey the geodesic equation, as we see immediately by rescaling the evolution parameter $\lambda \to \lambda/m$ in (2.10). Using the fact that $P^0 = dt/d\lambda$, one gets

$$P^0 \frac{dP^\mu}{dt} + \Gamma^\mu_{\nu\rho} P^\nu P^\rho = 0. \tag{2.15}$$

This equation is also valid for massless particles, with $g_{\mu\nu} P^\mu P^\nu = 0$ and the components of P^μ being given by the energy and linear momentum of the particle, $P^\mu = (E, P^i)$.

The geodesic equation makes it possible to compute the evolution in time of the energy and momentum of particles, and gives one of the main observational consequences of an expanding universe, the redshift of energy/momentum of photons, and more generally the fact that *physical* linear momentum of all particles decreases as $1/a$ as the universe expands.

Before deriving this result, we introduce two different definitions of linear momentum. By lowering indices using the metric we define the *comoving momentum*,

$$P_\mu = g_{\mu\nu} P^\nu. \tag{2.16}$$

The spatial components of this (co)vector represent the variables canonically conjugated to comoving coordinates, and we will soon show that they are kept invariant by the expansion. Using the mass shell condition (2.14) for $k = 0$ (the general case can be treated in close analogy using χ as a radial variable),

$$-E^2 + a^2(t) \sum_i (P^i)^2 = -m^2. \tag{2.17}$$

We can also introduce the *physical* momentum, which measures the rate of change in physical distances,

$$p^i = a(t)P^i, \tag{2.18}$$

so that

$$-E^2 + p^2 = -m^2, \quad p = \left(\sum_i p^{i2} \right)^{1/2}. \tag{2.19}$$

Let us now compute how the momentum components P^i change with the expansion. Using (2.12) we have

$$E\left(\frac{dP^i}{dt} + 2\frac{1}{a}\frac{da}{dt}P^i \right) = 0, \tag{2.20}$$

from which we see that

$$P^i \sim a^{-2} \tag{2.21}$$

$$P_i \sim \text{constant} \tag{2.22}$$

$$p^i \sim a^{-1}. \tag{2.23}$$

In the particular case of photons, this result gives a simple description of the observed redshift of wavelength. Suppose that a photon is in fact emitted by some astrophysical source (galaxy, star, etc.) at some value of time t_{em} with energy E_{em} and is then received and absorbed by a detector at t_{obs}. The energy which is measured is then $E_{obs} = E_{em} a(t_{em})/a(t_{obs})$, or in terms of the radiation wavelength,

$$\lambda_{obs}/\lambda_{em} \equiv 1 + z = a(t_{obs})/a(t_{em}), \tag{2.24}$$

which defines the redshift z. The redshift of a given object is related to the scale factor at emission time through

$$1 + z = \frac{a_0}{a}, \tag{2.25}$$

where a_0 is the scale factor today (usually set to unity in the case of a flat universe). This result gives an interpretation of the Hubble law, the linear dependence of the redshift on the distance between us and the source. For small redshifts we can expand (2.25) at first order around present time t_{obs}:

$$1 + z = 1 + \frac{1}{a}\frac{da}{dt}\Big|_{t_{obs}} (t_{obs} - t) + \mathcal{O}((t_{obs} - t)^2). \tag{2.26}$$

On the other hand, if we keep only first-order terms, the time difference $t_{obs} - t$ is also the physical distance d of the source in units $c = 1$; see (2.6). Defining the Hubble constant

$$H_0 = \frac{1}{a}\frac{da}{dt}\Big|_{t_{obs}}, \tag{2.27}$$

we find

$$cz \sim H_0 d. \tag{2.28}$$

The Hubble constant has dimension $[t]^{-1}$ and its order of magnitude is of a few tenths of km s^{-1} Mpc^{-1}. It is usually expressed in terms of the dimensionless parameter h as

$$H_0 = 100\, h\, \text{km s}^{-1}\, \text{Mpc}^{-1}. \tag{2.29}$$

Since the seminal paper of Hubble in 1929 (Hubble, 1929), the experimental determination of H_0 has been continuously refined till today, and it is currently known to be $h \sim 0.7$, with a few percent uncertainty. A nice review of experimental measurements of h can be found in Weinberg, 2008. The value of h will also appear here and there in the following, and can be slightly different depending upon the experimental datasets which are used. The redshift–distance relationship at higher z will be further considered in the next section, once a proper notion of distance in an expanding metric has been discussed.

Coming back to the energy redshift, consider now the case of a massive particle, such as (at least some) neutrino mass eigenstates. In this case we have

$$E = \sqrt{m^2 + p_{em}^2 \frac{a(t_{em})^2}{a(t)^2}}. \tag{2.30}$$

Later in this chapter we will see that background neutrinos were *emitted* when the scale factor was a factor of 10^{-10} smaller than today, and with a typical momentum and energy $p_{em} \sim$ MeV; thus they were highly relativistic at that time. Today their linear momentum is therefore on the order of $p \sim a(t_{em})$ MeV $\sim 10^{-4}$ eV. Even in the case of the lighter mass scheme compatible with neutrino flavour oscillation data, $m_1 \sim 0$, $m_2 \sim 10^{-2}$ eV and $m_3 \sim 0.05$ eV, we see that at least two neutrino states are nonrelativistic today.

The dynamics of the expansion is completely encoded in the time-dependent function $a(t)$ which appears in the metric components $g_{\mu\nu}$. These are the dynamical variables in general relativity and satisfy the Einstein equations

$$G_{\mu\nu} \equiv R_{\mu\nu} - \frac{1}{2}g_{\mu\nu}R = 8\pi G T_{\mu\nu} - \Lambda g_{\mu\nu}, \tag{2.31}$$

where $R_{\mu\nu}$ and R are the Ricci tensor and scalar, respectively:

$$R_{\mu\nu} = \frac{\partial \Gamma^\sigma_{\mu\nu}}{\partial x^\sigma} - \frac{\partial \Gamma^\sigma_{\mu\sigma}}{\partial x^\nu} + \Gamma^\sigma_{\rho\sigma}\Gamma^\rho_{\mu\nu} - \Gamma^\sigma_{\rho\nu}\Gamma^\rho_{\mu\sigma}, \quad R = R_{\mu\nu}g^{\mu\nu}. \tag{2.32}$$

Λ is the Einstein cosmological constant and G the Newton constant. In natural units, G defines a mass scale, the Planck mass scale:

$$m_{Pl} = G^{-1/2} = 1.22 \times 10^{19} \text{ GeV}. \tag{2.33}$$

We will be using both G and m_{Pl} in the following.

Finally, $T_{\mu\nu}$ is the stress–energy tensor of all matter species which fill the universe. This tensor is symmetric, $T_{\mu\nu} = T_{\nu\mu}$, and is covariantly conserved,

$$\nabla_\mu T^{\nu\mu} = \frac{\partial T^{\nu\mu}}{\partial x^\mu} + \Gamma^\nu_{\mu\rho}T^{\rho\mu} + \Gamma^\mu_{\mu\rho}T^{\nu\rho} = 0, \tag{2.34}$$

where ∇_μ denotes the connection, or covariant derivative, associated with the metric $g_{\mu\nu}$. In many applications it is a good approximation to describe a gas of particles such as electrons, photon, or neutrinos as a perfect fluid with no viscosity. In this case the corresponding stress–energy tensor takes the form

$$T^\mu_\nu = (\rho + P)u^\mu u_\nu + P\delta^\mu_\nu, \tag{2.35}$$

where ρ is the fluid energy density, P the pressure and u^μ the fluid 4-velocity. In particular, if the fluid has zero velocity in the comoving frame, $u^\mu = (1, 0, 0, 0)$, then

$$T^0_0 = -\rho, \quad T^i_j = P\,\delta^i_j. \tag{2.36}$$

Information on the particular fluid under consideration is encoded into the equation of state, the functional relationship relating pressure and energy density, $P = P(\rho)$. For example, for relativistic particles (usually denoted as "radiation"), $P = \rho/3$, whereas for nonrelativistic particles, with very small kinetic energy with respect to their rest mass (dust or simply "matter"), $P = 0$. At least in these two relevant cases, the equation of state is a simple linear relationship

$$P = w\rho. \tag{2.37}$$

The cosmological constant term can be effectively seen as a further contribution to the total stress–energy tensor, with the definition

$$T_{\mu\nu,\Lambda} \equiv -\frac{\Lambda}{8\pi G} g_{\mu\nu}, \tag{2.38}$$

so that $\rho_\Lambda = \Lambda/(8\pi G) = -P_\Lambda$ and $w = -1$ in this case. For a positive value of ρ_Λ, this contribution provides a negative pressure term in the Einstein equations.

Apart from a few (quite relevant) cases, namely the inflationary stage and the late accelerating stage, the stress–energy receives a contribution from a "gas" of particles, i.e., incoherent excitations of quantum fields, which can be treated to some extent as classical particles with suitable statistics (Bose–Einstein or Fermi–Dirac). This is, for example, the case of photons, electrons/positrons, neutrinos and baryons. In these cases the stress–energy tensor can be defined in terms of the distribution of particles in phase space, in the spirit and exploiting the methods of statistical mechanics. To elaborate on this point, we start defining the distribution function in the single-particle phase space (the so-called μ–space), $f(P_\mu, x^\nu)$, which gives the number of particles at a given point (x^i, P_j, P_0) and at a given time t. For a homogeneous and isotropic background the distribution function depends only on the energy, the time and the modulus of momentum. Integrating over the physical volume, we can define the particle current density as

$$n_\mu(t) = g \int \frac{d^4 P}{(2\pi)^4} (-\det(g))^{-1/2} P_\mu \, 2(2\pi)\theta(P_0)\delta(g^{\mu\nu} P_\mu P_\nu + m^2) \, f(P, t) \tag{2.39}$$

and the stress–energy tensor as

$$T_\nu^\mu(t) = g \int \frac{d^4 P}{(2\pi)^4} (-\det(g))^{-1/2} P^\mu P_\nu \, 2(2\pi)\theta(P_0)\delta(g^{\mu\nu} P_\mu P_\nu + m^2) \, f(P, t), \tag{2.40}$$

where $\det(g)$ is the determinant of the metric, g is the number of internal degrees of freedom, such as helicity, and integration is over the covariant 4-momenta P_μ. The Dirac delta function enforces the mass shell condition ("real" particles), in other words, gives the energy as a function of momentum, whereas the θ function ensures positivity of the energy. Integrating over the energy, and using $(-\det(g))^{-1/2} d^3 P =$

$a^{-3} d^3 P = d^3 p$ (we recall that p^i denote the physical momentum), we obtain

$$n^\mu(t) = g \int \frac{d^3 p}{(2\pi)^3} \frac{P^\mu}{E} f(p, t) \tag{2.41}$$

$$T_\nu^\mu(t) = g \int \frac{d^3 p}{(2\pi)^3} \frac{P^\mu P_\nu}{E} f(p, t), \tag{2.42}$$

where the energy is now expressed in terms of p as in (2.19).

Notice that from isotropy, only the time component n^0 is nonzero, giving the particle number density

$$n(t) = g \int \frac{d^3 p}{(2\pi)^3} f(p, t). \tag{2.43}$$

Similarly, all nondiagonal components of T_ν^μ vanish and for the diagonal terms, again using isotropy, we get

$$T_0^0 = -\rho = -g \int \frac{d^3 p}{(2\pi)^3} E f(p, t) \tag{2.44}$$

$$T_j^i = P \delta_j^i = \delta_j^i g \int \frac{d^3 p}{(2\pi)^3} \frac{p^2}{3E} f(p, t). \tag{2.45}$$

We will come back to these expressions later on in this chapter. For the moment, let us only remark that for massless particles $E = p$, so that we get again the result $P = \rho/3$, whereas for nonrelativistic particles $E \sim m$ and

$$\rho \sim mn \tag{2.46}$$

$$P \sim \rho \int \frac{d^3 p}{(2\pi)^3} \frac{p^2}{3m^2} f(p, t) \left(\int \frac{d^3 p}{(2\pi)^3} f(p, t) \right)^{-1} \ll \rho. \tag{2.47}$$

Having specified the form of the stress–energy tensor, we can now specialize Einstein equations to the FRW metric and compute the dynamics of the expansion. The covariant conservation of $T^{\mu\nu}$ in the case of an ideal fluid reads, using (2.34) with $\nu = 0$[1],

$$\dot\rho + 3\frac{\dot a}{a}(\rho + P) = 0, \tag{2.48}$$

which is easy to solve for an equation of state of the form (2.37):

$$\rho(a) = a^{-3(1+w)}. \tag{2.49}$$

---

[1] Setting $\nu = i$ in (2.34) gives a trivial identity, $0 = 0$.

Here are some relevant cases:

$$\begin{aligned}
\rho_R &\sim a^{-4}, & w &= 1/3 & \text{(radiation)} \\
\rho_M &\sim a^{-3}, & w &= 0 & \text{(matter)} \\
\rho_\Lambda &\sim \text{const.}, & w &= -1 & \text{(cosmological constant).}
\end{aligned} \tag{2.50}$$

Computing the Einstein equations for a FRW metric is lengthy but in principle straightforward. The reader not familiar with these computations is advised to go through the explicit calculation, as well as referring to, e.g., Weinberg, 2008, where a detailed sketch is presented. We will not repeat them here but simply report the result for the 0–0 component, first deduced by Friedmann (Friedmann, 1922, 1924),

$$\frac{\dot{a}^2}{a^2} + \frac{k}{a^2} = \frac{8\pi G}{3}\rho + \frac{\Lambda}{3}, \tag{2.51}$$

where ρ is the total energy density, summed over all species which fill the universe. Along with (2.48), this equation is already sufficient to solve for the scale factor as a function of time. Indeed, the nondiagonal components of the Einstein and stress–energy tensors vanish as a consequence of the isotropy of the metric, and the diagonal space–space components are not adding further information, as they can be obtained by combining the time derivative of (2.51) with (2.48). This is a consequence of the Bianchi identities $\nabla_\mu G^{\nu\mu} = 0$. In particular, after differentiating the Friedmann equation and using (2.48), one gets an equation for the acceleration rate of the expansion,

$$\frac{\ddot{a}}{a} = -\frac{4\pi G}{3}(\rho + 3P) + \frac{\Lambda}{3}. \tag{2.52}$$

Notice that the cosmological constant term can also be written in terms of $\rho_\Lambda = -P_\Lambda$ and absorbed as a further contribution of the form $(\rho + 3P)$ in the bracket.

The Friedmann equation relates the expansion rate (also called the Hubble parameter),

$$H \equiv \frac{\dot{a}}{a}, \tag{2.53}$$

to the total energy density. If there is a single dominant contribution to ρ with a definite equation of state parameter w, the solution for the function $a(t)$ is very simple. Using (2.49) in (2.51), we obtain

$$a(t) = \begin{cases} t^{\frac{2}{3(1+w)}}, & w \neq -1 \\ e^{\overline{H}t}, & w = -1. \end{cases} \tag{2.54}$$

If the universe is *radiation-dominated*, the main contribution comes from relativistic particles with $w = 1/3$, leading to $a \sim \sqrt{t}$. For a universe dominated by nonrelativistic particles, i.e., *matter-dominated*, one gets $a \sim t^{2/3}$. A cosmological constant drives the scale factor toward an exponential solution, with an asymptotically constant Hubble parameter $H = \overline{H}$.

In terms of the critical density

$$\rho_{cr} \equiv \frac{3H^2}{8\pi G}, \tag{2.55}$$

the Friedmann equation can be rewritten in the form

$$\Omega(a) - 1 = \frac{k}{H^2 a^2}, \tag{2.56}$$

where $\Omega(a)$ is the total energy density in units of ρ_{cr}, $\Omega(a) = \rho/\rho_{cr}$. We see that the spatial curvature is fixed by the value of ρ. A universe filled with a critical energy density, $\Omega = 1$, is spatially flat. It has positive curvature if ρ is larger, whereas it is a Lobachevski space (negative curvature) for $\rho < \rho_{cr}$. It is also customary to look at the curvature term in (2.56) as a particular contribution to the energy density and define $\Omega_k \equiv -k/(H^2 a^2)$, so that

$$\Omega(a) + \Omega_k(a) = 1. \tag{2.57}$$

In (2.56) we have introduced the explicit dependence of Ω on the scale factor to distinguish this time-dependent variable (except for the case $\Omega = 1$) from its value computed *today*, which is often denoted by Ω in the literature, although sometimes an index "0" is used, Ω_0. In the following we will define the present contributions of radiation, matter, curvature and cosmological constant in units of the critical density today ($a = 1$),

$$\rho_{cr,0} = \frac{3H_0^2}{8\pi G} = 1.878 \times 10^{-29} h^2 \mathrm{g\ cm}^{-3}, \tag{2.58}$$

as

$$\Omega_R = \frac{\rho_{R,0}}{\rho_{cr,0}} = \frac{8\pi G \rho_{R,0}}{3H_0^2}, \quad \Omega_M = \frac{\rho_{M,0}}{\rho_{cr,0}}, \quad \Omega_k = -\frac{k}{H_0^2}, \quad \Omega_\Lambda = \frac{\rho_\Lambda}{\rho_{cr,0}}, \tag{2.59}$$

in terms of which the Friedmann equation for an expanding universe reads

$$H(a) = H_0 \left(\Omega_R a^{-4} + \Omega_M a^{-3} + \Omega_k a^{-2} + \Omega_\Lambda \right)^{1/2} \tag{2.60}$$

or, using redshift (see (2.25)),

$$H(z) = H_0 \left(\Omega_R(1+z)^4 + \Omega_M(1+z)^3 + \Omega_k(1+z)^2 + \Omega_\Lambda \right)^{1/2}. \tag{2.61}$$

A sketch of the evolution of the various contributions to $H(a)$, from photons, neutrinos, cold dark matter, baryons and cosmological constant is shown in Fig. 2.1.

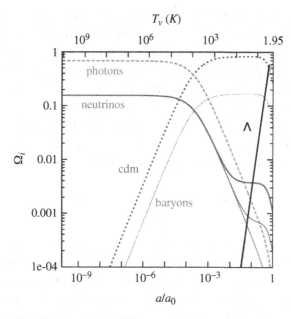

Figure 2.1 Evolution of the fractional energy density Ω of photons, three neutrino species (one massless and two massive, 0.05 and 0.009 eV), cold dark matter (cdm), baryons and a cosmological constant Λ, as functions of the scale factor a/a_0 or of the neutrino temperature T_ν. Notice the change in the behaviour of the two massive neutrino contributions when they become nonrelativistic particles.

For sufficiently small redshifts, the expansion rate can be Taylor-expanded at first order around the present time and parameterized in terms of H_0 and the deceleration parameter, defined as

$$q = -\frac{\ddot{a}}{a}\bigg|_0 \frac{1}{H_0^2}. \tag{2.62}$$

From (2.52), q can be written as

$$q = \frac{\Omega}{2}\left(1 + 3\frac{P}{\rho}\right)_0, \tag{2.63}$$

where in the bracket one should sum over all components which contribute to the stress–energy tensor today.

2.1.2 Distances in the universe

Definition of distances in an expanding universe is more involved than with a static background. If we receive today the light from a distant source, such as a galaxy or a quasar, to determine how far this source is from our observation point, it is necessary to take into account the cosmological expansion between the time the light was emitted and today. This along with the fact that a photon, and more generally an arbitrary particle species (such as a neutrino), will experience the stretching of physical distances during its journey renders this definition not unique. Actually, the particular operational procedures which are used to define distances will correspond to different results (luminosity distance, angular distance, etc.).

We recall that comoving coordinates are unaffected by the expansion. At a given time t, the physical distance between two systems placed at points labelled by comoving coordinates x^i and x'^i is, for a spatially flat metric,

$$d(t) = a(t) \sqrt{\sum_i (x^i - x'^i)^2}. \tag{2.64}$$

As this distance changes with time, unfortunately this definition might be of some help only if we were able to find an experimental procedure to measure it *instantaneously*, which is forbidden by causality, or as long as the relative speed is very small with respect to the speed of light. This amounts to saying that the scale factor has changed by a little, and expanding around t and recalling the definition of the Hubble parameter, we have that the relative velocity of the two systems is

$$v(t) \sim H(t)d(t). \tag{2.65}$$

If t is the present time, $H(t) = H_0$, this is again the Hubble law, where now the redshift z of the emission/absorption lines measured in the detector can be described as due to the Doppler effect caused by the relative velocity of the source and the detector, $z = v$ (c is as usual set to unity). From this result we see that the effects related to the expansion cannot be neglected for large redshifts (of order unity or bigger). In this case, distances cannot be defined without taking into account the whole expansion history between emission and detection.

Let us start by defining the so-called comoving particle horizon $\chi_p(t)$, defined as the comoving distance travelled by light between a very early initial time t_i and some time t. As photons propagate along null geodesics we get in the FRW metric – see (2.7) –

$$\chi_p(t) = \int_0^r \frac{dr'}{\sqrt{1 - kr'^2}} = \int_{t_i}^t \frac{dt'}{a(t')} = \int_0^{a(t)} \frac{da'}{a'^2 H(a')} = \int_z^\infty \frac{dz'}{H(z')}, \tag{2.66}$$

where we assume $a(t_i) \simeq 0$. This distance corresponds to the largest distance scale from which an observer can receive information at a given time t. Similarly, if we receive today light from a source with redshift z, the signal has travelled a comoving distance given by

$$\chi = \int_{a(t_{em})}^{a_0} \frac{da'}{a'^2 H(a')} = \int_0^z \frac{dz'}{H(z')}, \tag{2.67}$$

where t_{em} is the emission time, and can be obtained in terms of the redshift if we know the expansion history of the universe, by inverting the function $a(t) = a_0/(1 + z(t))$.

As we will see later, photons of the CMB are emitted from a thin surface, when they last interact with charged particles, the last scattering surface, placed at redshift $z_{LS} \sim 10^3$. If the energy density is dominated by dust, the comoving distance of this surface is therefore

$$\chi_{LS} = \frac{1}{H_0} \int_0^{z_{LS}} dz' \frac{1}{(1 + z')^{3/2}} = \frac{2}{H_0} \left(1 - \frac{1}{\sqrt{1 + z_{LS}}} \right), \tag{2.68}$$

whereas it increases if we include the contribution of Ω_Λ (for $\Omega_\Lambda = 1$ one would get $\chi_{LS} = z_{LS}/H_0$).

In the case of a particle with mass m, such as a neutrino, one can simply work out the analogue of (2.67). Using the mass shell condition, one finds

$$d\chi = \frac{\sqrt{E^2 - m^2}}{aE} dt \tag{2.69}$$

and using the modulus of the comoving momentum[2] $y = a\sqrt{E^2 - m^2}$,

$$\chi = \int_{a(t_{em})}^{a(t_{obs})} \frac{dt}{a(t)} \frac{y}{\sqrt{y^2 + m^2 a(t)^2}} = \int_0^z \frac{dz'}{H(z')} \frac{y}{\sqrt{y^2 + m^2/(1 + z)^2}}. \tag{2.70}$$

As an example, the reader could compute the distance corresponding to the cosmological neutrino background, showing that, although located at higher redshift with respect to the CMB's last scattering surface (and older times), it is indeed closer for neutrino masses larger than $m_\nu \sim 10^{-4}$ eV.

Coming back to photon propagation, we mentioned already that the notion of distance depends on the experimental procedure used to measure it. Suppose that we are interested in pointlike sources, such as far galaxies or stars, of which we know the absolute luminosity L; i.e., we consider a *standard candle*, the total energy emitted per unit time, as seen by an observer at rest in the source frame. In a Minkowski spacetime, the physical distance d of the source can be evaluated by measuring the energy flux Φ received at the detector:

$$\Phi = \frac{L}{4\pi d^2}, \quad \Rightarrow d = \sqrt{L/(4\pi \Phi)}. \tag{2.71}$$

This relationship is used to define the *luminosity distance* in an expanding universe:

$$d_L \equiv \sqrt{L/(4\pi \Phi)}. \tag{2.72}$$

To compute this expression we have to take into account that

---

[2] We will use the symbol y in the following to avoid confusion with the modulus of the spatial part of P^μ.

- When the signal is received, the photons have propagated over a physical distance $a(t_{obs})\chi$, with χ the comoving distance travelled by photons to reach us. For isotropic emission, they are homogeneously distributed over a thin spherical shell with surface $4\pi a(t_{obs})^2 r^2(\chi)$, where $r(\chi)$ is given by (2.8).
- The energy of photons is smaller than their energy at the emission time, by a factor $a(t_{em})/a(t_{obs})$ due to redshift.
- The arrival rate is also lower by the same factor. In fact, we know that

$$\chi = \int_{t_{em}}^{t_{obs}} \frac{dt}{a(t)} \tag{2.73}$$

and radiation emitted at $t_{em} + \delta t$ will be received at the time $t_{obs} + \delta t'$ such that

$$\chi = \int_{t_{em}+\delta t}^{t_{obs}+\delta t'} \frac{dt}{a(t)}. \tag{2.74}$$

For small δt we then have

$$\frac{\delta t}{a(t_{em})} = \frac{\delta t'}{a(t_{obs})}. \tag{2.75}$$

Thus, the energy received per unit time and unit surface, when expressed in terms of the absolute luminosity, is

$$\Phi = \frac{La^2(t_{em})}{4\pi a^4(t_{obs})r^2(\chi)}, \tag{2.76}$$

which gives

$$d_L = a^2(t_{obs})r(\chi)/a(t_{em}) = a(t_{obs})r(\chi)(1+z). \tag{2.77}$$

For a spatially flat metric this expression simplifes (neglecting radiation, an excellent approximation at low redshifts up to $z \leq 1$, where d_L is actually used, because today $\Omega_R \ll \Omega_M$; see later) to

$$d_L = (1+z)\frac{1}{H_0} \int_0^z dz' \frac{1}{(\Omega_M(1+z')^3 + \Omega_\Lambda)^{1/2}}, \tag{2.78}$$

with $a(t_{obs}) = 1$ having been set. Expanding at small redshifts at first order in z, using that for $k = 0$, $\Omega_M + \Omega_\Lambda = 1$, we get again the Hubble law. At second order, we obtain

$$d_L \sim \frac{z}{H_0} \left(1 + \frac{1}{2}(1-q)z \right), \tag{2.79}$$

where we have written $\Omega_M = 1 - \Omega_\Lambda$ in terms of the deceleration parameter using (2.63).

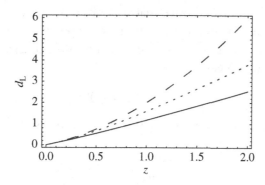

Figure 2.2 The luminosity distance, in units of c/H_0, for different spatially flat cosmologies. Full line corresponds to Einstein–de Sitter ($\Omega_M = 1$), $\Omega_\Lambda = 1$ (dashed), $\Omega_M = 0.25$ and $\Omega_\Lambda = 0.75$ (dotted).

For larger values of z, one should consider the exact expression (2.77). The behaviour of d_L for different values of Ω_M and Ω_Λ is shown in Fig. 2.2. Measuring d_L for a particular class of sources with quite well understood absolute luminosity, the Type Ia supernovae, has led to one of the main breakthroughs in cosmology of the last decades, giving evidence for a recent accelerated expansion stage (a negative q) and a large value of an energy component very close to a cosmological constant with $\Omega_\Lambda \sim 0.75$ today.

For non-spatially-flat metrics, the Friedmann equation imposes that the scale factor and the curvature parameter k are related as follows:

$$k = (\Omega - 1)a_0^2 H_0^2 \tag{2.80}$$

If we set k to one of its canonical values, $k = 0, \pm 1$, the scale factor today is fixed in terms of H_0. Instead, setting $a_0 = 1$ today determines the value of k. Using this relation to express the product $a_0 H_0$ in terms of $\Omega_k = 1 - \Omega$, the luminosity distance can be expressed as

$$d_L = \frac{1+z}{\sqrt{\Omega_k} H_0} \sinh\left(\Omega_k^{1/2} \int_0^z \frac{dz'}{(\Omega(z) + \Omega_k(1 + z')^2)^{1/2}}\right), \tag{2.81}$$

$\Omega(z)$ receiving contribution from all species (radiation, matter, cosmological constant, etc.). For example, for an *Einstein–de Sitter* cosmology with only dust dominating at late times, $\Omega(z) = \Omega_M(1 + z)^3$.

Whereas for pointlike sources with known luminosity, the luminosity distance described so far is used, observations of *extended* objects in the sky of known physical transverse length when they emit radiation require the definition of angular distance. Of course, this quantity is of practical use only when we can experimentally resolve this length, which will subtend a definite angular scale Θ. Let us call l its value, and suppose it is placed at a comoving distance $\chi(t_{em})$ from us.

To compute l in terms of Θ, notice that it is given by the distance between the two *simultaneous* photon emission events at the endpoints. From the FRW metric, setting the angular coordinate ϕ to be a constant for the geodesic paths of photons which reach us today, we get

$$l = a(t_{\mathrm{em}})r(\chi(t_{\mathrm{em}}))\Theta. \tag{2.82}$$

Angular distance is defined by analogy with the expression it takes in a static background,

$$d_{\mathrm{A}} = \frac{l}{\Theta} = a(t_{\mathrm{em}})r(\chi(t_{\mathrm{em}})) = \frac{a_0 r(\chi(z))}{1+z}. \tag{2.83}$$

Although for astrophysical sources it is quite a difficult task to define the transverse profile and define a sharp reference edge to be used as the variable l, CMB anisotropies provide a beautiful application of the angular distance observable. We will see in Chapter 5 that the angular correlation function of photon temperature fluctuations is characterized by a series of peaks. The angle subtending the first peak corresponds to a known transverse physical distance, the so–called sound horizon at recombination $d_{\mathrm{s}}(z_{\mathrm{LS}})$, where we recall that z_{LS} is on the order of 10^3. This quantity represents the largest distance scale over which sound waves can propagate in the plasma of photons and charged particles (electrons, baryons) till photon decoupling occurs. Its precise definition is given in Section 5.1.5. It is roughly on the order of the particle horizon at photon decoupling, $d_{\mathrm{s}}(z_{\mathrm{LS}}) \sim H^{-1}(z_{\mathrm{LS}})$, with corrections depending on the relative density of photons, baryons and dark matter. For given values of these densities and of z_{LS}, the distance $d_{\mathrm{s}}(z_{\mathrm{LS}})$ can be accurately computed, and finally divided by the observed angular scale of the peak in order to provide a measurement of $d_{\mathrm{A}}(z_{\mathrm{LS}})$. Because the latter quantity depends on H_0, Ω_k and Ω_Λ through the function $r(\chi(z))$, this method provides stringent constraints on spatial curvature and on the cosmological constant amplitude. When these results are used in combination with other measurements (e.g., of supernovae luminosity), they support the case for a flat (or very close to critical) universe. One may wonder why this is the case. The inflationary paradigm provides a natural interpretation of this fact.

2.2 Statistical mechanics in the expanding universe

2.2.1 The relativistic Boltzmann equation

The programme of statistical mechanics is to deduce the standard results of thermodynamics from a microscopic point of view. When the system under investigation is under thermodynamic equilibrium conditions, one can either consider a macroscopic description based on equilibrium thermodynamics, which involves macroscopic quantities such as energy density, number density and pressure, related by the equation of state, or take a microscopic point of view, where all properties of the system are obtained starting from the distribution function of particles in phase space. These two approaches have been already outlined in Section 2.1, where we obtained two different ways of writing the stress–energy tensor for a perfect fluid.

On the other hand, when external conditions are such that equilibrium is not guaranteed, we cannot exploit the powerful methods of equilibrium thermodynamics, nor assume that a gas of particles will be distributed according to standard Bose–Einstein or Fermi–Dirac functions. In this case, the tool which can be used to describe the evolution of a system in time is kinetic theory, pioneered by Boltzmann in his famous paper of 1872 (Boltzmann, 1872). His visionary (at that time) approach consists in finding the time evolution of the particle distribution function as the solution of an integro-differential equation, which is now known as the Boltzmann equation. The aim of this section is to review this tool, adapted to the case of a relativistic treatment in an expanding universe.

Why should we use kinetic theory to describe (at least some) features of the evolution of the universe? The reason is that during the expansion, equilibrium is not always guaranteed. Consider a fluid composed of thermal excitations of a particular quantum field (photons, neutrinos, charged leptons, etc.). Equilibrium is established by interaction processes such as scattering (which redistribute particle momenta and are crucial to achieve kinetic equilibrium) and interactions where the number of particles of a given species is not conserved, such as pair annihilations, which enforce chemical equilibrium among different species. Because the universe is expanding, equilibrium is maintained if the rate of microscopic interactions is higher than the expansion rate, given by H. In this case the evolution of the system is *quasi–static*, and can be considered as a sequence of equilibrium states. Consider the case of two-body processes of the form $a + b \leftrightarrow c + d$. The interaction rate for the species a is given by

$$\Gamma_a \sim \sigma v\, n_b, \tag{2.84}$$

where σ is the cross section evaluated at the typical energy of the interacting particles, v the $a - b$ relative velocity (we will see in the following that what is relevant is the *thermal average* of σv) and n_b the b number density (the target number density). Equilibrium is thus maintained if

$$\Gamma_a \gg H. \tag{2.85}$$

For $\Gamma_a \ll H$ the species a is *decoupled*, and all a particles simply follow geodesic motion. Typically, in the early universe, density and mean energy are sufficiently high to ensure that (2.85) is fulfilled. As expansion proceed, if Γ_a decreases faster than H, interaction processes become more and more inefficient, till at some value of redshift z_D,

$$\Gamma_a(z_D) \sim H(z_D), \qquad (2.86)$$

which defines the *decoupling* redshift z_D for the species a. For $z \gg z_D$ we expect the species to be in equilibrium. Its distribution will therefore be a standard Fermi–Dirac or Bose–Einstein, which depends on a time-dependent temperature $T(z)$ and chemical potential $\mu(z)$ that decrease smoothly as the expansion proceeds. This is a result of the kinetic theory, which we will justify soon. If the species decouples at z_D, we can therefore define a decoupling temperature $T_D \equiv T(z_D)$, so that below T_D particles a lose thermal contact with others species. Of course, decoupling is not an instantaneous phenomenon, but it lasts some redshift interval around z_D. In many cases, however, the *instantaneous decoupling limit* already provides quite an accurate result for the eventual distribution after decoupling. We will see an example of this when we describe neutrino decoupling. In general, the final distribution of decoupled species depends on the details of the decoupling phase, on the fact that decoupling takes place when they are still relativistic (*hot* relics) like neutrinos, or already nonrelativistic (*cold* relics) like cold dark matter (CDM), and on the fact that they carry conserved charges or not, such as baryon number, lepton number or electric charge. Although in some cases this distribution can be found by knowing its expression at T_D where equilibrium conditions are realized to some extent – this is typically the case for hot relics – in general, its behaviour is obtained by solving the corresponding Boltzmann equation, which we now discuss.

Consider the distribution function of some species in phase space $f(x^\mu, P^\nu)$, with P^μ the energy–momentum 4–vector, which satisfies the mass shell condition (2.14). In the absence of particle interactions, the value of f does not change as we follow a point in phase space along a geodesic motion. If we define the Liouville operator as

$$\mathcal{L}(f) \equiv \frac{df}{d\lambda}(x^\mu(\lambda), P^\nu(\lambda)), \qquad (2.87)$$

we therefore have that the *collisionless* Boltzmann equation is written

$$\mathcal{L}(f) = \frac{\partial f}{\partial x^\mu}\frac{dx^\mu}{d\lambda} + \frac{\partial f}{\partial P^\nu}\frac{dP^\nu}{d\lambda} = \frac{\partial f}{\partial x^\mu}P^\mu - \frac{\partial f}{\partial P^\nu}\Gamma^\nu_{\rho\sigma}P^\rho P^\sigma = 0, \quad (2.88)$$

where we have used the geodesic equation (2.10). In a FRW metric, homogeneity and isotropy requires that the distribution function depends on time and either energy E or the modulus of linear momentum P, once we use the mass shell condition. In this case, using the expression for the Christoffel symbols (2.12) (we consider a spatially flat metric), we have, for $f(t, P)$,

$$E\left(\frac{\partial f(t, P)}{\partial t} - 2HP\frac{\partial f(t, P)}{\partial P}\right) = 0 \tag{2.89}$$

or changing the variables and using the physical momentum $p = Pa$,

$$E\left(\frac{\partial f(t, p)}{\partial t} - Hp\frac{\partial f(t, p)}{\partial p}\right) = 0. \tag{2.90}$$

Finally, in terms of the modulus of the comoving momentum $y \equiv (\sum_i P_i^2)^{1/2}$,

$$\frac{\partial f(y, t)}{\partial t} = 0, \tag{2.91}$$

so that any function which depends upon the comoving momentum only is a solution of the collisionless Boltzmann equation.

In the presence of particle scatterings, annihilations, etc., the Boltzmann equation is modified to account for the change of particle density in a phase space volume due to these processes. Using physical momentum

$$\frac{1}{E}\mathcal{L}(f) = \frac{\partial f(t, p)}{\partial t} - Hp\frac{\partial f(t, p)}{\partial p} = C(f(p, t); f_i), \tag{2.92}$$

where by f_i we have denoted the distributions of all species i which interact with the one we are interested in. The right-hand side here defines the *collisional integral* C, given by the rate of all processes which produce particles (whose distribution we want to follow in time) with momentum p minus the rate of all interactions where these particles are instead destroyed. Of course, one should consider the analogous equations for all involved f_i, so that in general the problem is to solve a set of coupled equations for all coupled species.

The fact that the collisional integral is depending only on the distribution functions f, f_i is valid assuming the Boltzmann hypothesis of *Stosszahlansatz*, or molecular chaos, namely that particle momenta are uncorrelated with their positions. Consider for example again the case of binary collisions $a + b \leftrightarrow c + d$. The molecular chaos ansatz means that in a volume element d^3x the numbers of pairs of particles a and b with momentum p_a and p_b are given by

$$\left[f_a(p_a, t)d^3x\, d^3p\right]\left[f_b(p_b, t)d^3x\, d^3p\right]. \tag{2.93}$$

If we did not assume (2.93), the collisional integral would depend on the two-particle correlation functions f_{ab}, which are in general independent of the f's. In this case the Boltzmann equation (2.92) should be complemented by an analogous equation for f_{ab}, which would be linked to the three-point function f_{abc}, and so on.

Let us now write the expression for **C** for the particular case of two–particle into two–particle interactions. For the species a we have

$$\mathbf{C}(f_a; f_b, f_c, f_d)$$

$$= \frac{1}{E_a} \int d\pi(p_b) d\pi(p_c) d\pi(p_d) (2\pi)^4 \delta^{(4)}(p_a + p_b - p_c - p_d)$$

$$\times \left[\left| \mathcal{M}_{cd,ab} \right|^2 f_c(p_c, t) f_d(p_d, t)(1 \pm f_a(p_a, t))(1 \pm f_b(p_b, t)) \right.$$

$$\left. - \left| \mathcal{M}_{ab,cd} \right|^2 f_a(p_a, t) f_b(p_b, t)(1 \pm f_c(p_c, t))(1 \pm f_d(p_d, t)) \right]. \quad (2.94)$$

Notice that integration is made using the relativistic invariant measure

$$d\pi(p) \equiv \frac{d^3 p}{(2\pi)^3 \, 2E(p)} \quad (2.95)$$

and that the Dirac delta functions give conservation of energy and physical momentum. By $\mathcal{M}_{cd,ab}$ and $\mathcal{M}_{ab,cd}$ we have denoted the invariant amplitudes for the processes $c + d \to a + b$ and $a + b \to c + d$, respectively. In particular, if the interactions are invariant under time reversal, the two amplitudes have the same modulus, and the expression above slightly simplifies. Finally, the factors in brackets of the form $(1 \pm f)$ are due to the Pauli blocking effect for fermions (minus sign) and stimulated emission for bosons (positive sign). If the system is very dilute and the particle chemical potentials are very small, these factors reduce to unity, as in the case of particles obeying classical statistics.

One of the relevant properties of the collisional integral is that if we denote by $\mathcal{Q}$ a quantity which is conserved by interactions, then

$$\int \frac{d^3 p_a}{(2\pi)^3} \mathcal{Q} \mathbf{C} = 0. \quad (2.96)$$

For example, $\mathcal{Q}$ might be the energy or momentum, electric charge, etc.

Let us consider the case in which the *only* process involving a is the scattering $a + b \to a + b$, for which the particle numbers n_a and n_b are conserved. In this situation, one can choose a $\mathcal{Q}$ not depending on time or particle momentum, such as a conserved charge q_a or q_b, and prove that Eq. (2.96) is really satisfied. Indeed, if we integrate (2.94) over $d^3 p_a$ (assuming for simplicity that all quantum statistical factors can be neglected so that $(1 \pm f) = 1$), we obtain (up to a numerical factor $2(2\pi)^3$)

$$\int d\pi(p_a) d\pi(p'_a) d\pi(p_b) d\pi(p'_b)(2\pi)^4 \delta^4(p_a + p_b - p'_a - p'_b)$$

$$\times \left[\left| \mathcal{M}_{p'_a p'_b, p_a p_b} \right|^2 f_a(p'_a, t) f_b(p'_b, t) - \left| \mathcal{M}_{p_a p_b, p'_a p'_b} \right|^2 f_a(p_a, t) f_b(p_b, t) \right].$$

$$(2.97)$$

The unitarity of the S matrix implies

$$\int d\pi(p'_b)d\pi(p'_a)\left|\mathcal{M}_{p'_a\,p'_b,\,p_a\,p_b}\right|^2 = \int d\pi(p'_b)d\pi(p'_a)\left|\mathcal{M}_{p_a\,p_b,\,p'_a\,p'_b}\right|^2 \quad (2.98)$$

independent of time-reversal invariance (we recall that there is only one interaction process in this simple example). So we see that (2.97) can also be written as

$$\int d\pi(p_a)d\pi(p'_a)d\pi(p_b)d\pi(p'_b)(2\pi)^4\delta^4(p_a + p_b - p'_a - p'_b)$$

$$\times \left|\mathcal{M}_{p'_a\,p'_b,\,p_a\,p_b}\right|^2 \left[f_a(p'_a, t)f_b(p'_b, t) - f_a(p_a, t)f_b(p_b, t)\right] = 0, \quad (2.99)$$

where the last equality follows by interchanging momenta, $p_a \leftrightarrow p'_a$ and $p_b \leftrightarrow p'_b$, and using again (2.98). Hence, in that case, if we integrate (2.92) over $d^3 p_a$, both sides must vanish. This proves that

$$\int \frac{d^3 p_a}{(2\pi)^3}\left(\frac{\partial f(t, p_a)}{\partial t} - H p_a\frac{\partial f(t, p_a)}{\partial p_a}\right) = \dot{n}_a + 3Hn_a = 0, \quad (2.100)$$

where we have integrated the second term by parts and assumed that distribution functions vanish for $p \to \infty$ more rapidly than any power. Equation (2.100) expresses the covariant conservation of the number density $n_a^\mu = (n_a, 0, 0, 0)$ (defined in (2.41)) in the FRW model, justified by the covariant conservation of the particle charge current density $q_a n_a^\mu$.

In case there are more particle species and processes involved, but still, say, q_a is a conserved charge, then summing over all processes and using unitarity, one still obtains that the collisional integral for the a particle vanishes and $q_a n_a^\mu$ is covariantly conserved.

In a similar way it is easy to show that, choosing $\mathcal{Q} = E$ and summing over all species which are mutually interacting, we get

$$\sum_a \int \frac{d^3 p_a}{(2\pi)^3} E(p_a)\,\mathbf{C}(p_a; p_i) = 0, \quad (2.101)$$

which is equivalent to the covariant conservation of the total stress–energy tensor summed over all interacting species, see (2.44), (2.45) and (2.48):

$$\sum_a \int \frac{d^3 p_a}{(2\pi)^3} E(p_a)\left(\frac{\partial f(t, p_a)}{\partial t} - H p_a\frac{\partial f(t, p_a)}{\partial p_a}\right)$$

$$= \sum_a (\dot{\rho}_a + 3H(P_a + \rho_a)) = \sum_a \int \frac{d^3 p_a}{(2\pi)^3} E(p_a)\,\mathbf{C}(p_a; p_i) = 0, \quad (2.102)$$

where again we have used integration by parts of the second term on the l.h.s. If a species is only self–interacting, there is no transfer of energy and momentum to other particles, so that its $T^{\mu\nu}$ is covariantly conserved.

We know from standard kinetic theory that for $H = 0$ (no expansion) the collisional integral is zero under thermodynamic equilibrium conditions. This is simply due to the fact that at equilibrium the rate of processes producing particles a is exactly compensated for by all processes where a is destroyed, so that the net number per given momentum $f(p, t)$ does not change. The reader should be aware that this statement might generate confusion, as there are in fact too many collisions, yet $\mathbf{C} = 0$!

Is it still true that equilibrium distributions satisfy the Boltzmann equation for an expanding universe? The fact that $\mathbf{C} = 0$ at equilibrium can be shown easily, for example, for the case of binary interactions and recalling that equilibrium distribution functions in terms of the comoving momentum take the standard form

$$f_i = \left[\exp\left(\frac{E_i - \mu_i}{T_i} \right) \pm 1 \right]^{-1}. \tag{2.103}$$

It is left to the reader to show that indeed $\mathbf{C} = 0$ if all species share the same temperature (kinetic equilibrium) and moreover chemical potentials satisfy the chemical equilibrium condition

$$\mu_a + \mu_b = \mu_c + \mu_d. \tag{2.104}$$

Notice that this result also holds if the temperature and chemical potentials are time-dependent parameters.

If $\mathbf{C} = 0$, then equilibrium distributions will be solutions of the Boltzmann equation provided they are also in the kernel of the Liouville operator. If we rewrite (2.103) in the form

$$f(p, t) = \left[\exp\left(\beta E(p) - \xi \right) \pm 1 \right]^{-1} \tag{2.105}$$

with $\xi = \mu/T$ and $\beta = 1/T$, and allowing these parameters to be time-dependent, we see from (2.90) that this expression is a solution of the Boltzmann collisionless equation if

$$E(p)\dot{\beta} - H \frac{p^2}{E(p)}\beta - \dot{\xi} = 0, \tag{2.106}$$

which because ξ and β are independent reduces to

$$\xi = \text{const.} \tag{2.107}$$

$$E(p)^2 \dot{\beta} - H p^2 \beta = 0. \tag{2.108}$$

In the case of massless particles $E = p$ we can find an exact solution to (2.107) and (2.108):

$$\xi = \text{const.} \tag{2.109}$$

$$\beta \propto a. \tag{2.110}$$

One could also have found this result by observing that for relativistic species the equilibrium distribution (2.103) depends only on the comoving momentum, provided $Ta = \text{const.}$ Furthermore, if some species decouples when relativistic (a *hot* relic), it will retain a Fermi–Dirac or Bose–Einstein distribution with a decreasing T also after decoupling. In fact, if (2.103) holds till decoupling, it will also be a solution of the collisionless Boltzmann equation after decoupling, being a function only of the comoving momentum. This is why the CMB distribution today is such a perfect black body although photons decoupled at redshift $z \sim 10^3$. Nevertheless, their distribution today is the heritage of the early epoch when they were strongly coupled with charged particles. We will see that this is also the case for neutrinos, although in this case there is a tiny nonthermal distortion expected in their distribution because their decoupling takes place shortly before $e^{\pm}$ pairs annihilate.

Let us now discuss nonrelativistic particles. For $m \neq 0$ the second relation (2.108) admits no nontrivial solutions. This can be proven by observing that this condition for an arbitrary timelike 4–momentum $P^{\mu} = (E, p^i/a)$ is equivalent to the statement that there are spatially constant time-like Killing vectors in FRW, which is not the case. Rigorously speaking, for massive particles there are no equilibrium solutions to the Boltzmann equation in FRW. In practice, the shape of the distribution can be very close to equilibrium if interaction rates are sufficiently larger than H. When $T \gg m$ we can neglect the rest mass and if the species is strongly coupled its distribution is Fermi–Dirac or Bose–Einstein with $T \sim a^{-1}$. As temperature decreases to $T \sim m$, (2.103) is no longer a solution of the Boltzmann equation, but one can look for an approximate equilibrium form

$$f(p, t) = \left[\exp\left(\frac{E_i - \mu_i}{T_i} \right) \pm 1 \right]^{-1} (1 + \delta(p, t)) \tag{2.111}$$

with $|\delta(p, t)| \ll 1$. It can be shown that for a high interactions rate $\Gamma \gg H$ we have indeed – see (Bernstein, 1988) for a detailed discussion –

$$|\delta(p, t)| \sim \frac{H}{\Gamma} \ll 1. \tag{2.112}$$

Consider for example the case of electrons and positrons, which interact quite efficiently with photons through Compton scattering and pair processes $e^+ e^- \leftrightarrow \gamma\gamma$. For $T \ll m_e$ the Compton cross section is given by the Thomson formula,

$\sigma_T = 8\pi\alpha^2/(3m_e^2) \sim 0.67 \times 10^{-24}$ cm^2. Because the equilibrium distribution of photons is a solution of the corresponding Boltzmann equation for a photon temperature which scales as $T \sim a^{-1} = 1 + z$, we have that $T \sim m_e$ corresponds to a redshift $z \sim m_e/T_0 \sim 10^{10}$. As we will discuss later in this chapter, at this value of the redshift the universe is radiation-dominated, down to the radiation–matter *equality* point $z_{eq} \sim 10^3$. Recalling (2.50), the Hubble parameter then reads

$$H \sim H_0 \left(\frac{1+z}{1+z_{eq}}\right)^2 (1+z_{eq})^{3/2}, \quad z > z_{eq}. \tag{2.113}$$

On the other hand, the photon number density is obtained by integrating the black-body distribution and summing over the two polarization states. Neglecting a factor of order one due to photon reheating by $e^{\pm}$ annihilations, which we will discuss at length soon, we have

$$n_\gamma \sim 2 \int \frac{d^3 p}{(2\pi)^3} \frac{1}{e^{p/T} - 1} = \frac{1}{\pi^2}(1+z)^3 \int y^2 dy \frac{1}{e^{y/T_0} - 1}$$

$$= \frac{2\zeta(3)}{\pi^2}(1+z)^3 T_0^3, \tag{2.114}$$

where $\zeta(3) \sim 1.202$, so that

$$\Gamma \sim c\,\sigma_T \frac{2\zeta(3)}{\pi^2}(1+z)^3 T_0^3. \tag{2.115}$$

Plugging numerical values into (2.114) and (2.115), we get that at $T \sim m_e$ the ratio H/Γ is extremely small and we can estimate the amount of departure from equilibrium conditions for $e^{\pm}$ as

$$|\delta(p,t)|(T \sim m_e) \sim \frac{\pi^2}{2\zeta(3)\,c\,\sigma_T\,T_0^2\,m_e} \frac{H_0}{(1+z_{eq})^{-1/2}} \sim 10^{-17} \tag{2.116}$$

so that $e^{\pm}$ are distributed with a perfect equilibrium Fermi–Dirac function. Because pair processes also have a high rate compared with H, chemical equilibrium holds. From (2.104) and the fact that photons have zero chemical potential, we get

$$\xi_e + \xi_{e^+} = 0. \tag{2.117}$$

Finally, we can write for $T \le m_e$

$$f_{e^{\pm}} \sim \exp\left(-\frac{m_e}{T} \mp \xi_e - \frac{p^2}{2m_e T}\right). \tag{2.118}$$

In many relevant applications of the Boltzmann equation to cosmology, it is a good approximation to assume that kinetic equilibrium holds thanks to fast scattering processes. In these cases it is convenient to integrate (2.92) over phase space and get the equation ruling the evolution of the number densities. As before, we consider the example of two body–two body processes. We also specialize the result to the particular case where all particles involved either are nonrelativistic or, in case they are relativistic, have small (or vanishing as for photons) chemical potential. With these assumptions we can highly simplify the Boltzmann equation and neglect all effects due to quantum statistics. On the other hand, this case is sufficiently general to cover the main examples we are going to consider in the following. Using integration by parts on the l.h.s. and using the scale factor as the parameter of evolution, we have

$$
Ha\frac{dn_a}{da} + 3Hn_a
$$

$$
= \int d\pi(p_a)d\pi(p_b)d\pi(p_c)d\pi(p_d)(2\pi)^4\delta^{(4)}(p_a + p_b - p_c - p_d)
$$

$$
\times |\mathcal{M}_{ab,cd}|^2 \left[e^{-(E_c+E_d)/T} e^{\mu_c/T} e^{\mu_d/T} - e^{-(E_a+E_b)/T} e^{\mu_a/T} e^{\mu_b/T} \right], \quad (2.119)
$$

where for simplicity we have also assumed time reversal invariance so that the squared modulus of the invariant amplitude for direct and inverse processes is the same. Notice that for the involved particles the number density is simply given in terms of the chemical potential as

$$
n_i \sim e^{\mu_i/T} \int \frac{d^3p}{(2\pi)^3} e^{-E_i/T}. \quad (2.120)
$$

Using energy conservation, (2.119) can also be cast in the form

$$
a^{-2}\frac{d}{da}\left(n_a a^3\right) = \frac{\langle \sigma|v|\rangle n_b}{H} n_a \left(\exp\left(\frac{\mu_c + \mu_d - \mu_a - \mu_b}{T}\right) - 1 \right), \quad (2.121)
$$

where we have defined

$$
\langle \sigma|v|\rangle \equiv \int d\pi(p_a)d\pi(p_b)d\pi(p_c)d\pi(p_d)(2\pi)^4\delta^{(4)}(p_a + p_b - p_c - p_d)
$$

$$
\times |\mathcal{M}_{ab,cd}|^2 e^{-(E_a+E_b)/T} \left(\int \frac{d^3p}{(2\pi)^3} e^{-E_a/T} \int \frac{d^3p}{(2\pi)^3} e^{-E_b/T} \right)^{-1}, \quad (2.122)
$$

which is the product of the cross section σ and relative velocity for the process $a + b \rightarrow c + d$, averaged over the thermal particle distributions in momentum.

A look at the prefactor on the r.h.s. shows that the rate of change of the a particle number per comoving volume $n_a a^3$ is related to the ratio of the interaction rate $\Gamma_a = \langle \sigma|v|\rangle n_b$ and the Hubble parameter, as we already discussed at the beginning of this section. When this ratio is large, the species a is in chemical equilibrium, the number density scales as a^{-3} and the Boltzmann equation is satisfied because of (2.104). This chemical equilibrium condition can also be rewritten using (2.120) in the form

$$\frac{n_c n_d}{n_a n_b} = \frac{\int d^3 p \, e^{-E_c/T} \int d^3 p \, e^{-E_d/T}}{\int d^3 p \, e^{-E_a/T} \int d^3 p \, e^{-E_b/T}}, \tag{2.123}$$

which is also called the Saha equation. This equation is very useful to get the order of magnitude of the time or redshift at which some key events take place. The first main example is the recombination stage, namely when electrons and protons recombine to form neutral hydrogen, through the two-body process

$$p + e^- \leftrightarrow H + \gamma. \tag{2.124}$$

Similarly, it is also used to understand the time at which deuterium starts forming, i.e., the onset of primordial nucleosynthesis, due to proton–neutron fusion:

$$p + n \leftrightarrow {}^2H + \gamma. \tag{2.125}$$

In both cases we will see that eventually chemical equilibrium is no longer satisfied as these phenomena proceed; so for example the eventual yield of ^{2}H cannot be grasped by simply using the Saha condition. Nevertheless, the latter provides a typically quite accurate estimate for the time scale when nucleosynthesis (or recombination) starts.

2.2.2 When equilibrium holds

Under equilibrium conditions we can easily compute the stress–energy tensor for a given species i. Recalling (2.44) and (2.45) and summing over particles and antiparticles,

$$\rho_i = g_i \int \frac{d^3 p}{(2\pi)^3} E(p) \left(\frac{1}{e^{E(p)/T_i - \xi_i} \pm 1} + \frac{1}{e^{E(p)/T_i + \xi_i} \pm 1} \right) \tag{2.126}$$

$$P_i = g_i \int \frac{d^3 p}{(2\pi)^3} \frac{p^2}{3E(p)} \left(\frac{1}{e^{E(p)/T_i - \xi_i} \pm 1} + \frac{1}{e^{E(p)/T_i + \xi_i} \pm 1} \right), \tag{2.127}$$

where g_i is the number of internal degrees of freedom such as helicity and colour. Notice that we have assumed opposite chemical potentials for particles and antiparticles, assuming chemical equilibrium with photons. Even if the species is not *directly* coupled to photons, such as neutrinos (at least at lowest order in perturbation theory), this condition is guaranteed by processes such as $i + \bar{i} \leftrightarrow e^- + e^+$ or $i + \bar{i} \leftrightarrow p + \bar{p}$. In particular, for $\xi_i = 0$ for relativistic particles, one finds

$$\rho_i = 3P_i = \begin{cases} \frac{\pi^2}{30} g_i T_i^4, & \text{boson} \\ \frac{7}{8} \frac{\pi^2}{30} g_i T_i^4, & \text{fermion.} \end{cases} \tag{2.128}$$

If $\xi_i \neq 0$ we can also work out analytically the result for fermions in the relativistic limit,

$$\rho_i = 3P_i = \frac{7}{8}\frac{\pi^2}{30}g_i T_i^4 \left(1 + \frac{30\xi_i^2}{7\pi^2} + \frac{15\xi_i^4}{7\pi^4}\right). \tag{2.129}$$

Similarly, in the same limit, one finds for the particle–antiparticle asymmetry of a fermionic species

$$n_i - n_{\bar{i}} = \frac{g_i}{6}T_i^3 \left(\xi_i + \frac{\xi_i^3}{\pi^2}\right), \tag{2.130}$$

whereas the number density can be expressed in a closed form only for $\xi_i \ll 1$, in which case one gets

$$n_{i,\bar{i}} = \frac{3\zeta(3)}{4\pi^2}g_i T_i^3 \pm \frac{g_i}{12}T_i^3\xi_i + \mathcal{O}(\xi_i^2). \tag{2.131}$$

For relativistic bosons, $T_i \gg m_i$, because the chemical potential cannot be greater than the particle mass,[3] we have $\xi \ll 1$, and in this limit

$$n_{i,\bar{i}} \sim \frac{\zeta(3)}{\pi^2}g_i T_i^3 \tag{2.132}$$

$$n_i - n_{\bar{i}} \sim \frac{g_i T_i^3}{3}\xi_i, \tag{2.133}$$

where the last result only applies in the case $\mu_i \ll m_i$ and no Bose condensate forms; otherwise a large fraction of the particle/antiparticle asymmetry could be accommodated there.

For nonrelativistic particles and small chemical potentials, $\mu_i \ll m_i$, the distribution is given by the Maxwell–Boltzmann function

$$f_i \sim g_i \exp\left(-\frac{m_i}{T_i} \pm \xi_i - \frac{p^2}{2m_i T_i}\right) \tag{2.134}$$

so that

$$n_i \sim g_i \left(\frac{m_i T_i}{2\pi}\right)^{3/2} e^{-m_i/T_i \pm \xi_i} \tag{2.135}$$

$$\rho_i \sim m_i n_i + \frac{3}{2}T_i n_i \tag{2.136}$$

$$P_i \sim T_i n_i \ll \rho_i. \tag{2.137}$$

Notice that if some particles have a positive chemical potential, the density of their antiparticles is smaller by a factor $\exp(-2\xi_i)$.

---

[3] The distribution function should be positive definite. When ξ approaches the value of the minimal energy (the particle mass), a Bose condensate develops.

Let us consider now the entropy associated with the particle fluids. We recall that for a homogeneous system, such as fluids in a FRW universe, the specific entropy or entropy density, $s = S/V$, is obtained from the grand potential, $\Omega = -PV$, as

$$s = -\frac{1}{V}\frac{\partial\Omega}{\partial T}\bigg|_{\mu,V} = \frac{\partial P}{\partial T}\bigg|_{\mu}, \tag{2.138}$$

which gives

$$s = g\int\frac{d^3p}{(2\pi)^3}\frac{\exp\left(\frac{E-\mu}{T}\right)}{\left[\exp\left(\frac{E-\mu}{T}\right)\pm 1\right]^2}\frac{E-\mu}{T^2}\frac{p^2}{3E}. \tag{2.139}$$

This expression can also be written as

$$s = -g\int\frac{d^3p}{(2\pi)^3}\left(\frac{\partial}{\partial p}\frac{1}{\exp\left(\frac{E-\mu}{T}\right)\pm 1}\right)\frac{E-\mu}{3T}p, \tag{2.140}$$

and integrating by parts, one obtains

$$s = \frac{\rho - \mu n + P}{T}. \tag{2.141}$$

For relativistic species, because the chemical potential scales as the temperature, the specific entropy behaves as $s \sim a^{-3}$, so that the entropy per comoving volume is conserved. This also holds for particles which decouple when still relativistic, because their distribution keeps the equilibrium shape after decoupling. We will see an application of this result to neutrinos in the following. The total entropy density in relativistic species can be written as the contribution of one polarization state for the photons, multiplied by an effective number of "entropy relativistic degrees of freedom," g_s:

$$s_R = \frac{4}{3}\rho_R \equiv g_s\frac{2\pi^2}{45}T^3. \tag{2.142}$$

For vanishing chemical potentials,

$$g_s = \sum_{i,\text{boson}}g_i\left(\frac{T_i}{T}\right)^3 + \frac{7}{8}\sum_{j,\text{fermion}}g_j\left(\frac{T_j}{T}\right)^3. \tag{2.143}$$

In the definition of g_s we have allowed for different temperatures T_i with respect to the photon temperature T. We will see that T_i may differ from T, for example, in the case of neutrinos after the $e^{\pm}$ annihilation stage.

The parameter g_s is, in general, a function of T. At mass thresholds $T \sim m_i$ the ith species becomes nonrelativistic and its contribution becomes negligible. Correspondingly, g_s decreases, and all species which at this stage are still in thermal equilibrium with i particles get reheated by the entropy released by the rapid $i - \bar{i}$ annihilations. Their temperature behaves in this phase as $T \sim g_s^{-1/3} a^{-1}$, as predicted by entropy conservation.

For nonrelativistic species s becomes exponentially small if $\xi = 0$; see (2.135)–(2.137). On the other hand, for species which carry some conserved charge, there will be an associated nonzero value of the chemical potential. In particular, the constant value of the charge per comoving volume, Q, can be defined by normalizing the particle–antiparticle asymmetry to the entropy density of relativistic species s_R (which scales as a^{-3}):

$$Q \equiv \frac{n_i - n_{\bar{i}}}{s_R} \sim \frac{g_i}{s_R} \left(\frac{m_i T}{2\pi} \right)^{3/2} e^{-m_i/T} \left(e^{\xi_i} - e^{-\xi_i} \right). \tag{2.144}$$

Examples of $Q \neq 0$ occur for the baryon or lepton numbers, B and L, which are conserved by all fundamental interactions in the low-energy regime but do not vanish in our universe; see Chapter 3. For a nonrelativistic species ($m_i \gg T$) with an excess of particles over antiparticles ($Q > 0$) but a very small charge-to-entropy ratio[4] ($Q \ll 1$), we infer from (2.144) that

$$\frac{m_i}{T} - \xi_i \sim - \log \left(Q \frac{2^{5/2} \pi^{7/2}}{45} g_s \right) + \frac{3}{2} \log \left(\frac{m_i}{T} \right), \tag{2.145}$$

so that ξ_i grows as the temperature decreases. Correspondingly, the antiparticle density decreases exponentially as $\exp(-2m_i/T)$ because of annihilation processes into lighter particles. The relic particle density is then fixed by the conservation of Q,

$$n_i \sim Q g_s T^3. \tag{2.146}$$

The specific entropy of such relics compared to that of relativistic particles reads

$$\frac{s_i}{s_R} \sim -Q \log \left(Q \frac{2^{5/2} \pi^{7/2}}{45} g_s \right) + \frac{3}{2} Q \log \left(\frac{m_i}{T} \right). \tag{2.147}$$

This ratio is suppressed by the small value of Q, but grows logarithmically as the expansion proceeds.

2.3 The expansion stages

2.3.1 Inflation

The hot Big Bang model described so far suffers from some difficulties which cannot be overcome if one assumes that the history of the universe has been always

---

[4] If Q is very large, the mean kinetic energy for fermions is given by the value of the Fermi energy, and thus it is not correct to assume that for $m_i \gg T$ particles are nonrelativistic. For bosons, a large Q might lead instead to the formation of a Bose condensate.

characterized by decelerated expansion. We recall that this is always the case for ordinary matter because pressure (and energy density) is positive and thus $\ddot{a} < 0$; see (2.63). Some of these problems rely on the particular theoretical framework one has in mind for the dynamics of the early universe (the monopole problem, the gravitino problem, etc.). Others are even more serious, as they are not related to any theoretical prejudice and require some extremely well-tuned initial conditions which sound very unnatural.

The *flatness problem* is related to the fact that the curvature's contribution to the universe's expansion, parameterized by Ω_k, increases quickly with time if $\ddot{a} < 0$. Following Eq. (2.56), the Friedmann equation gives

$$|\Omega - 1| = \frac{|k|}{a^2 H^2} \sim |k| a^{1+3w}, \tag{2.148}$$

with w the total pressure-to-density ratio. For decelerated expansion, occurring for instance during radiation and matter domination, one has $1 + 3w > 0$. Then the value $\Omega = 1$ is a *past* attractor, so the FRW metric is closer and closer to the flat limit as we go back in time. To explain why the density today is remarkably close to the critical one, $|\Omega_k| \ll 1$, one should fine-tune the curvature radius in the early universe down to unnaturally small values.

The *horizon problem* comes from the fact that if the expansion was always decelerated in the early universe, the particle horizon (2.66) at the time of decoupling z_{LS} (evaluated with an initial time arbitrarily close to the singularity) would be too small. Indeed, this scale should be seen on the last scattering surface under an angle of approximately $1°$. Larger angular scales should correspond to patches which were causally disconnected at photon decoupling, so it is paradoxical that the temperature is the same on the whole 4π solid angle, up to tiny fluctuations on the order of $\delta T \sim 10^{-5} T_0$.

It was first proposed by Starobinsky, 1980 and Guth, 1981 to solve these problems by assuming a stage of accelerated expansion prior to radiation domination. Let us sketch the main argument.

In a decelerated universe, if photons are causally disconnected at some redshift z, this implies that they were *never* in contact before, because the comoving particle horizon is a monotonically growing function of time, or in terms of the scale factor

$$\chi_p = \int_0^a \frac{da'}{a'^2 H(a')} \sim a^{(1+3w)/2}, \quad 1 + 3w > 0. \tag{2.149}$$

However, suppose that the comoving horizon is receiving its main contribution from some earlier epoch, rather that at recent times. This is only possible if $w \leq -1/3$, i.e., if expansion is accelerated, because in this case the integrand in (2.149) grows as we approach the Big Bang, $a = 0$.

Let us denote as a_{in} and a_{end} the scale factors at which this accelerated phase starts and ends, respectively. Let us also assume that the expansion is then of the quasi-de Sitter type, with $w \sim -1$, as in a cosmological-constant-dominated Universe. This stage would contribute to the particle horizon by an amount

$$\chi_p = \int_{a_{in}}^{a_{end}} \frac{da'}{a'^2 H} = \frac{1}{\overline{H} a_{in}} - \frac{1}{\overline{H} a_{end}} \sim \frac{1}{\overline{H} a_{in}}, \qquad (2.150)$$

with $\overline{H}$ the (almost) constant value of the Hubble parameter during inflation. The last equality holds if inflation lasts long enough, so that $a_{end} \gg a_{in}$. In this case we see that the comoving particle horizon is a constant, implying that the corresponding physical scale grows like the scale factor. If we start with a small initial patch at a_{in}, inside which causality is established by physical processes, this single patch blows up and might embed the whole present observable universe, or even a much bigger region. Then all observed CMB photons may originate from a single causally connected region, and there is no more horizon problem. The flatness problem also disappears because accelerated expansion with $w < -1/3$ implies that Ω is driven extremely close to one during inflation, as shown by (2.148). Another crucial property of inflation is that physical scales grow faster than the Hubble radius during any accelerated expansion stage, whereas during radiation or matter domination they tend to re-enter within the Hubble scale. Thus the largest cosmological scales that we can observe today, which are on the order of the present Hubble scale, might have been within a single causally connected region already at the beginning of inflation.

The flatness and horizon problems can be solved only if there has been a sufficient amount of accelerated expansion. In particular, a comoving wavelength equal to the Hubble scale at the beginning of inflation should not re-enter inside this scale earlier than today. Hence the minimum duration of inflation is given by the condition that the comoving wavenumber equal to $k_* = a_{in} \overline{H}$ is also equal to $a_0 H_0$,

$$\overline{H} a_{in} = H_0 a_0. \qquad (2.151)$$

The values of the Hubble parameter at a_{in} and a_{end} are almost the same for a quasi-de Sitter expansion $\overline{H}$ and can be related to H_0 using the fact that when inflation ends, the universe enters radiation domination, and switches to matter domination at the equality point a_{eq} (see the next section). Using $H \sim a^{-2}$ during radiation domination and $H \sim a^{-3/2}$ during matter domination, we have

$$\overline{H} \frac{a_{end}^2}{a_{eq}^2} \frac{a_{eq}^{3/2}}{a_0^{3/2}} = H_0 \qquad (2.152)$$

(for simplicity, we neglected the impact of a recent stage of cosmological constant domination). Finally, substituting this expression into (2.151), we get

$$\frac{a_{\text{end}}}{a_{\text{in}}} \sim \frac{a_0}{a_{\text{end}}} \left(\frac{a_{\text{eq}}}{a_0} \right)^{1/2}. \tag{2.153}$$

The number of e-folds N between two cosmological times is defined as the logarithm of the amount of expansion between these times. Hence $N \equiv \log(a_{\text{end}}/a_{\text{in}})$ represents the number of e-folds during inflation. The previous condition states that the minimum number of e-folds during inflation is given by the number of e-folds between the end of inflation and today. We can have a feeling for the minimum value of N by specifying the value T of the temperature of relativistic species at the beginning of radiation domination, because in a very crude approximation $T \sim (a_0/a_{\text{end}}) T_0$. This value cannot be smaller than the primordial nucleosynthesis scale $T \sim$ MeV, because an inflationary stage at this epoch would result in a much higher expansion rate. Nuclear processes would then be very inefficient, preventing helium production. This condition gives $N > 27$. More sophisticated arguments can be invoked to narrow this bound, but they usually depend on which extension of the standard model of particle physics is assumed. Note that during inflation, the contribution of curvature to the expansion, given by $\Omega_k = 1 - \Omega$, is reduced by a factor e^{-2N}.

Inflationary models are typically based on the idea that the dominant energy density during this stage is the potential energy of a (fundamental or effective) scalar field. If the motion of this field is sufficiently slow, its kinetic energy may remain subdominant for an extended period of time. Any small region where spatial variations of the field can be neglected will then undergo a stage of accelerated expansion and give rise to a large, homogeneous, causally connected, and almost flat region, which could encompass our whole observable universe. Furthermore, inflation provides an explanation for the production of primordial perturbations (Mukhanov and Chibisov, 1981; Guth and Pi, 1982; Hawking, 1982; Starobinsky, 1982; Bardeen *et al.*, 1983; Brandenberger *et al.*, 1983). Indeed, quantum perturbations of the scalar field responsible for inflation (called the *inflaton*) play naturally the role of a local time-shifting function $\delta t(\vec{x})$, affecting the later evolution of all quantities. These primordial perturbations can seed CMB anisotropies, and grow in order to form structures through gravitational instability. The simplest inflationary models predict that by the end of inflation and on cosmologically observable scales, the universe will be filled with effectively classical stochastic perturbations that are

- coherent in phase space,
- described in terms of a single time-shifting function $\delta t(\vec{x})$,
- obeying to Gaussian statistics with a nearly scale-invariant spectrum in Fourier space.

These last two characteristics will be discussed more in detail in sections 5.1.3 and 5.1.4. It is remarkable that *all* of these properties coincide with what is needed to explain the observed CMB anisotropies and matter fluctuations in our universe. This excellent agreement provides a further strong argument in favour of inflation.

The scalar field is generically assumed to be coupled with ordinary matter fields, those appearing in the standard model of particle physics. At the end of inflation, when the kinetic energy of the inflaton field becomes comparable to its potential energy, it can decay into particles through such couplings. The decay products keep interacting, and at some point may reach thermal equilibrium. After such a *reheating* stage, the universe enters into a radiation-dominated epoch.

For a much more detailed discussion of inflationary models, we advise the reader to go through the books (Linde, 1990; Kolb and Turner, 1994; Mukhanov, 2005; Weinberg, 2008), or the reviews Mukhanov *et al.* (1992); Lyth and Riotto (1999).

2.3.2 Radiation and matter domination

Radiation domination

When inflation is over and the inflaton field has reheated the universe, the energy density becomes dominated by the radiation density $\rho_R \sim T^4$. In fact, primordial nucleosynthesis requires a period of radiation domination down to at least $T \sim$ MeV. The proportion of light nuclei produced by nucleosynthesis is strongly sensitive to the Hubble parameter evolution, and the behaviour $H(T) \propto \sqrt{\rho_R} \sim T^2/m_{Pl}$ provides the best agreement between theory predictions and data.

The total energy density and pressure of radiation can be expressed in terms of the photon energy density for each polarization degree of freedom as follows:

$$\rho_R = g_* \frac{\pi^2}{30} T^4 = 3 P_R, \tag{2.154}$$

where g_* is the effective number of degrees of freedom contributing to ρ_R. For particles with relativistic Fermi–Dirac or Bose–Einstein distributions and vanishing chemical potentials, the results of Section (2.2.2) lead to

$$g_* = \sum_{i,\text{boson}} g_i \left(\frac{T_i}{T}\right)^4 + \frac{7}{8} \sum_{j,\text{fermion}} g_j \left(\frac{T_j}{T}\right)^4, \tag{2.155}$$

where we have considered the general case of particles not necessarily sharing the temperature of photons, $T_i \neq T$. Like the parameter g_s entering into the total specific entropy of relativistic species, g_* is in general a function of T. At mass thresholds $T \sim m_i$ the ith species becomes nonrelativistic and, if it is in chemical equilibrium with lighter species, its contribution becomes negligible, vanishing as $\exp(-m_i/T)$. Correspondingly g_* decreases quite rapidly by $\Delta g_* = (30\rho_i)/(\pi^2 T^4)$.

Using the Friedmann equation and neglecting the spatial curvature term, the Hubble parameter during the radiation-dominated regime reads

$$H(T) = \sqrt{\frac{4\pi^3 G}{45} g_* T^2}. \tag{2.156}$$

As we said, radiation dominates during the earlier stages after inflation. During this period, several phenomena with relevant observable signatures take place. Others are still in the domain of theoretical speculations, though in some cases they are quite strongly motivated on the basis of our present understanding of fundamental interactions. Let us go through them briefly, before discussing some of them in more detail in the following:

- *Baryon asymmetry generation.* The present value of the baryon density can be understood if some baryon-violating interactions took place in the early universe. The temperature range for such a mechanism is still unknown but is likely to exceed the electroweak transition temperature, $T \sim 100\,\text{GeV}$, at least in the framework of the *baryogenesis through leptogenesis* scenario discussed in Chapter 3.
- *Electroweak phase transition.* At $T \sim 100\,\text{GeV}$ the SM gauge symmetry $SU(2)_L \times U(1)_Y$ is spontaneously broken down to $U(1)_Q$. As a consequence all fermions and $W^\pm$ and Z^0 gauge bosons become massive. Fermion and baryon numbers are also violated above this temperature, by nonperturbative effects that we will shortly review in Chapter 3.
- *Quark–gluon transition.* At high temperatures quarks and gluons are not bound into hadrons. When the temperature drops below a value on the order of the nonperturbative QCD scale, $\Lambda_{QCD} \sim 200\,\text{MeV}$, they *hadronize*, and become confined in baryons and mesons. This transition is not fully understood theoretically.
- *Neutrino decoupling and primordial nucleosynthesis.* At $T \sim \text{MeV}$, weak interactions are no longer efficient in maintaining neutrinos in equilibrium with electrons, positrons and photons, because the corresponding interaction rates become smaller than the Hubble parameter. This phase will be described at its zero approximation level in the next section and then in more detail in Chapter 4. Soon after, chemical equilibrium between neutrons and protons is also lost because of the inefficiency of weak processes, and the neutron-to-proton density ratio reaches an almost constant asymptotic value, only affected by neutron decay. Nuclear reactions burn a relevant fraction of free nucleons into light nuclei, starting from deuterium formation at $T \sim 0.08\,\text{MeV}$, and then ^{4}He formation. Only light elements are produced, because the expansion prevents the formation of a substantial fraction of nuclei heavier than ^{7}Li. After nucleosynthesis, the universe contains mainly hydrogen and ^{4}He nuclei, with a mass

fraction

$$Y_p = \frac{m_{\text{He}} n_{\text{He}}}{m_{\text{N}} n_{\text{B}}} \sim 4 \frac{n_{\text{He}}}{n_{\text{B}}}, \qquad (2.157)$$

where n_{B} is the baryon density and m_{N} the average nucleon mass. Standard nucleosynthesis predicts $Y_p \sim 0.25$, whereas the yields of ^{2}H, ^{3}He and ^{7}Li are much smaller.

Primordial nucleosynthesis is in fact one of the main observational tools we can use to constrain the evolution of the universe till the MeV scale. Its theoretical description is rooted in a very robust basis (weak and nuclear processes at the MeV scale or lower), and on the other hand observations of primordial nuclei can provide remarkably detailed data, although still affected by systematics. We will say much more on this in Chapter 4.

Matter domination

The radiation-dominated epoch ends when nonrelativistic matter starts to be the dominant contribution to the total energy density. Because the temperature is decreasing as expansion goes on, massive particles sooner or later will become nonrelativistic, when T is on the order of their rest mass. However, as long as equilibrium holds and a species with mass m is in equilibrium with lighter species, their distribution always keeps the standard Fermi–Dirac or Bose–Einstein shape, which for $T \leq m$ is exponentially suppressed by the factor $\exp(-m/T)$. In this case these particles simply disappear from the plasma and release their entropy to lighter species.

As an example, consider the heaviest charged leptons $\tau^{\pm}$. At temperatures below $m_\tau \sim 1.78$ GeV, τ's will efficiently annihilate through, for example, electromagnetic processes such as $\tau^+ \tau^- \rightarrow \gamma\gamma, \mu^+\mu^-, e^+e^-$, etc., whereas the inverse processes are less and less kinematically allowed, because the typical kinetic energy T is lower than m_τ. This imbalance between direct and inverse processes produces the steep exponential decrease of $\tau^{\pm}$ number density, and the corresponding energy density

$$\rho_\tau \sim m_\tau (m_\tau T)^{3/2} e^{-m_\tau/T}, \qquad T \ll m_\tau \qquad (2.158)$$

soon becomes negligible with respect to that of light particles. For example, ρ_τ is only a subpercent fraction of the photon energy density at $T = m_\tau/10$.

The matter-dominated epoch is driven by those nonrelativistic species which have a non-negligible number density, which decreases as a^{-3} for temperatures smaller than their mass. Their distribution, therefore, should not develop the rapidly decreasing exponential term $\exp(-m/T)$. There are, in fact, two possible way to get this, and obtain a *relic abundance* of massive particles:

- It could be that a given species is no longer in chemical equilibrium with lighter species. In this case the collisional integral in the Boltzmann equation becomes negligible and the number density simply dilutes inversely proportionally to the comoving volume. In other words, the typical interaction rates when T is on the order of the particle mass is substantially smaller than the Hubble rate, as we discussed in Section 2.2.2. An example of this possibility is provided by the weakly interacting massive particle (WIMP) scenario for cold dark matter. In this case the annihilation rate into light particles freezes out for temperatures not much smaller than the WIMP mass.

- On the other hand, the massive particles might be in equilibrium, but carry a quantum number which is conserved by all interactions in the plasma. This is equivalent to saying that their distribution has a chemical potential ξ associated with this charge,

$$n \sim (mT)^{3/2} e^{-m/T+\xi}, \qquad (2.159)$$

which grows as temperature decreases and cancels the kinematic effect of the $\exp(-m/T)$ factor. As far as we know, this second scenario applies to baryons (protons and neutrons at low energy), but it has also been invoked for dark matter in the *asymmetric dark matter* scenario (see, e.g., Hooper *et al.*, 2009; Kaplan *et al.*, 2009). If an initial asymmetry between baryons and antibaryons is produced in the early universe, say more baryons than antibaryons, and all interactions at late times do not violate baryon number conservation, proton and neutron distributions develop a large (positive) chemical potential ξ at low temperature, and their number density is not exponentially suppressed. On the other hand, the number density of antiparticles is strongly depleted because their chemical potential $-\xi$ is negative (we are assuming that chemical equilibrium holds, and thus particles and antiparticles have opposite ξ's). In other words, when the temperature reaches the nucleon mass scale $T \sim$ GeV, all antibaryons annihilate, and only a relic fraction of baryons (corresponding to the initial asymmetry) remain. This topic will be discussed in detail in Chapter 3. Let us define for the moment only the parameter η_B, which will be extensively used in the following:

$$\eta_B \equiv \frac{n_B - n_{\overline{B}}}{n_\gamma} \sim \frac{n_B}{n_\gamma}. \qquad (2.160)$$

After all antibaryons have disappeared, this parameter remains constant as long as the photon density scales as a^{-3}. We will see soon that this is the case after the $e^\pm$ annihilation stage, which takes place at $T \sim m_e$. Below this temperature η_B is a way to quantify the baryon density. Observations from primordial nucleosynthesis and CMB anisotropies are in agreement with a value on the order of $\eta_B \sim 6 \times 10^{-10}$. If we use the expression (2.132) for n_γ today and the value of the critical density $\rho_{cr,0}$ in (2.58), η_B can be simply related to the baryon fractional density today,

$$\Omega_B = \frac{m_N n_B}{\rho_{cr,0}} = \eta_B \frac{m_N n_{\gamma,0}}{\rho_{cr,0}} \sim 0.365 \times 10^8 \eta_B \, h^{-2} \sim 0.02 \, h^{-2}. \qquad (2.161)$$

As we will see in the following, an important parameter is the *equality point*, i.e., the value of the scale factor a_{eq}, or redshift z_{eq}, at which $\rho_R = \rho_M$. It represents the onset of the matter-dominated epoch. Because this parameter plays a crucial role in cosmology and leaves an imprint on the spectrum of matter fluctuations and CMB anisotropies, it is worth computing it in some detail.

We have already anticipated that data from primordial nucleosynthesis and CMB temperature fluctuations agree on a present baryon energy density on the order of $\Omega_B h^2 \sim 0.02$. This is in fact a subleading contribution to the total matter energy density today, which receives a larger contribution from dark matter, $\Omega_C h^2 \sim 0.11$, so that the total matter density is $\Omega_M h^2 \sim 0.13$. To get a first order of magnitude, let us compute the redshift of equality, assuming that radiation consists only of photons. From the known values of the photon temperature $T_0 = 2.725$ K and of the critical density $\rho_{cr,0}$ – see (2.58) – we get

$$\Omega_\gamma h^2 = 2.47 \times 10^{-5}. \qquad (2.162)$$

Because the radiation energy density scales as $(1 + z)^4$, whereas the matter density evolves as $(1 + z)^3$, the equality point in this case is given by

$$\Omega_\gamma h^2 (1 + z_{eq}) = \Omega_M h^2 \Rightarrow 1 + z_{eq} \sim 4 \times 10^4 \Omega_M h^2. \qquad (2.163)$$

At equality, the photon temperature was then equal to $9.4 \, \Omega_M h^2$ eV.

Note that the value of z_{eq} is insensitive to the dark matter mass, as long as it is much larger than the eV scale. However, to obtain the correct value of z_{eq}, we should now include in the computation the three standard neutrinos. All other particles (charged leptons, etc.) are already nonrelativistic and make a negligible contribution to the energy budget.

Actually, we do not yet know the neutrino mass spectrum, but we have an upper bound from laboratory measurements, $m_\nu \leq 2$ eV. We will show soon that at $T \sim$ eV neutrinos have already decoupled from the thermal bath and their temperature is lower than photon temperature by a factor $T_\nu/T \sim (4/11)^{1/3}$. Putting these numbers together, we see that at equality, neutrinos were still relativistic and should be counted as *radiation* to compute the correct value of z_{eq}. Using (2.155) and assuming zero chemical potential for neutrinos, we get

$$\Omega_R h^2 = \Omega_\gamma h^2 \left(1 + \frac{7}{8} 3 \left(\frac{4}{11}\right)^{4/3}\right). \qquad (2.164)$$

Hence the neutrino contribution to ρ_R is on the same order of ρ_γ, and leads to a small decrease in z_{eq},

$$1 + z_{eq} = \frac{\Omega_M h^2}{\Omega_R h^2} = 2.4 \times 10^4 \Omega_M h^2 \sim 0.3 \times 10^4. \qquad (2.165)$$

Finally the photon temperature at equality appears to be close to $T_{eq} \sim 0.7\,eV$, whereas the neutrino temperature at the same time is about $0.5\,eV$.

Using (2.61), it is easy to show that as long as possible contributions from spatial curvature and from a cosmological constant can be neglected, the Hubble parameter can be expressed in terms of H_0, Ω_M, and z_{eq}:

$$H(z) = H_0 \sqrt{\Omega_M} (1+z)^{3/2} \left(1 + \frac{1+z}{1+z_{eq}} \right)^{1/2}. \qquad (2.166)$$

For $z < z_{eq}$, the universe enters the matter-dominated era, during which two important phenomena take place:

- When the photon temperature is on the order of $0.1\,eV$, i.e., around a redshift $z \sim 10^3$, almost all electrons and protons recombine to form neutral hydrogen. Matter becomes transparent to radiation, and photons can move unperturbed along null geodesics. This is the epoch at which the image of the CMB that we see today is emitted. Precise measurements of CMB anisotropies provide extremely useful information on the composition of the universe at this time. The recombination process is sketched in Section 2.4.1, whereas Chapter 5 describes the physics of CMB anisotropies, with of course particular emphasis on the role played by neutrinos.
- When the universe is matter-dominated, structures on large scales start forming. The initial tiny perturbations seen in the CMB are efficiently amplified by gravitational instability and eventually lead to galaxy and cluster patterns. As long as perturbations remain small, structure formation can be modeled with linear perturbation theory. Semianalytical methods or numerical approaches should be used when matter density inhomogeneities become large. These topics are covered by several books and reviews including Peebles, 1980; Padmanabhan, 1993; Dekel and Ostriker, 1999; Mo *et al.*, 2010. We will summarize theoretical approaches to structure formation in Chapter 6, insisting on the role played by neutrinos in the growth of structures.

2.3.3 Λ *(or dark energy) domination*

In the late nineties, two independent experimental groups found an unexpected result: the expansion of the universe today and at small redshift, $z \leq 1$, is accelerated (Riess *et al.*, 1998; Perlmutter *et al.*, 1999). This evidence comes from a measurement at high redshift, up to $z \sim 1$, of the luminosity distance d_L defined in (2.81). At these redshifts, in fact, d_L is sensitive to the deceleration parameter q and can distinguish among different expansion histories $H(z)$. As mentioned before, measuring d_L requires sources which can be trusted as standard candles. Moreover, they should be bright enough to remain detectable at high redshifts. Fortunately, there are excellent candidates which fit both requirements, namely Type Ia supernovae. These objects result from explosive events which take place in binary systems where one of the two stars is a white dwarf. When the degenerate star accretes matter from its companion and reaches the limiting Chandrasekar mass, it becomes unstable and produces a thermonuclear explosion. Because the mass scale of such supernovae is fixed by the Chandrasekar bound, their luminosity can vary only within a small range. Furthermore, this luminosity is correlated with the time rise and fall of the event. Observing the light-curve in time gives a reliable way to measure the absolute supernova magnitude.

Observations show that distant supernovae (with typically $z \sim 1$) look fainter than would be expected if the expansion rate corresponded to a matter-dominated regime, as illustrated by Fig. 2.3. This observation can be accommodated by assuming that the universe expansion is accelerated at late times (i.e., small redshift). Equation (2.63) shows that the sign of the acceleration parameter q is sensitive to the mean equation of state of the several components contributing to the Friedmann equation. A negative q (i.e., positive $\ddot{a}$) is a clear sign that the dominant component has $P \leq -\rho/3$. In particular, a cosmological constant with $P = -\rho$ is a very good fit to the data. In addition, the combination of supernova data with measurements of the scale of the peak in the angular correlation function of CMB anisotropies points to a nearly flat universe. Hence the current standard model of cosmology, called ΛCDM, assumes that the universe has zero spatial curvature, and a nonzero cosmological constant Λ, and is filled today with photons, baryons, CDM and neutrinos (assumed to be approximately massless in the minimal ΛCDM version). Under the ΛCDM assumption, current supernovae data are best fitted by a fractional density of matter (baryons and CDM) $\Omega_M = 0.28$, and hence a cosmological constant density $\Omega_\Lambda = 0.72$, with a 15% statistical uncertainty and less than 20% systematic error.

The fact that today the universe is dominated by Ω_Λ raises two problems that are somehow related. Is there any way to understand from a theoretical point of view why the value of Λ is on the order of H_0^2? Furthermore, why are the matter and cosmological constant contributions on the same order of magnitude today? Actually, although Λ is a constant, matter energy density scales as $(1 + z)^3$, so ρ_Λ and ρ_M should be very fine-tuned in order to be almost the same today. This problem is usually called the *coincidence problem*.

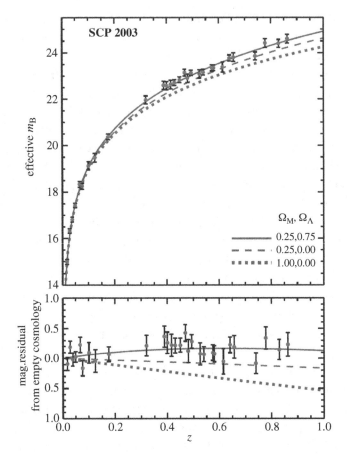

Figure 2.3 The results of the Supernova Cosmological Project (from Knop *et al.*, 2003). The effective apparent magnitude for 42 high-redshift Type Ia supernovae is plotted versus z and compared with different cosmological models. Reproduced by permission of the AAS.

Although explaining such a small value of the cosmological constant is still an open question (Weinberg, 1989), the coincidence problem might be alleviated by assuming that the species responsible for the acceleration of the universe is in fact evolving with redshift and such that P is close to $-\rho$ today, but not necessarily in the past. This can be achieved with an equation of state $P = w\rho$ with w close to minus one, or more generally with a more complicated or time-dependent equation of state. Such an unknown species is generically called *dark energy* (DE). In this book, we will not discuss DE models and the several tuning problems that they have to face, but when scrutinizing cosmological bounds on neutrino properties, we will occasionally assume that the acceleration is caused by DE with a free parameter w instead of a pure cosmological constant. This assumption is relevant only for late cosmological evolution: it can affect the duration of the accelerated stage for fixed $\Omega_M = 1 - \Omega_{DE}$, as well as the evolution of cosmological perturbations during this recent stage.

2.4 A first look at photon and neutrino backgrounds

2.4.1 Photon decoupling and the formation of the cosmic microwave background

The CMB plays an important role in constraining neutrino physics. Let us introduce the basics of CMB physics at the level of homogeneous cosmology, in order to prepare for the discussion of CMB anisotropies in Chapter 5.

Photons of the CMB are perhaps the best example of a blackbody distribution, with a temperature reaching $T_0 = 2.725 \pm 0.002\,\mathrm{K}$ today (Mather *et al.*, 1999). This is the signal that in the past photons were in thermodynamic equilibrium, thanks to fast electromagnetic processes involving charged particles at sufficiently high energy. As the temperature decreased, however, free electrons and nuclei (mainly protons and a small fraction of ^{4}He) started forming neutral atoms, and matter became neutral, up to a tiny fraction of free electrons. At *recombination* the universe started to be transparent to radiation, and the Compton scattering rate of photons over free electrons became negligible compared to the expansion rate. If we approximatively describe recombination as taking place in a very short redshift interval around a central value, the CMB photons we see today originate from the *last scattering surface*, corresponding to a redshift $z_{LS} = 1/a_{LS} - 1$ such that the photon scattering rate is on the order of the Hubble parameter. The limit of *instantaneous photon decoupling* is not entirely correct, and in fact the finite width of the last scattering surface is responsible for detectable effects on CMB anisotropies on very small angular scales. Once decoupled, photons satisfy a Boltzmann equation with a zero collisional term. Because at z_{LS} their distribution is a Bose–Einstein distribution

$$f_\gamma(p) = \left[\exp\left(\frac{p\,a_{LS}}{T_{LS}a_{LS}} \right) - 1 \right]^{-1} \tag{2.167}$$

and we saw in Section 2.2.1 that for massless particles any function of the comoving momentum $(p\,a)$ is a solution of the collisionless Boltzmann equation, after last scattering the photon distribution keeps its equilibrium shape,

$$f_\gamma(p) = \left[\exp\left(\frac{p\,a}{T_{LS}a_{LS}} \right) - 1 \right]^{-1}, \tag{2.168}$$

with a decreasing temperature $T = T_{LS}a_{LS}/a$.

To compute the value of z_{LS} one should first understand the mechanism of recombination and get the typical redshift at which it takes place. We first need to compute the evolution of the fraction X_e of free electrons. We will then infer the typical photon scattering rate, and compare it with $H(T)$ to find the last scattering epoch. In fact, neutral hydrogen is not directly formed by capturing an electron on the lower energy level $1s$, because the emitted photon would have just the right energy to ionize another atom, and this would produce no decrease in X_e. Recombination proceeds mainly through electron captures in the $2s$ excited state, which then relax to ground states by emitting two lower-energy photons. A detailed description of recombination requires following free electrons and thermal photons, as well as the density of each excited hydrogen level (at least $2s$ and $2p$). The corresponding kinetic equations are typically solved numerically. This description is beyond the scope of this book, and the interested reader might want to go through the standard papers (Peebles, 1968 and Seager *et al.*, 1999), or the semi-analytic description of Mukhanov, 2005 and Weinberg, 2008. Here we are interested only in getting an estimate of z_{LS}, and for this it is enough to apply the Saha condition described in Section 2.2.1.

If we define the number density of free electrons/protons as $n_e = n_p$ and of neutral hydrogen atoms as n_H, equation (2.123) applied to the case of the electromagnetic reaction

$$e + p \leftrightarrow \gamma + H \tag{2.169}$$

reads

$$\frac{n_p n_e}{n_H} = \left(\frac{m_e T}{2\pi}\right)^{3/2} \exp\left(-\frac{m_e + m_p - m_H}{T}\right). \tag{2.170}$$

We have neglected the proton/hydrogen atom mass difference in the prefactor. It is instead crucial to keep it in the exponential, which depends on the hydrogen binding energy $B_H = m_e + m_p - m_H \sim 13.6$ eV. Notice that we have used the nonrelativistic limit for both protons and electrons. Indeed, we expect recombination to take place at energies at most on the order of B_H, which is much smaller than the electron mass. Defining the ionized fraction as

$$X_e \equiv \frac{n_e}{n_e + n_H}, \tag{2.171}$$

we can write (2.170) in the form

$$\frac{X_e^2}{1 - X_e} = \frac{1}{n_e + n_H} \left(\frac{m_e T}{2\pi}\right)^{3/2} \exp\left(-\frac{B_H}{T}\right). \tag{2.172}$$

Using the definition of the ^{4}He mass fraction in (2.157), we have $n_p + n_H = n_B(1 - Y_p)$. Introducing also the baryon–to–photon density ratio η_B defined in

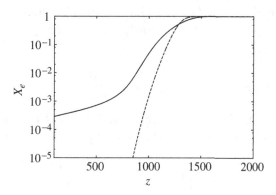

Figure 2.4 The free electron fraction X_e versus redshift for $Y_p = 0.25$ and $\eta_B = 6 \times 10^{-10}$. The dashed line shows the prediction of the Saha equation (2.173), whereas the solid line is given by RECFAST (Wong *et al.*, 2007), a numerical algorithm providing an accurate approximation to the solution of the full system of Boltzmann equations at all relevant hydrogen atom levels.

(2.160), we can cast the Saha equation in the form

$$\frac{X_e^2}{1 - X_e} = \frac{1}{\eta_B(1 - Y_p)} \left(\frac{m_e}{T}\right)^{3/2} \frac{\sqrt{\pi}}{2^{5/2}\zeta(3)} \exp\left(-\frac{B_H}{T}\right). \qquad (2.173)$$

Around the temperature $T \sim B_H$, the r.h.s. is still on the order of 10^{15}, which means that the fraction X_e is very close to unity and all electrons are still free. This is due to the large number of photons compared to baryons, i.e., to the very low value of η_B. Indeed, even if at $T \sim B_H$ the mean photon energy is as low as B_H, the high-energy tail of the distribution still contains a large number of more energetic photons which would immediately dissociate any formed hydrogen atom. The behaviour of X_e as predicted by the Saha equation (2.173) is shown in Fig. 2.4 for $Y_p = 0.25$ and $\eta_B = 6 \times 10^{-10}$. We see that X_e decreases quite rapidly around $z \sim 10^3$, corresponding to $T \sim 0.3$ eV.

Although the Saha condition succeeds in giving a correct estimate of the recombination redshift, it fails in giving the asymptotic value of X_e, which would be predicted to be negligibly small. To find this value, one should solve the kinetic equations which govern the recombination stage. The point is that the equilibrium condition is not applicable to the number density of hydrogen atoms in the ground state, because this population is not in equilibrium with excited states when recombination starts. The predicted value for X_e levels off at redshifts on the order of a few hundred, and remains small but not completely negligible, $X_e \sim 10^{-3}$; see Fig. 2.4.

We can now compute the redshift of the last photon scattering surface. Compton processes become inefficient when matter starts recombining and the free electron density falls. Using the (nonrelativistic) limit of the Compton cross section, given by the Thomson formula $\sigma_T = 8\pi\alpha^2/(3m_e^2) \sim 0.67 \times 10^{-24}$ cm^2, we have for the photon scattering rate

$$\Gamma_\gamma \sim \sigma_T n_B X_e \sim 1.7 \times 10^{-31} X_e \left(\frac{\Omega_B h^2}{0.022}\right) (1+z)^3 \, \text{cm}^{-1}. \tag{2.174}$$

This should be compared with the Hubble parameter $H(T)$, for which we can use the expression (2.166). The condition $\Gamma_\gamma \sim H(T)$ for $\Omega_B h^2 = 0.022$ reads

$$4 \times 10^{-3} X_e(z)(1+z)^{3/2} \left(1 + \frac{1+z}{1+z_{\text{eq}}}\right)^{-1/2} \left(\frac{0.13}{\Omega_M h^2}\right)^{1/2} \sim 1. \tag{2.175}$$

The l.h.s. is still quite large at equality because $X(z_{\text{eq}}) = 1$, but it falls quickly below one as X_e decreases to its final value of order 10^{-3}. Hence last scattering takes place during recombination, around $z_{\text{LS}} \sim 10^3$.

2.4.2 *The cosmic neutrino background*

In the early universe, the three left-handed neutrinos ν_α and their CP conjugated states are thermally excited in the primeval plasma of particles. They are maintained in kinetic and chemical equilibrium with charged leptons, baryons and photons by weak interactions. In this regime the neutrino distribution is of the Fermi–Dirac type, with a negligible contribution of the mass to the energy

$$f_{\nu_\alpha}(p) = \frac{1}{e^{\frac{p}{T}-\xi_\alpha} + 1}$$

$$f_{\bar{\nu}_\alpha}(p) = \frac{1}{e^{\frac{p}{T}+\xi_\alpha} + 1}, \tag{2.176}$$

with p the physical momentum and T the photon temperature, whose evolution is given by the conservation of entropy in the baryon–lepton–neutrino–photon fluid. We have seen that the temperature decreases as a^{-1}, except near the times when particles disappear from the thermal bath and release their entropy to lighter particles. The neutrino–antineutrino asymmetry, energy density, pressure and entropy density take the standard forms seen in Section 2.2.2,

$$n_{\nu_\alpha} - n_{\bar{\nu}_\alpha} = \frac{T^3}{6}\xi_\alpha \left(1 + \frac{\xi_\alpha^2}{\pi^2}\right) \tag{2.177}$$

$$\rho_{\nu_\alpha} + \rho_{\bar{\nu}_\alpha} = \frac{7\pi^2}{120}T^4 \left(1 + \frac{30\,\xi_\alpha^2}{7\pi^2} + \frac{15\,\xi_\alpha^4}{7\pi^4}\right) = 3(P_{\nu_\alpha} + P_{\bar{\nu}_\alpha}) \tag{2.178}$$

$$s_{\nu_\alpha} + s_{\bar{\nu}_\alpha} = \frac{7\pi^2}{90}T^3 \left(1 + \frac{15\,\xi_\alpha^2}{7\pi^2}\right), \tag{2.179}$$

whereas the neutrino or antineutrino number density is given by the polylogarithmic function $n_{\nu_\alpha, \bar{\nu}_\alpha} = -2Li_3(-e^{\pm \xi_\alpha})$, which for small ξ_α reduces to

$$n_{\nu_\alpha, \bar{\nu}_\alpha} = \frac{3\zeta(3)}{4\pi^2} T^3 \pm \frac{\xi_\alpha}{12} T^3 + \mathcal{O}(\xi_\alpha^2). \tag{2.180}$$

We will see in the following chapters that neutrino chemical potentials are expected to be very small on the basis of theoretical arguments (see Chapter 3). Furthermore, their values can be *experimentally constrained* using primordial nucleosynthesis. In this Section we will set $\xi_\alpha = 0$, and postpone the discussion of finite ξ_α's to chapter 4.

As the temperature decreases, the neutrino interaction rate falls faster than the Hubble rate, and eventually becomes smaller than $H(T)$. Weak rates are then inefficient in mantaining neutrinos at equilibrium, and they decouple from the electromagnetic plasma (baryons, leptons, photons). To describe neutrino decoupling in detail, one should use the set of Boltzmann equations governing the evolution of their distribution function. This approach will be presented in Chapter 4. However, to estimate the value of T at which decoupling takes place, we can simply compare the interaction rate with $H(T)$. A posteriori we will see that the decoupling temperature is in the MeV range, when the universe is still radiation-dominated and the thermal bath is populated by electron–positron pairs, photons and (nonrelativistic) baryons. The leading processes contributing to equilibrium are scattering over electrons/positrons and pair conversions, $e^+ e^- \leftrightarrow \nu_\alpha \bar{\nu}_\alpha$. As long as the temperature T is smaller than the W and Z boson masses, the corresponding cross section times velocity for charged and neutral current interactions is on the order of

$$\langle \sigma v \rangle \sim G_F^2 T^2. \tag{2.181}$$

The weak interaction rate for neutrinos is therefore given by

$$\Gamma_\nu \sim G_F^2 T^2 n_e \sim G_F^2 T^5, \tag{2.182}$$

where n_e is the electron/positron (target) density, which is on the order of T^3 as long as electrons are still relativistic. On the other hand, the Hubble parameter in the radiation-dominated epoch is given by

$$H(T) = \sqrt{\frac{8\pi G}{3} g_* \frac{\pi^2}{30} T^4} \sim \sqrt{g_*} \frac{T^2}{m_{\text{Pl}}}. \tag{2.183}$$

The decoupling temperature $T_{\nu D}$ is thus given by

$$G_F^2 T_{\nu D}^5 = \sqrt{g_*} \frac{T_{\nu D}^2}{m_{\text{Pl}}} \Rightarrow T_{\nu D} = \left(\frac{\sqrt{g_*}}{G_F^2 m_{\text{Pl}}}\right)^{1/3} \sim g_*^{1/6} \, \text{MeV}. \tag{2.184}$$

We see that $T_{\nu D}$ is very weakly dependent on the number of relativistic degrees of freedom g_* and is of order MeV. At this temperature, only electrons/positrons, photons and neutrinos themselves contribute to g_*, and therefore

$$g_* = 2 + \frac{7}{8}4 + \frac{7}{8}6 = \frac{43}{4} = 10.75. \tag{2.185}$$

A more refined computation of the decoupling temperature can be performed using the kinetic equation (Dolgov, 2002). The relevant Boltzmann equations in terms of the variables $x = ma$ and $y = pa$, with m some mass scale (usually taken 1 MeV), read

$$Hx\frac{\partial f_{\nu_e}}{\partial x} = -\frac{80G_F^2(\tilde{g}_L^{\prime 2} + g_R^{\prime 2})m^9}{3\pi^3 x^5}yf_{\nu_e} \tag{2.186}$$

$$Hx\frac{\partial f_{\nu_{\mu,\tau}}}{\partial x} = -\frac{80G_F^2(g_L^{\prime 2} + g_R^{\prime 2})m^9}{3\pi^3 x^5}yf_{\nu_{\mu,\tau}}. \tag{2.187}$$

To get these expressions, one uses the neutrino interaction rate amplitudes as in Tables 1.5 and 1.6, and approximates the particle distributions as Boltzmann functions. This approximation is typically quite good, with an accuracy of 10%. Because during radiation domination $H \sim x^{-2}$, it is easy to find that the solutions are functions of the combination y/x^3, or yT^3. In particular the (momentum-dependent) decoupling temperatures are found to be $T_{\nu_e D} = 2.7y^{-1/3}$ MeV and $T_{\nu_{\mu,\tau}D} = 4.5y^{-1/3}$ MeV. Taking the average values of momenta, $y \simeq 3$, one therefore finds $T_{\nu_e D} = 1.87$ MeV and $T_{\nu_\mu,\nu_\tau D} = 3.12$ MeV (see also Dolgov and Zeldovich, 1981; Enqvist *et al.*, 1992a). The slightly lower value for the ν_e species is simply due to the fact that at the MeV scale the thermal bath is not flavour blind (there are only electrons/positrons), and ν_e can interact also through charged current processes in addition to neutral currents, so they remain in equilibrium slightly later.

The limit in which we consider neutrino decoupling as an instantaneous event hapenning at $T_{\nu D}$ is called the *instantaneous decoupling limit*. In fact, decoupling takes place over an extended range of time. Because neutrinos have different momenta and weak cross sections grow with s (the squared total energy in the two–particle center of mass of the system), more energetic neutrinos will be kept in equilibrium longer than low-energy neutrinos. This in turn implies that when $e^\pm$ annihilate, some thermal distortions will be imprinted in the neutrino distribution with respect to a standard Fermi–Dirac function. In Chapter 4 we will see that these effects are quite small and change the neutrino energy density at the percent level.

After decoupling, neutrinos propagate freely, and their distribution remains unchanged but for the effect of redshift of physical momentum. They are an example of hot relics; i.e., they decouple while they are relativistic particles. In the instantaneous decoupling approximation, we can assume the distribution at $T_{\nu D}$ to be a Fermi–Dirac one, which written in terms of the comoving momentum reads

$$f_{\nu_\alpha}(p) = \left[\exp\left(\frac{p \, a_D}{T_{\nu D} a_D}\right) + 1 \right]^{-1}, \tag{2.188}$$

with a_D the scale factor at $T_{\nu D}$. The neutrino distribution at later times is therefore again a Fermi–Dirac function (the same result holds if we include the constant chemical-potential–to–temperature ratio ξ_α),

$$f_{\nu_\alpha}(p) = \left[\exp\left(\frac{p \, a}{T_{\nu D} a_D}\right) + 1 \right]^{-1}. \tag{2.189}$$

In particular, the distribution in terms of physical momenta is entirely specified by a "temperature" parameter which scales as $T = T_{\nu D} a_D / a$, where quotation marks have been used to warn the reader that though the distribution inherited from the decoupling phase is Fermi–Dirac, neutrinos are no longer in equilibrium for $T \leq T_{\nu D}$. This temperature scaling is equivalent to the conservation of the neutrino entropy per comoving volume, $s_{\nu,\bar\nu_\alpha} a^3 = \text{constant}$.

The neutrino-to-photon temperature ratio after neutrino decoupling can be inferred from the conservation of entropy for the electromagnetic plasma of $e^\pm$ and photons. As long as electron/positron pairs are relativistic, the photon temperature simply scales as a^{-1} because the number of degrees of freedom g_s is constant. When the temperature drops below a value T_{ann} on the order of the electron mass m_e, electromagnetic interactions are still very efficient, but the $e^\pm$ pair annihilation $e^+ e^- \to \gamma\gamma$ cannot be efficiently compensated for by inverse pair production. In other words, equilibrium prescribes that $e^\pm$ distributions are suppressed by the term $\exp(-m_e/T)$, and practically all electrons and positrons disappear, apart from a tiny relic electron density related by charge conservation and electric neutrality to the baryon number parameter η_B. As a result, photons are heated, and their temperature in this phase is not decreasing as the inverse scale factor. Neutrinos are basically left undisturbed by pair annihilations, because the analogous weak processes $e^+ e^- \to \nu_\alpha \bar\nu_\alpha$ have a low rate compared to the Hubble parameter, and only a tiny fraction of $e^\pm$ goes into neutrinos (see Chapter 4 for a detailed discussion). Notice that this fraction would be exactly zero in the instantaneous decoupling limit, because $T_{\nu D} > T_{\text{ann}}$.

To compute the evolution of the photon temperature during this stage, consider the entropy density of coupled $e^\pm$ and photons,

$$s_{e^\pm, \gamma} = \frac{2\pi^2}{45} T^3 g_s(T), \tag{2.190}$$

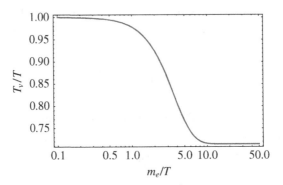

Figure 2.5 The neutrino-to-photon temperature ratio versus the nondimensional *time* variable m_e/T.

where $g_s(T)$ has been defined in (2.143) and in the present case reads

$$g_s(T) = 2 + \frac{45}{\pi^4} \int_0^\infty \left(\sqrt{x^2 + \frac{m_e^2}{T^2}} + \frac{x^2}{3\sqrt{x^2 + \frac{m_e^2}{T^2}}} \right) \frac{x^2 dx}{\exp \sqrt{x^2 + \frac{m_e^2}{T^2}} + 1}.$$

$$(2.191)$$

For $T \gg m_e$, electrons and positrons are still relativistic, and

$$g_s \to 2 + \frac{7}{8}(2 + 2) = \frac{11}{2}. \tag{2.192}$$

In the opposite limits electrons and positrons have negligible entropy and

$$g_s \to 2. \tag{2.193}$$

Because the entropy per comoving volume $s_{e^\pm,\gamma} a^3$ is conserved, and $T_\nu \sim a^{-1}$, we have

$$g_s(T) \frac{T^3}{T_\nu^3} = \text{constant}. \tag{2.194}$$

At neutrino decoupling $T_\nu = T$. After the $e^\pm$ annihilation stage, when $T \ll m_e$, g_s drops and we get

$$\frac{11}{2} = 2 \frac{T^3}{T_\nu^3} \Rightarrow \frac{T_\nu}{T} = \left(\frac{4}{11} \right)^{1/3}, \quad T \ll m_e. \tag{2.195}$$

The behavior of the ratio T_ν/T is shown in Fig. 2.5 versus the adimensional variable $x = m_e/T$.

If we use the known value of the CMB temperature today, $T_0 = 2.725\,\mathrm{K}$, we see that neutrinos are as cold as $T_{\nu,0} = 1.945\,\mathrm{K}$, or $1.676 \times 10^{-4}\,\mathrm{eV}$ in natural units.

Using (2.180) and assuming negligible chemical potentials, the number density of cosmic neutrinos is given for each flavour by

$$n_{\nu,0} = n_{\bar{\nu},0} \sim 56\,\text{cm}^{-3}, \tag{2.196}$$

yielding a remarkably large flux with respect to other astrophysical neutrino sources, including solar neutrinos. Unfortunately their distribution peaks at a very low energy, making their direct detection impossible or at least very challenging (see the related discussion in Chapter 7).

After $e^{\pm}$ annihilation, the relativistic degrees of freedom populating the universe are photons and neutrinos only, so that the corresponding energy density can be written as

$$\rho_R = \rho_\gamma \left(1 + \frac{7}{8} \left(\frac{4}{11} \right)^{4/3} 3 \right). \tag{2.197}$$

This result holds only if

- there are only three light neutrino species and no other relativistic particles;
- neutrino distributions are standard Fermi–Dirac functions with zero chemical potentials;
- we trust the instantaneous decoupling limit.

A convenient way to parameterize any deviation from these assumptions consists in defining the *effective number of neutrino species* N_{eff},

$$\rho_R = \rho_\gamma \left(1 + \frac{7}{8} \left(\frac{4}{11} \right)^{4/3} N_{\text{eff}} \right). \tag{2.198}$$

The name of this parameter is historically linked to the fact that in the 1980s, the number of active (light) neutrino generations was still debated. After LEP precision electroweak measurements on the Z resonance and its invisible width, we know that this number is three (Beringer *et al.*, 2012), but before these experimental results were available, cosmology was already able to put some bounds, mainly from primordial nucleosynthesis, since the first analysis of Shvartsman, 1969; Steigman *et al.*, 1977. Measuring N_{eff} (in the broader sense of measuring the radiation density) is still of great interest today, and the set of observable quantities which can be used to this end is even enlarged. In fact, beside nucleosynthesis, we will see in Chapters 5 and 6 that the spectra of CMB anisotropies and of matter fluctuations can provide powerful constraints on ρ_R and therefore on N_{eff}. Of course, the question is not any more to assess the number of light-flavour neutrino species, but rather to check if there is any evidence for other as yet unidentified light particles contributing to ρ_R, or if the neutrino distributions in phase space have some nontrivial features, such as chemical potentials. In other words, $N_{eff} \neq 3$ might be due to fermionic or bosonic degrees of freedom which have nothing to do with neutrinos, or simply show that our naive expectation of a simple Fermi–Dirac distribution for the three neutrinos should be questioned.

From the value of $T_{\nu,0}$ we see that at least two of the three neutrino mass eigenstates are nonrelativistic today, because $T_{\nu,0} < (\delta m^2)^{1/2}$, with $\delta m^2 \sim 10^{-5}$ eV^2 the squared mass difference involved in the solar neutrino problem; see Chapter 1. Their contribution to the present energy density can easily be worked out. Using the present value of the critical density and (2.196) we have (summing only over neutrinos which are nonrelativistic today)

$$\Omega_\nu = \frac{\sum_i m_{\nu i} n_{\nu,0}}{\rho_{cr,0}} = \frac{\sum_i m_{\nu i}}{eV} \frac{1}{94.1(93.1)\,h^2}, \qquad (2.199)$$

where standard Fermi–Dirac distributions with zero chemical potential have been assumed, and the number in brackets accounts for the effects of noninstantaneous neutrino decoupling (see Chapter 4). If the lightest neutrino eigenstate is still relativistic today, its contribution to Ω_ν is in any case subdominant. So, for simplicity, the above formula can be used as a good approximation in all cases, summing over all three species.

Before neutrino oscillation experiments, the mass of ν_μ and ν_τ was poorly constrained. If we look into the Particle Data Group (Beringer *et al.*, 2012), we still find the following upper bounds from laboratory experiments only: $m_{\nu_\mu} \leq 0.17$ MeV and $m_{\nu_\tau} \leq 18$ MeV. In the past, the constraints coming from the fact that the neutrino density should not overclose the universe, $\Omega_\nu < 1$, and therefore $\sum_i m_{\nu i} \leq 94 h^2$ eV, were the strongest information on the sum of neutrino masses. This result is usually referred to as the Cowsik–McClelland bound (Cowsik and McClelland, 1977), although it was also previously considered in Gerstein and Zeldovich, 1966 (see also Szalay and Marx, 1976). Today, all neutrino masses are constrained to be at most on the order of 1 eV, so that neutrinos can contribute only a small fraction of the total energy density of the universe. Nevertheless, their effect at late times during structure formation might be very significant, as we will discuss at length in Chapter 6.

So far we have only considered left–handed neutrinos and CP-conjugated states, so for each flavour we have counted only two degrees of freedom. If neutrinos are Majorana particles, right-handed partners might not exist at all, or could be very massive particles playing a role only at very high temperatures; see Chapter 3. In this case they would have already disappeared from the thermal bath in the primordial nucleosynthesis epoch.

On the other hand, if neutrinos have a Dirac mass term, right-handed neutrinos are also excited through the mass term, which mixes chirality states. However, if only left-handed weak currents are at work, the production rate of right-handed states is suppressed with respect to that of left-handed ones by a factor $(m_\nu/E)^2$ in the relativistic limit. Thus

$$\Gamma_R \sim G_F^2 T^5 \frac{m_\nu^2}{T^2}, \tag{2.200}$$

which compared with the Hubble parameter and imposing $\Gamma_R \geq H(T)$ shows that these processes are at equilibrium only for $T > \text{MeV}^3/m_\nu^2$; i.e., $T > 10^9$ GeV for $m_\nu \sim$ eV. However, for such high temperatures, much higher than the W, Z boson masses, the weak cross section has different behaviour and decreases as the inverse squared center of mass energy,

$$\sigma \sim G_F^2 m_{W,Z}^4/T^2, \quad T > m_{W,Z}. \tag{2.201}$$

If we use this expression we find $T \leq 10^2 \, \text{MeV}(m_\nu/\text{eV})^{2/3}$, which is well below the range where (2.201) can be applied. We conclude that Γ_R is always smaller than the universe's expansion rate, so that the total neutrino density is the same as in the case of Majorana neutrinos.

Another contribution to the ν_R production at high temperatures may come from direct decay of gauge bosons, with a rate also much lower than $H(T)$ for neutrino masses smaller than keV (see, e.g., Dolgov, 2002). The possibility of right-handed neutrino production by right-handed weak currents, for instance in extensions of the Standard Model to gauge symmetries also having an $SU(2)_R$ factor, will be discussed later on. Let us anticipate that in this case they would contribute to N_{eff} by an amount comparable to that for left-handed neutrinos, unless the right-handed gauge bosons are sufficiently massive, because cross sections scale as $M_{W_R}^{-4}$. Because primordial nucleosynthesis is a very sensitive probe of N_{eff}, it is possible to deduce lower bounds on M_{W_R}.

3

Neutrinos in the early ages

A Tale of Three Numbers: **1**, **3** *and* ∞.[1] This might be a good subtitle for this chapter, which is about the problem of baryogenesis in the early universe.

We have already mentioned in Chapter 1 that observations of the light nuclear yields produced during primordial nucleosynthesis and the features of the CMB anisotropies single out a definite value for **1** parameter, the baryon-to-photon density ratio $\eta_B \sim 6 \times 10^{-10}$ at low temperatures, much below the nucleon mass. If the universe's expansion were starting with symmetric initial conditions, i.e., a zero initial baryon number per comoving volume, and no baryon-violating interactions were at work at all stages of this expansion, η_B would be expected to be much smaller, as we will show in the following. Moreover, we should detect a comparable number of antibaryons in the solar system or on larger scales, such as our galaxy (30 kpc) or the Local Group (3 Mpc). However, this is not what we do observe. Baryogenesis is a collection of several theoretical ideas on how the tiny value of η_B might be dynamically produced at some stage of the universe's history. The number of models which have been proposed is quite large, although maybe not really ∞! Despite their different particular properties, they all share some common features, because they should all fullfill three basic conditions which were first put forward by Sakharov quite long ago (Sakharov, 1967).

Interestingly, among all models, a very elegant and simple solution is closely related to neutrinos, and thus it is of particular relevance to the subject of this book. It is based on the idea that an asymmetry produced in the lepton number if neutrinos are Majorana particles can later be converted into a baryon asymmetry by some particular configurations of the electroweak gauge and Higgs fields called sphalerons. This scenario of *baryogenesis through leptogenesis* (Fukugita and Yanagida, 1986) is indeed one of the most promising possibilities, and it also has the appealing feature of linking to some extent the high-energy-scale physics which drives lepton

---

[1] After Charles Dickens.

and baryon number production to the low-energy and measurable properties of neutrinos, such as mixing matrix and masses.

The basics of this scenario will be discussed in Section 3.4. We start by arguing in Section 3.1 that the value of η_B is only compatible with a baryon asymmetric universe, and analyze the three Sakharov conditions in Sections 3.2 and 3.3.

As we mentioned, there is quite a long list of models which have been proposed to produce successful baryogenesis. Under the spell of leptogenesis only, there are a few hundred papers in the literature which deal with variations of the standard scenario or discuss intriguing new possibilities: supersymmetric leptogenesis, soft leptogenesis, Dirac leptogenesis, triplet scalar and triplet fermion leptogenesis, less hierarchical leptogenesis, resonant leptogenesis, leptogenesis in the framework of the v-MSM of Asaka and Shaposhnikov, 2005, preheating leptogenesis. We have chosen to give an introduction to the basics of the leptogenesis *standard model* only. Discussing all models would deserve a separate companion volume and is outside of the mainstream of our tale. The reader interested in a more complete overview might refer to one of the several review papers which are available in the literature, such as Dolgov, 1992; Riotto and Trodden, 1999; Davidson *et al.*, 2008; Boyarsky *et al.*, 2009a, or to the books Kolb and Turner, 1994; Fukugita and Yanagida, 2003.

3.1 The baryon number of the universe

Observations show that matter is far more abundant than antimatter in the visible universe. All structures we see, stars, galaxies and clusters are made of nucleons and electrons. If the universe contained similar structures composed of antinucleons and positrons, we would see a very intense γ ray emission, originating from the boundaries between matter and antimatter domains, because of annihilation of $p\bar{p}$ pairs into π mesons, including π^0, which then decay into photon pairs. More precisely, antimatter appears to be very rare inside our galaxy, because in cosmic rays we observe 10^4 times more protons than antiprotons, and 10^5 times more helium than anti-helium nuclei. The existence of larger antimatter domains (of galactic or cluster size) appears very unlikely, because statistically we would expect that some of them would intersect our last scattering surface. This would lead to strong patterns in CMB maps, which are not observed (Cohen and De Rújula, 1997; Kinney *et al.*, 1997).

The idea that the universe is symmetric in its baryon and antibaryon content, i.e., has zero baryon number B per comoving volume, is also difficult to reconcile with the baryon-to-photon density ratio, as measured by both primordial nucleosynthesis, mainly from the observed cosmological ^{2}H abundance – see Chapter 4 – and CMB anisotropies. Both kind of observations, which refer to different epochs of the expansion history, single out a value on the order of

$$\eta_B \equiv \frac{n_B}{n_\gamma} \sim 273.97 \times 10^{-10} \, \Omega_B h^2 \sim 6 \times 10^{-10}. \tag{3.1}$$

Despite its smallness, this result is indeed too large!

We have already seen in Section 2.2.2 that a convenient way to express the baryon number per comoving volume is to normalize the asymmetry density to the entropy density for relativistic species, because we know that the latter scales as a^{-3}:

$$B = \frac{n_B - n_{\overline{B}}}{s_R}. \tag{3.2}$$

Suppose now that $B = 0$, so that $n_B = n_{\overline{B}}$. At very high temperatures, exceeding the nucleon mass scale, protons and neutrons are maintained in thermal equilibrium with other species including photons by strong and electromagnetic interactions. As the temperature drops below m_N, $N\overline{N}$ annihilation processes strongly dilute the nucleon density and their equilibrium distribution in phase space develops the exponential suppression factor $\exp(-m_N/T)$. We can compute the relic abundance of baryons using the standard criterion of comparing the annihilation rate to the Hubble expansion rate H. Because at low energies the annihilation cross section is on the order of $\sigma \sim m_\pi^{-2}$, with m_π the pion mass, and the average relative velocity is given by $v \sim 1.6\sqrt{T/m_N}$, in a radiation-dominated expansion the freeze-out temperature is

$$m_\pi^{-2} \sqrt{\frac{T}{m_N}} 2 \left(\frac{m_N T}{2\pi}\right)^{3/2} e^{-m_N/T} \sim \sqrt{\frac{8\pi}{3 m_{Pl}^2} g_* \frac{\pi^2}{30}} T^2 \tag{3.3}$$

or

$$\frac{m_N}{T} \sim \log\left(\frac{m_N m_{Pl}}{m_\pi^2}\right) + \log\left(\frac{6\sqrt{10}}{(2\pi)^3 \sqrt{g_*}}\right). \tag{3.4}$$

The first term on the r.h.s. is ~ 48, which alone would give $T \sim 20$ MeV, below the $\mu^\pm$ annihilation temperature, whereas the second one has a negative sign but is much smaller, for g_* receiving contribution from standard particles only ($e^\pm$, photons and decoupled neutrinos in this temperature range). We finally get $T \sim 21$ MeV. Plugging this value into the nucleon-to-photon density ratio, we obtain

$$\frac{n_B}{n_\gamma} = \frac{n_{\overline{B}}}{n_\gamma} \sim 10^{-17}, \tag{3.5}$$

which is eight orders of magnitude smaller than what is observed; see (3.1).

On the other hand, if some unknown physical mechanism segregated matter from antimatter, it should have been efficient when the baryon-to-photon density ratio was given by (3.1), i.e., using the nonrelativistic equilibrium distribution for baryons at $T \sim 37$ MeV, or $t \sim 0.7 \times 10^{-3}$ s. At this time the physical particle horizon, the largest causally connected distance, was quite small:

$$a(t)\chi_p(t) = \frac{c}{H_0} \frac{a(t)^2}{\sqrt{a_{eq}}} \sim 10^2 \, \text{km}. \tag{3.6}$$

The largest matter in segregated islands of matter/antimatter is that contained in the largest causally connected volume and can be estimated as

$$M \sim m_N \eta_B n_{\gamma,0} \left(\frac{37\,\text{MeV}}{T_0}\right)^3 \frac{4\pi}{3} \left(a(t)\chi_p(t)\right)^3 \sim 10^{-7} M_\odot. \tag{3.7}$$

The fact that we see no antimatter on much larger mass scales, such as the local group or the Virgo cluster, thus remains unexplained.

We are therefore led to the conclusion that the universe has a nonzero and positive baryonic charge B (more nucleons than antinucleons). At high temperatures the nucleon and antinucleon densities are comparable for small values of B, because nucleons are relativistic and the ratio μ_B/T is small. Once the temperature drops below m_N, basically all antinucleons disappear by the effect of annihilations and we are left with a relic baryon density, which is simply given by

$$\frac{n_B - n_{\bar{B}}}{s_R} \sim \frac{n_B}{s_R}. \tag{3.8}$$

Therefore, the value of B today is simply related to η_B by a numerical factor, $B \sim (n_{\gamma,0}/s_{R,0})\,\eta_B$. Notice that, differently from η_B, B is a constant parameter if baryon number is conserved by particle interactions, whereas η_B undergoes dilution when some entropy is released to photons by annihilating species, as for example during the $e^\pm$ annihilation stage. In this latter case, the value of η_B decreases by a factor of $4/11$, because the product $T a$ increases by a factor of $(11/4)^{1/3}$, as we have seen in Chapter 2.

Before Sakharov pointed out the necessary conditions for dynamically producing baryon asymmetry, the value of η_B was assumed as an initial condition at the Big Bang epoch, although quite fine-tuned to a small value. Moreover, at that time there were no hints of a possible violation of baryon number in fundamental interactions. We now consider how, in fact, B can be generated during the evolution of the universe, starting from symmetric conditions, and can be preserved until today, a much more theoretically appealing possibility.

3.2 Sakharov conditions

In a celebrated paper, Sakharov discussed the necessary and sufficient conditions for the production of a nonzero baryon asymmetry. They are the following (Sakharov, 1967):

(i) there should be B-violating interactions;
(ii) these interactions should also violate charge conjugation symmetry C *and* CP, the combined charge conjugation and parity symmetry;
(iii) a baryon asymmetry can be produced only if there is some departure from thermodynamic equilibrium.

The first condition is quite obvious. If all interactions preserved the baryon number, the Hamiltonian H would commute with B, so that if $B = 0$ at some initial time, it will remain so at any later time.

The requirement of both C and CP violation is also quite easy to understand. Consider some arbitrary process from an initial state i to a final state f where some baryon number is produced. If C were a symmetry for this process, then the rate at which it occurs $\Gamma(i \to f)$ would be the same for the reaction involving antiparticles, $\Gamma(\bar{i} \to \bar{f})$, so that no baryon number would be effectively produced. Let us recall that the CPT theorem, which holds on the basis of quite general assumptions for a local relativistic quantum field theory, implies that because the combined transformation of C, P and time reversal T is always an exact symmetry, CP invariance is equivalent to T invariance. Therefore, if p_i (p_f) and s_i (s_f) are the initial (final) particle momenta and spins, then CP symmetry implies that

$$\Gamma(i \to f; p_i, s_i; p_f, s_f) = \Gamma(f \to i; -p_f, -s_f; -p_i, -s_i). \qquad (3.9)$$

Because the baryon numbers produced in the $i \to f$ and $f \to i$ processes are opposite, integrating over both momenta and spins again, we get that no total baryon number can be produced.

Let us finally consider the third condition. One way to see that no baryon number can develop in thermal equilibrium is to recall that the mean value of B, as of all physical quantities, in a statistical mixture described by the density matrix ϱ is given by

$$\langle B \rangle = \frac{\text{Tr}(\varrho B)}{\text{Tr}(\varrho)} \qquad (3.10)$$

Because at equilibrium $\varrho = \exp(-\beta H)$ and H commutes with the CPT operator Θ, whereas $B\Theta = -\Theta B$, we have, using the cyclicity property of the trace,

$$\langle B \rangle = \frac{\text{Tr}\left(\Theta\Theta^{-1}e^{-\beta H} B\right)}{\text{Tr}(\varrho)} = \frac{\text{Tr}\left(\Theta e^{-\beta H}\Theta^{-1} B\right)}{\text{Tr}(\varrho)}$$

$$= -\frac{\text{Tr}\left(\Theta e^{-\beta H} B\,\Theta^{-1}\right)}{\text{Tr}(\varrho)} = -\langle B \rangle \qquad (3.11)$$

and $\langle B \rangle = 0$.

This result can be also understood as follows. At equilibrium the particle distribution function is given by the standard Fermi–Dirac or Bose–Einstein function, which depends on energy and chemical potential only. If there are baryon-violating interactions, the chemical potential associated to baryon number vanishes, so that $f(p) = [\exp(\beta E(p)) \pm 1]^{-1}$. Because from CPT symmetry particles and antiparticles have the same mass, they share the very same distribution in phase space, and therefore the asymmetry is zero.

A crucial role in this reasoning is played by unitarity of the scattering S matrix, i.e., the fact that the sum over all possible transitions to and from some initial state should sum to one. We already used this condition in Chapter 2; see Eq. (2.98). Consider the collisional integral $\mathbf{C}$ which rules the evolution of a particular state i, an index which collectively denotes the particle species, momentum, spin, etc. Summing over all possible processes of the kind $i + j_1 + \cdots + j_n \leftrightarrow k_1 + \cdots + k_m$, we have

$$\mathbf{C} = \sum_{j_i,\ldots,j_n,k_1,\ldots,k_m} \left[\left|\mathcal{M}_{k_1\ldots k_m,ij_1\ldots j_n}\right|^2 f_{k_1} \cdots f_{k_m} (1 \pm f_i)(1 \pm f_{j_1}) \ldots (1 \pm f_{j_n}) \right.$$

$$\left. - \left|\mathcal{M}_{ij_1\ldots j_n,k_1\ldots k_m}\right|^2 f_i f_{j_1} \cdots f_{j_n} (1 \pm f_{k_1}) \ldots (1 \pm f_{k_m}) \right]. \qquad (3.12)$$

If CP is violated, time reversal T is also not a symmetry, so that direct and inverse squared invariant amplitudes are in general different, and we cannot factor them out in each term of (3.12). At equilibrium $\mathbf{C}$ vanishes. We want to show that equilibrium distributions satisfy $\mathbf{C} = 0$ if all particles share the same temperature and

$$\mu_i + \mu_{j_1} + \cdots + \mu_{j_n} = \mu_{k_1} + \cdots + \mu_{k_m}. \qquad (3.13)$$

To this end we have to invoke unitarity (see Dolgov, 1979; Toussaint et al., 1979; Weinberg, 1979). This condition tells us that (compare with the simple case considered in Eq. (2.98) where quantum statistical effects are neglected)

$$\sum_{k_1,\ldots,k_m} \left|\mathcal{M}_{ij_1\ldots j_n,k_1\ldots k_m}\right|^2 (1 \pm f_{k_1}) \ldots (1 \pm f_{k_m})$$

$$= \sum_{k_1,\ldots,k_m} \left|\mathcal{M}_{k_1\ldots k_m,ij_1\ldots j_n}\right|^2 (1 \pm f_{k_1}) \ldots (1 \pm f_{k_m}), \qquad (3.14)$$

so that using this to rewrite the second term in (3.12), we get

$$\mathbf{C} = \sum_{j_i,\ldots,j_n,k_1,\ldots,k_m} \left|\mathcal{M}_{k_1\ldots k_m,ij_1\ldots j_n}\right|^2 \left[f_{k_1} \cdots f_{k_m} (1 \pm f_i)(1 \pm f_{j_1}) \ldots (1 \pm f_{j_n}) \right.$$

$$\left. - f_i f_{j_1} \cdots f_{j_n} (1 \pm f_{k_1}) \ldots (1 \pm f_{k_m}) \right]. \qquad (3.15)$$

Again, when equilibrium distributions are inserted, $\mathbf{C}$ vanishes, because in this case the two terms in the brackets cancel.

3.3 C, CP, *B*, out of equilibrium and all that

We now consider how all Sakharov conditions can be explicitly realized. These conditions are already fulfilled in the framework of the electroweak Standard Model (SM) but to an extent which, with our the present understanding, is unable to explain the experimental result of Eq. (3.1). Baryogenesis, much like the neutrino mass puzzle, seems to require some new physics beyond our present wisdom of fundamental interactions. Perhaps the two problems are related.

3.3.1 C and CP violation

In the SM, charge conjugation is maximally violated because SM is a chiral gauge theory. Whereas left-handed particles interact with $SU(2)_L$ gauge bosons, right-handed fermions are singlets with respect to $SU(2)_L$ transformations. Particles and antiparticles also have different weak hypercharges; see Chapter 1. We also know that CP symmetry is broken in the SM, although the amplitudes of violating effects are much smaller, as first detected in 1964 observing neutral K meson decays (Christenson *et al.*, 1964) and currently well established also in B meson decays; see, e.g., Beringer *et al.*, 2012 for a summary of this topic. We have mentioned in Chapter 1 that all CP-breaking effects in the quark sector can be understood in terms of the phase δ which appears in the CKM matrix, and that any CP-violating observable quantity should be proportional to the Jarlskog parameter $J \sim 10^{-5}$ – see Eq. (1.28) – which is invariant under arbitrary phase redefinitions of quark fields.

Provided the other two Sakharov criteria are satisfied, B violation and out-of-equilibrium conditions, the fact that C and CP are known to be broken is good news and might lead to the optimistic idea that baryogenesis can be explained in terms of well-known SM physics. Unfortunately, this does not seem to be the case. Apart from the problem of how departures from thermodynamic equilibrium can be realized, the amount of CP violation is too small to account for the observed value of η_B.

A back-of-the-envelope way to understand this is the following. From our short review of Chapter 1 it should be clear that if quark fields were degenerate in mass, one could absorb the CKM matrix in the charged weak current $J^{\mu}_{W,Q}$ (1.25) by a suitable redefinition of quarks by a unitary transformation. This is analogous to the case of the lepton charged weak current for massless (and thus degenerate) neutrinos considered in Section 1.2; see (1.20). This means that all CP-violating effects ϵ_{CP} should be proportional to quark mass differences – actually to squared mass differences – because we can always change the overall sign of the fermion mass term by independently redefining the right and left components,

$$\epsilon_{CP} \sim J \frac{(m_t^2 - m_c^2)(m_t^2 - m_u^2)(m_c^2 - m_u^2)(m_b^2 - m_s^2)(m_b^2 - m_d^2)(m_s^2 - m_d^2)}{E^{12}},$$

(3.16)

with E some energy scale. Because B-violating processes in the framework of the SM are effective only at high energies, on the order of the electroweak breaking scale $E_{EW} \sim v \sim 10^2$ GeV or above – see the next section – we find

$$\epsilon_{CP} \sim 10^{-20},$$

(3.17)

which is much smaller than the order of magnitude of η_B.

It seems that this argument already suggests that the production of a large enough baryon asymmetry requires some physics beyond the SM. For example, very popular models for baryogenesis are represented by Grand Unified Theories (GUT). In these theories, as we will discuss in the next section, there is room for B-violating processes. Furthermore, larger CP–violating effects can be obtained in view of the enlarged number of Higgs scalars and gauge boson fields.

A further interesting possibility is related to lepton physics, in particular to our beloved neutrinos. One may wonder how leptons might be relevant to producing a baryon asymmetry, because they do not carry any baryon number. We will see, in fact, that sphalerons reshuffle baryon and lepton numbers but keep unchanged their difference $B - L$, so if some finite value for L is produced in the early universe, it will give rise to a baryon asymmetry too. This is the *baryogenesis through leptogenesis* scenario, which we will discuss in detail. For the moment, we point out only that lepton number L is not a symmetry as soon as we introduce both Dirac and Majorana neutrino mass terms, as in the seesaw models. Moreover, if we have n_s sterile states N_s, we can introduce Yukawa terms into the Lagrangian density, which couple these states to left-handed neutrinos and the Higgs field Φ, whose expectation value eventually produce the neutrino Dirac mass term after spontaneous symmetry breaking:

$$\mathcal{L}_Y^R = Y_{\alpha s}^\nu \overline{L_\alpha} i\sigma_2 \Phi^* N_s + \text{h.c.}$$

(3.18)

Because the coupling matrix Y^ν in general has complex entries, interaction processes mediated by this term violate CP symmetry.

3.3.2 Baryon and lepton number violation

The SM lagrangian is invariant under the two global abelian symmetries corresponding to baryon and lepton numbers. This fact is just *accidental*, meaning that it is not the effect of some symmetry requirement, as for gauge transformations.

Rather, if one writes down all possible renormalizable terms which are invariant under the SM group, one finds that no one breaks B or L. At any order in perturbation theory, there are no processes where B or L can change. However, there are nonperturbative effects which may give rise to some violation. This is the case for instanton and sphaleron configurations of the gauge and Higgs fields, which we are going to discuss in this section. Before doing that, let us first briefly mention how the situation changes if one enlarges the SM gauge symmetry to some GUT.

The idea at the basis of GUT is that at energy scales above some energy threshold M_{GUT} the symmetry group of fundamental interactions is a larger group G, either simple such as $SU(5)$ (Georgi and Glashow, 1974) or $SO(10)$ (Fritzsch and Minkowski, 1975), or semi-simple such as $SU(4) \times SU(2)_L \times SU(2)_R$ (Pati and Salam, 1974), or finally containing a $U(1)$ factor such as $SU(3)_C \times SU(2)_L \times SU(2)_R \times U(1)_{B-L}$ (Mohapatra and Pati, 1975), which contains the SM gauge symmetry group as a proper subgroup. As in the standard electroweak symmetry-breaking mechanism, G is broken down to the SM group, possibly through a chain of several intermediate breaking steps at decreasing energy scales,

$$G \to G' \to G'' \to \cdots \to SU(3)_C \times SU(2)_L \times U(1)_Y. \qquad (3.19)$$

Differently from the SM, both quarks and leptons of a given generation can be accomodated in the same irreducible representation of G. For example, in the case of $SU(5)$, a single generation of fermions is assigned to the $\bar{\mathbf{5}} \oplus \mathbf{10}$, with $\bar{\mathbf{5}} = (q_{\alpha R}^{DC}, L_{\alpha L})$ and $\mathbf{10} = (Q_{\alpha L}, q_{\alpha R}^{UC}, l_{\alpha R}^{C})$. The case of $SO(10)$ is even more elegant, because all left-handed quarks and leptons and their conjugate states are accomodated in a single representation, the 16-dimensional spinorial representation of $SO(10)$, which also naturally include the SM sterile state ν_R^C.

Because now gauge transformation can turn, say, a quark into a lepton, there are many interaction terms where both B and L are violated. These processes are mediated by some extra gauge vector bosons, the leptoquarks. In the case of $SU(5)$, there are 12 such fields which carry both $SU(3)_C$ and $SU(2)_L$ charges, in addition to the eight gluons and the four $SU(2)_L \times U(1)_Y$ $W^{\pm}$, W^0 and B, for a total of 24 gauge bosons, the dimension of the adjoint representation of $SU(5)$. For the $SO(10)$ case there are even more new gauge bosons, the adjoint representation having dimension 45.

Baryon-number-violating terms lead dangerously to proton instability, via processes such as $p \to e^+ \pi^0$. The fact that there is an extremely high lower limit on the lifetime of these decays, for example, $\tau_{p \to e^+ \pi^0} \geq 8.2 \times 10^{33}$ years (Beringer *et al.*, 2012), means that the typical mass of these gauge bosons must be very high. Because their mass is proportional via gauge coupling to the breaking scale M_{GUT}, this means that if we believe in GUT models, they underwent spontaneous symmetry breaking in the very early universe, when the temperature was on the order of $M_{GUT} \gtrsim 10^{15}$ GeV.

If B- and L-violating interactions jeopardize proton stability, they are good news for baryogenesis. In fact, in the early stages of the expansion, where the typical temperature of the plasma was very high, these interactions were not suppressed by the high values of leptoquark masses, so they might lead to some baryon or lepton asymmetry. Actually, the GUT baryogenesis was among the first schemes proposed to explain the value of η_B today, based on the *out-of-equilibrium decay* of gauge or heavy Higgs scalars. We will describe the basic structure of this scenario in the following.

The main problem of theories like $SU(5)$ is that although B and L are not symmetries of the model, nevertheless their combination $B - L$ is always conserved. Even if some initial B asymmetry is produced at very early times, it will be completely erased by the lower-energy sphaleron processes, which wash out any initial asymmetry in $B + L$, whereas they cannot change the value of their difference $B - L$. If $B - L$ is initially zero, no baryon and lepton asymmetry is left over today.

Interestingly, the case of $SO(10)$ GUT or left–right symmetric models such as $SU(3)_C \times SU(2)_L \times SU(2)_R \times U(1)_{B-L}$ is different. In fact, $B - L$ in this class of models is a gauge charge, whose corresponding symmetry is spontaneously broken at some intermediate step on the way down to the SM. At the $B - L$ breaking scale the ν_R^C neutrinos take a Majorana mass term, so we cannot assign a definite value of L to these states if they also have standard Dirac mass coupling with their left-handed partners. This is the main starting point of leptogenesis. Of course we will have much more to say about this in the next sections.

We close here our brief summary of GUT baryogenesis, which is somewhat beyond the subject and aims of this book. The interested reader may have a look at, e.g., Kolb and Turner, 1994; Riotto and Trodden, 1999.

Let us now come back to the issue of B and L violation in the SM.

It was first noted by 't Hooft that the combination of the nontrivial vacuum structure of nonabelian gauge theories and anomalous violation of baryon and lepton currents leads to B- and L-violating processes at the nonperturbative level, mediated by instantons ('t Hooft, 1976). Let us recall that although for massless fermions both left-handed and right-handed currents are classically conserved, the triangle diagram (Adler, 1969; Bell and Jackiw, 1969) leads to a quantum anomaly. For example, in quantum electrodynamics,

$$\partial_\mu J_L^\mu = \partial_\mu \overline{\psi}_L \gamma^\mu \psi_L = -\frac{e^2}{16\pi^2} F^{\rho\sigma} \tilde{F}_{\rho\sigma} \tag{3.20}$$

$$\partial_\mu J_R^\mu = \partial_\mu \overline{\psi}_R \gamma^\mu \psi_R = \frac{e^2}{16\pi^2} F^{\rho\sigma} \tilde{F}_{\rho\sigma}, \tag{3.21}$$

where $\tilde{F}_{\rho\sigma} = \epsilon_{\rho\sigma\mu\nu}F^{\mu\nu}/2$. This means that the vector current $J^\mu = J_L^\mu + J_R^\mu$ is still conserved (fortunately, because it is related to conservation of electric charge!) whereas the axial one is anomalous:

$$\partial_\mu J_5^\mu = \partial_\mu \overline{\psi}\gamma^\mu\gamma_5\psi = \frac{e^2}{8\pi^2}F^{\rho\sigma}\tilde{F}_{\rho\sigma}. \tag{3.22}$$

The same result holds for nonabelian gauge theories such as the SM. However, differently from quantum electrodynamics, SM is a *chiral* gauge theory and left- and right-handed fermions couple differently to gauge fields, and this has a crucial consequence. Consider the case $SU(2)_L$. Because only left-handed fermions couple to gauge fields this time, $\partial_\mu J_R^\mu$ vanishes and the vector current is also anomalous,

$$\partial_\mu \overline{\psi}\gamma^\mu\psi = -\frac{g^2\kappa}{16\pi^2}F_a^{\rho\sigma}\tilde{F}_{a\rho\sigma}, \tag{3.23}$$

where κ is a numerical coefficient depending on the number and type of virtual particles in the triangle Feynman diagram. The charge associated with this current,

$$Q = \int d^3x\, J^0, \tag{3.24}$$

counts the number of particles minus antiparticles in a given state, summed over the two possible chiralities. Using the Gauss theorem, we get from (3.23)

$$Q(t_f) - Q(t_i) = -\frac{g^2\kappa}{16\pi^2}\int_{t_i}^{t_f}\int d^3x\, F_a^{\rho\sigma}\tilde{F}_{a\rho\sigma} \tag{3.25}$$

so that if the r.h.s. is nonzero, the value of Q is not conserved.

If ψ in (3.23) is a quark field, and we sum over all quarks,

$$J^{(B)\mu} = \frac{1}{3}\sum_q \overline{q}\gamma^\mu q, \tag{3.26}$$

we immediately recognize the baryon current, and the associated charge as the baryon number B. Similarly summing over all leptons, we get the lepton current $J^{(L)\mu}$ and lepton number L. Both are anomalous in the SM and one finds in particular

$$\partial_\mu J^{(B)\mu} = \partial_\mu J^{(L)\mu} = \frac{N_f}{32\pi^2}\left(-g^2 F_a^{\rho\sigma}\tilde{F}_{a\rho\sigma} + g'^2 B^{\rho\sigma}\tilde{B}_{\rho\sigma}\right), \tag{3.27}$$

where N_f is the number of generations and $F_a^{\rho\sigma}$ and $B^{\rho\sigma}$ are the $SU(2)_L$ and $U(1)_Y$ field strength, respectively; see Chapter 1. Notice that because both lepton and baryon currents have the same anomaly (for the same number of quark and lepton generation), whereas $B + L$ can change, the difference $B - L$ is still a conserved charge and cannot vary in any SM process, even at the nonperturbative level.

The fact that, indeed, there are processes where $B + L$ changes is related to the nontrivial structure of vacua in nonabelian gauge theories. If we integrate Eq. (3.27), we can relate the variation of, say, B to the Chern–Simons numbers

$$B(t_f) - B(t_i) = N_f \left(N_{CS}(t_f) - N_{CS}(t_i) \right), \tag{3.28}$$

where

$$N_{CS} = -\frac{g^2}{16\pi^2} \int d^3x \, 2\epsilon^{ijk} \text{Tr} \left(\partial_i A_j A_k + \frac{2}{3} ig A_i A_j A_k \right) \tag{3.29}$$

and A_i denotes the gauge field summed over the $SU(2)_L$ index.

The Chern–Simons numbers are integers, $0, \pm 1, \pm 2$, etc., which label the infinite set of vacua of a nonabelian gauge theory. Classically, any ground state should be time-independent and minimize the energy density. Choosing the gauge $A_0 = 0$, this means that A_i should correspond to a pure gauge, i.e.,

$$A_i = \frac{i}{g} \left(\partial_i U(x) \right) U(x)^{-1}, \tag{3.30}$$

with $U(x)$ a gauge transformation. Moreover, the Higgs field should be at the minimum of the potential

$$\Phi = U(x) \begin{pmatrix} 0 \\ v \end{pmatrix}. \tag{3.31}$$

If we choose the trivial transformation $U(x) = \mathbb{I}$, we get $A_i = 0$ and in this case the computation of the Chern–Simons number gives of course, zero. If any arbitrary time-independent gauge transformation could be continuously transformed into the trivial one, then all vacua would be identical. However, this is not true for non-abelian gauge theories, because the $U(x)$ can be decomposed into homotopy classes labelled by integer (positive or negative) *winding numbers n*. Two transformations $U(x)$ and $U'(x)$ belong to the same class if they can be continuously deformed one into the other. If we take a representative $U^{(n)}(x)$ in each of these classes, we can then write down an infinite set of topologically nonequivalent vacua,

$$A_i^n = \frac{i}{g} \left(\partial_i U^{(n)}(x) \right) \left(U(x)^{(n)} \right)^{-1}, \quad \Phi^{(n)} = U^{(n)}(x) \begin{pmatrix} 0 \\ v \end{pmatrix}, \tag{3.32}$$

which have different Chern–Simons numbers $N_{CS} = n$.

Coming back to Eq. (3.28), we see that baryon and lepton number violation is related to the probability of transition from configuration of gauge fields to another with different values of N_{CS}. This probability is very small at low energies. In fact, the SM vacuum states are separated by a potential barrier on the order of $8\pi v/g \sim 10$ TeV (see, e.g., Klinkhamer and Manton, 1984). At zero temperature the only way for this transition to occur is by a tunnel effect. This is exactly the role played by the instanton configuration studied by 't Hooft, who found that the probability of a variation of N_{CS} by one unit and thus of B and L violation processes is very strongly suppressed by the exponential factor $\exp(-16\pi^2/g^2) \sim 10^{-160}$ ('t Hooft, 1976).

Basically, in everyday life baryon- and lepton-violating processes never occur. However, in the early universe, when the temperature and density conditions were very different, one naively expects that the potential barrier between states with different N_{CS} could have been classically overcome if the kinetic energy were sufficiently large (Dimopoulos and Susskind, 1978; Kuzmin *et al.*, 1985). The transition rate in this case is determined by a different solution of classical gauge and the Higgs equation of motion, which has been called *sphaleron* (Manton, 1983; Klinkhamer and Manton, 1984) from the Greek word $\sigma\phi\alpha\lambda\epsilon\rho\acute{o}\varsigma$ meaning "ready to fall". It corresponds to an unstable static solution whose mass M_{sph} is given by the height of the potential barrier and which interpolates between two contiguous vacua. The probability of exciting such a configuration should be proportional to the Boltzmann factor $\exp(-M_{sph}/T)$. Indeed, by taking into account the fact that the Higgs vev v is decreasing as the temperature increases, and eventually vanishes at the electroweak phase transition, the probability per unit time and volume for a $\Delta N_{CS} = 1$ process (Kuzmin *et al.*, 1985; Arnold and McLerran, 1987, 1988; Ringwald, 1988; Carson *et al.*, 1990) is found to be

$$\Gamma \propto \exp\left(-\frac{8\pi m_W(T)}{g^2 T}\right), \tag{3.33}$$

where we have written the Higgs vev in terms of the (temperature-dependent) W boson mass. Above the electroweak transition at $T_{EW} \sim 100$ GeV, v is zero, the gauge bosons are massless and the barrier disappears. There are no more sphalerons and the transitions can occur much more efficiently. In this case, a rough estimate of the transition rate gives (Khlebnikov and Shaposhnikov, 1988)

$$\Gamma \sim \left(\frac{g^2}{4\pi}T\right)^4. \tag{3.34}$$

A more precise calculation involving quantum corrections is presented in Bodeker, 1998. This leads to very fast B- and L-violating processes till very high energies. Using the standard criterion (recall that Γ is rate per unit volume),

$$\frac{\Gamma}{T^3} \gtrsim H(T) = \left(\frac{8\pi}{3} g_* \frac{\pi^2}{30}\right)^{1/2} \frac{T^2}{m_{Pl}}, \tag{3.35}$$

one finds that the rate is higher than H up to $T \sim 10^{12}$ GeV.

Which processes correspond to a given ΔN_{CS}? From (3.28) we read that baryon and lepton numbers both change by $N_f \Delta N_{CS}$. For three fermion generations a transition where N_{CS} changes by one involves nine left-handed quarks, three per color, and three left-handed leptons, as for example

$$\emptyset \leftrightarrow 2u + d + 2c + s + t + 2b + e^- + \nu_\mu + \tau^-. \tag{3.36}$$

3.3.3 *Relating baryon and lepton numbers*

Because sphaleron processes reshuffle the baryon and lepton numbers through $(B + L)$-violating transitions, while keeping constant the value of $B - L$, a naive expectation would be that any initial value for $B + L$ is driven exactly to zero as long as sphaleron processes are in equilibrium. If we denote by B_0 and L_0 the initial values for B and L on some high energy scale, we can write the trivial relations

$$B_0 = \frac{B_0 + L_0}{2} + \frac{B_0 - L_0}{2}, \quad L_0 = \frac{B_0 + L_0}{2} - \frac{B_0 - L_0}{2}, \tag{3.37}$$

so that if $B_0 + L_0$ asymmetry were completely washed out one would get at low energy the final values

$$B = \frac{B_0 - L_0}{2} = -L. \tag{3.38}$$

Although on the right order of magnitude, this result is not entirely correct. The precise calculation of the final baryon and lepton numbers depends crucially on the value of the temperature T compared to the Higgs vev v. In the following, we present only a simplified calculation leading to the correct results in the limit $T \gg v$. In this limit, baryons and leptons can be described in terms of elementary particles. We refer the reader to Khlebnikov and Shaposhnikov, 1996 for a calculation valid at any temperature.

We have seen in Chapter 2 that, for small values of $\mu_i \ll T$, the asymmetry for the relativistic particle species i can be written as

$$n_i - n_{\bar{i}} \simeq \begin{cases} \frac{g}{6} T^2 \mu_i, & \text{fermion} \\ \\ \frac{g}{3} T^2 \mu_i, & \text{boson.} \end{cases} \tag{3.39}$$

We should take into account the constraints which all interaction processes being at equilibrium and conserved charges Q impose on the chemical potentials of particles of the SM (Khlebnikov and Shaposhnikov, 1988; Harvey and Turner, 1990). We recall that both these kind of constraints result in some linear relations among the values of μ_i, with i ranging over the particle involved in the specific process (see, e.g., (2.104)) or possessing a nonzero Q charge. The various constraints on the μ_i are the following:

- At high temperatures the $SU(3)_C \times SU(2)_L \times U(1)_Y$ gauge interactions are all in equilibrium, so that all particles in the same group representation have the same chemical potential, and furthermore, the chemical potential of gauge bosons is zero.
- The rapid 12–particle interactions induced by sphaleron processes lead to

$$\sum_{\alpha} \left(3\mu_{Q_{\alpha L}} + \mu_{L_{\alpha L}} \right) = 0, \tag{3.40}$$

where the sum is over the generation index α.
- Left- and right-handed quarks interact via $SU(3)_C$ instanton processes, which lead to the relation

$$\sum_{\alpha} \left(2\mu_{Q_{\alpha L}} - \mu_{q_{\alpha R}^U} - \mu_{q_{\alpha R}^D} \right) = 0. \tag{3.41}$$

- The Yukawa interactions give relations among the left-handed and right-handed fermion and Higgs chemical potentials:

$$\mu_{Q_{\alpha L}} - \mu_{\Phi} - \mu_{q_{\alpha R}^D} = 0$$

$$\mu_{Q_{\alpha L}} + \mu_{\Phi} - \mu_{q_{\alpha R}^U} = 0$$

$$\mu_{L_{\alpha L}} - \mu_{\Phi} - \mu_{l_{\alpha R}} = 0. \tag{3.42}$$

We remark that in general, although for heavy quarks and leptons the Yukawa coupling, which are weighted by the fermion mass squared, are strong enough to be in equilibrium in a wide temperature range, this might not be not the case for lighter particles. We will assume for simplicity that all (3.42) hold independently of α. The reader interested in a more detailed analysis of this issue and the role of flavour in baryogenesis/leptogenesis scenarios can have a look at Davidson *et al.*, 2008.
- Hypercharge neutrality implies

$$\sum_{\alpha} \left(\mu_{Q_{\alpha L}} + 2\mu_{q_{\alpha R}^U} - \mu_{q_{\alpha R}^D} - \mu_{L_{\alpha L}} - \mu_{l_{\alpha R}} \right) + 2\mu_{\Phi} = 0. \tag{3.43}$$

Using all these relations, and assuming that the chemical potentials are flavour-blind, we can express all μ's in terms of one of them only, for example μ_{L_L}:

$$\mu_{\Phi} = \frac{4N_f}{6N_f + 3} \mu_{L_L}, \quad \mu_{q_R^U} = \frac{2N_f - 1}{6N_f + 3} \mu_{L_L}, \quad \mu_{q_R^D} = -\frac{6N_f + 1}{6N_f + 3} \mu_{L_L}$$

$$\mu_{Q_L} = -\frac{1}{3} \mu_{L_L}, \quad \mu_{l_R} = \frac{2N_f + 3}{6N_f + 3} \mu_{L_L}, \tag{3.44}$$

where we have considered the more general case of N_f fermion generations. If we now make use of (3.39), the total baryon and lepton numbers are

$$B = N_f \left(2\mu_{Q_L} + \mu_{q_R^U} + \mu_{q_R^D} \right) = -\frac{4}{3} N_f \mu_{L_L}$$

$$L = N_f \left(2\mu_{L_L} + \mu_{l_R} \right) = \frac{14 N_f + 9}{6 N_f + 3} N_f \mu_{L_L} \tag{3.45}$$

or

$$B = \frac{8 N_f + 4}{22 N_f + 13} (B - L)$$

$$L = -\frac{14 N_f + 9}{8 N_f + 4} B, \tag{3.46}$$

which for $N_f = 3$ gives $L \sim -1.8\,B$. We recall that these results are only valid in the limit $T \gg v$. However, the results at lower temperature differ only by a small amount (Khlebnikov and Shaposhnikov, 1996).

3.3.4 The out-of-equilibrium decay scenario

In this section we describe a way to produce out-of-equilibrium conditions which was first proposed in the framework of GUT baryogenesis – see, e.g., Kolb and Turner, 1994 – and as we will see is also at the basis of the leptogenesis scenario. We note just in passing that this scheme is far from being unique. For example, a departure from equilibrium can be produced at phase transitions in the early universe. One very popular example is the electroweak breaking stage, which combined with the B- and L-violating effects of sphalerons, was advocated as a possible mechanism for baryogenesis in the framework of the SM, provided they are of the first type (Kuzmin *et al.*, 1985), i.e., involving latent heat. In this case the nature of the transition depends upon one crucial parameter, the mass of the Higgs boson. It is a strong first-order transition for relatively low values $m_H \lesssim 40$ GeV, which are ruled out from LEP data – see Beringer *et al.*, 2012 – so this scenario is not viable, unless one considers SM extensions as its supersymmetric version. See Riotto and Trodden, 1999 for a review.

The out-of-equilibrium decay scenario goes as follows. Consider some particle species X with mass m_X and its antiparticle $\overline{X}$, which are coupled to light degrees of freedom, such as the fermionic particles of the SM or the scalar Higgs, via baryon (or lepton)-number-violating interaction terms. To illustrate the main features of the model, it is enough to consider the simplest case, where the X particle can decay into two different channels f_1 and f_2 with baryon numbers B_1 and B_2. The partial decay rates can be written as

$$\Gamma(X \to f_1) = \frac{1}{2}(1 + \epsilon)\Gamma_D, \quad \Gamma(\overline{X} \to \overline{f}_1) = \frac{1}{2}(1 + \overline{\epsilon})\Gamma_D$$

$$\Gamma(X \to f_2) = \frac{1}{2}(1 - \epsilon)\Gamma_D, \quad \Gamma(\overline{X} \to \overline{f}_2) = \frac{1}{2}(1 - \overline{\epsilon})\Gamma_D, \quad (3.47)$$

where Γ_D is the total decay rate and we have used CPT, which tells us that X and $\overline{X}$ have the same width. Notice that $\epsilon - \overline{\epsilon}$ parameterizes the strength of C and CP violation.

At high temperatures, $T \gg m_X$, X and $\overline{X}$ are assumed to be excited in the primordial plasma, and to be in thermodynamic equilibrium with other relativistic species via scattering processes, $X\overline{X}$ pair annihilations and pair productions, decays and inverse decays $f_{1,2} \to X, \overline{X}$, whose rates in this temperature regime are on the order of αT, with α some coupling constant depending on the particular process. Because there are baryon-violating processes and equilibrium holds, no baryon asymmetry is initially present, so that $n_X = n_{\overline{X}}$.

As the temperature drops below m_X, the $X, \overline{X}$ densities exponentially decrease as $\exp(-m_X/T)$ if interactions are strong enough to maintain them in equilibrium. Because annihilation processes are *self-quenching*, being proportional to n_X, typically the main processes which regulate their abundance are decay and inverse decay, whose rates for a renormalizable theory are[2]

$$\Gamma_D = g^2 m_X \qquad (3.48)$$

$$\Gamma_{ID} \sim \Gamma_D \left(\frac{m_X}{T}\right)^{3/2} e^{-m_X/T}. \qquad (3.49)$$

If Γ_D is much larger than the Hubble parameter when $T \sim m_X$, decays are quite efficient in keeping X and $\overline{X}$ in equilibrium, and their densities rapidly decrease. On the other hand, suppose that decay processes are quite slow, i.e.,

$$\Gamma_D \lesssim H(T \sim m_X). \qquad (3.50)$$

In this case the $X, \overline{X}$ density per comoving volume remain frozen and soon become much larger than the equilibrium value, as shown in Fig. 3.1. As T continues to decrease, the inverse decays become quite negligible, because the mean kinetic energy in the plasma is smaller than the X mass and the process is kinematically suppressed, as in (3.49). The $X, \overline{X}$ eventually decay at a temperature T_D such that $\Gamma_D \sim H(T_D)$, which is to say, when the universe is at least as old as the particle lifetime. Decays take place in out of equilibrium conditions, because $n_{X,\overline{X}} \gg n_{X,\overline{X}}^{eq}$,

---

[2] If X only couple via gravitational interactions, the decay rate by dimensional argument is on the order of $\Gamma_D \sim m_X^3/m_{Pl}^2$. We will not consider this case in the following.

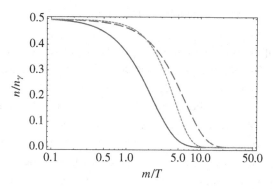

Figure 3.1 The number density of a scalar particle normalized to photon density for equilibrium (solid line) and for a small decay width $\Gamma/H(m) = 0.1$ for matter (long-dashed)- and radiation (short-dashed)-dominated universes. The value of n_X remains much higher than the equilibrium value till X particles eventually decay at $T \sim T_D$; see text. In this example $T_D \sim 0.2\, m_X$ or $T_D \sim 0.3\, m_X$ for matter- and radiation-dominated expansion, respectively.

and this is precisely the departure from equilibrium which allows for production of baryon asymmetry.

Equation (3.50) translates into a lower bound on the X particle mass. Assuming that the universe is radiation-dominated, we have, using (3.48),

$$m_X \gtrsim \frac{g^2}{\sqrt{g_*}} m_{\mathrm{Pl}}. \tag{3.51}$$

If X is a gauge boson of a GUT theory, then typically the coupling is on the order of $g^2 \sim 10^{-2} - 10^{-1}$, and we get $m_X \gtrsim (10^{17} - 10^{18})/\sqrt{g_*}$ GeV. In the case of a scalar particle decaying in fermion pairs, or a heavy fermion producing a scalar-fermion pair, the relevant interaction term is the Yukawa term. For example, if X is a heavy neutrino coupled to the Higgs and the left-handed lepton doublet, we have $g^2 \sim m_D^2/v^2$, with m_D on the order of the charged lepton Dirac masses; see Chapter 2. Therefore, $m_X \gtrsim (10^7 - 10^{15})/\sqrt{g_*}$ GeV. We see that in both cases the X particles should be very heavy, well above the electroweak breaking scale. In fact, these bounds were soon interpreted as a further hint in favour of GUT baryogenesis, because the GUT breaking scale is typically on the order of 10^{16} GeV, so that heavy gauge bosons are indeed very massive.

The value of g_* which enters Eq. (3.51) depends of course on the theoretical framework which is considered. In the SM, at very high energy scale, summing over all degrees of freedom (lepton, quarks, gauge bosons and Higgs scalar), we have $g_* \sim 10^2$. It is typically larger in theories which extend the SM group or in supersymmetric models. In any case, the main conclusion of this discussion, that only very heavy particles can provide out-of-equilibrium decay scenarios, is quite unchanged. In different cases, the lower limit might vary by one order of magnitude or so depending on g_*.

Let us now compute the final baryon asymmetry produced in our simple model. In general, besides decays, one should also consider the inverse decay processes, as well as baryon-violating scatterings $f_1 \leftrightarrow f_2$ mediated by the heavy X, $\overline{X}$ particles. These two contributions should be taken into account if the out-of-equilibrium condition is mildly satisfied, $\Gamma \sim H(m_X)$, and in general, they tend to wash out the effect of decays and reduce the final baryon asymmetry (in fact, if equilibrium holds, $\Gamma \gtrsim H(m_X)$, the baryon numbers produced by all these processes exactly balance and no baryon asymmetry is obtained). In this case the time evolution of B can be found by only solving a set of Boltzmann equations for the species involved. We will discuss one example in the next section, devoted to leptogenesis.

We assume here that T_D is sufficiently smaller than m_X and we can neglect the inverse decay processes and baryon-violating scatterings. The average baryon number density produced in X decays is

$$n_X(T_D) \frac{\Gamma(X \to f_1)B_1 + \Gamma(X \to f_2)B_2}{\Gamma_D} = n_X(T_D) \left(\frac{B_1 + B_2}{2} + \epsilon \frac{B_1 - B_2}{2} \right).$$

$$(3.52)$$

Similarly, for the C-conjugated states, we get

$$n_{\overline{X}}(T_D) \left(-\frac{B_1 + B_2}{2} - \overline{\epsilon} \frac{B_1 - B_2}{2} \right).$$

$$(3.53)$$

Summing the two contributions, the final total baryon number density normalized to the specific entropy of relativistic species s_R after decays is

$$B = \frac{n_X(T_D)}{2 s_R} (\epsilon - \overline{\epsilon})(B_1 - B_2).$$

$$(3.54)$$

Notice that B vanishes if $\epsilon - \overline{\epsilon} = 0$ (C and CP symmetry) or $B_1 = B_2$ (B is a conserved quantum number). If at decay the energy density is still dominated by relativistic particles, then we can neglect the entropy release from the decaying heavy X, $\overline{X}$ and $s_R \sim g_s T_D^3$. Because $n_X(T_D) \sim n_\gamma(T_D) \sim T_D^3$, we obtain

$$B \sim \frac{(\epsilon - \overline{\epsilon})(B_1 - B_2)}{g_s}.$$

$$(3.55)$$

On the other hand, if the universe is dominated by the X energy density before decays, the value of s_R can be computed as follows. In the limit of instantaneous decay at T_D, the reheating temperature T_{RH} of relativistic species at decay time is, from conservation of energy,

$$m_X n_X(T_D) = g_* \frac{\pi^2}{30} T_{RH}^4, \tag{3.56}$$

so that

$$s_R = \frac{4}{3} g_s \frac{\pi^2}{30} T_{RH}^3 = \frac{4}{3} n_X(T_D) \frac{m_X}{T_{RH}}, \tag{3.57}$$

where we have assumed that $g_* = g_s$, i.e., that all relativistic species share the same temperature; see Eqs. (2.154) and (2.143). The reheating temperature can be obtained from the condition that at T_D the decay rate equals the Hubble parameter,

$$\Gamma_D \sim \left(\frac{8\pi}{3m_{Pl}^2} \rho_X(T_D) \right)^{1/2} = \left(\frac{8\pi}{3m_{Pl}^2} g_* \frac{\pi^2}{30} T_{RH}^4 \right)^{1/2}, \tag{3.58}$$

and we find

$$T_{RH} \sim \frac{0.78}{g_*^{1/4}} (\Gamma_D m_{Pl})^{1/2}. \tag{3.59}$$

Finally, we obtain

$$B \sim \frac{3}{4} \frac{T_{RH}}{m_X} (\epsilon - \bar{\epsilon})(B_1 - B_2) \sim \frac{1}{g_*^{1/4}} \left(\frac{\Gamma_D m_{Pl}}{m_X^2} \right)^{1/2} (\epsilon - \bar{\epsilon})(B_1 - B_2). \tag{3.60}$$

3.4 Basics of leptogenesis

As discussed in the previous section, baryogenesis requires new physics beyond the SM, which must account for new sources of CP violation, and a mechanism leading to a departure from thermal equilibrium. A nice mechanism based on heavy sterile neutrinos involved in the seesaw model was first proposed in Fukugita and Yanagida, 1986 and it is now popular under the name of *baryogenesis through leptogenesis* or simply *leptogenesis*. The Yukawa couplings of such particles provide the source of CP violation, and their values can be tuned to yield decay rates which are small enough to ensure out-of-equilibrium decay conditions. The lepton number violation is encoded in the Majorana nature of these neutrinos. Sphaleron processes are then charged to convert the lepton asymmetry into a baryon asymmetry. Below, we will focus on this standard realization of leptogenesis (see Chen, 2007 and in particular Davidson *et al.*, 2008 for a much more detailed discussion).

3.4.1 Standard leptogenesis and Majorana neutrinos

As we have already discussed in Section 1.3.3, the simplest implementation of seesaw mechanism is represented by *Type I models*, where a certain number n_s of sterile Majorana neutrinos $N_s \equiv v_{sR} + v_{sR}^C$ are introduced. In this case, the most general Dirac/Majorana Lagrangian involving the Yukawa coupling for leptons and right-handed neutrinos with Higgs doublet Φ reads

$$\mathcal{L}_Y^R = -Y_{\alpha s}^v \, \overline{v_{\alpha L}} \, N_s \, \overline{\phi}_0 + Y_{\alpha s}^v \, \overline{l_{\alpha L}} \, N_s \, \phi_- - Y_{\alpha s}^{v*} \, \overline{N_s} \, v_{\alpha L} \, \phi_0$$

$$+ Y_{\alpha s}^{v*} \, \overline{N_s} \, l_{\alpha L} \, \phi_+ - \frac{1}{2} M_s \, \overline{N_s} \, N_s. \tag{3.61}$$

We recall that for energies much higher than the electroweak vacuum expectation value v, all SM particles are massless.

At tree level, N_s can decay via the two channels $N_s \to \phi_0 + v_{\alpha L}$ and $N_s \to \phi_+ + l_{\alpha L}$. Because they are Majorana spinors, they can also decay in the CP conjugated channels $N_s \to \overline{\phi}_0 + v_{\alpha L}^C$ and $N_s \to \phi_- + l_{\alpha L}^C$. The total width then yields

$$\Gamma_{D_s} = \sum_\alpha \left[\Gamma(N_s \to \phi_0 + v_{\alpha L}) + \Gamma(N_s \to \phi_+ + l_{\alpha L}) + \Gamma(N_s \to \overline{\phi}_0 + v_{\alpha L}^C) \right.$$

$$\left. + \Gamma(N_s \to \phi_- + l_{\alpha L}^C) \right] = \frac{M_s}{8\pi} (Y^{v\dagger} Y^v)_{ss}. \tag{3.62}$$

Let N_1 be the lightest right-handed neutrinos. Assuming that their interactions till they are in equilibrium wash out any lepton number asymmetry previously generated in the decay of other N_s at temperatures $T \gg M_1$, the final asymmetry depends on the dynamics of N_1 only. If out-of-equilibrium conditions are fulfilled,

$$\Gamma_{D_1} < H(T \sim M_1), \tag{3.63}$$

when N_1 eventually decay, a lepton asymmetry is generated because the CP asymmetry which arises from the interference of the tree level and one-loop diagrams, as shown in Fig. 3.2. In particular, an imaginary part develops in the one-loop contribution, because there are branch cuts corresponding to intermediate on-shell particles, which are much lighter than the heavy neutrino states

$$\varepsilon_1 \equiv \frac{\Delta_1}{\Gamma_{D_1}} \simeq \frac{1}{8\pi} \frac{1}{(Y^{v\dagger} Y^v)_{11}} \sum_{s \neq 1} \text{Im} \left\{ (Y^{v\dagger} Y^v)_{1s}^2 \right\} \left[f \left(\frac{M_s^2}{M_1^2} \right) + g \left(\frac{M_s^2}{M_1^2} \right) \right], \tag{3.64}$$

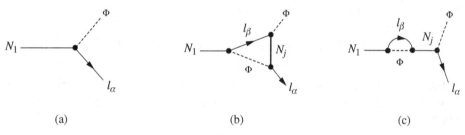

Figure 3.2 Diagrams with right-handed neutrinos from Eq. (3.61) contributing to the lepton number asymmetry. The asymmetry (3.64) is produced by the interference of the tree-level process (a) and the one-loop vertex correction (b) and self-energy (c) diagrams.

where

$$\Delta_1 = \sum_\alpha \left[\Gamma(N_1 \to \phi_0 + \nu_{\alpha L}) + \Gamma(N_1 \to \phi_+ + l_{\alpha L}) \right.$$

$$\left. - \Gamma(N_1 \to \overline{\phi_0} + \nu_{\alpha L}^C) - \Gamma(N_1 \to \phi_- + l_{\alpha L}^C) \right]. \tag{3.65}$$

The $f(x)$ in Eq. (3.64) comes from the loop integration in diagram (b) of Fig. 3.2 and it is given by

$$f(x) \equiv \sqrt{x} \left[1 - (1+x) \ln\left(\frac{1+x}{x} \right) \right]. \tag{3.66}$$

Diagram (c), which is the nondiagonal one-loop self-energy, for $|M_s - M_1| \gg |\Gamma_{D_s} - \Gamma_{D_1}|$ gives the term

$$g(x) \equiv \frac{\sqrt{x}}{1-x}. \tag{3.67}$$

For hierarchical neutrino masses, $M_1 \ll M_2, \dots M_{n_s}$, the asymmetry takes the simple form

$$\varepsilon_1 \simeq -\frac{3}{8\pi} \frac{1}{(Y^{\nu\dagger} Y^\nu)_{11}} \sum_{s \neq 1} \mathrm{Im}\left\{ (Y^{\nu\dagger} Y^\nu)^2_{1s} \right\} \frac{M_1}{M_s}. \tag{3.68}$$

If the decay temperature is well below M_1, we can neglect the effect of inverse decays and L-violating scatterings. This holds as long as

$$\frac{\Gamma_{D_1}}{H(T \sim M_1)} \simeq \frac{m_{\mathrm{Pl}}}{(1.66)(8\pi)\sqrt{g_*}} \frac{(Y^{\nu\dagger} Y^\nu)_{11}}{M_1} < 1, \tag{3.69}$$

which leads to a constraint on m_{ν_1}, defined as

$$m_{\nu_1} \equiv (Y^{\nu\dagger} Y^\nu)_{11} \frac{v^2}{2 M_1} \simeq 20\sqrt{g_*} \frac{v^2}{m_{\rm Pl}} \frac{\Gamma_{D_1}}{H}\bigg|_{T=M_1} \lesssim 10^{-3} \text{ eV}. \qquad (3.70)$$

For the electroweak SM we have $g_* \simeq 10^2$, whereas in supersymmetric models this value is almost doubled. In both cases one gets the same order of magnitude of (3.70).

In general, one expects a certain amount of washout of lepton number asymmetry produced by the N_1 decay. This can be parameterized by a coefficient κ which weights the final lepton asymmetry

$$L = \kappa \frac{\varepsilon_1}{g_*}. \qquad (3.71)$$

In the limit of small κ or strong washout regime, all L-violating interactions are in equilibrium, whereas $\kappa = 1$ corresponds to $T_D \ll M_1$.

To get a more detailed prediction for the leptogenesis model, one should solve a set of Boltzmann equations.[3] The main processes to be considered are

- the decay channels of N_1 and the inverse processes:

$$N_1 \leftrightarrow \phi_0 + \nu_{\alpha L} \quad N_1 \leftrightarrow \phi_+ + l_{\alpha L}$$
$$N_1 \leftrightarrow \overline{\phi_0} + \nu^C_{\alpha L} \quad N_1 \leftrightarrow \phi_- + l^C_{\alpha L}; \qquad (3.72)$$

- 2–2 scattering $|\Delta L| = 1$ processes, both in the s and t channels:

$$N_1 \nu_{\alpha L} \leftrightarrow q^U_L q^{UC}_R, \quad N_1 l_{\alpha L} \leftrightarrow q^{UC}_L q^D_R$$
$$N_1 \nu^C_{\alpha L} \leftrightarrow q^{UC}_L q^U_R, \quad N_1 l^C_{\alpha L} \leftrightarrow q^U_L q^{DC}_R \qquad (3.73)$$

and

$$N_1 q^{UC}_L \leftrightarrow \nu^C_{\alpha L} q^{UC}_R, \quad N_1 q^U_L \leftrightarrow l^C_{\alpha L} q^D_R$$
$$N_1 q^U_L \leftrightarrow \nu_{\alpha L} q^U_R, \quad N_1 q^{UC}_L \leftrightarrow l_{\alpha L} q^{DC}_R; \qquad (3.74)$$

- 2–2 scattering with $|\Delta L| = 2$ through the exchange of a virtual N_1:

$$(\nu_{\alpha L} \phi_0, l_{\alpha L} \phi_+) \leftrightarrow (\nu^C_{\beta L} \overline{\phi_0}, l^C_{\beta L} \phi_-) \qquad (3.75)$$

or

$$\nu_{\alpha L} \nu_{\beta L} \leftrightarrow \overline{\phi_0} \overline{\phi_0} \quad \nu_{\alpha L} l_{\beta L} \leftrightarrow \overline{\phi_0} \phi_- \quad l_{\alpha L} l_{\beta L} \leftrightarrow \phi_- \phi_-$$
$$\nu^C_{\alpha L} \nu^C_{\beta L} \leftrightarrow \phi_0 \phi_0 \quad \nu^C_{\alpha L} l^C_{\beta L} \leftrightarrow \phi_0 \phi_+ \quad l^C_{\alpha L} l^C_{\beta L} \leftrightarrow \phi_+ \phi_+. \qquad (3.76)$$

---

[3] Actually, for even more precise results, based on a full quantum description, it is necessary to switch to the Kadanoff–Baym equations (Anisimov *et al.*, 2011).

Following Buchmüller *et al.* (2002), we write the relevant Boltzmann equations for the number of N_1 neutrinos and the $B - L$ number per comoving volume, N_{N_1} and N_{B-L} (Luty, 1992; Plümacher, 1997), as

$$\frac{dN_{N_1}}{dt} = -(\Gamma_D + \Gamma_S)(N_{N_1} - N_{N_1}^{eq}) \tag{3.77}$$

$$\frac{dN_{B-L}}{dt} = -\varepsilon_1 \Gamma_D (N_{N_1} - N_{N_1}^{eq}) - \Gamma_W N_{B-L}. \tag{3.78}$$

The terms proportional to Γ_D account for decays and inverse decay processes (3.72), with the inverse decay rate given by the detailed balance condition $\Gamma_{ID} = (n_{N_1}^{eq}/n_l) \Gamma_D$, $n_{N_1}^{eq}$ being the equilibrium value for the N_1 number density.

The term Γ_S is due to $2 \leftrightarrow 2$ scattering processes, divided into the t-channel (3.73) and the s-channel (3.74):

$$\Gamma_S = 2\Gamma_{\phi,t}^{(N_1)} + 4\Gamma_{\phi,s}^{(N_1)}. \tag{3.79}$$

Finally, Γ_W contains inverse decays, $\Delta L = 1$ and $\Delta L = 2$ processes, and represents the washout damping term

$$\Gamma_W = \left(\frac{1}{2}\Gamma_{ID} + 2\Gamma_{\phi,t}^{(l)} + \Gamma_{\phi,s}^{(l)}\frac{N_{N_1}}{N_{N_1}^{eq}}\right) + 2\,\Gamma_N^{(l)} + 2\Gamma_{N,t}^{(l)}, \tag{3.80}$$

with Γ_N and $\Gamma_{N,t}$ denoting the rates of processes (3.75) and (3.76), respectively.

We recall that all interaction rates are thermally averaged over a kinetic equilibrium distributions. They are related to reaction densities γ_i by $\Gamma_i^{(X)} = \gamma_i/n_X^{eq}$, which can be obtained from the reduced cross sections (Luty, 1992) $\hat{\sigma}_i(s/M_1^2)$:

$$\gamma_{(i)}(x) = \frac{M_1^4}{64\pi^4}\frac{1}{x}\int_{(m_a^2+m_b^2)/M_1^2}^{\infty} dy\,\hat{\sigma}_{(i)}(y)\,\sqrt{y}\,K_1(x\,\sqrt{y}). \tag{3.81}$$

We denote by m_a and m_b the masses of the two particles in the initial state, whereas K_1 is a Bessel function.

Because all rates are expressed as functions of the temperature of relativistic species, it is convenient to replace time in Eqs. (3.77) and (3.78) with the evolutionary parameter $z = M_1/T$. We have $dt/dz = 1/(Hz)$ and assuming that g_* remains constant during leptogenesis, the kinetic equations for leptogenesis now are

$$\frac{dN_{N_1}}{dz} = -(D + S)(N_{N_1} - N_{N_1}^{eq}) \tag{3.82}$$

$$\frac{dN_{B-L}}{dz} = -\varepsilon_1 D(N_{N_1} - N_{N_1}^{eq}) - W N_{B-L}, \tag{3.83}$$

where $(D, S, W) \equiv (\Gamma_D, \Gamma_S, \Gamma_W)/(Hz)$.

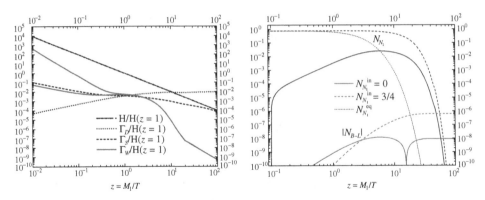

Figure 3.3 Left figure: The rates normalized to the expansion rate at $z = 1$. The two branches for Γ_W at small z correspond to two different estimates of the washout rate Γ_W. Right figure: The evolution of the N_1 abundance and the $B - L$ asymmetry for $\varepsilon_1 = -10^{-6}$ and $\overline{m} = 0.05$ eV. The two curves for N_1 density correspond to the cases of a zero or a thermal value. The parameters are chosen as $M_1 = 10^{10}$ GeV, $m_{\nu_1} = 10^{-3}$ eV and $\overline{m} = 0.05$ eV. Reprinted from Buchmüller *et al.*, 2002, with permission from Elsevier.

These equations should be solved numerically. They depend on three mass scales, the N_1 mass M_1, $\overline{m}^2 = \sum_{i=1}^{3} m_{\nu_i}^2$ and m_{ν_1}; see Eq. (3.70).

Successful leptogenesis requires a departure from thermal equilibrium for the decaying heavy Majorana neutrinos, and only partially effective washout processes. This means that $(\Gamma_D + \Gamma_S)(z \sim 1) < H(z \sim 1)$ and $\Gamma_W(z \sim 1) < H(z \sim 1)$. These conditions are fulfilled for the typical values $M_1 = 10^{10}$ GeV, $m_{\nu_1} = 10^{-3}$ eV, $\overline{m} = 0.05$ eV, for which the three rates and the Hubble parameter are shown in Fig. 3.3, where we also show the evolution of the $B - L$ asymmetry for these parameters. Both plots are from Buchmüller *et al.* (2002). Notice that even assuming a vanishing initial density for N_1, Yukawa interactions bring the heavy neutrinos into thermal equilibrium quite rapidly.

For strongly hierarchical right-handed neutrino masses, when the asymmetry ε_1 depends on the decay of the lightest right-handed neutrino, leptogenesis becomes very predictive for the lepton asymmetry (Buchmüller *et al.*, 2002, 2005; Nir, 2007), provided that N_1 decays at temperature $T \gtrsim 10^{12}$ GeV. In particular, if $M_1/M_2 \ll 1$, there is an upper bound on ε_1 called the "Davidson–Ibarra" bound (Davidson and Ibarra, 2002),

$$|\varepsilon_1| \leq \frac{3}{16\pi} \frac{M_1(m_{\nu_3} - m_{\nu_2})}{v^2} \equiv \varepsilon_1^{\mathrm{DI}}, \qquad (3.84)$$

which is obtained by expanding ε_1 up to leading order in M_1/M_2. Because $|m_{\nu_3} - m_{\nu_2}| \leq \sqrt{\Delta m_{32}^2} \sim 0.05$ eV, one gets the following lower bound on M_1:

$$M_1 \geq 2 \times 10^9 \text{ GeV}. \tag{3.85}$$

For degenerate light neutrinos, the leading terms in the expansion of ε_1 in M_1/M_2 and M_1/M_3 vanish. Computing the next-to-leading-order terms, one can obtain a looser bound (Hambye *et al.*, 2004):

$$|\varepsilon_1| \lesssim \text{Max}\left(\varepsilon^{\text{DI}}, \frac{M_3^3}{M_1 M_2^2}\right). \tag{3.86}$$

Several alternatives to standard leptogenesis have been proposed. A general remark is that the standard (hierarchical) model relies on very heavy right-handed neutrinos, as shown by the bound (3.85) on the lightest mass M_1. This means that such models are very difficult to test in colliders (and also, in the specific context of supergravity, that they would lead to a problem of overproduction of gravitinos after inflation). As mentioned previously, the sterile neutrino mass scale can be lowered if one invokes a degenerate rather than hierarchical spectrum of right-handed neutrinos. This can be assumed still within the context of the Type I seesaw model. A very small difference between the mass of two Majorana neutrinos may lead to a resonant CP asymmetric decay of sterile neutrinos (Pilaftsis, 1997; Pilaftsis and Underwood, 2004), or to resonant neutrino oscillations (Akhmedov *et al.*, 1998). Such oscillations may occur in the context of the νMSM scenario already introduced in Section 1.3.3 (Asaka and Shaposhnikov, 2005; Shaposhnikov, 2008). The right-handed neutrino mass scale can then be lowered considerably (even below the EW scale in the νMSM case: hence this model can explain both the smallness of active neutrino masses and the baryon asymmetry of the universe without involving new physics significantly above the EW scale). More radical departures from the standard leptogenesis model invoke, for instance, the Type II or Type III seesaw models, or nonthermal production of right-handed neutrinos at reheating, or Dirac neutrinos with highly suppressed Yukawa couplings. We refer the reader to the review of Davidson *et al.*, 2008 for a summary of such alternatives.

3.4.2 Leptogenesis and neutrino oscillation: Two right-handed neutrinos

In general, there are no simple relations between the amount of CP violation invoked in leptogenesis models and low-energy CP-violating processes, such as neutrino oscillation and neutrinoless double beta decay. This is due to the presence of extra phases and mixing angles in the heavy neutrino sector. In some cases, however, it is still possible to establish some relationship, which is strongly dependent upon the particular form which is adopted for the form of the neutrino Dirac mass term.

One example is the 3×2 seesaw model studied in (Frampton *et al.*, 2002). Consider two right-handed neutrinos $N_{1,2}$ and the Lagrangian density term

$$\mathcal{L} = \frac{1}{2} N_i^T \, C^\dagger \, M_{ij} \, N_j - \overline{N}_i D_{i\alpha}(-i\Phi^T \sigma_2) L'_{\alpha L} + \text{h.c.}, \qquad (3.87)$$

with the Majorana and Dirac mass matrices given by

$$M = \begin{pmatrix} M_1 & 0 \\ 0 & M_2 \end{pmatrix} \qquad D = \begin{pmatrix} a & a' & 0 \\ 0 & b & b' \end{pmatrix}. \qquad (3.88)$$

As already discussed in Section 1.3.3, the effective mass matrix for light neutrinos due to the seesaw formula is given by

$$\hat{L} = D^T M^{-1} D = \begin{pmatrix} \dfrac{a^2}{M_1} & \dfrac{aa'}{M_1} & 0 \\ \dfrac{aa'}{M_1} & \dfrac{a'^2}{M_1} + \dfrac{b^2}{M_2} & \dfrac{bb'}{M_2} \\ 0 & \dfrac{bb'}{M_2} & \dfrac{b'^2}{M_2} \end{pmatrix}, \qquad (3.89)$$

where phases have been arranged in such a way that a, b, b' are real and $a' = |a'|e^{i\delta}$. The matrix $\hat{L}$ can be diagonalized in terms of a unitary transformation U,

$$\frac{1}{2} \sum_{\alpha\beta} v'^T_\alpha \, C^\dagger \, \hat{L}_{\alpha\beta} \, v'_\beta = \frac{1}{2} \sum_{ij} v_i^T \, C^\dagger \, \left(U^T \hat{L} U \right)_{ij} v_j, \qquad (3.90)$$

where v_i stands for mass eigenvector light Majorana neutrinos. When we assume that $a^2/M_1 \ll b^2/M_2$ and make the choices $|a'| = \sqrt{2}a$ and $b = b'$, the three mass eigenvalues are

$$m_{v_1} = 0, \quad m_{v_2} = \frac{2a^2}{M_1}, \quad m_{v_3} = \frac{2b^2}{M_2}. \qquad (3.91)$$

The unitary matrix U takes the simple expression

$$U = \begin{pmatrix} 1/\sqrt{2} & 1/\sqrt{2} & 0 \\ -1/2 & 1/2 & 1/\sqrt{2} \\ 1/2 & -1/2 & 1/\sqrt{2} \end{pmatrix} \times \begin{pmatrix} 1 & 0 & 0 \\ 0 & \cos\theta & \sin\theta \\ 0 & -\sin\theta & \cos\theta \end{pmatrix}, \qquad (3.92)$$

where $\theta \simeq m_{v_2}/\sqrt{2}m_{v_3}$. Both the experimental values of oscillation angles in the light active neutrino sectors and the corresponding squared mass differences can be quite well reproduced for the chosen values of a' and b'.

If the lighter state N_1 produces the baryon asymmetry by the out-of-equilibrium decays, using the results of Section 3.4.1, one finds that the final value of B is proportional to

$$B \propto \xi_H = \mathrm{Im}\left(DD^\dagger\right)^2_{12} = \mathrm{Im}\left(a'b\right)^2 = |a'|^2 b^2 \sin(2\delta). \tag{3.93}$$

On the other hand, at low energy the CP violation in neutrino oscillations is governed by the quantity

$$\xi_L = \mathrm{Im}\left(h_{12}h_{23}h_{31}\right), \quad \text{where} \quad h = \hat{L}\hat{L}^\dagger. \tag{3.94}$$

From the expression for h entries which can be computed by the explicit form of $\hat{L}$, one gets that the sign of the baryon number asymmetry, which is positive from observations, is related to the sign of the CP violation in neutrino oscillation

$$\xi_L = -\frac{a^4 b^4}{M_1^3 M_2^3}\left(2 + \frac{|a'|^2}{a^2}\right)\xi_H \propto -B. \tag{3.95}$$

4

Neutrinos in the MeV age

Why me? Why now? These are the kind of questions that cosmological neutrinos could ask themselves about the strange coincidence of relevant facts at the MeV range of temperatures. The four known forces of Nature all play a role in this very interesting epoch. When the universe was from one-tenth of a second to a few minutes old, neutrinos experienced decoupling from electromagnetic plasma while flavour neutrino oscillations became effective, and they witnessed electron–positron annihilations and in the meantime were involved in the business of fixing the initial conditions for the primordial production of light nuclei. All these processes, which in principle could have occurred in the early universe well separated in time, depend upon the values of a bunch of unrelated parameters such as the Fermi constant, the neutrino mixing angles and squared mass differences, the electron mass and the binding energy of nuclei, in particular that of deuterium. The fact that all these events take place almost simultaneously means that they cannot be understood by a back-of-the-envelope calculation. Once more neutrinos put out a challenge to physicists.

In this chapter we first consider in Section 4.1 the process of relic neutrino decoupling in more detail than in Chapter 2, going beyond the instantaneous decoupling approximation in which neutrinos simply no longer interact with other particles below a certain temperature $T_{\nu D}$. To this end, one should solve the Boltzmann integro-differential equations for the neutrino momentum distributions, with essentially no approximations. This gives the shape of relic neutrino spectra and their contribution to radiation, parameterized in terms of N_{eff}. We will see that the process of neutrino decoupling is almost, but not completely, independent of $e^{\pm}$ annihilations.

In Section 4.2 we discuss the cosmological consequences of the present experimental evidence for flavour neutrino oscillations, which we reviewed in Chapter 1. First, we describe the main features that distinguish flavour oscillations in the expanding universe with a very small baryon asymmetry from the standard case in matter discussed in Section 1.3.5. The formalism of neutrino density matrices is then introduced, in order to cope with a situation where both neutrino oscillations and interactions are effective, and it is later applied to a refined calculation of relic neutrino decoupling and to the case of nonzero neutrino–antineutrino asymmetries. The addition of extra sterile neutrino species also leads to interesting cosmological consequences that are described afterwards.

The final part of this chapter is devoted to primordial nucleosynthesis, one of the observational pillars of the hot Big Bang model and also known simply as Big Bang nucleosynthesis (BBN). An overview of the two phases of BBN (the freeze-out of the neutron-to-proton density ratio and the onset of nuclear fusion reactions) is presented in Section 4.3, followed by a short discussion of the light element observations and a comparison of theory versus data in the standard case. The main bounds on neutrino properties, understood in a broad sense in order to include the possibility of extra radiation, always encoded in the parameter N_{eff}, are finally reviewed in Section 4.4.

Of course, there are many issues which are not discussed here. The interested reader can find more details on the subject of relic neutrino decoupling in the review (Dolgov, 2002), whereas for more details on the formalism of flavour oscillations in the expanding universe one can start with Prakash *et al.*, 2001. For primordial nucleosynthesis, excellent reviews exist, such as Sarkar, 1996; Olive *et al.*, 2000; Steigman, 2007; Iocco *et al.*, 2009; and Pospelov and Pradler, 2010.

4.1 Neutrino decoupling

In Chapter 1 we have already introduced the basic properties of the cosmological neutrino background, which essentially arise from the fact that neutrinos decouple from the electromagnetic plasma when they are relativistic. We found the decoupling temperature to be $T_{\nu D} \sim 2 - 3$ MeV, with a slight dependence on the neutrino flavour, from a mere comparison of the expansion rate of the universe with the weak interaction rate. In this section we will study in more detail the decoupling process and calculate to what extent the spectrum of relic neutrinos preserves an equilibrium form as in Eq. (2.189), under the simplifying assumption of no neutrino oscillations.

In fact, a single decoupling temperature does not exist at all, and the values of $T_{\nu D}$ that we obtained in the instantaneous decoupling approximation should be regarded as a fair approximation. From Table 1.7, the cross section of a process such as the e^+e^- annihilation into neutrinos depends on the squared total energy s,

$$\sigma_{e^+e^- \to \bar{\nu}_e \nu_e} \simeq \frac{G_F^2 s}{6\pi} \left(\tilde{g}_L^{l2} + g_R^{l2} \right), \tag{4.1}$$

where we have considered the limit $s \gg m_e^2$. It is clear that neutrinos and antineutrinos with larger momenta will be kept in thermal contact with the electromagnetic components for a longer time. This fact would not modify the final spectrum of neutrinos if all particles in contact with them were relativistic during the full decoupling period[1] and their temperature was always falling as a^{-1}. However, this is not the case for electrons and positrons, because the effect of their mass becomes important for temperatures $T \sim m_e$, relatively close to $T_{\nu D}$. As soon as the processes $e^+ e^- \leftrightarrow \gamma\gamma$ are more favoured in the annihilation direction, the temperature shared by electrons, positrons and photons T falls at a rate lower than a^{-1}. Photons are then heated by the massive annihilation of electrons and positrons. More energetic neutrinos which could be still interacting with $e^{\pm}$ would also be *hotter* than assumed in the instantaneous decoupling approximation.

The average neutrino heating was first estimated in Dicus *et al.*, 1982; Herrera and Hacyan, 1989; Raha and Mitra, 1991, assuming that their spectrum is kept in equilibrium. However, this approximation does not hold because neutrinos will acquire distortions in their momentum spectra from relic $e^+ e^- \rightarrow \nu\bar{\nu}$ processes, whose size was calculated in later works (Dodelson and Turner, 1992; Dolgov and Fukugita, 1992; Hannestad and Madsen, 1995; Dolgov *et al.*, 1997).

We define the neutrino nonthermal distortions with respect to a standard Fermi–Dirac function for the physical momenta

$$f_{eq}(p) = \left[\exp\left(\frac{p}{T_\nu}\right) + 1 \right]^{-1} \tag{4.2}$$

as

$$\delta_{\nu_\alpha}(t, p) = \frac{f_{\nu_\alpha}(t, p)}{f_{eq}(p)} - 1, \tag{4.3}$$

which depends on the neutrino flavor, because electron neutrinos and antineutrinos interact with $e^{\pm}$ through both charged and neutral weak processes. The distortions also vary with time until neutrinos with a given momentum are no longer interacting.

A proper calculation of the neutrino nonthermal distortions involves solving the Boltzmann equations for the evolution of particle distributions in phase space as introduced in Eq. (2.92), which in this case read

$$\left(\frac{\partial}{\partial t} - Hp \frac{\partial}{\partial p} \right) f_{\nu_\alpha}(t, p) = \mathbf{C}[f_{\nu_\alpha}; f_{\nu_\beta}, f_{e^{\pm}}]. \tag{4.4}$$

---

[1] At this period the abundance of nonrelativistic particles such as protons and neutrons is tiny compared with that of electrons, positrons or photons, and so they play a negligible role in neutrino decoupling.

The weak interactions which are relevant for neutrino decoupling are neutrino self-interactions and scattering or annihilation processes involving electrons and positrons. The corresponding collisional integral **C**, written in terms of two-fermion into two-fermion interactions only, is then, $\nu_\alpha + b \leftrightarrow c + d$ as given in Eq. (2.94),

$$
\mathbf{C}\left[f_{\nu_\alpha}; f_b, f_c, f_d\right]
$$

$$
= \frac{1}{p_{\nu_\alpha}} \int d\pi(p_b) d\pi(p_c) d\pi(p_d) (2\pi)^4 \delta^{(4)}(p_{\nu_\alpha} + p_b - p_c - p_d)
$$

$$
\times S \left|A_{\nu_\alpha b \to cd}\right|^2 \left[f_c f_d (1 - f_{\nu_\alpha})(1 - f_b) - f_{\nu_\alpha} f_b (1 - f_c)(1 - f_d)\right], \quad (4.5)
$$

where each distribution function depends on time and momentum. Neutrinos are relativistic particles in the MeV temperature range, so we have already substituted the modulus of momentum for their energy. This is, of course, not the case for electrons and positrons. The matrix elements $\mathcal{A}$ of all processes are listed in Tables 1.5 and 1.6, and S is the symmetrization factor, which includes $1/2!$ for each pair of identical particles in initial and final states and a factor of 2 if there are two identical particles in the initial state.

The full evolution of the neutrino distributions can be obtained solving numerically Eqs. (4.4) as shown in Hannestad and Madsen, 1995; Dolgov *et al.*, 1997; Dolgov *et al.*, 1999, and we will discuss the results later on. Let us first derive an approximate analytical expression for the final distortion $\delta_{\nu_e}(p)$ following (Dolgov and Fukugita, 1992; Dolgov, 2002). For the annihilation process $\nu_e(1) + \bar{\nu}_e(2) \leftrightarrow e^+(3) + e^-(4)$ it can easily be shown that

$$
S |\mathcal{A}_{12 \to 34}|^2 = 128 G_F^2 [\tilde{g}_L^{l2}(p_1 \cdot p_4)^2 + g_R^{l2}(p_1 \cdot p_3)^2 + \tilde{g}_L^l g_R^l m_e^2 (p_1 \cdot p_2)]. \quad (4.6)
$$

We now approximate the Fermi distribution functions with the Boltzmann limit, both for neutrinos with a momentum-dependent distortion,

$$
f_{\nu_e}(t, p) = \exp(-p/T_\nu)\left[1 + \delta_{\nu_e}(t, p)\right], \quad (4.7)
$$

and for $e^\pm$, which have an equilibrium distribution with photon temperature:

$$
f_e(t, p) = \exp(-E_e/T) \simeq \exp(-E_e/T_\nu)\left[1 + \frac{E_e}{T_\nu}\frac{\Delta T}{T}\right]. \quad (4.8)
$$

Here $\Delta T = T - T_\nu$ depends on time.

The corresponding statistical factor in this limit in the collisional integral of Eq. (4.5) is $f_3 f_4 - f_1 f_2$ and depends only on the total energy $E_3 + E_4$. The integral over $d^3p_3 d^3p_4$ can be carried out using the δ^4 and we obtain

$$\mathbf{C} \simeq \frac{G_F^2}{\pi^3} p_1 \int d\xi\, dp_2\, p_2^3 \sqrt{1 - \frac{4m_e^2}{s}(1-\xi)^2}$$

$$\times \exp\left(\frac{-p_1 - p_2}{T_\nu}\right) \left[\frac{p_1 + p_2}{T_\nu}\frac{\Delta T}{T} - \delta_{\nu_e}(t, p_1) - \delta_{\nu_e}(t, p_2)\right]$$

$$\times \left[\frac{1}{3}(\tilde{g}_L^{l2} + g_R^{l2})\left(1 - \frac{m_e^2}{s}\right) + 2\tilde{g}_L^l g_R^l \frac{m_e^2}{s}\right], \tag{4.9}$$

where $s = p_1 p_2(1-\xi)$. We anticipate that the final nonthermal distortions δ_ν will be small, so we neglect the terms proportional to δ_ν in the collisional integral. In addition, we will consider the limit $s \gg m_e^2$. Taking into account these approximations, the last integral over $d\xi\, dp_2$ can be solved, and we obtain the following evolution equation for the distortion:

$$\left(\frac{\partial}{\partial t} - Hp\frac{\partial}{\partial p}\right)\delta_{\nu_e}(t, p) \simeq \frac{16G_F^2}{3\pi^3}(\tilde{g}_L^{l2} + g_R^{l2})\frac{\Delta T}{T}T_\nu^3 p(p + 4T_\nu). \tag{4.10}$$

This equation can be integrated if we note that the temperature parameter T_ν evolves as $\dot{T}_\nu = -HT_\nu$, which gives

$$\delta_{\nu_e}(t, p/T_\nu) \simeq \frac{16G_F^2}{3\pi^3}(\tilde{g}_L^{l2} + g_R^{l2})\frac{p}{T_\nu}\left(\frac{p}{T_\nu} + 4\right) \times \int_{T_\nu}^{T_{\nu,i}} \frac{d\lambda}{H}\lambda^4\frac{\Delta T}{T}. \tag{4.11}$$

Inserting the expression for H in a radiation-dominated universe, Eq. (2.156), we obtain

$$\delta_{\nu_e}(p/T_\nu) \simeq 0.031\frac{p}{T_\nu}\left(\frac{p}{T_\nu} + 4\right) \times \int_{T_\nu}^{T_{\nu,i}} d\lambda\, \lambda^2\frac{\Delta T}{T}\ \text{MeV}^{-3}. \tag{4.12}$$

As shown in Dolgov and Fukugita, 1992, the factor $p/T_\nu + 4$ is modified with an extra term $7(p/T_\nu - 4)/4$ when the terms from elastic scattering of neutrinos with electrons and positrons are also taken into account. Finally, it can be checked that a good approximation for ΔT as calculated from Eq. (2.194) for temperatures in the range $[3, 0.5]$ MeV is

$$T_\nu^2\frac{\Delta T}{T} \simeq 6 \times 10^{-3}\ \text{MeV}^2. \tag{4.13}$$

Inserting this expression into Eq. (4.12) and the upper and lower values of this temperature range, we obtain an estimate for the spectral distortion of neutrinos,

$$\delta_{\nu_e}(p/T_\nu) \approx 4.6 \times 10^{-4}\frac{p}{T_\nu}\left(\frac{11}{4}\frac{p}{T_\nu} - 3\right). \tag{4.14}$$

Thus the final value of δ_ν is at the percent level for the typical average momentum $p \sim 3T_\nu$, increasing for larger neutrino momenta as $(p/T_\nu)^2$. Although this is just a rough estimate of the distortion, due to the approximations we have used, we will see that it is not very far from the correct results.

As we commented before, the full evolution of the neutrino spectral distortions from Eqs. (4.4) was calculated in a series of works once it was found that the various terms in the collisional integral can be reduced from nine to two dimensions as described in the Appendices of Refs. (Hannestad and Madsen, 1995; Dolgov *et al.*, 1997). The system becomes numerically tractable, and the evolution of $\delta_{\nu_\alpha}(t, p)$ was obtained with the required accuracy, including Fermi–Dirac statistics and a finite electron mass. As in many situations in the expanding universe, the equations are simplified when we use the following dimensionless variables instead of time, neutrino momenta and photon temperature:

$$x \equiv ma, \qquad y \equiv pa, \qquad z \equiv Ta, \tag{4.15}$$

where m is an arbitrary mass scale (for instance 1 MeV or m_e) and the scale factor a can be normalized so that $a(t) \rightarrow 1/T$ at high temperatures. In terms of these variables, the kinetic equations for the neutrino distributions are

$$Hx\partial_x f_{\nu_\alpha}(x, y_1) = \mathbf{C}\left[f_{\nu_\alpha}(x, y_1); f_{\nu_\beta}(x, y_i), f_{e^\pm}(x, z)\right], \tag{4.16}$$

where the collision term is a sum over all possible scattering and annihilation processes, each contributing with a two-dimensional integral. We do not write it here, but its full expression can be found, e.g., in Dolgov *et al.*, 1997. Because muon and tau neutrinos have the same interactions at MeV temperatures, one must solve the equation for f_{ν_e} and for $f_{\nu_x} \equiv f_{\nu_\mu, \nu_\tau}$, unless we take into account flavour oscillations as described later. Moreover, $f_{\nu_\alpha} = f_{\bar{\nu}_\alpha}$ in the absence of initial neutrino chemical potentials.

The kinetic equations for neutrinos are supplemented by the continuity equation for the total energy density ρ_R of the relativistic plasma, the three neutrino states and the electromagnetic components γ and $e^\pm$ (which are always in equilibrium),

$$\frac{d\rho_R}{dt} = -3H\left(\rho_R + P_R\right), \tag{4.17}$$

where P_R is the total pressure. This equation is more adequate to find the evolution of the photon temperature $T = mz/x$ than the conservation of entropy density in Eq. (2.194) because it takes into account the effect of relic neutrino heating. It can be rewritten as

$$x\frac{d\bar{\rho}_R}{dx} = \bar{\rho}_R - 3\bar{P}_R. \tag{4.18}$$

Here barred quantities indicate dimensionless energy densities $\bar{\rho} \equiv \rho(x/m)^4$ and pressure $\bar{P} \equiv P(x/m)^4$. After substituting the corresponding expressions, one finds an equation for dz/dx,

$$\frac{dz}{dx} = \frac{\dfrac{x}{z}F_1(x/z) - \dfrac{1}{2z^3}\displaystyle\int_0^\infty dy\, y^3 \left(\dfrac{df_{\nu_e}}{dx} + 2\dfrac{df_{\nu_x}}{dx}\right)}{\dfrac{x^2}{z^2}F_1(x/z) + F_2(x/z) + \dfrac{2\pi^4}{15}}, \tag{4.19}$$

where the functions F_i are given by

$$F_1(\tau) \equiv \int_0^\infty d\omega\, \omega^2 \frac{\exp(\sqrt{\omega^2 + \tau^2})}{(\exp(\sqrt{\omega^2 + \tau^2}) + 1)^2}$$

$$F_2(\tau) \equiv \int_0^\infty d\omega\, \omega^4 \frac{\exp(\sqrt{\omega^2 + \tau^2})}{(\exp(\sqrt{\omega^2 + \tau^2}) + 1)^2}. \tag{4.20}$$

The system of Eqs. (4.16) and Eq. (4.19) has been solved either using a discretization in a grid of dimensionless momenta y_i as in Hannestad and Madsen, 1995; Dolgov *et al.*, 1997 or with an expansion of the nonthermal distortions in moments as in Esposito *et al.*, 2000a; Mangano *et al.*, 2002.

Let us now describe the main results for the neutrino spectral distortions, following (Mangano *et al.*, 2002), where the quantum electrodynamics (QED) corrections were also included. These arise from finite-temperature QED corrections to the electromagnetic plasma. In fact, electromagnetic interactions modify the $e^\pm$ and γ dispersion relations, and thus the energy density and pressure (Heckler, 1994). More precisely, the energy density is lowered so that the $e^\pm$ annihilation phase releases less entropy with respect to the noninteracting particle limit calculation, modifying, among others, the equations for dz/dx. Because most of this energy ends up into photons, this decrease results in a final smaller value of z.

We present in Fig. 4.1 the evolution of the distortion of the neutrino distribution as a function of x for a particular neutrino momentum ($y = 10$). At high temperatures or $x \lesssim 0.2$, neutrinos are in good thermal contact with $e^\pm$ and their distributions only change keeping an equilibrium shape with the photon temperature $[\exp(y/z(x)) + 1]^{-1}$ (the T_γ line in the figure). In the intermediate region $0.2 \lesssim x \lesssim 4$, weak interactions become less effective in a momentum-dependent way, leading to distortions in the neutrino spectra which are larger for ν_e's than for the other flavours, as expected. Finally, at high values of x neutrino decoupling is complete, and the distortions reach their asymptotic values, which are shown in Fig. 4.2. For the particular neutrino momentum in Fig. 4.1, the final value of the δ_ν is 4.4% (ν_e) and 2% ($\nu_{\mu,\tau}$). The dependence of the nonthermal distortions in momentum is easily visible, following a polynomial function which actually is quite close to the estimate in Eq. (4.14).

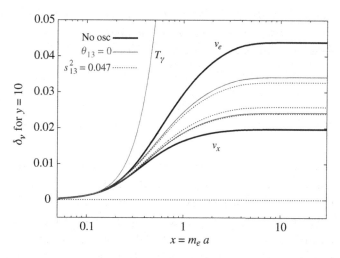

Figure 4.1 Evolution of the distortion of the ν_e and $\nu_x = \nu_{\mu,\tau}$ spectrum for a particular comoving momentum ($y = 10$). The line labeled with T_γ corresponds to the distribution of a neutrino which would be in full thermal contact with the electromagnetic plasma. The thick solid lines correspond to the case without neutrino oscillations, whereas the thinner ones include nonzero mixing. In the case with $\theta_{13} \neq 0$ one can distinguish the distortions for ν_μ (middle line) and ν_τ (lower line). (Reprinted from Mangano *et al.*, 2005, with permission from Elsevier.)

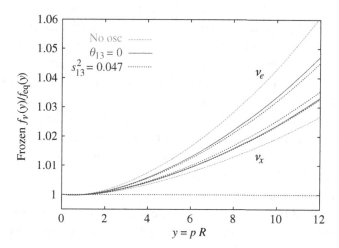

Figure 4.2 Frozen distortions of the flavour neutrino spectra as a function of the comoving momentum. The outer dashed lines correspond to the case without neutrino oscillations, whereas the inner lines include nonzero mixing. In the case with $\theta_{13} \neq 0$ one can distinguish the distortions for ν_μ (middle line) and ν_τ (lower line). (Reprinted from Mangano *et al.*, 2005, with permission from Elsevier.)

Table 4.1 *Frozen values of z_{fin}, the neutrino energy densities $\delta\bar{\rho}_{\nu_\alpha} \equiv \delta\rho_{\nu_\alpha}/\rho_{\nu_0}$, N_{eff} and ΔY_p in the absence of flavour neutrino mixing*

Case	z_{fin}	$\delta\bar{\rho}_{\nu_e}(\%)$	$\delta\bar{\rho}_{\nu_{\mu,\tau}}(\%)$	N_{eff}	ΔY_p
With QED corrections	1.3978	0.94	0.43	3.046	1.71×10^{-4}
No QED	1.3990	0.95	0.43	3.035	1.47×10^{-4}

These distortions in the relic neutrino spectra are small and their direct observation is at present unfeasible, considering that direct detection of the cosmological neutrino background represents an extremely challenging goal, as discussed in Chapter 7. However, it is interesting to note that the δ_ν's are much larger than the measured departure from equilibrium of CMB photons, and in any case should be included in the calculation of any observable related to relic neutrinos. For instance, the integrated effect of the spectral distortions on the radiation energy density, described with the asymptotic effective number of neutrinos N_{eff}, can be calculated easily from Eq. (2.198) as

$$N_{\text{eff}} = \left(\frac{z_0}{z_{\text{fin}}}\right)^4 \left(3 + \frac{\delta\rho_{\nu_e}}{\rho_{\nu_0}} + \frac{\delta\rho_{\nu_\mu}}{\rho_{\nu_0}} + \frac{\delta\rho_{\nu_\tau}}{\rho_{\nu_0}}\right), \qquad (4.21)$$

where $z_0 = (11/4)^{1/3} \simeq 1.40102$. In Table 4.1 we present the results for the dimensionless photon temperature z_{fin}, the change in the neutrino energy densities with respect to ρ_{ν_0} (the energy density in the instantaneous decoupling limit) and the small modification in the prediction for the primordial abundance of ^{4}He, to be discussed later in Section 4.3. Inserting the values in the first row of this table into Eq. (4.21), one obtains that in order to take into account the effect of neutrino heating one must use $N_{\text{eff}} = 3.046$, whereas the total contribution of radiation to the cosmological energy density is $\Omega_R = 1.6918\,\Omega_\gamma$, as long as the relic neutrinos are relativistic. We will see later in Section 4.2 that the unavoidable presence of flavour oscillations does not modify N_{eff}, but the individual flavour distortions are different (Mangano *et al.*, 2005).

Finally, the effect of the distortions on the contribution of massive neutrinos to the present energy density can be calculated with the results shown in Fig. 4.2. The change in neutrino densities shifts the present energy density to

$$\Omega_\nu = \frac{\sum_i m_{\nu i}}{93.1\,h^2\,\text{eV}}, \qquad (4.22)$$

where the sum includes all neutrinos which are nonrelativistic today; compare this result with Eq. (2.199).

4.2 Neutrino oscillations in the expanding universe

The experimental evidence for flavour neutrino oscillations that we reviewed in Section 1.4.1 naturally leads us to consider their implications for the early universe. However, we have seen that relic neutrinos of all flavours possess, in the approximation of instantaneous decoupling from the rest of the plasma, the same momentum distributions. In this case, the effect of oscillations would be simply to exchange equal neutrino spectra, without further consequences.

There exist, however, a few well-motivated situations where one expects unequal neutrino distributions, examples being the small flavour-dependent distortions that we have calculated in the previous section or the possible presence of nonzero neutrino asymmetries which are not the same for different flavours. In addition, the existence of sterile neutrinos could lead to interesting cosmological implications if they were exclusively or largely populated via their mixing with active neutrinos. All these cases justify the inclusion of this section, where we explain when flavour oscillations could be effective in the early universe and which is the proper formalism to find their evolution, which involves the neutrino density matrices. Afterwards, we will present the main cosmological implications of oscillations with and without sterile neutrinos.

4.2.1 Effective matter potentials

As shown in Section 1.3.5, oscillations are modified if neutrinos propagate in a flavour-asymmetric medium. In the case where the particles in the background are unpolarized electrons, we found that the effective potential from charged current interactions felt by electron neutrinos is $V_{CC} = \sqrt{2}\, G_F\, n_e$. If the medium also contains a comparable density of positrons, as in the early universe at MeV temperatures, it can easily be shown with equations similar to those in Section 1.3.5 that

$$V_{CC} = \sqrt{2}\, G_F\, (n_{e^-} - n_{e^+}). \tag{4.23}$$

This dependence on the difference between the number densities of fermions and antifermions applies also to the neutral current effective potential, felt by all neutrino flavours.

In the very early universe there were almost equal numbers of baryons and antibaryons, with a present value of the baryon asymmetry fixed by observations to be $\eta_B = (n_B - n_{\bar{B}})/n_\gamma \sim 6 \times 10^{-10}$. Because charge neutrality in the universe implies that the electron density should match the proton density, it turns out that the first-order matter effect in Eq. (4.23) does not dominate at MeV temperatures.

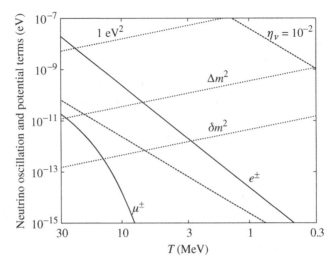

Figure 4.3 Evolution of the vacuum oscillation term (dotted lines) for the two mass differences δm^2 and Δm^2 and a putative squared mass difference of 1 eV², as well as matter potential terms in the relevant temperature range. Matter effects from $e^\pm$ at second order (solid lines) are much larger than the first-order potential term (short-dashed lines), suppressed by η_B. For comparison, the diagonal part of the first-order potential term from background neutrinos with a large asymmetry $\eta_\nu = 10^{-2}$ is also shown; see Section 4.2.4.

Let us consider the W gauge boson propagator in Eq. (1.29) with Δ the momentum exchanged, but now expand it to second order in the limit $q^\mu \ll m_W$,

$$\frac{-g_{\mu\nu} + q_\mu q_\nu / m_W^2}{q^2 - m_W^2} \approx \frac{g_{\mu\nu}}{m_W^2} + \frac{1}{m_W^4} \left(g_{\mu\nu} q^2 - q_\mu q_\nu \right). \qquad (4.24)$$

The effective potential in Eq. (4.23) is obtained from the first term in this expansion, whereas if p is the neutrino momentum the second-order terms lead to

$$V_{CC}^T = -\frac{8\sqrt{2} G_F p}{3 m_W^2} (\rho_{e^+} + \rho_{e^-}), \qquad (4.25)$$

i.e., a matter potential that depends on the sum of the $e^\pm$ energy densities (Nötzold and Raffelt, 1988). Because for relativistic particles $n \propto T^3$ and $\rho \propto T^4$, one would expect that the second-order potential would become much smaller than the one in Eq. (4.23) with the cosmological expansion, but this occurs only at temperatures below 1 MeV, as shown in Fig. 4.3. Thus at high temperatures the V_{CC}^T term is the most relevant.

When we discussed the neutrino effective potentials in Section 1.3.5, we showed that flavour neutrino oscillations are in general suppressed by matter effects until the vacuum and background terms become equal in magnitude. In the early universe at MeV temperatures, this condition should be fulfilled for the vacuum and the charged-lepton background terms

$$-\frac{\Delta m^2}{2p}\cos 2\theta, \qquad -\frac{8\sqrt{2}G_F p}{3m_W^2}(\rho_{l^-} + \rho_{l^+}). \qquad (4.26)$$

Note that the vacuum terms grow with the cosmological expansion because of their dependence on p^{-1}, as shown in Fig. 4.3 for the two relevant squared-mass differences for active neutrinos, Δm^2 and δm^2, using the values in Table 1.9.

Comparing the two terms in this equation for the corresponding mixing parameters, one can estimate when the onset of flavour oscillations is expected, i.e., the temperature at which the lines in Fig. 4.3 cross. Tau leptons are too heavy to have a significant density at MeV temperatures, whereas the energy density of $\mu^\pm$ is exponentially suppressed, leading to $\nu_\tau - \nu_\mu$ mixing driven by Δm^2 and θ_{23} at $T \simeq 15$ MeV (Dolgov *et al.*, 2002), when weak interactions are fully effective. For flavour neutrino oscillations involving ν_e's, the crucial parameters are Δm^2 and θ_{13} and one easily finds that both terms in Eq. (4.26) are equal at a temperature

$$T_c \simeq 19.9 \left(\frac{p}{T}\right)^{-1/3} \left(\frac{|\Delta m^2|}{\text{eV}^2}\right)^{1/6} \text{MeV} \qquad (4.27)$$

for $\cos 2\theta_{13} \simeq 1$ and $e^\pm$ taken as relativistic particles. For $|\Delta m^2| = 2.3 \times 10^{-3}$ eV^2 and an average neutrino momentum, one finds $T_c \simeq 5$ MeV, which approximately agrees with the corresponding crossing point in Fig. 4.3. Finally, for the last combination of neutrino mixing parameters (δm^2 and θ_{12}), one finds $T_c \simeq 2.7$ MeV. Both values of T_c are slightly higher than the neutrino decoupling temperature and precede the onset of primordial nucleosynthesis.

These values for T_c are just a rough estimate of the temperatures at which the flavour oscillations *could* be effective in the early universe, whereas their implications will depend on the initial differences in the neutrino flavour spectra. Note that we have not yet considered the presence of neutrino interactions or the dependence on unknown features of the mixing parameters, such as the sign of Δm^2. In particular, real collisions should be included if $T \gtrsim 1$ MeV, but this demands a proper formalism beyond the Boltzmann equations, which we now discuss.

4.2.2 Density matrix formalism

In Chapter 1 we saw that refraction effects modify the flavour oscillations of neutrinos and found that neutrino mixing in matter can differ from that in vacuum. The evolution of the neutrino flavour transitions was obtained in Section 1.3.5 with the Schrödinger equation, but this simple formalism is insufficient for a situation such as the early universe at MeV temperatures. Here we have to deal simultaneously with time-dependent vacuum and matter terms *and* neutrino interactions in the primeval plasma. In this case one has to exploit the density matrix formalism, which is more appropriate to describe mixed quantum states for neutrinos and possible loss of coherence due to real collisions. We will follow the derivation of Sigl and Raffelt, 1993 (for earlier works see Dolgov, 1981; Barbieri and Dolgov, 1991; Raffelt *et al.*, 1993), but skipping most of the technical aspects.

Let us consider the momentum expansion of a left-handed[2] massless neutrino field

$$\psi_L(x) = \int \frac{d^3p}{(2\pi)^3} \left(u_{\vec{p}} a_{\vec{p}} + v_{-\vec{p}} b^\dagger_{-\vec{p}} \right) \exp(i\,\vec{p}\cdot\vec{x}). \tag{4.28}$$

Here the Dirac spinors $u_{\vec{p}}$ and $v_{\vec{p}}$ refer to massless negative-helicity particles and positive-helicity antineutrinos[3], respectively, whereas $a_{\vec{p}}(b^\dagger_{\vec{p}})$ is an annihilation (creation) operator for negative (positive)-helicity neutrinos (antineutrinos) of three-momentum $\vec{p}$. For n neutrino states, either active (i.e., weakly interacting) or sterile, these operators are column vectors with anticommutation relations $\{a_i(\vec{p}), a^\dagger_j(\vec{p}')\} = \{b_i(\vec{p}), b^\dagger_j(\vec{p}')\} = \delta_{ij}(2\pi)^3\delta^3(\vec{p} - \vec{p}')$.

From all possible bilinears involving the operators a and b, it can be shown that the only ones that do not violate lepton number by two units, or whose expectation values do not oscillate around zero, are $a^\dagger a$ and $b^\dagger b$ (Sigl and Raffelt, 1993). With the additional assumption of spatial homogeneity, their expectation values contribute only for equal momenta. Therefore, a homogeneous ensemble of neutrinos can be characterized by $n \times n$ *density matrices* $\varrho_{\vec{p}}$ (or matrices of densities) defined for neutrinos and antineutrinos with given three-momentum $\vec{p}$, respectively, as

$$\langle a^\dagger_\beta(\vec{p}) a_\alpha(\vec{p}') \rangle = (2\pi)^3\delta^3(\vec{p} - \vec{p}')(\varrho_{\vec{p}})_{\alpha\beta}$$

$$\langle b^\dagger_\alpha(\vec{p}) b_\beta(\vec{p}') \rangle = (2\pi)^3\delta^3(\vec{p} - \vec{p}')(\bar\varrho_{\vec{p}})_{\alpha\beta}. \tag{4.29}$$

Here the reversed order of the indices in the definition of $\bar\varrho$ guarantees that both density matrices transform in the same way under a unitary transformation $\psi' = U\psi$. The diagonal elements of $\varrho_{\vec{p}}$ and $\bar\varrho_{\vec{p}}$ are the usual distribution functions (occupation numbers) for the corresponding neutrino species, whereas the off-diagonal ones encode phase information and vanish for zero mixing.

The evolution in time of the neutrino and antineutrino density matrices is given by[4] (Raffelt *et al.*, 1993; McKellar and Thomson, 1994)

$$i\frac{d\varrho_{\vec{p}}}{dt} = [\Omega^0_{\vec{p}}, \varrho_{\vec{p}}] + [\Omega^{\text{int}}_{\vec{p}}, \varrho_{\vec{p}}] + \mathbf{C}[\varrho_{\vec{p}}, \bar\varrho_{\vec{p}}] \tag{4.30}$$

---

[2] In the limit of relativistic neutrinos we can ignore the right-handed fields.
[3] Recall that for massless spinors, helicity and chirality coincide.
[4] As in the case of the Boltzmann kinetic equations, in the early universe these expressions include the derivative term $(\partial/\partial t - Hp\,\partial/\partial p)\varrho_{\vec{p}}$ and can be rewritten in terms of the dimensionless variables defined in Eq. (4.15).

and similarly for the antineutrino matrices $\bar{\varrho}_{\vec{p}}$. The first term on the r.h.s. describes oscillations in vacuum with

$$\Omega_{\vec{p}}^0 = -\overline{\Omega}_{\vec{p}}^0 = \frac{M^2}{2p}, \qquad (4.31)$$

where $p = |\vec{p}|$ and M is the neutrino mass matrix, diagonal in the mass basis. The matter potential term in Eq. (4.30) is defined as

$$\Omega_{\vec{p}}^{\text{int}} = \sqrt{2}\, G_F \left[\mathsf{L} - \frac{8p}{3m_W^2} \mathsf{E} \right] + \sqrt{2}\, G_F \left[\varrho - \bar{\varrho} - \frac{8p}{3m_Z^2} (\mathsf{U} + \bar{\mathsf{U}}) \right] \qquad (4.32)$$

$$\overline{\Omega}_{\vec{p}}^{\text{int}} = \sqrt{2}\, G_F \left[\mathsf{L} + \frac{8p}{3m_W^2} \mathsf{E} \right] + \sqrt{2}\, G_F \left[\varrho - \bar{\varrho} + \frac{8p}{3m_Z^2} (\mathsf{U} + \bar{\mathsf{U}}) \right]. \qquad (4.33)$$

The first two terms in this equation correspond to the matter potentials felt by neutrinos in a background of charged leptons. Both are diagonal matrices in the flavour basis and proportional to the difference L of number densities or the sum E of energy densities of charged leptons and antileptons. We have seen in the previous subsection that the second kind of terms are dominant at MeV temperatures in an almost charge-symmetric universe. Note that the diagonal components of L and E vanish in the corresponding entries for sterile neutrino states.

The last two terms in Eqs. (4.32) and (4.33) are the refractive contributions of background neutrinos, caused by neutrino self-interactions, and depend on

$$\varrho = \int \frac{d^3 p}{(2\pi)^3} \varrho_{\vec{p}} \qquad \mathsf{U} = \int \frac{d^3 p}{(2\pi)^3} p\varrho_{\vec{p}}, \qquad (4.34)$$

with similar definitions for antineutrinos. Again, zeroes appear whenever either of the two flavour indices in $\varrho_{\alpha\beta}$ corresponds to an inert (sterile) neutrino state. As first realized by Pantaleone, 1992, these matter terms caused by neutrino self-interactions can have nondiagonal elements in the flavour basis, sometimes leading to unexpected behavior of neutrino transitions in media where the neutrino density is large. We will see one example when considering cosmological flavour neutrino asymmetries. These terms are also quite important in media with high density such as Type-II supernovae; see, e.g., the review Duan *et al.*, 2010.

The collision integral $\mathbf{C}[\varrho_{\vec{p}}, \bar{\varrho}_{\vec{p}}]$ in Eq. (4.30) includes all possible neutrino interaction processes and is a generalization of the expression that we described when discussing the usual Boltzmann kinetic equation. Different types of neutrino interactions will lead to separate contributions to the collision term: charged-current (CC) and effective neutral-current (NC) interactions with a medium and neutral-current neutrino self-interactions (S), as shown in Sigl and Raffelt, 1993:

$$C[\varrho_{\vec{p}}, \bar{\varrho}_{\vec{p}}] = \left(\frac{d\varrho_{\vec{p}}}{dt} \right)_{\text{CC}} + \left(\frac{d\varrho_{\vec{p}}}{dt} \right)_{\text{NC}} + \left(\frac{d\varrho_{\vec{p}}}{dt} \right)_{\text{S}}, \qquad (4.35)$$

and an analogous expression for the antineutrino collision terms. The CC term includes the contribution of processes such as $e^- + p \leftrightarrow n + \nu_e$, which make a negligible contribution in the early universe. Instead, the weak processes that contribute to the NC term are those that in the interaction hamiltonian are bilinear in the left-handed neutrino field ψ. After a suitable Fierz transformation, they can be written as an effective NC interaction with an external medium consisting of charged leptons $l^\pm$. It can be shown that this leads to

$$\left(\frac{d\varrho_{\vec{p}}}{dt}\right)^l_{\text{NC}} = \frac{1}{2} \int \frac{d^3 p'}{(2\pi)^3} \left\{ W^l(p', p)(1 - \varrho_{\vec{p}})G^l \varrho_{\vec{p}'} G^l - W^l(p, p')\varrho_{\vec{p}} G^l(1 - \varrho_{\vec{p}'}) \right.$$

$$\left. + W^l(-p', p)(1 - \varrho_{\vec{p}})G^l(1 - \bar{\varrho}_{\vec{p}'})G^l - W^l(p, -p')\varrho_{\vec{p}} G^l \bar{\varrho}_{\vec{p}'} + \text{h.c.} \right\}.$$

(4.36)

The non-negative transition probabilities $W^l(p', p)$ are Wick contractions of medium operators whose expressions can be found in Sigl and Raffelt, 1993, and G^l is a hermitian $n \times n$ matrix of dimensionless coupling constants, which in the flavour basis reads $G^l = \text{diag}(g_1^l, \ldots, g_n^l)$. One can easily see that this collision term resembles that in Eq. (2.94), but with a matrix form. The first two terms in this integral are due to neutrino scattering off the medium, the positive (negative) term being a gain (loss) term corresponding to scattering processes. The third and fourth expressions in the integral account for pair processes, with the positive term being a gain term from pair creations by the medium, whereas the negative one is a loss term from pair annihilations. The final collision term in Eq. (4.30) arises from neutrino self-interactions, which, as in the case of NC interactions, lead to a similar but lengthy expression that can be found in Sigl and Raffelt, 1993.

The general evolution equations for the neutrino and antineutrino density matrices constitute a set of nonlinear integro-differential equations. Note that if there is only one flavour present, the oscillation terms in Eq. (4.30) vanish and the collision integrals reduce to the usual Boltzmann collision terms. On the other hand, the density matrix formalism enhances the complexity of the evolution equations, which become more difficult to solve numerically even with powerful computer tools. Actually, to our knowledge, the complete set of equations for $\varrho_{\vec{p}}$ and $\bar{\varrho}_{\vec{p}}$ has not been solved yet without approximations. Instead, this problem has been attacked in many papers under various simplifying assumptions, above all regarding the collision terms.

Let us now consider the application of the density matrix formalism to a two-flavour situation involving the states ν_e and ν_β, where ν_β can be an active or a sterile neutrino species. The usual relation between flavour and mass eigenstates is

$$\nu_e = \cos\theta \, \nu_1 + \sin\theta \, \nu_2$$

$$\nu_\beta = -\sin\theta \, \nu_1 + \cos\theta \, \nu_2,$$

(4.37)

with θ the vacuum mixing angle. A convenient expansion of the corresponding 2×2 density matrices for neutrinos and antineutrinos is

$$\varrho_{\vec{p}}(t) = \frac{1}{2}\left[P_{\vec{p}}^0(t) + \vec{\sigma} \cdot \vec{P}_{\vec{p}}(t)\right]$$

$$\bar{\varrho}_{\vec{p}}(t) = \frac{1}{2}\left[\overline{P}_{\vec{p}}^0(t) + \vec{\sigma} \cdot \overline{\vec{P}}_{\vec{p}}(t)\right]. \tag{4.38}$$

Here, σ_i are the Pauli matrices whereas $\vec{P}_{\vec{p}}$ ($\overline{\vec{P}}_{\vec{p}}$) is the usual polarization vector in flavour space of a neutrino (antineutrino) mode with momentum $\vec{p}$. The z direction indicates flavour, so the P_z component of the polarization vector is related to the diagonal terms of the density matrix that give us the flavour occupation numbers, for instance, $\varrho_{ee} = (P^0 + P_z)/2$. Correspondingly, the $P_{x,y}$ components vanish for zero neutrino mixing.

The evolution equations for P^0, $\overline{P}^0$ and the polarization vectors can be found from Eq. (4.30) for the density matrices. Neglecting the collision terms for the moment, P^0 and $\overline{P}^0$ are constant and one obtains

$$\left(\frac{\partial}{\partial t} - Hp\frac{\partial}{\partial p}\right)\vec{P}_{\vec{p}} = \vec{V} \times \vec{P}_{\vec{p}} = \left(\frac{\Delta m^2}{2p}\vec{B} + \vec{V}_e + \vec{V}_\nu\right) \times \vec{P}_{\vec{p}}$$

$$\left(\frac{\partial}{\partial t} - Hp\frac{\partial}{\partial p}\right)\overline{\vec{P}}_{\vec{p}} = \overline{\vec{V}} \times \overline{\vec{P}}_{\vec{p}} = \left(-\frac{\Delta m^2}{2p}\vec{B} + \overline{\vec{V}}_e + \overline{\vec{V}}_\nu\right) \times \overline{\vec{P}}_{\vec{p}}, \tag{4.39}$$

where $\Delta m^2 = m_2^2 - m_1^2$ and $\vec{B} = (\sin 2\theta, 0, -\cos 2\theta)$. If we assume that $e^\pm$ are the only charged leptons abundant in the primeval plasma, the corresponding potential term that dominates is

$$\vec{V}_e = -\overline{\vec{V}}_e = -\frac{8\sqrt{2}G_F p}{3m_W^2}(\rho_{e^+} + \rho_{e^-})\hat{z}, \tag{4.40}$$

i.e., along the flavour direction ($\hat{z}$ is a unit vector in the z-direction).

The equations of motion (4.39) for the polarization vectors are the usual spin-precession formula for neutrino oscillations in vacuum (see, e.g., Stodolsky, 1987). The vector $\vec{B}$ gives us an *effective magnetic field* around which $\vec{P}_{\vec{p}}$ precesses with rate $\Delta m^2/2p$. Therefore, $\vec{P}_{\vec{p}}$ plays the role of an angular momentum vector whereas $\vec{M}_{\vec{p}} = (\Delta m^2/2p)\vec{P}_{\vec{p}}$ plays the role of its associated magnetic dipole moment. The effect of charged leptons in the background is to reduce the effective mixing angle for all momentum modes, suppressing flavour oscillations for temperatures higher than T_c in Eq. (4.27). Even in the absence of interactions, this spin-precession picture (or geometrical representation) can be a useful tool for representing and calculating the effect of matter on neutrino oscillations (see, e.g., Chapter 9 of Giunti and Kim, 2007).

The specific form of the refraction terms coming from neutrinos in the medium depends on whether the second neutrino state ν_β is an active or a sterile species. Let us first consider the active neutrino case, with ν_β either ν_μ or ν_τ. The effective potential terms for neutrinos and antineutrinos are, from Eq. (4.33),

$$\vec{V}_\nu = \sqrt{2}\, G_{\rm F} \left[\vec{P} - \vec{\overline{P}} - \frac{8p}{3m_Z^2} \left(\vec{U} + \vec{\overline{U}} \right) \right]$$

$$\vec{\overline{V}}_\nu = \sqrt{2}\, G_{\rm F} \left[\vec{P} - \vec{\overline{P}} + \frac{8p}{3m_Z^2} \left(\vec{U} + \vec{\overline{U}} \right) \right], \qquad (4.41)$$

where we have introduced the total polarization vector $\vec{P} = \int d^3p/(2\pi)^3\, \vec{P}_{\vec{p}}$ and $\vec{U} = \int p\, d^3p/(2\pi)^3\, \vec{P}_{\vec{p}}$, with similar expressions for antineutrinos. In contrast to the refraction terms coming from $e^\pm$, the neutrino potentials are not necessarily along the z direction. Thus, the vacuum oscillations are not obviously suppressed in a medium dense in neutrinos, differently from a standard background medium, as we will discuss later in Section 4.2.4.

If ν_β is a sterile neutrino state, it does not interact with the medium, so that the potential term is (Nötzold and Raffelt, 1988)

$$\vec{V}_\nu = \sqrt{2}\, G_{\rm F} \left[n_\gamma \left(2\eta_{\nu_e} + \eta_{\nu_\mu} + \eta_{\nu_\tau} \right) - \frac{8p}{3m_Z^2} \left(2\rho_{\nu_e} + \rho_{\nu_\mu} + \rho_{\nu_\tau} \right) \right] \hat{z}$$

$$\vec{\overline{V}}_\nu = \sqrt{2}\, G_{\rm F} \left[n_\gamma \left(2\eta_{\nu_e} + \eta_{\nu_\mu} + \eta_{\nu_\tau} \right) + \frac{8p}{3m_Z^2} \left(2\rho_{\nu_e} + \rho_{\nu_\mu} + \rho_{\nu_\tau} \right) \right] \hat{z}, \qquad (4.42)$$

where we have included the contribution of $\nu_{\mu,\tau}$ (assumed unmixed with ν_e) and defined the neutrino flavour asymmetries as $\eta_{\nu_\alpha} \equiv (n_{\nu_\alpha} - n_{\overline{\nu}_\alpha})/n_\gamma$. Therefore, in this case the neutrino background term is parallel to $\hat{z}$ and plays a role similar to that of the standard matter term. In this situation one should also include the subdominant asymmetric potential term, proportional to $\eta_e - \eta_n/2$ $(-\eta_n/2)$ for $\nu_e - \nu_s$ $(\nu_\mu - \nu_s)$ oscillations.

The evolution equations for P^0, $\overline{P}^0$ and the polarization vectors should be completed with the corresponding collision terms from Eq. (4.35). However, we saw that these terms involve multidimensional integrals that are not so easy to solve. Therefore, while waiting for a calculation of the effect of neutrino oscillations in the early universe with the full collision integrals in matrix form, here we describe some approximate expressions for the terms involving real interactions of neutrinos used in previous works on active–sterile oscillations.[5] In particular, we follow

---

[5] For similar equations in the active–active case, see McKellar and Thomson, 1994; Hannestad, 2002.

(Prakash *et al.*, 2001) and add collision terms to Eq. (4.39) of the form

$$\left(\frac{\partial}{\partial t} - Hp\frac{\partial}{\partial p}\right)\vec{P}_{\vec{p}} = \vec{V} \times \vec{P}_{\vec{p}} - \frac{1}{2}D_p\,(P_{\vec{p}}^x\,\hat{x} + P_{\vec{p}}^y\,\hat{y}) + R_p\,\hat{z}$$

$$\left(\frac{\partial}{\partial t} - Hp\frac{\partial}{\partial p}\right)P_{\vec{p}}^0 = R_p,\tag{4.43}$$

where D_p is the decoherence or damping function and R_p is the repopulation or refilling function (Barbieri and Dolgov, 1991; Enqvist *et al.*, 1992b). Both functions are in general dynamical quantities because the collision rates depend on the neutrino distribution functions and hence on the neutrino density matrices.

The decoherence function D_p quantifies the loss of quantum coherence due to neutrino interactions with the particles in the primeval plasma. Essentially one assumes that neutrino wavefunctions collapse into weak-interaction eigenstates each time they interact in a process that distinguishes among flavours (Raffelt *et al.*, 1993). This damping term affects the components of the polarization vector that are orthogonal to the flavour direction, i.e., the off-diagonal elements of the neutrino density matrix. It is related to the total collision rate Γ for the active neutrinos as $D_p = \Gamma_p/2$. A simple expression for D_p can be obtained by assuming thermodynamical equilibrium and small lepton number, as well as neglecting Pauli blocking effects (McKellar and Thomson, 1994; Bell *et al.*, 1999),

$$\Gamma_\alpha(p) = \langle \Gamma_\alpha\rangle\frac{p}{\langle p\rangle} = \kappa_\alpha\frac{180\zeta(3)}{7\pi^4}G_F^2 T^5\,\frac{p}{T},\tag{4.44}$$

where $\langle\Gamma_\alpha\rangle$ is the thermally averaged total collision rate for ν_α and $\langle p\rangle = 7T\pi^4/180\zeta(3) \simeq 3.15T$ is the average momentum for a Fermi–Dirac spectrum. The κ coefficients are numbers of order unity; $\kappa_e \simeq 4$ and $\kappa_{\mu,\tau} \simeq 2.9$, according to Enqvist *et al.*, 1992b. More accurate values were calculated in Dolgov *et al.*, 2000a from the thermal averaging of the complete weak rates, including Fermi statistical factors, leading to $\kappa_e \simeq 3.56$ and $\kappa_{\mu,\tau} \simeq 2.5$. This approximate expression for D_p does not depend on P^0 nor on $\vec{P}$, and this greatly simplifies the numerical solution of the equations.

The repopulation function $R(p)$ is approximately given by

$$R_p = \Gamma_\alpha(p)\left\{f_{\text{eq}}(p, \mu_{\nu_\alpha}) - \frac{1}{2}\left(P_{\vec{p}}^0 + P_{\vec{p}}^z\right)\right\},\tag{4.45}$$

where $f_{\text{eq}}(p, \mu_{\nu_\alpha})$ is the Fermi–Dirac momentum distribution with a possible chemical potential. This expression can be obtained from the general collision term, with a positive (negative) contribution from all processes that can generate (deplete) neutrinos with momentum p. The main approximation used is that all distribution functions except that for ν_α present an equilibrium form, whereas the ν_α spectrum is approximately a Fermi–Dirac (McKellar and Thomson, 1994; Bell *et al.*, 1999). The expression for the repopulation function can be interpreted as follows: all weak processes with ν_α's tend to change their actual distribution functions toward the equilibrium form. The interested reader can find more details on the decoherence and repopulation functions in Bell *et al.*, 1999; Dolgov, 2002.

4.2.3 *Flavour oscillations and relic neutrino distortions*

We found in Section 4.1 that a detailed calculation of the full process of relic neutrino decoupling showed that cosmological neutrinos present an energy spectrum very close to an equilibrium Fermi–Dirac form. However, small spectral distortions arise which are flavour-dependent, those of electron neutrinos and antineutrinos being roughly twice as large as the distortions for the other flavours, as shown in Table 4.1. The potential effect of flavour oscillations on neutrino decoupling in the early universe was already noted long ago by Langacker *et al.*, 1987. Here we describe the formalism and results of recent analyses in some detail, because they represent to date the *standard* calculation of relic neutrino decoupling.

We have already described all the elements needed for a proper calculation of neutrino decoupling in the presence of flavour oscillations. The set of equations to be solved is those for the neutrino density matrix, Eq. (4.30), modified to take into account the expansion of the universe, together with the covariant conservation equation for the total energy–momentum tensor, Eq. (4.18). Earlier works simplified this task, using integrated kinetic equations and Maxwell–Boltzmann statistics (Hannestad, 2002). These approximations were relaxed in Mangano *et al.*, 2005 and a full calculation was performed with a density matrix formalism.

The neutrino density matrix for a mode with momentum p that we need in this case is a 3×3 matrix,

$$\varrho_p(t) = \begin{pmatrix} \varrho_{ee} & \varrho_{e\mu} & \varrho_{e\tau} \\ \varrho_{\mu e} & \varrho_{\mu\mu} & \varrho_{\mu\tau} \\ \varrho_{\tau e} & \varrho_{\tau\mu} & \varrho_{\tau\tau} \end{pmatrix}, \tag{4.46}$$

where, as usual, the diagonal elements correspond to the number density of the different flavours, whereas the off-diagonal terms are nonzero in the presence of neutrino mixing. In the absence of a neutrino asymmetry (or of additional couplings flipping neutrinos into antineutrinos and vice versa), antineutrinos follow the same evolution as neutrinos, and a single density matrix is enough to describe the system.

The equations of motion for ϱ_p of Eq. (4.30) were numerically solved in Mangano *et al.*, 2005, using the comoving variables in Eq. (4.15) and a discretization in neutrino momenta.

The matter potential term is as in Eq. (4.32) with zero neutrino asymmetry, ignoring the usual refractive term $\sqrt{2}G_F L$, which is negligible at high temperatures compared to the E term and at $T \simeq 1$ MeV compared to the vacuum oscillation term. Some approximations were mode in order to calculate the collision term of order G_F^2 in Eq. (4.30). For the off-diagonal terms that break quantum coherence, the simple damping functions reported in Eq. (4.44) were used. For the diagonal ones, collision integrals similar to Eq. (4.5) were included, replacing the neutrino distribution functions with the corresponding diagonal elements of ϱ_p. Given the order of the effects considered, one should also include the finite-temperature QED corrections to the electromagnetic plasma as in Mangano et al., 2002.

Because only the three flavour neutrinos are included, the vacuum oscillation term is proportional to M^2, the mass-squared matrix in the flavour basis that is related to the diagonal one in the mass basis $M_m^2 = \text{diag}(m_1^2, m_2^2, m_3^2)$ via the neutrino mixing matrix U defined in Chapter 1; see Eq. (1.56). In the following we will assume vanishing CP-violating phases.

There are five oscillation parameters ($\Delta m^2, \delta m^2, \theta_{12}, \theta_{23}, \theta_{13}$), whose best-fit values and uncertainties were discussed in Chapter 1. The recent measurement of θ_{13} from the Double CHOOZ (Abe et al., 2012), Daya-Bay (An et al., 2012) and RENO (Ahn et al., 2012) experiments was also mentioned there. The first four parameters are already known with good precision, such that the results concerning neutrino decoupling remain unchanged within the currently favoured regions for oscillations parameters. Also, θ_{13} is now better determined and is found to be different from zero with high statistical significance. To illustrate the role of this mixing angle in the flavour dynamics during neutrino decoupling, we will consider two limiting cases, $\theta_{13} = 0$ and $\sin^2 \theta_{13} = 0.047$. The second value is almost the present 3σ upper bound. The results for a vanishing θ_{13} are shown just for comparison. Notice that the results for the present best-fit value are quite close to what we will find for the $\sin^2 \theta_{13} = 0.047$ case.

If we go back to Fig. 4.1, we can follow the evolution of the neutrino spectral distortions δ_ν as a function of x for a particular neutrino momentum, with and without flavour oscillations. The evolution without oscillations was already described in Section 4.1, so here we emphasize the role of neutrino mixing. We have seen that oscillations can be effective when the vacuum oscillation term, growing as x^2, becomes larger than the background potential proportional to the energy density of $e^\pm$, decreasing as x^4. Thus, for $x \lesssim 0.3$ this last term dominates, suppressing flavour oscillations so that the neutrino distortions grow as in the absence of mixing. Then the $e^\pm$ potential adiabatically disappears, leading to the usual MSW-type evolution and a convergence of the flavour neutrino distortions. For the example shown in Fig. 4.1, if $\theta_{13} = 0$ the final value of δ_{ν_e} at $y = 10$ is reduced to 3.4%, whereas $\nu_{\mu,\tau}$ increases to 2.4%. For $\sin^2 \theta_{13} = 0.047$, one finds 3.2%

Table 4.2 *Frozen values of* z_{fin}, *the neutrino energy densities* $\delta\bar{\rho}_{\nu_\alpha} \equiv \delta\rho_{\nu_\alpha}/\rho_{\nu_0}$, N_{eff} *and* ΔY_p *including flavour neutrino oscillations*

Case	z_{fin}	$\delta\bar{\rho}_{\nu_e}(\%)$	$\delta\bar{\rho}_{\nu_\mu}(\%)$	$\delta\bar{\rho}_{\nu_\tau}(\%)$	N_{eff}	ΔY_p
$\theta_{13} = 0$	1.3978	0.73	0.52	0.52	3.046	2.07×10^{-4}
$\sin^2\theta_{13} = 0.047$	1.3978	0.70	0.56	0.52	3.046	2.12×10^{-4}

for electron neutrinos and a different distortion for muon (2.6%) and tau (2.4%) neutrinos.

The effect of neutrino oscillations is evident in the asymptotic nonthermal distortions depicted in Fig. 4.2, reducing the difference between the different δ_{ν_α}. The corresponding momentum distributions for the neutrino mass eigenstates $\nu_{1,2,3}$, relevant to any cosmological epoch when neutrinos are nonrelativistic, can easily be found from the flavour ones through the relation

$$f_{\nu_i}(y) = \sum_{\alpha=e,\mu,\tau} |U_{\alpha i}|^2 f_{\nu_\alpha}(y) \tag{4.47}$$

which, for $f_{\nu_\mu} \sim f_{\nu_\tau}$ and the best-fit values of the neutrino mixing angles, gives the simple relations

$$f_{\nu_1}(y) \simeq 0.7 f_{\nu_e}(y) + 0.3 f_{\nu_x}(y)$$

$$f_{\nu_2}(y) \simeq 0.3 f_{\nu_e}(y) + 0.7 f_{\nu_x}(y)$$

$$f_{\nu_3}(y) \simeq f_{\nu_x}(y). \tag{4.48}$$

Finally, in Table 4.2, we present the results for the dimensionless photon temperature z_{fin}, the change in the neutrino energy densities with respect to ρ_{ν_0} and the change in the prediction for the primordial abundance of ${}^4\text{He}$ (see later in Section 4.3). The main conclusion is that flavour oscillations do not modify the contribution of neutrino heating to the total relativistic energy density, leaving the value of $N_{\text{eff}} = 3.046$ unchanged.

4.2.4 Flavour oscillations and relic neutrino asymmetries

A major cosmological implication of neutrino oscillations among active states at MeV temperatures is the evolution of putative flavour neutrino–antineutrino asymmetries. In analogy with the baryon parameter η_B, a difference between the number density of neutrinos and antineutrinos of flavour $\alpha = e, \mu, \tau$ can be parameterized by the ratio

$$\eta_{\nu_\alpha} = \frac{n_{\nu_\alpha} - n_{\bar{\nu}_\alpha}}{n_\gamma}. \tag{4.49}$$

Based on the equilibration of baryon and lepton numbers per comoving volume, B and L, by sphalerons in the very early universe discussed in Chapter 3, all η_{ν_α} should be on the same order as the cosmological baryon number η_B, i.e., a few times 10^{-10}. In such a case, cosmological neutrino asymmetries would be too small to have any observable consequence.

However, values of these parameters which are orders of magnitude larger than η_B are not excluded by observations. Actually, large η_{ν_α} are predicted in theoretical models where the generation of lepton asymmetry took place after the electroweak phase transition or the electroweak washing out of preexisting asymmetries is not effective (Dolgov, 1992; Casas *et al.*, 1999; McDonald, 2000). In any case, their possible high values would raise fundamental questions about both electroweak physics and the early universe (see, e.g., Semikoz *et al.* 2009; Schwarz and Stuke 2009), so it is interesting to discuss the observational implications of a large cosmological lepton asymmetry stored in neutrinos.

The presence of a cosmological neutrino–antineutrino asymmetry when neutrinos are still interacting implies that their spectrum presents a Fermi–Dirac form characterized by neutrino degeneracy parameters $\xi_\alpha \equiv \mu_{\nu_\alpha}/T_\nu$, as in Eq. (2.105). Moreover, the neutrino energy density, pressure and entropy density then take the standard form listed in Eqs. (2.177)–(2.179). For convenience, we again write the neutrino energy density

$$\rho_{\nu_\alpha} + \rho_{\bar{\nu}_\alpha} = \frac{7\pi^2}{120}T_\nu^4\left(1 + \frac{30\,\xi_\alpha^2}{7\pi^2} + \frac{15\,\xi_\alpha^4}{7\pi^4}\right). \tag{4.50}$$

Because this expression contains only even powers of ξ_α, the additional terms from nonzero asymmetries are always positive – this is simply the effect of a high mean kinetic energy due to Fermi energy in either neutrino or antineutrino sectors – thus enhancing the contribution of radiation content. If we express this contribution to ρ_R in terms of the effective parameter N_{eff}, as defined in Eq. (2.198), we get

$$N_{\text{eff}} = 3 + \sum_{\alpha=e,\mu,\tau}\left[\frac{30}{7}\left(\frac{\xi_\alpha}{\pi}\right)^2 + \frac{15}{7}\left(\frac{\xi_\alpha}{\pi}\right)^4\right], \tag{4.51}$$

where we have neglected the small effect of nonthermal neutrino distortions. Thus, any nonvanishing flavour neutrino asymmetry leads to $N_{\text{eff}} > 3$ without the introduction of additional relativistic particles. Neutrino degeneracy parameters of order $\xi_\alpha \gtrsim 0.3$ are needed to have a contribution of the asymmetries to N_{eff} at least at the same level as the effect of nonthermal distortions discussed in Section 4.2.3.

The present cosmological observations are not sensitive to neutrino asymmetry if $|\eta_\nu| \lesssim 10^{-2}$. Only higher values lead to a significant enhancement of N_{eff} or to changes in the production of light elements during Big Bang nucleosynthesis. We will see in Section 4.4.2 that the primordial abundance of ^{4}He is strongly sensitive to the presence of an electron neutrino asymmetry and sets a stringent BBN bound on η_{ν_e} which does not apply to the other flavours, leaving a total neutrino asymmetry of order unity unconstrained (Kang and Steigman, 1992; Hansen *et al.*, 2002). However, this conclusion relies on the absence of neutrino oscillations that would mix the asymmetries among different flavours. Therefore, it is an important issue to calculate the evolution of the η_{ν_α}'s in the epoch before BBN with three-flavour neutrino oscillations, as we describe next.

To this end, we need again to solve the set of equations of motion for the 3×3 neutrino (antineutrino) density matrices, Eq. (4.30), together with the continuity equation for the total energy density ρ_R, Eq. (4.18). In this case, however, neutrinos and antineutrinos do not share the same density matrix and the evolution of both must be calculated. The system to solve for neutrinos is now

$$i \frac{d\varrho_{\vec{p}}}{dt} = [\Omega_{\vec{p}}, \varrho_{\vec{p}}] + C[\varrho_{\vec{p}}, \bar{\varrho}_{\vec{p}}], \tag{4.52}$$

where the first term on the r.h.s. describes coherent flavour oscillations,

$$\Omega_{\vec{p}} = \frac{\mathsf{M}^2}{2p} + \sqrt{2}\, G_F \left[-\frac{8p}{3m_w^2} \mathsf{E} + \varrho - \bar{\varrho} \right], \tag{4.53}$$

and similarly for the antineutrino matrices $\bar{\varrho}_{\vec{p}}$ with the corresponding sign changes, as in Eqs. (4.31) and (4.33). We only consider the dominant two matter-potential terms from the charged leptons and neutrinos in the background, proportional to E and $\varrho - \bar{\varrho}$, respectively.

For the present values of neutrino oscillation parameters, we saw from Eq. (4.27) that one expects oscillations involving electron neutrinos to be effective before $T = 1$ MeV, thus before the onset of BBN. However, this conclusion does not take into account the presence of the neutrino background term, which for the values of the neutrino asymmetries considered here, $|\eta_\nu| \gtrsim 10^{-2}$, is much larger than the charged lepton term, as clearly shown in Fig. 4.3. Although this would in principle suggest that flavour oscillations are even more suppressed, the fact that $\varrho - \bar{\varrho}$ is not diagonal in the flavour basis leads to some counterintuitive evolution. It turns out that the effect of the large background term from neutrinos, rather than to suppress oscillations, is to synchronize them (Samuel, 1993; Pastor *et al.*, 2002). Although the $\varrho - \bar{\varrho}$ term dominates, neutrinos and antineutrinos with different momenta oscillate with a certain common frequency and remain sensitive to the charged-lepton contribution. Therefore, as found in a series of analyses (Abazajian *et al.*, 2002; Dolgov *et al.*, 2002; Wong, 2002), flavour neutrino asymmetries show a behavior similar (although not exactly the same) to that with only the E matter term, an important result for their impact on the primordial production of light nuclei. These works treated the collision terms, $C[\varrho_{\vec{p}}, \bar{\varrho}_{\vec{p}}]$ in a schematic way,

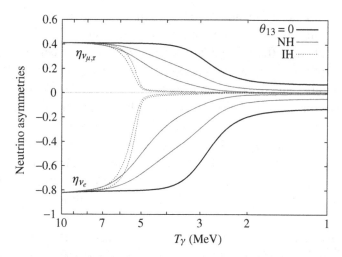

Figure 4.4 Evolution of the flavour neutrino asymmetries versus photon temperature T_γ with $\eta_{\nu_e}^{in} = -0.82$ and zero total asymmetry. The outer solid curves correspond to vanishing θ_{13}, whereas the inner ones were calculated in the normal neutrino mass hierarchy (NH) for two values of $\sin^2 \theta_{13}$: from left to right, 0.04 and 0.02. The same two values of $\sin^2 \theta_{13}$ apply to the cases shown as dotted lines, but in the inverted hierarchy (IH) case. (Reprinted from Mangano *et al.*, 2012, with permission from Elsevier.)

including only damping terms, as in Eq. (4.44), but not the corresponding repopulation functions for the diagonal elements of $\varrho_{\vec{p}}$ and $\bar{\varrho}_{\vec{p}}$.

A better approximation for the collision terms, similar to the one described in the previous subsection, was considered in Pastor *et al.*, 2009 and developed in Mangano *et al.*, 2011, 2012. In particular, the collisional integrals for the diagonal elements of the density matrices allow a proper calculation of the transfer of entropy from neutrinos to the electromagnetic plasma through processes such as $\nu\bar{\nu} \rightarrow e^+ e^-$.

Let us now describe one example of evolution of the flavour neutrino asymmetries shown in Fig. 4.4. The initial condition that was assumed for flavour neutrinos is a Fermi–Dirac spectrum with degeneracy parameters $\xi_x \equiv \xi_\mu = \xi_\tau$ and ξ_e (opposite sign for antineutrinos), as guaranteed by efficient $\nu_\tau - \nu_\mu$ mixing driven by Δm^2 and θ_{23} at $T \simeq 15$ MeV (Dolgov *et al.*, 2002), when weak interactions are fully effective. Equivalently, one can choose as free parameters the initial electron neutrino asymmetry $\eta_{\nu_e}^{in}$ produced at some early stage and the total neutrino asymmetry (or lepton number L) $\eta_\nu = \eta_{\nu_e}^{in} + 2\eta_{\nu_x}^{in}$.

Electron neutrinos and antineutrinos enter the oscillation game through flavour conversions driven by the parameters Δm^2 and θ_{13} at $T_c \sim 5$ MeV, as we saw in Section 4.2.1. If $\Delta m^2 > 0$ (normal neutrino mass hierarchy, NH), both terms in

Eq. (4.26) have the same sign and neutrino oscillations follow an MSW conversion when the vacuum term overcomes the matter potential at $T \simeq T_c$. In this case, the amplitude of this effect depends upon the value of θ_{13}, being larger as this mixing angle grows. This can be seen in Fig. 4.4 for one particular case with zero total lepton number and different choices of θ_{13}. The conversion for nonzero θ_{13} is more evident for the inverted mass hierarchy, because of the resonant character of the MSW transition for $\Delta m^2 < 0$. Indeed, for IH and $\sin^2 \theta_{13} \gtrsim 5 \times 10^{-3}$ the sum of the two terms in Eq. (4.26) vanishes and equipartition of the total lepton asymmetry among the three neutrino flavours is quickly achieved. For negligible θ_{13} flavour oscillations are not effective until $T \lesssim 3\,\mathrm{MeV}$ (outer lines in Fig. 4.4), as expected from Eq. (4.26), with the relevant squared mass difference given this time by δm^2.

The time when flavour oscillations in the presence of neutrino asymmetries become effective is important not only to establish the electron neutrino asymmetry at the onset of BBN but also to determine whether weak interactions with e^+e^- can still keep neutrinos in good thermal contact with the surrounding plasma. Oscillations redistribute the asymmetries among the flavours, but only if they occur early enough would interactions preserve Fermi–Dirac spectra for neutrinos, in such a way that a chemical potential μ_{ν_α} is well defined for each η_{ν_α} and the relations in Eqs. (2.105) and (4.51) remain valid. For instance, if the initial values of the flavour asymmetries have opposite signs, neutrino conversions will tend to reduce the asymmetries which in turn will decrease N_{eff}. On the other hand, if flavour oscillations take place at temperatures close to neutrino decoupling this would not hold, and an extra contribution of neutrinos to radiation is expected with respect to the value in Eq. (4.51), as emphasized in Pastor *et al.*, 2009.

This effect is easy to understand. Suppose all neutrino interactions are switched off and consider two-flavour mixing, say ν_e and ν_μ. Starting from initial thermal distributions and some value ξ_e^{in} and ξ_μ^{in}, efficient flavour conversion would mix neutrino distributions but does not change their total density in phase space,

$$f_{\nu_e}(p) + f_{\nu_\mu}(p) = \frac{1}{e^{\frac{p}{T_\nu} - \xi_e^{\mathrm{in}}} + 1} + \frac{1}{e^{\frac{p}{T_\nu} - \xi_\mu^{\mathrm{in}}} + 1}. \tag{4.54}$$

On the other hand, if neutrino scatterings/annihilations are at work, the ν_e and ν_μ distributions would rapidly adjust to a Fermi–Dirac shape with the same chemical potential ξ:

$$f_{\nu_e}(p) + f_{\nu_\mu}(p) = \frac{2}{e^{\frac{p}{T_\nu} - \xi} + 1}. \tag{4.55}$$

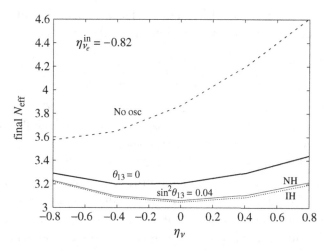

Figure 4.5 Final contribution of neutrinos to the total radiation energy density, parameterized by N_{eff}, as a function of the total neutrino asymmetry for a particular value of the initial electron neutrino asymmetry ($\eta_{\nu_e}^{\text{in}} = -0.82$). From top to bottom, the various lines correspond to the following cases: no neutrino oscillations, $\theta_{13} = 0$, $\sin^2 \theta_{13} = 0.04$ for normal (solid line) and inverted (dotted line) neutrino mass hierarchy. (Reprinted from Mangano *et al.*, 2012, with permission from Elsevier.)

The value of ξ is given by the conservation of the total lepton asymmetry – see (2.130) –

$$\xi + \frac{\xi^3}{\pi^2} = \frac{\xi_e^{\text{in}} + \xi_\mu^{\text{in}}}{2} + \frac{\left(\xi_e^{\text{in}}\right)^3 + \left(\xi_\mu^{\text{in}}\right)^3}{2\pi^2}, \tag{4.56}$$

which can be analytically solved using the Cardano formula. For $\xi_e^{\text{in}} \pm \xi_\mu^{\text{in}} \simeq 0$, we get $\xi \simeq (\xi_e^{\text{in}} + \xi_\mu^{\text{in}})/2$. Computing the total energy density in the two cases of Eqs. (4.54) and (4.55), one finds that it is always higher for the distribution (4.54), unless $\xi_e = \xi_\mu$. In other words, interaction processes act as a sink and transfer the excess in energy density due to chemical potentials to the electromagnetic plasma, reducing in this way the value of N_{eff}.

The final value of N_{eff} from neutrino asymmetries is given in Fig. 4.5. Here the initial electron neutrino asymmetry was fixed to $\eta_{\nu_e}^{\text{in}} = -0.82$ as in Fig. 4.4, but the total asymmetry was varied in the range $-0.8 \leq \eta_\nu \leq 0.8$. In the absence of neutrino mixing the final value of N_{eff} is that given by Eq. (4.51), directly related to the chemical potentials, and for this particular range it can be as large as $N_{\text{eff}} \simeq 4.6$. Instead, when oscillations are included, the three flavour asymmetries are modified and the contribution of neutrinos is greatly reduced. For $\sin^2 \theta_{13} = 0.04$ and both normal and inverted mass hierarchies, the final flavour asymmetries are given by $\eta_{\nu_\alpha} \simeq \eta_\nu/3$. In such a case, we expect neutrinos almost to follow Fermi–Dirac spectra and N_{eff} as given in Eq. (4.51). For instance, for $\eta_\nu = 0.8$, one has $\xi_{\nu_{e,\mu,\tau}} \simeq$ 0.38 and a total contribution to the radiation energy density $N_{\text{eff}} \simeq 3.2$.

4.2.5 Active–sterile oscillations

We have reviewed in Chapter 1 the main experiments that provide results on flavour neutrino oscillations. We saw that a minimal scenario with three flavour or active neutrinos can account for practically all data on neutrino oscillations, but there exist some anomalies that could be better explained if the number of light neutrinos is extended with the addition of one or more sterile states. Here we consider whether such active–sterile oscillations will be effective in the early universe.

Let us first consider the main possibilities for cosmological production of sterile neutrinos. Because these particles are insensitive to weak interactions, they do not follow the behaviour of active neutrinos and are not expected to be present at MeV temperatures. Even if they could interact through other kind of interactions, significantly weaker than the standard weak ones, as predicted by extended particle physics models, they would have a thermal spectrum at very high temperatures, but their density would have been strongly diluted by many subsequent particle–antiparticle annihilations. Therefore, barring nonthermal production from additional physics beyond the SM, such as the existence of a relic ν_s population from the decay of a heavy particle like a gauge singlet boson (Kusenko, 2006; Shaposhnikov and Tkachev, 2006; Petraki and Kusenko, 2008), the main way of obtaining a significant abundance of sterile neutrinos is through their mixing with the active ones.

In principle, the cosmological evolution of the active–sterile neutrino system should be found by solving the corresponding Boltzmann kinetic equations for the density matrices, as described in Section 4.2.2. There exists a vast literature on this subject, where different analyses have considered several approximations, in general solving equations similar to those in Eq. (4.43). Early references include Barbieri and Dolgov, 1990; Enqvist *et al.*, 1990; Kainulainen, 1990; Barbieri and Dolgov, 1991; Enqvist *et al.*, 1991, 1992b; for a more complete list see, e.g., Dolgov, 2002. However, we do not need a full calculation to understand the key points of ν_s production via oscillations in certain limits.

Although in general one should consider, at least, a 4×4 mixing of three active neutrinos and one sterile species (with four masses, six mixing angles and three CP-violating phases), let us first assume an admixture of one sterile state to electron neutrinos as defined in Eq. (4.37). We showed in Section 4.2.1 that in the early universe one expects that neutrino oscillations could be effective when the vacuum oscillation term becomes larger than the main matter-potential term from charged leptons, both given in Eq. (4.26). A good approximation is that this happens when the temperature drops below a certain value T_c given in Eq. (4.27). This expression holds for values of the squared-mass difference Δm^2 of the active–sterile neutrino system on the same order as the values shown in Fig. 4.3 or smaller. Comparing the value of T_c with the decoupling temperature of active neutrinos, $T_{\nu D} \sim 2$ MeV, one obtains that for values

$$\Delta m^2 \lesssim 1.3 \times 10^{-7} \, \text{eV}^2, \tag{4.57}$$

active–sterile oscillations are effective *after* neutrino decoupling. In such a case, the total comoving number density of active and sterile neutrinos will be constant because active neutrinos are no longer interacting with the rest of the primeval plasma. Correspondingly, distortions in the momentum spectra of neutrinos are expected, whereas the combined contribution of the active–sterile system to the radiation content will amount to $N_{\text{eff}} = 1$. Instead, for much higher values of Δm^2, oscillations are effective *before* neutrino decoupling, when weak interactions are frequent. Thus, the sterile neutrinos will become populated whereas the energy spectrum of the active ones will be kept in equilibrium. In this case, their total contribution to radiation can be as large as $N_{\text{eff}} = 2$, depending on the specific value of the mixing angle and on the sign of Δm^2.

For active–sterile oscillations with $\Delta m^2 > 1$ eV^2, the value of T_c exceeds the range of temperatures shown in Fig. 4.3. Because it is interesting to consider values as high as a few keV^2, one can write it as

$$T_c \approx 150 \left(\frac{|\Delta m^2|}{\text{keV}^2} \right)^{1/6} \text{MeV}. \tag{4.58}$$

This expression is approximate because one should take into account the proper evolution of other particle species at temperatures much higher than $T \gtrsim 30$ MeV, and in particular close to the QCD transition phase at $T \sim 200$ MeV.

The evolution of the active–sterile neutrino system depends on the sign of Δm^2. For negative values there could be a cancellation of the two terms in Eq. (4.26) for both helicity states, leading to resonant oscillations (resonant production, RP). Instead, for $\Delta m^2 > 0$, one has the so-called nonresonant production of sterile neutrinos (NRP). In both cases, for Δm^2 higher than the value in Eq. (4.57), the actual growth of the sterile neutrino population depends on the interplay between oscillations and interactions. If the mean free path of the active neutrinos is much shorter than the matter oscillation length, the probability for conversion of an active neutrino into a sterile state becomes very small (Stodolsky, 1987). The system is completely incoherent at very high temperatures and frozen (Quantum Zeno effect; see, e.g., Bell *et al.*, 1999).

In the NRP case, it was shown in Barbieri and Dolgov, 1990 that the production probability of sterile neutrinos is

$$\Gamma_s = \langle \sin^2 2\theta_m \sin^2(\omega_{\text{osc}}t)\Gamma_a \rangle, \tag{4.59}$$

where θ_m is the mixing angle in matter and ω_{osc} the frequency of oscillations in the medium. Here Γ_a is the production rate of active neutrinos in the plasma, described in Section 4.2.2, and the averaging is done over the thermal background. If the oscillation frequency is very high, one can substitute $\sin^2(\omega_{\text{osc}}t) \simeq 1/2$, and if θ_m is small one obtains

$$\Gamma_s \approx \theta_m^2 \Gamma_a \tag{4.60}$$

for a small number density of ν_s and active neutrinos close to equilibrium. Therefore, a thermal or close-to-thermal population of sterile neutrinos is expected, provided that such a production rate of ν_s is higher than the expansion rate of the universe, a condition that holds unless either Δm^2 or θ is very small. For instance, Dolgov and Villante (2004) found that equilibrium of sterile neutrinos is completely established only for

$$\sin^4 2\theta_{es}\,\Delta m^2 \gtrsim 3 \times 10^{-5}\,\text{eV}^2. \tag{4.61}$$

This is modified for mixing with nonelectronic neutrinos, being just a rough estimate of the solution of the quantum kinetic equations.

For RP of sterile neutrinos at temperatures below T_c the resonance propagates from low to high values of the neutrino comoving momentum y, covering the whole momentum distribution while the active neutrinos are repopulated by interactions. The thermalization of ν_s is thus significantly enhanced, even for quite small values of the mixing angle.

To illustrate this discussion with an actual calculation of the active–sterile system with the kinetic equations in the two-flavour approximation (see Section 4.2.2), among the many published analyses we have chosen that described in Hannestad *et al.*, 2012a. Their results for the final extra contribution of sterile neutrinos to radiation, in the case of zero initial lepton asymmetry, are shown as isocontours of ΔN_{eff} in Fig. 4.6 as a function of the mixing parameters $\delta m_s^2 \equiv \Delta m^2$ (in eV^2 units) and $\sin^2 2\theta_s \equiv \sin^2 2\theta$. In the NRP case (top panel), one can clearly see that the same final N_{eff} corresponds to constant values of $\sin^4 2\theta\,\Delta m^2$. For RP (bottom panel), ν_s's are more efficiently brought into equilibrium, even for quite small values of the mixing angle.

It is interesting to note that for the active–sterile parameters needed to solve the oscillation anomalies described in Section 1.4.2, the thermalization of sterile neutrinos is achieved; i.e., $N_{\text{eff}} = 4$. An example of the best-fit values of a particular calculation in the $3 + 1$ neutrino model is indicated in the plot. Therefore, it seems

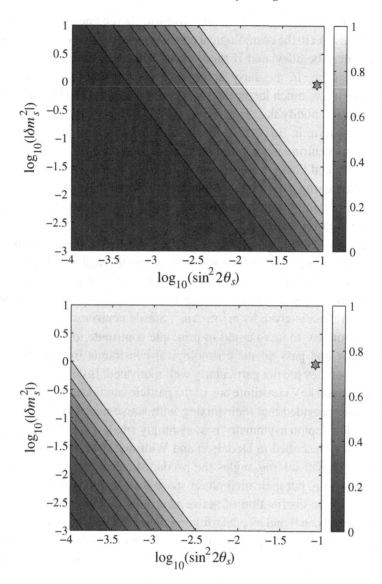

Figure 4.6 Isocountours of the final value of ΔN_{eff} in the $\sin^2 2\theta_s - \delta m_s^2$ plane for vanishing lepton asymmetry and $\delta m_s^2 > 0$ (top panel) and $\delta m_s^2 < 0$ (bottom pannel). The star denotes the best-fit mixing parameters as in the $3+1$ global fit in Giunti and Laveder, 2011c: $(\delta m_s^2, \sin^2 2\theta_s) = (0.9 \text{ eV}^2, 0.089)$. (Adapted from Hannestad *et al.*, 2012a. Courtesy of S. Hannestad, I. Tamborra, and T. Tram.)

that such extra radiation is guaranteed in these situations unless oscillations are suppressed. Moreover, we will see that this $3+1$ case may be compatible with BBN data, as well as CMB and large-scale structure (LSS) observations to be discussed in Chapters 5 and 6 (but the $3+2$ scheme can lead to too much radiation).

In any case, some tension remains in all active–sterile schemes with masses on the eV scale with respect to the cosmological constraints on the sum of neutrino masses. This tension may be alleviated if there exists some mechanism that suppresses the production of ν_s in the early universe, as in the case of a nonzero initial lepton asymmetry η_ν much larger than η_B (Foot and Volkas, 1995). Such a lepton asymmetry would modify the resonance conditions through its contribution to the matter-potential term in Eq. (4.42), in such a way that conversions are restricted for values of the neutrino momentum close to the maximum of the active neutrino distribution (Shi and Fuller, 1999). The behaviour is different in the RP (blocking of ν_s production until ν_a decoupling) and NRP (delayed thermalization) cases, but the final contribution of sterile neutrinos to N_{eff} can be significantly suppressed (see, e.g., Fig. 4 in Hannestad *et al.*, 2012a).

The results shown in Fig. 4.6 do not include the case of $|\Delta m^2| > 10$ eV^2, but the corresponding behaviour of the sterile neutrino thermalization can easily be predicted: larger squared-mass differences lead to a final ν_s spectrum with an equilibrium form unless the mixing angle is very small. For heavier masses the sterile neutrino mass is given by $m_s^2 \simeq \Delta m^2$. Sterile neutrinos in the intermediate mass range (from eV to keV) could in principle contribute to a small fraction of the dark matter and pass all the cosmological constraints from CMB and LSS observations, but they are not particularly well motivated. Instead, sterile neutrinos with masses around keV constitute one of the particle candidates for the dark matter of the universe, provided that their mixing with active neutrinos is very small. If the cosmological lepton asymmetry is vanishingly small ($L \ll 10^{-6}$), the NRP of such keV ν_s was described in Dodelson and Widrow, 1994; Dolgov and Hansen, 2002. For very small mixing angles the produced sterile neutrinos do not reach thermal equilibrium, but their final phase space distribution is well approximated by the Fermi–Dirac distribution of active neutrinos multiplied by a suppression factor that depends on θ and m_s, determining their relic density. For a larger initial asymmetry, the final phase space distribution has a more complicated shape, and the relic density is enhanced for the same value of the mass and mixing angle (Shi and Fuller, 1999). A detailed calculation of the sterile neutrino phase-space distribution and relic density was performed in these two cases by Asaka *et al.*, 2007 and Laine and Shaposhnikov, 2008. The region in (m_s, θ) space leading to the correct relic density (if all or most of DM is assumed to be in the form of sterile neutrinos) in the NRP and RP cases is displayed in Fig. 4.7. The cosmological bounds will be discussed in Section 6.5.4.

The evolution of active–sterile neutrino oscillations with small mass differences $\Delta m^2 \lesssim 10^{-7}$ eV^2, where weak interactions are irrelevant, has been calculated with exquisite numerical precision in a number of papers, both for NRP (Kirilova and Chizhov, 1998a), also with a pre-existing lepton asymmetry

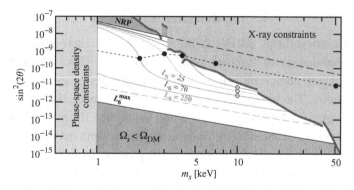

Figure 4.7 Parameter space of keV sterile neutrinos. Neutrinos of mass m_s and mixing angle θ can account for the total DM density if they are in the white region, bounded from above by the NRP case (for negligible lepton asymmetry L) and from below by the BBN bound on L. Between these bounds, neutrinos can be produced resonantly (RP), and the various lines correspond to different possible values of $L_6 \equiv 10^6 L$. A lower limit on m_s can be inferred from the Tremaine–Gunn phase space density bound (Tremaine and Gunn, 1979; Boyarsky *et al.*, 2009b), whereas upper limits are derived from a compilation of X-ray observations. (Reprinted with permission from Boyarsky *et al.*, 2009c. Copyright 2011 by the American Physical Society.)

(Kirilova and Chizhov, 1998b), and for RP (Kirilova and Chizhov, 2000). More recent analyses also included the effects of an initial abundance of sterile neutrinos (Kirilova and Panayotova, 2006).

All calculations described so far correspond to the two-neutrino limit of one active and one sterile state, but a proper calculation should also include the unavoidable presence of mixing among the active neutrinos. A four-flavour density matrix has not been performed, but a few analyses did solve simplified kinetic equations taking into account active neutrino mixing (such as Dolgov and Villante, 2004). In Section 4.4.4 we will discuss the main results.

Finally, we mention the possibility of generation of a very large lepton (neutrino) asymmetry by active–sterile neutrino oscillations, whose possible exponential rise was noticed in Barbieri and Dolgov, 1991. Its origin comes from the fact that η_ν appears with opposite sign in the matter-potential terms of neutrinos and antineutrinos. The dynamical generation of η_ν from active–sterile oscillations was described in Foot *et al.*, 1996 and developed in a series of papers (see Section 12 of Dolgov, 2002 for discussion and references). It requires very low values of the mixing angles: $\sin^2 2\theta \lesssim (2-4) \times 10^{-5}$ for eV^2 according to numerical results of Foot and Volkas, 1997; Dolgov and Villante, 2004. Such a large lepton asymmetry, in addition to affecting the active–sterile oscillations, could have important cosmological consequences, as we reviewed in Section 4.2.4.

4.3 Big Bang nucleosynthesis

Big Bang nucleosynthesis (BBN) is the epoch of the early universe when the primordial abundances of light elements, mainly ^{2}H, ^{3}He, ^{4}He and ^{7}Li, were produced. Heavier nuclei are only synthesized in stars or as a consequence of stellar explosions. The basic framework of the BBN emerged in the decade after the seminal paper by Alpher *et al.*, 1948 and since then it has become one of the observational pillars of the hot Big Bang model, providing maybe the most robust probe of the earliest phases of the universe's expansion.

We start this section with an overview of BBN dynamics, describing the transition from a hot and dense plasma of electrons, positrons, photons and nucleons in kinetic and chemical equilibrium to a universe with frozen abundances of light elements. For more details, we refer the reader to many excellent reviews on BBN (e.g., Sarkar, 1996; Olive *et al.*, 2000; Steigman, 2007; Iocco *et al.*, 2009; Pospelov and Pradler, 2010).

We have described in Section 4.1 the process of neutrino decoupling at a temperature of 2–3 MeV. At that moment, the densities of all nuclei are set by nuclear statistical equilibrium, a different name for what we called in Chapter 2 the Saha equation. Fast nuclear and electromagnetic interactions keep nuclear species in chemical equilibrium. Because there are too many photons per baryon, $\eta_B^{-1} \sim 10^9$, for temperatures of the typical order of magnitude of nuclear binding energy per nucleon (few MeV), baryon matter is all in the form of free neutrons and protons.

Consider for example the case of deuterium, which is formed via the proton–neutron fusion reaction

$$n + p \leftrightarrow {}^2\text{H} + \gamma. \tag{4.62}$$

Applying the Saha equation to this process we get – see Eq. (2.123) –

$$\frac{n_{^2\text{H}}}{n_p n_n} = \frac{3}{4} \left(\frac{2\pi \left(m_n + m_p - B_{^2\text{H}} \right)}{m_n m_p T} \right)^{3/2} \exp\left(\frac{B_{^2\text{H}}}{T} \right). \tag{4.63}$$

In the prefactor the deuterium binding energy $B_{^2\text{H}} \sim 2.2$ MeV can be neglected, whereas it is crucial to keep it in the exponential term. The $3/4$ factor is due to the different spin states, 2 for both proton and neutron and 3 for ^{2}H. Because at $T \sim$ MeV we have $n_n \sim n_p \sim n_B = \eta_B n_\gamma$, as we will see soon, it follows that

$$\frac{n_{^2\text{H}}}{n_B} \sim \eta_B \left(\frac{T}{m_\text{N}} \right)^{3/2} \exp\left(\frac{B_{^2\text{H}}}{T} \right), \tag{4.64}$$

with m_N the average nucleon mass. From this expression one finds that at $T \sim B_{^2\text{H}}$ only a fraction of order 10^{-12} of baryons are in the form of deuterium. Notice the analogy of this result with the proton–electron recombination history we considered in Chapter 2. In that case, too, because the large number of photons per proton/electron, we found that neutral hydrogen formation is delayed to temperatures lower than the hydrogen binding energy.

Higher-mass nuclei are even less abundant. Their density normalized to n_B is suppressed by higher powers of η_B. From the Saha equation for the equilibrium of $A - Z$ free neutrons, Z free protons, and the nuclei $N(A, Z)$ with binding energy $B(A, Z)$ it is easy to get the behaviour

$$\frac{n_{N(A,Z)}}{n_B} \sim A^{3/2} \eta_B^{A-1} \left(\frac{2\zeta(3)}{\pi^2}\right)^{A-1} \left(\frac{2\pi T}{m_N}\right)^{3(A-1)/2} \exp\left(\frac{B(A, Z)}{T}\right). \quad (4.65)$$

Soon after neutrinos decouple, charged-current weak interactions involving neutrons and protons also become too slow to guarantee $n - p$ chemical equilibrium. For temperatures below $T_D \sim 0.7$ MeV, the n/p density ratio departs from its equilibrium value and freezes out at the value $n/p = \exp(-\Delta m/T_D) \sim 1/6$, with $\Delta m = 1.29$ MeV the neutron–proton mass difference, and then is reduced only by neutron decays. At this stage, the photon temperature is already below the deuterium binding energy, but deuterium synthesis starts only when the photodissociation process become ineffective. This is the so-called *deuterium bottleneck*, which is overcome at $T_{BBN} \sim 0.07$ MeV, for which the ratio in (4.64) is of order one.

The onset of ^{2}H formation leads to a whole nuclear process network that produces heavier nuclei, until BBN eventually stops. As soon as deuterium forms, it is quite immediately burned into ^{4}He, which has the highest binding energy per nucleon among light nuclei. This nucleus represents the main BBN outcome, and its abundance can be quite accurately obtained by very simple arguments. Indeed, the final value $n_{^4He}$ is very weakly sensitive to the details of the nuclear network, and a very good approximation is to assume that all neutrons which have not decayed at T_{BBN} are eventually bound into helium nuclei (see, e.g., Kolb and Turner, 1994; Sarkar, 1996). This leads to the famous result for the helium mass fraction $Y_p \equiv 4 n_{^4He}/n_B$,

$$Y_p \sim \frac{2}{1 + \exp\left(\Delta m/T_D\right)\exp\left(t(T_{BBN})/\tau_n\right)} \sim 0.25, \quad (4.66)$$

with $t(T_{BBN})$ the value of time at T_{BBN} and τ_n the neutron lifetime.

A more accurate determination of the primordial abundances of ^{4}He and in particular of other light nuclei requires the solution of a set of coupled kinetic equations, supplemented by Einstein equations and covariant conservation of the total stress–energy tensor, as well as conservation of baryon number and electric charge. This is typically obtained from numerical codes from the pioneering works of Wagoner, 1969 and Kawano, 1988, 1992 to more recent ones such as

Table 4.3 *The nuclei whose abundances are usually tracked in BBN codes*

		$A - Z$							
Z	0	1	2	3	4	5	6	7	8
0		n							
1	H	^{2}H	^{3}H						
2		^{3}He	^{4}He						
3				^{6}Li	^{7}Li	^{8}Li			
4				^{7}Be		^{9}Be			
5				^{8}B		^{10}B	^{11}B	^{12}B	
6						^{11}C	^{12}C	^{13}C	^{14}C
7						^{12}N	^{13}N	^{14}N	^{15}N
8							^{14}O	^{15}O	^{16}O

PArthENoPE (Pisanti *et al.*, 2008), although nice semianalytical studies also exist (Mukhanov, 2004).

Let us summarize the general BBN setting, starting with some definitions. We consider N_{nuc} species of nuclides, whose number densities, n_i, are normalized with respect to the total number density of baryons n_{B},

$$X_i = \frac{n_i}{n_{\text{B}}} \qquad i = p,\, n,\, {}^2\text{H},\, {}^3\text{He},\, \ldots . \tag{4.67}$$

The list of nuclides which are typically considered in BBN analysis is reported in Table 4.3. To quantify the most interesting abundances, those of ^{2}H, ^{3}He, ^{4}He and ^{7}Li, it is usual to express them as

$$\frac{^2\text{H}}{\text{H}} = \frac{X_{^2\text{H}}}{X_p}, \qquad \frac{^3\text{He}}{\text{H}} = \frac{X_{^3\text{He}}}{X_p}, \qquad Y_p = 4X_{^4\text{He}}, \qquad \frac{^7\text{Li}}{\text{H}} = \frac{X_{^7\text{Li}}}{X_p}, \tag{4.68}$$

i.e., the ^{2}H, ^{3}He and ^{7}Li number densities normalized to hydrogen, and the ^{4}He mass fraction, Y_p. Indeed, Y_p is related only approximately to the *true* helium mass fraction, because the ^{4}He mass is not simply given by four times the atomic mass unit. The difference is quite small, on the order of 0.5%, because of the effect of ^{4}He binding energy. In any case, to refer to Y_p as the helium mass fraction is so common that we will adopt this notation as well.

In the temperature range of interest for BBN, few MeV $\gtrsim T \gtrsim 10\,\text{keV}$ electrons and positrons are kept in thermodynamical equilibrium with photons. Their distribution is a Fermi–Dirac function with a low chemical potential, because of the charge neutrality of the universe, $\mu_e / T \simeq \eta_{\text{B}}$. On the other hand, to follow the neutrino–antineutrino distribution in detail, it is necessary to write down evolution equations for their distribution in phase space as we did in Section 4.1, rather than simply using their energy density computed for a Fermi–Dirac distribution.

The set of differential equations ruling primordial nucleosynthesis is the following (see, e.g., Esposito *et al.*, 2000b)

$$\frac{\dot{n}_B}{n_B} = -3H \tag{4.69}$$

$$\dot{X}_i = \sum_{j,k,l} N_i \left(\Gamma_{kl \to ij} \frac{X_k^{N_k} X_l^{N_l}}{N_k! \, N_l!} - \Gamma_{ij \to kl} \frac{X_i^{N_i} X_j^{N_j}}{N_i! \, N_j!} \right) \tag{4.70}$$

$$n_B \sum_j Z_j X_j = n_{e^-} - n_{e^+} \simeq \frac{1}{3} T^2 \mu_e. \tag{4.71}$$

Equation (4.69) is the total baryon number conservation per comoving volume. The set of N_{nuc} Boltzmann equations (4.70) describes the density evolution of each nuclide species and Eq. (4.71) states the cosmological charge neutrality in terms of the electron chemical potential, with Z_j the electric charge of nucleus j. Additional required equations are the expression of the Hubble parameter during radiation domination, Eq. (2.156), the covariant conservation of the total energy density, Eq. (2.48), and the Boltzmann equations (4.4) for the neutrino spectra or their generalization for density matrices to account for neutrino flavour oscillations.

Because electromagnetic and nuclear scattering keep the nonrelativistic baryons in kinetic equilibrium, their energy density ρ_B and pressure P_B are given by equilibrium values

$$\rho_B = \left[M_u + \sum_i \left(\Delta M_i + \frac{3}{2} T \right) X_i \right] n_B \tag{4.72}$$

$$P_B = T n_B \sum_i X_i, \tag{4.73}$$

with ΔM_i and M_u the ith nuclide mass excess and the atomic mass unit, respectively.

In Eq. (4.70), i, j, k, l denote nuclear species, N_i the number of nuclides of type i entering a given reaction (and analogously N_j, N_k, N_l), whereas the Γ's stand for the reaction rates. For example, in the case of decay of the species i, $N_i = 1$, $N_j = 0$ and $\sum_{kl} \Gamma_{i \to kl}$ is the inverse lifetime of the nucleus i. For two–body collisions, $N_i = N_j = N_k = N_l = 1$ and $\Gamma_{ij \to kl} = \langle \sigma_{ij \to kl} \, v \rangle$, the thermal averaged cross section for the reaction $i + j \to k + l$ times the $i - j$ relative velocity.

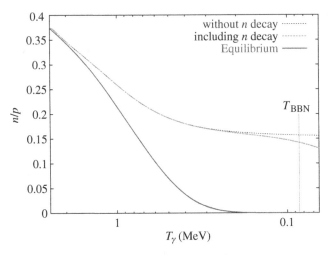

Figure 4.8 Evolution of the neutron-to-proton number density ratio at MeV temperatures, with and without neutron decays, compared to its equilibrium value. T_{BBN} indicates the approximate onset of ^{2}H synthesis.

4.3.1 Neutron–proton chemical equilibrium

Neutrons and protons are kept in chemical equilibrium by weak interactions, through the charged-current processes we already saw in Chapter 1,

$$
\begin{aligned}
&\text{(a)} \quad \nu_e + n \to e^- + p \quad &&\text{(d)} \quad \bar{\nu}_e + p \to e^+ + n \\
&\text{(b)} \quad e^- + p \to \nu_e + n \quad &&\text{(e)} \quad n \to e^- + \bar{\nu}_e + p \\
&\text{(c)} \quad e^+ + n \to \bar{\nu}_e + p \quad &&\text{(f)} \quad e^- + \bar{\nu}_e + p \to n,
\end{aligned}
\tag{4.74}
$$

which enforce their number density ratio to follow the equilibrium value $n_n/n_p \equiv n/p = \exp(-\Delta m/T)$, as dictated by the Saha equation. For $T \gtrsim \Delta m \simeq \text{MeV}$, $n_n \sim n_p \sim n_B$. Shortly before the onset of BBN, processes (a)–(f) become too slow, chemical equilibrium is lost and the ratio n/p freezes out for temperatures below the decoupling temperature $T_D \sim 0.7 \text{ MeV}$, as shown in Fig. 4.8. The value of T_D is only indicative, and is obtained by requiring that the interaction rates equal the Hubble parameter. Residual free neutrons are partially depleted by decay until deuterium starts forming at T_{BBN} and neutrons get bound in nuclei, first in deuterium and eventually in ^{4}He.

The fact that processes (a)–(f) fix the neutron fraction, and thus Y_p, means that to get an accurate theoretical prediction for Y_p requires a careful treatment of the weak rates. Large improvements on this issue have been obtained in the last two decades, which we briefly summarize in the following. Extensive analysis can be found in, e.g., Esposito *et al.*, 1999; Lopez and Turner, 1999.

At the lowest order, the calculation is rather straightforward, and is obtained by using V–A theory and in the limit of infinite nucleon mass (the Born limit); see (Weinberg, 1972). The latter approximation is justified in view of the typical energy scale of interest, of order $T \sim$ MeV, much smaller than m_N. For example, the neutron decay rate in the massless neutrino limit takes the form

$$\omega_B(n \to e^- + \bar{\nu}_e + p) = \frac{G_F^2}{2\pi^3}(C_V^2 + 3C_A^2) \int_{m_e}^{\Delta m} dE_e E_e^2 E_\nu p_\nu$$

$$\times [1 - f_{\bar{\nu}_e}(p_\nu, T_\nu)][1 - f_e(p_e, T)], \quad (4.75)$$

where $p_\nu = \Delta m - \sqrt{p_e^2 + m_e^2}$. It is just the expression of the inverse neutron lifetime, but for a neutron decaying in a bath of electrons and antineutrinos, which introduce the effect of the two Pauli blocking factors. The rates for all other processes in Eq. (4.74) can be obtained from (4.75) simply by changing (i) the statistical factors, (ii) the expression for neutrino energy in terms of the electron energy and (iii) electron energy integration limits.

Neglecting the small effects of distortion in neutrino–antineutrino distribution functions but taking into account the time evolution of the neutrino-to-photon temperature ratio T_ν/T, the thermal averaged rates we can compute numerically. In Fig. 4.9 we show the total Born rates, ω_B, for $n - p$ processes, where

$$\omega_B(n \to p) = \omega_B(n \to e^- + \bar{\nu}_e + p) + \omega_B(e^+ + n \to \bar{\nu}_e + p)$$

$$+ \omega_B(\nu_e + n \to e^- + p)$$

$$\omega_B(p \to n) = \omega_B(e^- + \bar{\nu}_e + p \to n) + \omega_B(\bar{\nu}_e + p \to e^+ + n)$$

$$+ \omega_B(e^- + p \to \nu_e + n). \quad (4.76)$$

The accuracy of the results in the Born approximation is, at best, on the order of 9%, as we estimated in Chapter 1 comparing the prediction of Eq. (1.47) for the neutron lifetime at very low temperatures with the experimental value $\tau_n^{\exp} = (880.1 \pm 1.1)$ s (Beringer *et al.*, 2012). Note, however, that this average is currently obtained by combining measurements of τ_n which are not compatible with each other, the lowest value being $\tau_n^{\exp} = (878.5 \pm 0.7 \pm 0.3)$ s (Serebrov *et al.*, 2005), so the situation should be cleared up by further experimental efforts.

The Born calculation can be improved by considering different types of corrections, which include electromagnetic radiative, finite-nucleon-mass and finite-temperature radiative corrections, as reviewed in Iocco *et al.*, 2009. Their effect on the weak rates discussed so far has been considered in detail in Esposito *et al.*, 1999; Lopez and Turner, 1999; Serpico *et al.*, 2004. The leading contributions are given by electromagnetic radiative corrections, which decrease monotonically

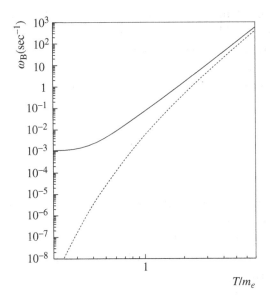

Figure 4.9 The total Born rates, $\omega_B(n \to p)$ (solid line) and $\omega_B(p \to n)$ (dashed line). (Reprinted from Iocco *et al.*, 2009, with permission from Elsevier.)

with increasing temperature for both $p \to n$ and $n \to p$ processes, and by finite-nucleon-mass effects. Their sum changes the Born estimate for a few percent correction at the freeze-out temperature T_D.

Actually, radiative corrections sensibly improve the agreement between the theoretical prediction for the neutron lifetime at zero temperature and its experimental value. The order 9% difference between the present value of τ_n^{exp} and the Born result is reduced to 0.5%. Therefore, we might be confident that all $n - p$ weak rates are quite accurately computed, say at less than the percent level (slightly worse if one uses the neutron lifetime of Serebrov *et al.*, 2005).

Let us now describe the influence of the nonthermal distortions calculated in Section 4.2.3 on the weak rates in Eq. (4.74). Naively, one would expect that distortions of the neutrino spectra at a percent level would result in an effect of similar size. However, this does not occur, for the following reason. An excess of neutrinos at the high-energy tail of the spectrum results in more destruction of neutrons in reaction (a) and their production in reaction (d). The nonequilibrium contribution to the second process is more efficient because the number density of protons at $T \approx 0.7$ MeV is six to seven times higher than that of neutrons. This results in an increase of the frozen neutron-to-proton ratio n/p, and in a corresponding increase of ^{4}He. On the other hand, an excess of neutrinos at low energies produces a decrease of n/p because reaction (d) is suppressed by threshold effects. Moreover, an overall increase of the neutrino energy density leads to a lower freezing temperature of the weak processes, which also decreases n/p. The combination of all effects causes only a minor change in the primordial abundance of ^{4}He, on the order of 10^{-4}, as listed in Table 4.2.

Finally, consider electromagnetic finite-temperature corrections. These are due to interaction processes with on-shell (real) photon or $e^{\pm}$ emission and absorption from the thermal bath. Apart from modifying the neutron–proton chemical equilibrium through the shape of the $e^{\pm}$ distribution function, these effects are also slightly changing the energy density and pressure of the electromagnetic component, $e^{\pm}$ and photons and thus the expression of the expansion rate H. It has been shown that at the time of BBN, all these effects are quite small, at the level of per mill on the ^{4}He abundance (Lopez and Turner, 1999).

4.3.2 The nuclear network

Nuclear processes during the BBN proceed in a hot but low-nucleon-density plasma with a significant population of free neutrons, which expands and cools down very rapidly. These conditions lead to peculiar *out-of-equilibrium* nuclear burning. In particular, the low density is responsible for the suppression of three–body reactions and an enhanced effect of the Coulomb barrier, which as a matter of fact inhibits any reaction with interacting nuclear charges $Z_1 Z_2 \gtrsim 6$. Along with the lack of tightly bound nuclides with $A = 5 - 8$ which could have worked as a bridge to the synthesis of heavier and more stable isotopes, such as ^{12}C, this is the reason BBN is producing low-mass nuclei only, mainly ^{4}He. We recall that in stars, in a much higher-density environment, ^{12}C is produced via the triple-α process, which is very strongly suppressed during BBN.

The most efficient categories of reactions are proton, neutron and deuterium captures (p, γ), (n, γ), (d, γ), charge exchanges (p, n) and proton and neutron stripping (d, n), (d, p)[6]. The leading nuclear reactions are graphically summarized in Fig. 4.10 and listed in Table 4.4, as identified in Smith *et al.*, 1993.

Apart from a few cases, such as the deuterium formation process R_1, and of course R_0, for which there are reliable theoretical calculations, nuclear rates are typically obtained by experiment. They have been critically reexamined and collected by the NACRE collaboration.[7] New and very accurate rate determinations have been also quite recently obtained in the low center of mass energy range, below MeV, one example being the determination of the ^{3}He$(\alpha, \gamma)^7$Be of Costantini *et al.*, 2009. This has been possible by a sensible reduction of backgrounds by performing experiments in underground laboratories. Recent reviews

---

[6] We use here the standard notation adopted by nuclear physicists. For example, $p(n, \gamma)d$ denotes the neutron–proton fusion reaction $n + p \leftrightarrow {}^2$H $+ \gamma$.

[7] http://pntpm.ulb.ac.be/Nacre/nacre.htm.

Table 4.4 *The BBN reactions of Fig. 4.10*

Symbol	Reaction	Symbol	Reaction
R_0	τ_n	R_8	$^3\mathrm{He}(\alpha, \gamma)^7\mathrm{Be}$
R_1	$p(n, \gamma)d$	R_9	$^3\mathrm{H}(\alpha, \gamma)^7\mathrm{Li}$
R_2	$^2\mathrm{H}(p, \gamma)^3\mathrm{He}$	R_{10}	$^7\mathrm{Be}(n, p)^7\mathrm{Li}$
R_3	$^2\mathrm{H}(d, n)^3\mathrm{He}$	R_{11}	$^7\mathrm{Li}(p, \alpha)^4\mathrm{He}$
R_4	$^2\mathrm{H}(d, p)^3\mathrm{H}$	R_{12}	$^4\mathrm{He}(d, \gamma)^6\mathrm{Li}$
R_5	$^3\mathrm{He}(n, p)^3\mathrm{H}$	R_{13}	$^6\mathrm{Li}(p, \alpha)^3\mathrm{He}$
R_6	$^3\mathrm{H}(d, n)^4\mathrm{He}$	R_{14}	$^7\mathrm{Be}(n, \alpha)^4\mathrm{He}$
R_7	$^3\mathrm{He}(d, p)^4\mathrm{He}$	R_{15}	$^7\mathrm{Be}(d, p)2\,^4\mathrm{He}$

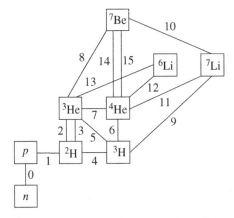

Figure 4.10 The most relevant reactions for primordial nucleosynthesis.

and analysis of the whole nuclear network can be found in, e.g., Cyburt, 2004; Serpico *et al.*, 2004; Iocco *et al.*, 2009.

In general, although improvements in the nuclear reaction rates would still sharpen the BBN predictions, it is fair to conclude that currently, the experimental errors on the leading processes of Table 4.4 propagate to uncertainties on the most interesting X_i which are below the statistical and systematic uncertainty on the experimental measurements of primordial abundances from astrophysical observations, which should be compared with theoretical predictions.

Let us briefly follow the BBN nuclear path. After the freeze-out of weak interactions, $p(n, \gamma)^2\mathrm{H}$ is the only reaction able to synthesize sensible amounts of nuclei, because it involves the only two nuclear species with nonvanishing abundances, protons and neutrons. As we have seen, the deuterium bottleneck is overcome at a temperature of $T_{\mathrm{BBN}} \sim 70$ keV. At this point, a large fraction of all the available neutrons is locked into deuterium nuclei. The effect of the bottleneck on the

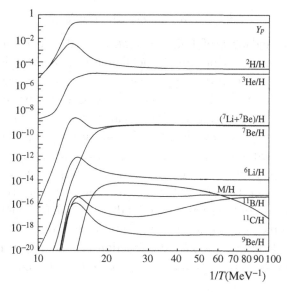

Figure 4.11 Evolution of the abundances of light elements produced during BBN for a baryon density $\Omega_B h^2 = 0.022$. The line M/H refers to the abundance of *metals* (nuclei with $Z > 4$) such as C, N and O. (Reprinted from Iocco *et al.*, 2009, with permission from Elsevier.)

following nuclear processes is twofold. On one hand, the delayed population of deuterium results in BBN taking place at relatively low temperatures, with consequences for the efficiency of all the following reactions. Furthermore, neutrons continue to decay till T_{BBN}, changing the n/p at the time of effective ^{2}H production.

Efficient production of deuterium marks the beginning of the nuclear phase of BBN. The formation of heavier elements such as ^{3}H, ^{3}He, ^{4}He proceeds through the interaction of deuterium nuclei with nucleons and other ^{2}H nuclei, followed by ^{3}H and ^{3}He nucleon and deuterium capture processes. ^{7}Li is produced via tritium and, for a sufficiently high baryon density, especially through ^{3}He radiative capture on ^{4}He. The latter path leads to ^{7}Be, which then decays into ^{7}Li by electron capture.

The evolution of light nuclide abundances as a function of the temperature of the plasma is shown in Fig. 4.11. They rise quite steeply once deuterium forms and, by the time the universe cools to a few tens of keV, they reach their final values. Notice also that all ^{7}Li is produced by ^{7}Be electron capture processes for the chosen value of $\Omega_B h^2 \simeq 0.022$. For lower baryon density, $\Omega_B h^2 \lesssim 0.01$, the main ^{7}Li production channel is rather ^{3}H$(\alpha, \gamma)^7$Li. From this figure one can also see that elements such as C, N, O, in astrophysicist jargon "the metals", are basically absent from the chemical composition of the early universe, an additional robust prediction of BBN.

4.3.3 Light-element observations

The primordial abundances of light elements are inferred from measurements performed in a large variety of astrophysical environments. Despite the increased number and precision of spectroscopic observations over the past two decades, at present the situation is still quite involved owing to the presence of relevant systematic errors. Unfortunately, these errors are to a large extent irreducible.

Our nearby universe is strongly evolved with respect to primordial chemical composition. The methods proposed and developed to infer primordial yields have focused either on very old and hence little evolved astrophysical regions, or on the capability to correct for the effect of the galactic evolution on the BBN abundances. For example, because of their very weak binding, ^{2}H nuclei contained in prestellar nebulae are burned out during their collapse. Hence, the post-BBN deuterium evolution is expected to be a monotonic function of time, and astrophysical deuterium measurements can be assumed to represent lower bounds of its primordial abundance.

Such a simple scheme cannot be applied to the more tightly bound ^{3}He nucleus, which can either be produced by ^{2}H burning or be destroyed in hotter regions. For this reason the inferred primordial ^{3}He abundance is intrinsically a model-dependent quantity.

Hydrogen burning in the stellar population has increased the amount of ^{4}He as well as *metals* such as C, N and O. Usually, ^{4}He is measured in old and very little evolved systems versus their metallicity, and extrapolating *linearly* to zero metallicity. Although this is the best approach to date, it leads to some systematic uncertainty.

Finally, ^{7}Li is a very weakly bound nuclide which has an extremely involved post-BBN chemical evolution. It is easily destroyed in the interiors of stars but can survive in the cooler outer layers of stars with shallow convective zones, where it can be measured by means of absorption spectra. However, the scenario is much more involved and likely the observed abundances are not reflecting the primordial BBN values.

$$^2H$$

It is believed that there are no astrophysical sources of deuterium because it is destroyed by stellar evolution processes. Any astrophysical observation can therefore provide a lower bound for its primordial abundance, such as those of the local interstellar medium (ISM) in the Milky Way (Linsky *et al.*, 2006). Inside the Local Bubble (<100 pc from the Sun) the deuterium-to-hydrogen ratio is roughly constant, ^{2}H/H $= (1.56 \pm 0.04) \times 10^{-5}$ (Wood *et al.*, 2004), although an unexpected scatter of a factor of 2 is observed.

The best method to date for deuterium measurements is the observation of absorption lines due to intervening hydrogen-rich clouds along the line of sight of background quasars at high redshifts, called Quasar Absorption Systems (QAS) (Adams, 1976). Conventional models of galactic chemical evolution predict a small contamination of the primordial ^{2}H/H in this case (Fields, 1996). However, a successful implementation of this method requires a suitable neutral hydrogen column density, low metallicity to reduce the chances of deuterium astration and finally, a low internal velocity dispersion of the gas for good resolution of the hydrogen/deuterium isotope shift (Pettini *et al.*, 2008). For this reason, there are only a few measurements which can be used for the determination of primordial deuterium. An average value of the ^{2}H/H ratio was obtained in Iocco *et al.*, 2009 using seven determinations from different quasar absorption systems,

$$^2\text{H/H} = (2.87 \pm 0.22) \times 10^{-5}, \tag{4.77}$$

following the more conservative approach of the results of (Pettini *et al.*, 2008).

A different proposed method for deuterium measurement is based on the fluctuations in the absorption of CMB photons by neutral gas at very high redshifts $z \approx 7\text{--}200$, and on the strength of the cross correlation of brightness-temperature fluctuations due to resonant absorption of CMB photons in the 21-cm line of neutral hydrogen and in the 92-cm line of neutral deuterium. Although technically challenging, this measurement could provide in the future a very clean determination of ^{2}H/H (Sigurdson and Furlanetto, 2006).

^3He

There is only one spectral transition which allows for neat detection of ^{3}He, namely the 3.46 cm spin-flip transition of ^{3}He$^+$, the analog of the widely used 21-cm line of hydrogen. This is a powerful tool for identification of this isotope, because there is no corresponding transition in ^{4}He$^+$. The emission is quite weak; hence ^{3}He has been observed outside the solar system only in a few HII regions and planetary nebulae (PN) in the Galaxy. The values found in PN are one order of magnitude larger than for protosolar material and local ISM determinations, in agreement with the expectation of a net stellar production of ^{3}He in at least some stars. From the correlation between the metallicity of the particular galactic environment and its distance from the center of the galaxy, one should measure a gradient in ^{3}He abundance versus metallicity and/or distance. However, such a correlation is not observed (Bania *et al.*, 2002). Because of this unexpected result, currently only an upper limit to the primordial abundance of ^{3}He is available, using the observations of a peculiar galactic HII region (Bania *et al.*, 2002):

$$^3\text{He/H} < (1.1 \pm 0.2) \times 10^{-5}. \tag{4.78}$$

A different approach is based on the observation that the ratio $(^2H + {}^3He)/H$ shows a high level of stability during galactic evolution. This is partially supported by observations and chemical evolution models; see for example Geiss and Gloeckler, 2007. In this case, using the value reported in (Geiss and Gloeckler, 1998) for protostellar material, namely $(^2H + {}^3He)/H = (3.6 \pm 0.5) \times 10^{-5}$, and the estimate of deuterium of Eq. (4.77), one finds

$$^3He/H = (0.7 \pm 0.5) \times 10^{-5}, \tag{4.79}$$

which is consistent with (4.78).

^4He

Primordial ^4He is enriched by stellar burning. This process is correlated with the metallicity (C, N and O yields) of the particular astrophysical environment, because heavier nuclei are also a product of nuclear processes inside sufficiently massive stars. This observation is at the basis of the most well-established technique for deriving the primordial value of ^4He mass fraction. Because the universe was born with negligible amounts of nuclei heavier than ^7Li, as we already mentioned in the previous section, Y_p can be derived by extrapolating the $Y_p - O/H$ and $Y_p - N/H$ correlations to zero metallicity for a sample of objects, and correcting for model-dependent star evolutionary effects (although all stars produce ^4He, heavier element abundances depend on their mass).

The key data for inferring ^4He primordial abundance are provided by observations of helium and hydrogen emission lines from the recombination of ionized hydrogen and helium in low-metallicity extragalactic HII regions (Steigman, 2007). Typically, dwarf irregular and blue compact galaxies, the least chemically evolved known galaxies, are used as the most natural environment. They are thought to contain very little helium synthesized in stars after the BBN, minimizing the chemical evolution problem; see, e.g., Izotov and Thuan, 2004.

Despite the conceptual simplicity of the method, the ^4He mass fraction so derived is affected by statistical and (mainly) systematic uncertainties. Statistical uncertainties are easier to deal with. They can be decreased by looking for very high signal-to-noise ratio helium spectra. On the other hand, all recent analyses of Y_p agree that the systematic error is the main factor responsible for the spread of the Y_p determinations, although different authors report different error budgets, sometimes analyzing the same objects. The use of new He$_I$ recombination coefficients (Porter *et al.*, 2007), has remarkably changed the predictions by pushing them up a few percent, as shown in Izotov *et al.*, 2007; Peimbert *et al.*, 2007. In particular, in Peimbert *et al.*, 2007, the uncertainty quoted for the value of Y_p seems to provide a realistic estimate (± 0.003) of the residual indetermination affecting the ^4He mass fraction. Based on this analysis of the error budget, we adopt as a reference value in the following the estimate of (Iocco *et al.*, 2009):

$$Y_p = 0.250 \pm 0.003. \tag{4.80}$$

It should be noted, however, that apart from the error issue, the central value of Y_p is also debated. Recently, new studies of metal-poor HII regions have appeared in the literature (Aver *et al.*, 2010, 2011, 2012; Izotov and Thuan, 2010). Although these groups both agree on a central value higher than the result of Eq. (4.80), a different estimate of possible systematic effects which dominate the total uncertainty budget is given in Izotov and Thuan, 2010 and Aver *et al.*, 2010, with the second quoting a larger error, on the order of 4%:

$$Y_p = 0.2565 \pm 0.0010 \, (\text{stat.}) \pm 0.0050 \, (\text{syst.}) \tag{4.81}$$

$$Y_p = 0.2561 \pm 0.0108. \tag{4.82}$$

As we will see, such a high value of Y_p may have a big impact on the determination of N_{eff} from BBN. We also mention that in Aver *et al.*, 2012, the Markov chain Monte Carlo method was exploited to determine the ^{4}He abundance and the uncertainties derived from observations of metal-poor nebulae, finding

$$Y_p = 0.2534 \pm 0.0083, \tag{4.83}$$

which is fairly well in agreement with the typical standard prediction of BBN, $Y_p \sim 0.250$ (see later), also because of the generous uncertainty estimate.

Finally, we briefly point out that other constraints on Y_p can be obtained by other indirect methods. For example, in Salaris *et al.*, 2004, Y_p was bounded from studies of galactic globular clusters, $Y_p \lesssim 0.250 \pm (0.006)_{\text{stat}} \pm (0.019)_{\text{sys}}$. CMB anisotropies are also sensitive to ^{4}He, through its role in the recombination and reionization histories. Although the combination of present CMB experiments, WMAP, ACBAR and QUaD, provides marginal detection of primordial helium, $Y_p = 0.326 \pm 0.075$ at 68% C.L. (Komatsu *et al.*, 2011), even with the PLANCK experiment, the error bars from CMB will be larger than the present systematic spread of the astrophysical determinations (see, e.g., Trotta and Hansen, 2004; Hamann *et al.*, 2008).

^7Li

The two stable isotopes of lithium, ^{6}Li and ^{7}Li, continue to puzzle astrophysicists and cosmologists who try to reconcile their primordial abundance as inferred from observations with the BBN predictions (Fields, 2011). From the astrophysical point of view the questions mainly concern the observation of lithium in cold interstellar gas and in all type of stars in which lithium lines are either detected or potentially detectable. A chance to link primordial ^{7}Li with the BBN abundance was first proposed by Spite and Spite, 1982, who showed that the lithium abundance in the warmest metal-poor dwarfs was independent of metallicity. The constant lithium abundance defining what is commonly called *the Spite plateau* suggested that this may be the lithium abundance in pre-galactic gas provided by the BBN.

Several technical and conceptual difficulties have been responsible for a long list of ^{7}Li determinations (see, e.g., Iocco *et al.*, 2009). A simple unweighted average and half-width of the distribution of data gives

$$^7\text{Li/H} = \left(1.86^{+1.30}_{-1.10}\right) \times 10^{-10} \qquad (4.84)$$

which, as we will see, disagrees by almost a factor of 2–3 with the theoretical predictions. Currently, this result is thought to be related to the fact that observed values are not primordial.[8]

4.3.4 Theory vs. data

In the standard BBN model, the only free parameter is the value of the baryon-to-photon number density, η_B, or equivalently, the baryon energy density parameter $\Omega_B h^2$, barring values of N_{eff} different than the reference value 3.046. We will come back to this important issue in the next section.

We present in Fig. 4.12 the dependence of the final value of the primordial abundances of the four main nuclides on $\eta_{10} \equiv 10^{10}\eta_B$, calculated using the numerical code PArthENoPE (Pisanti *et al.*, 2008), along with the experimental values of the abundances and their corresponding uncertainties, as discussed in the previous section. One can see that Y_p is less sensitive to the value of the baryon density (note the linear scale). On the other hand, the abundances of both ^{2}H and ^{3}He are steeply decreasing functions of η_B. The nonmonotonic behaviour of ^{7}Li is due to the fact that at lower η_B this nuclide is mainly produced by ^{3}H$(\alpha, \gamma)^7$Li, but as η_B grows, ^{3}He$(\alpha, \gamma)^7$Be, followed by ^{7}Be decay by electron capture, takes over and the lithium abundance starts growing again.

The theoretical expectations for ^{2}H and ^{4}He agree quite well with the experimental bands, but the ^{7}Li observations lie a factor of 2–3 below the BBN prediction. Apart from this problem, which is likely to be ascribed to the nonprimordial nature of ^{7}Li measurements, it is quite remarkable that a single parameter, η_B,

---

[8] The very recent status of the ^{7}Li *problem* was discussed at the topical workshop *Lithium in the cosmos*, http://www.iap.fr/lithiuminthecosmos2012/index.html, whose proceedings will be published in *Memorie della Società Astronomica Italiana – Supplementi*.

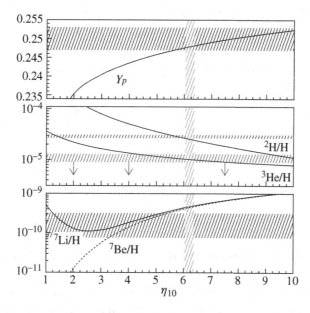

Figure 4.12 Values of the primordial abundances as a function of $\eta_{10} \equiv 10^{10}\eta_B$ in the standard BBN scenario. The hatched bands represent the experimental determination with 1σ statistical errors on Y_p, ^{2}H/H, and ^{7}Li/H, whereas the band with the arrows is the upper bound on ^{3}He/H from Eq. (4.78). The vertical band represents the WMAP 7-year result $\Omega_B h^2 = 0.02258 \pm 0.00057$ (68% C.L.). (Reprinted from Iocco et al., 2009, with permission from Elsevier.)

can fit nuclear abundances which span several orders of magnitude, and that its value is in fair agreement with a completely independent measurement from CMB anisotropies, shown as a vertical band (68% C.L. range) in Fig. 4.12.

The value of η_B is mainly fixed by deuterium. With present data the preferred value is $\Omega_B h^2 = 0.021 \pm 0.002$ at 95% C.L., compatible at 2σ with the WMAP 7-year result $\Omega_B h^2 = 0.02258 \pm 0.00057$ (Komatsu et al., 2011; Larson et al., 2011). As a test of consistency of BBN, it would be highly desirable to have a cleaner determination of ^{3}He, which is also strongly dependent on baryon density.

In Table 4.5 we report the values of abundances for the WMAP-7 99% C.L. range, evaluated using PArthENoPE (Pisanti et al., 2008). Notice the value of ^{3}He, which almost saturates the upper bound of Eq. (4.78).

4.4 Bounds on neutrino properties from Big Bang nucleosynthesis

Cosmological neutrinos influence the primordial production of light elements in two ways. First of all, they contribute to the radiation energy density that fixes the

Table 4.5 *The theoretical values of the main primordial abundances for $\Omega_B h^2$ in the WMAP-7 99% C.L. range, evaluated using* PArthENoPE

$\Omega_B h^2$	0.02258 ± 0.00171
Y_p	$0.2487^{+0.0007}_{-0.0008}$
$^2\mathrm{H/H} \times 10^5$	$2.56^{-0.28}_{+0.34}$
$^3\mathrm{He/H} \times 10^5$	1.02 ∓ 0.05
$^7\mathrm{Li/H} \times 10^{10}$	$4.65^{+0.75}_{-0.71}$

Source: Pisanti *et al.*, 2008.

expansion rate of the universe during BBN, a background effect parameterized by N_{eff}. Actually, this parameter encodes the contribution of all relativistic species in addition to photons, their number and distribution in phase space. It is mainly measured by the primordial abundance of ^{4}He. In a next-to-minimal BBN model, one can leave η_B and N_{eff} as free parameters.

A more direct role in BBN is played by electron neutrinos and antineutrinos, which are involved in the charged-current weak processes which rule the neutron/proton chemical equilibrium. Any change in the ν_e or $\bar{\nu}_e$ momentum distribution can shift the n/p ratio freeze-out temperature and in turn modify the primordial ^{4}He abundance. As an example, if there is an electron-antineutrino neutrino asymmetry parameterized by the ν_e chemical potential μ_e, the chemical equilibrium condition reads $n/p \sim \exp(-\Delta m/T - \mu_e/T)$.

Today, we know from oscillation experiments and tritium β-decay the neutrino mass range and mixing angles, and we have seen that flavour oscillations before the onset of BBN mix the three active neutrino distributions. The important news for BBN is that these pieces of information have greatly simplified the phenomenology. A plethora of cases once popular in the literature are now excluded. Among the ones which were of remarkable interest only a decade ago we can mention (i) a lower-than-three effective number of neutrinos due to a "massive ν_τ" (improper, ν_τ not being a mass eigenstate); (ii) a decaying ν_τ; (iii) the thermalization of right-handed neutrinos (for the Dirac mass case), which is inhibited by the smallness of the neutrino masses, which couple these sterile states to the active components. We do not treat these issues further and direct the interested reader to these historical topics to the review (Dolgov, 2002). Rather, we will focus on the BBN bounds on electromagnetic interactions and other nonstandard properties of neutrinos. The last part of this section is devoted to the impact on BBN of sterile neutrinos, which also shows a rich phenomenology.

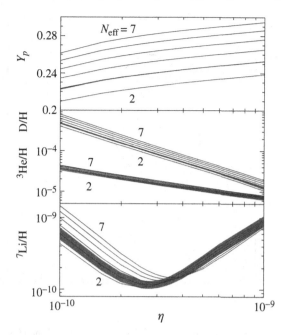

Figure 4.13 BBN abundance predictions as a function of the baryon-to-photon ratio $\eta = \eta_B$, varying the radiation content from $N_{\text{eff}} = 2$ to 7. (Reprinted from Cyburt *et al.*, 2002, with permission from Elsevier.)

4.4.1 Extra relativistic degrees of freedom

We have seen in the previous sections that the BBN outcome depends on the expansion rate at sub-MeV temperatures. In a universe dominated by radiation, this is given by Eq. (2.156), which as we saw can be parameterized in terms of the effective number of neutrino species N_{eff} as defined in Eq. (2.198). The effect of a varying N_{eff} on the primordial production of light nuclei is depicted in Fig. 4.13, where one can see that all abundances are modified, but especially the value of Y_p, because a larger N_{eff} produces a speed-up of universe expansion, which in turn leads to an earlier n/p freeze-out, a larger relic neutron abundance, and thus, a higher yield of ^{4}He. This fact was used in the pioneering analyses (Shvartsman, 1969; Steigman *et al.*, 1977) to get the first bounds on N_{eff} from BBN.

A theoretical analysis of BBN less conservative than in Section 4.3.4 can be performed by relaxing the hypothesis of a standard number of relativistic degrees of freedom. In such a case we have two free parameters, N_{eff} and η_B, and it is possible to obtain bounds on the largest (or smallest) amount of radiation present at the BBN epoch, in the form of decoupled relativistic particles, or nonstandard features of active neutrinos.

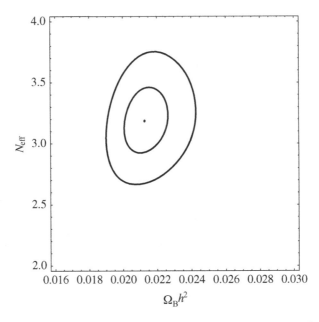

Figure 4.14 Allowed regions at 68% and 95% C.L. in the plane $(\Omega_B h^2, N_{\mathrm{eff}})$. The ^{2}H and Y_p are chosen as in Eqs. (4.77) and (4.80).

An example of this kind of study is shown in Fig. 4.14, where allowed regions at 68% and 95% C.L. are displayed in the plane $(\Omega_B h^2, N_{\mathrm{eff}})$, which is fairly compatible with the prediction of the standard cosmological model $N_{\mathrm{eff}} = 3.046$. In this case both ^{2}H and Y_p data have been used as from Eqs. (4.77) and (4.80).

Although the value of $\Omega_B h^2$ is mainly fixed by deuterium, the bound on N_{eff} is due to the assumed value of Y_p. Three active neutrinos and a baryon density of $\Omega_B h^2$ in the WMAP range give $Y_p \simeq 0.25$. If analysis of experimental data suggesting a higher ^{4}He mass fraction is confirmed, as discussed in Aver *et al.*, 2010, 2011, 2012; Izotov and Thuan, 2010, then the preferred value of N_{eff} shifts to higher values, corresponding to a *half* extra effective neutrino. Yet it is fair to conclude that, because of systematic errors on Y_p, to date there is no statistically significant evidence in favour of a nonstandard amount of relativistic species during BBN.

Because the reliability of bounds on N_{eff} depends on how accurate the astrophysical determinations of Y_p are, any complementary study that can provide independent limits on the radiation content is very welcome. An example that we will discuss in Chapter 5 is the analysis of CMB anisotropies and other late cosmological data with a free N_{eff}. Alternatively, one could also derive the BBN bounds on N_{eff} from deuterium data and $\Omega_B h^2$ fixed by WMAP observations, or consider very weak assumptions on the astrophysical determination of the helium abundance, as performed in Mangano and Serpico, 2011. These authors showed that a robust upper limit $N_{\text{eff}} \leq 4$ at 95% C.L. is found, assuming that the minimum effect of stellar processing is to keep constant (rather than increase, as expected) the helium content of a low-metallicity gas.

The BBN bound on the relativistic degrees of freedom is one of the most important cosmological tests that any model with nonstandard physics must pass. In most cases, such models predict a larger radiation content, which may contradict this bound on N_{eff}, an example being the case of extra sterile neutrino species, to be discussed later. It is also worth mentioning at least one case which features $N_{\text{eff}} < 3$: in a scenario where the reheating of the universe after inflation takes place at temperatures in the MeV range, it can happen that there is not enough time for neutrinos to acquire thermal spectra from interactions with $e^{\pm}$. Effectively, this translates into a smaller contribution of neutrinos to radiation than in the standard case. From a BBN analysis of this scenario, a bound on the lowest possible reheating temperature $T_{\text{RH}} \gtrsim 2\text{--}4\,\text{MeV}$ was obtained at 95% C.L. (Hannestad, 2004; Ichikawa *et al.*, 2005).

4.4.2 Relic neutrino asymmetries

In Section 4.2.4 we discussed a scenario where the total lepton number of the universe, which must be in the form of neutrinos, is much larger than the baryon asymmetry. We also saw how the effect of neutrino oscillations modifies the flavour distribution of the neutrino asymmetries, reaching frozen values at the beginning of BBN. Here we describe how one can constrain neutrino asymmetries with observations of primordial abundances.

Let us start by describing the effects of neutrino asymmetries η_{ν_α} on BBN neglecting, for the moment, neutrino mixing. This case provides a nice example of the two ways neutrinos can change the production of light elements. First of all, neutrino asymmetries make a positive extra contribution to their energy density, given in terms of the parameter N_{eff} in Eq. (4.51), which speeds up the expansion rate. This background effect does not depend on neutrino flavour and, as can be seen in Fig. 4.13, modifies all final abundances and in particular that of ^4He.

The second effect appears only if a nonzero $\nu_e\text{--}\bar{\nu}_e$ asymmetry exists. When equilibrium holds, the neutrino distribution depends upon the degeneracy parameter ξ_e. In such a case, the equilibrium value of the neutron-to-proton ratio fixed by β processes in Eq. (4.74) is

$$\frac{n}{p} \simeq \exp\left(-\frac{\Delta m}{T} - \xi_e\right). \tag{4.85}$$

Therefore, the freeze-out value of n/p decreases (increases) if ξ_e is positive (negative), leading to a primordial abundance of ^4He which is smaller (larger). If ξ_e were the only additional parameter, the corresponding range allowed by BBN would be quite narrow,

$$-0.01 \lesssim \xi_e \lesssim 0.07, \qquad (4.86)$$

i.e., compatible with $\xi_e = 0$ (see for example Kang and Steigman, 1992; Esposito *et al.*, 2000b; Di Bari, 2002). Adding an asymmetry in the other neutrino flavours, which provides an extra contribution to the number of relativistic degrees of freedom, can compensate for the effect of a positive ξ_e on ^4He. Indeed, there exists a *degeneracy* in the (ξ_e, N_{eff}) plane that can be removed only by using other data, such as CMB observations. For instance, the analysis of BBN and CMB/LSS data yielded the allowed regions (Hansen *et al.*, 2002)

$$-0.01 < \xi_e < 0.22, \quad |\xi_{\mu,\tau}| < 2.6, \qquad (4.87)$$

which clearly shows that as far as neutrino asymmetries is concerned, BBN is not flavour-blind.

The origin of this degeneracy can be easily understood. For positive ξ_e, neutron abundance at freeze-out of weak interactions decreases. However, we can shift the decoupling temperature T_D at earlier times by increasing the Hubble parameter through a larger N_{eff}. The same neutron density is obtained for different values of ξ_e and N_{eff}, but provided the combination

$$\frac{\Delta m}{T_D(N_{\text{eff}})} + \xi_e \qquad (4.88)$$

remains fixed; see Eq. (4.85).

We now consider the effect of oscillations. The first analyses (Dolgov *et al.*, 2002; Abazajian *et al.*, 2002; Wong, 2002) showed that oscillations lead to strong flavour conversions,[9] so that effective flavour equilibrium between all active neutrino species is established well before BBN, in particular for nonzero values of the mixing angle θ_{13}. In this case, if flavour oscillations enforce the condition that all ξ_α are almost the same during BBN, the stringent bound on ξ_e applies to all flavours. For the common value of the neutrino degeneracy parameter ξ, the corresponding predictions for the primordial abundances are shown in Fig. 4.15. The allowed range of ξ is then $-0.021 \leq \xi \leq 0.005$. This in turn also implies that if neutrinos indeed reach perfect kinetic and chemical equilibrium before they decouple, any excess in cosmic radiation density must be ascribed to extra relativistic degrees of

---

[9] Unless flavour oscillations are suppressed in exotic scenarios, for instance, in the presence of a hypothetical neutrino–Majoron coupling of order $g \sim 10^{-6}$ (Dolgov and Takahashi, 2004).

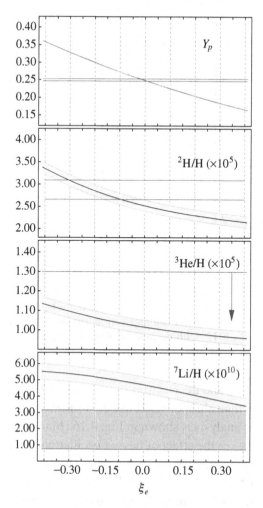

Figure 4.15 Light-element abundances as a function of a common neutrino degeneracy parameter. The gray $1\,\sigma$ error bands include the uncertainty of the WMAP determination of the baryon density, as well as the uncertainties from the nuclear cross sections of (Serpico *et al.*, 2004). (Reprinted from Iocco *et al.*, 2009, with permission from Elsevier.)

freedom, because the additional contribution to radiation density due to ξ is very small.

However, as emphasized in Section 4.2.4, a proper inclusion of collisions in the neutrino kinetic equations shows that equilibration among neutrinos and with the electromagnetic plasma is guaranteed only if the mixing angle θ_{13} is large enough (Pastor *et al.*, 2009). The role of neutrinos in fixing the BBN yields can be understood only by following the time evolution of their distributions (Mangano *et al.*, 2011, 2012), which depend upon the initial values of the total neutrino

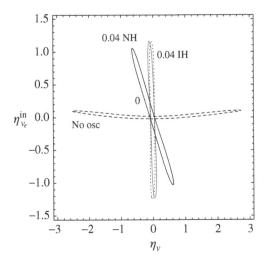

Figure 4.16 BBN contours at 95% C.L. in the $\eta_\nu - \eta_{\nu_e}^{\mathrm{in}}$ plane for several values of $\sin^2\theta_{13}$: 0 (solid line), 0.04 and normal mass hierarchy (NH) (almost vertical solid line), 0.04 and inverted mass hierarchy (IH) (dotted line). The case of no neutrino flavour oscillations is shown for comparison as the dashed contour. (Reprinted from Mangano *et al.*, 2012, with permission from Elsevier.)

asymmetry $\eta_\nu = \sum_\alpha \eta_{\nu_\alpha}$ and the electron neutrino asymmetry $\eta_{\nu_e}^{\mathrm{in}}$. Once these two parameters are given, the values of ^{2}H/H and Y_p follow.

An example of this analysis is shown in Fig. 4.16 (Mangano *et al.*, 2012). From this plot one can easily see the effect of flavour oscillations on the BBN constraints on the total neutrino asymmetry. With no neutrino mixing, the value of η_{ν_e} is severely constrained by ^{4}He data, as from Eq. (4.86), whereas the asymmetry for other neutrino flavours could be much larger. On the other hand, flavour oscillations imply that an initially large $\eta_{\nu_e}^{\mathrm{in}}$ can be compensated by an asymmetry in the other flavours with opposite sign. The most restrictive BBN bound on η_{ν_e} applies then to the total asymmetry, an effect that can be seen graphically in Fig. 4.16 as a *rotation* of the allowed region from a quasi-horizontal one for zero mixing to an almost vertical region for $\sin^2\theta_{13} = 0.04$, in particular for the inverted mass hierarchy. In all cases depicted in Fig. 4.16, the allowed values of the asymmetries are fixed by both deuterium and ^{4}He, the latter imposing that the value of η_{ν_e} at BBN must be very close to zero. For values of θ_{13} suggested by present experimental data, the combined effect of oscillations and collisions leads to an efficient mixing of all neutrino flavours before BBN. Therefore, the individual neutrino asymmetries have similar values, approximately $\eta_{\nu_\alpha} \simeq \eta_\nu/3$, and the BBN bound on the electron neutrino asymmetry applies to all flavours, and in turn to η_ν as considered in Serpico and Raffelt, 2005; Simha and Steigman, 2008; Iocco *et al.*, 2009. For

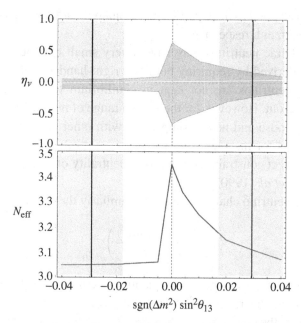

Figure 4.17 Region of total neutrino asymmetry (upper panel) and largest values of N_{eff} from primordial neutrino asymmetries (lower panel) compatible with BBN at 95% C.L., as a function of θ_{13} and the neutrino mass hierarchy. The vertical solid lines and shadowed regions indicate the best-fit and 2σ allowed range of θ_{13}, respectively, from the RENO reactor experiment (Ahn *et al.*, 2012). (Adapted from Mangano *et al.*, 2012. Reprinted with permission from Elsevier.)

$\sin^2\theta_{13} \sim 0.03$ the allowed region at 95% C.L. is $-0.2\,(-0.1) \lesssim \eta_\nu \lesssim 0.15\,(0.05)$ for neutrino masses following a normal (inverted) mass hierarchy. In the IH this result approximately holds for any value of $\sin^2\theta_{13} \gtrsim 0.005$ due to the resonant character of the flavour conversions. This can be seen from Fig. 4.17, where we show the allowed regions of the total neutrino asymmetry and the largest value of N_{eff} from neutrino asymmetries compatible with BBN, as a function of the mixing angle θ_{13} and the mass hierarchy. The Daya Bay and RENO results imply that neutrino chemical potentials cannot produce a value of N_{eff} larger than 3.2.

4.4.3 Nonstandard neutrino electromagnetic properties and interactions

The most general structure of effective neutrino electromagnetic interactions is

$$\mathcal{L}_{\text{int}} = -e_\nu \, \bar{\nu} \, \gamma_\mu \, \nu \, A^\mu - a_\nu \, \bar{\nu} \, \gamma_\mu \, \gamma_5 \, \nu \, \partial_\lambda F^{\mu\lambda} - \frac{1}{2} \bar{\nu} \, \sigma_{\alpha\beta} \, (\mu + \epsilon \, \gamma_5) \, \nu \, F^{\alpha\beta}, \quad (4.89)$$

where $F^{\alpha\beta}$ is the electromagnetic field tensor and $\sigma_{\alpha\beta} = [\gamma_\alpha, \gamma_\beta]$. The form factors $\{e_\nu, a_\nu, \mu, \epsilon\}$ are functions of the transferred squared momentum q^2, and in the limit

$q^2 \to 0$ they correspond to the electric charge, anapole moment, and magnetic and electric dipole moment, respectively.

In principle, Dirac neutrinos may have a very small electric charge e_ν. BBN bounds may be derived by requiring both that right-handed partners are not populated and that neutrinos are not kept in equilibrium too long after the weak interaction freeze-out. However, for the known range of neutrino masses, the BBN bounds are very loose and never competitive with other astrophysical or laboratory constraints, as for instance from red giants, $e_\nu \lesssim 2 \times 10^{-14}$ (Raffelt, 1999). Actually, the indirect constraint coming from neutrality of matter is even stronger, $e_\nu \lesssim 10^{-21}$ (Foot *et al.*, 1990).

The so-called neutrino charge radius (and similarly the anapole radius)

$$\langle r^2 \rangle = \frac{6}{e} \left(\frac{\partial e_\nu(q^2)}{\partial q^2} \right)_{q^2=0} \tag{4.90}$$

is a well-defined (gauge-independent) quantity (Bernabéu *et al.*, 2004). The SM expectations are in the range $\langle r^2 \rangle \simeq 1-4$ nb. Differently from the case of the magnetic moment, the charge radius does not couple neutrinos to on-shell photons, so stellar cooling arguments are not very sensitive to $\langle r^2 \rangle$. For Dirac neutrinos the channel $e^- e^+ \to \nu_R \bar{\nu}_R$ is still effective, provided that new physics violates the cancellation between vector and axial contribution that otherwise applies in the SM. If this cancellation does not take place, the corresponding bounds from the Supernova SN1987A (Grifols and Massó, 1989) and nucleosynthesis (Grifols and Massó, 1987) are of the same order of magnitude,

$$|\langle r^2 \rangle| \lesssim \begin{cases} 2 \text{ nb,} & \text{SN1987A} \\ 7 \text{ nb,} & \text{BBN.} \end{cases} \tag{4.91}$$

These bounds do not apply to Majorana neutrinos. In this case, even in the ν_τ-sector where BBN may have a sensitivity comparable to or better than laboratory experiments, a change of one order of magnitude above the SM level in the channel $e^- e^+ \to \nu_\tau \bar{\nu}_\tau$ due to some new physics effect, would change Y_p only at the 0.1% level. (For a further discussion of this point, see Hirsch *et al.*, 2003).

Massive neutrinos can have magnetic moments, although for Majorana particles only off-diagonal elements in flavour space are nonvanishing. There are two possible processes of interest for BBN: (i) the thermalization of Dirac neutrinos via, e.g., the process $\nu_L e \to \nu_R e$ or (ii) a radiative decay $\nu_L^i \to \nu_L^j \gamma$, the latter being possible also for Majorana neutrinos. In the absence of primordial magnetic fields, the BBN bound on the diagonal elements coming from the thermalization of right-handed neutrinos is not as restrictive as the one coming from the red giant cooling argument, $\mu \lesssim 3 \times 10^{-12} \mu_B$, from plasmon decay $\gamma^* \to \nu\bar{\nu}$ (Raffelt, 1999). Here

μ_B is the Bohr magneton. The radiative decay rate for a transition $i \to j$ is

$$\Gamma_{ij}^{\gamma} = \frac{|\mu_{ij}|^2 + |\epsilon_{ij}|^2}{8\pi} \left(\frac{m_i^2 - m_j^2}{m_i} \right)^3 \simeq 5.3 \left(\frac{\mu}{\mu_B} \right)^2 \left(\frac{(m_i^2 - m_j^2)/m_i}{\text{eV}} \right)^3 \text{s}^{-1}$$

(4.92)

which leads to a lifetime definitely too long to be relevant for BBN. In any case, the cosmological bounds on the neutrino lifetime in Mirizzi *et al.*, 2007 exclude any effect of the radiative neutrino decay at the BBN epoch.

Besides anomalous electromagnetic interactions, neutrinos might undergo non-standard interactions (NSI) with charged leptons. We have already discussed this issue in Chapter 1. Because during BBN the only charged leptons are electrons/positrons, it is enough to consider the low-energy four-fermion inter-actions' lagrangian density

$$\mathcal{L}_{\text{NSI}} = -2\sqrt{2}G_F \left[\epsilon_{\alpha\beta}^L (\overline{\nu}_{\alpha L} \gamma^\mu \nu_{\beta L})(\overline{e}_L \gamma_\mu e_L) + \epsilon_{\alpha\beta}^R (\overline{\nu}_{\alpha L} \gamma^\mu \nu_{\beta L})(\overline{e}_R \gamma_\mu e_R) \right]$$

(4.93)

with the NSI parameters $\epsilon_{\alpha\beta}^L$ and $\epsilon_{\alpha\beta}^R$ constrained by laboratory measurements to be at most of $\mathcal{O}(1)$, as we reviewed in Chapter 1. It was found in Mangano *et al.*, 2006a that, for NSI parameters within the ranges allowed by present laboratory data, nonstandard neutrino–electron interactions do not essentially modify the density of relic neutrinos nor the bounds on neutrino properties from cosmological observables. This depends on the fact that a large modification of the neutrino spectra would only be achieved if the decoupling temperature is shifted to values below the electron mass. The presence of neutrino–electron NSI within laboratory bounds may enhance the entropy transfer from electron–positron pairs into neutrinos, up to a value of $N_{\text{eff}} = 3.12$ (and correspondingly $\Delta Y_p \simeq 6 \times 10^{-4}$), which are at most three times the corrections due to nonthermal distortions that appear for standard weak interactions, but still probably too small to be detectable in the near future.

Neutrinos could be also coupled to a scalar or pseudoscalar particle with a Yukawa interaction term

$$\mathcal{L}_{\nu\phi} = -\phi \, \lambda_{\alpha\beta} \, \overline{\nu}_{\alpha L} \, \Gamma \, \nu_{\beta L}$$

(4.94)

with α, β flavour indices and $\Gamma = 1$, γ_5 for scalar and pseudoscalar ϕ, respectively. A famous example is the Majoron model (Chikashige *et al.*, 1981; Gelmini and Roncadelli, 1981; Georgi *et al.*, 1981; Schechter and Valle, 1982b), where the Majoron ϕ is the Goldstone boson associated to the breaking of the lepton number symmetry. In the early universe, a large enough λ would make it possible to populate thermally the species ϕ. For $m_\phi \ll 1$ MeV, it was found in Chang and Choi, 1994 that the coupling matrix entries $\lambda_{\alpha\beta}$ should be smaller than 10^{-5}, if one considers one additional boson ($\Delta N_{\text{eff}} = 3/7$) to be incompatible with the observations. To date, this is still viable, however, and annihilation processes $\bar{\nu}\nu \leftrightarrow \phi\phi$ may be responsible for a "neutrinoless universe" well after the BBN epoch, provided that $m_\phi \ll m_\nu$ (Beacom *et al.*, 2004). In this case, we would have wasted our time in writing this book.

Relatively less attention has been paid to the case where the associated (pseudo) boson is massive. In Cuoco *et al.*, 2005, the authors consider MeV-scale ϕ particles produced at early epochs via additional couplings with other SM particles and later decaying into neutrinos in out-of-equilibrium conditions, with a rate

$$\Gamma(\phi \to \nu_\alpha \bar{\nu}_\beta) = |\lambda_{\alpha\beta}|^2 \frac{m_\phi}{8\pi}. \tag{4.95}$$

The resulting neutrino distribution would be the superposition of the canonical Fermi–Dirac function and a peak corresponding to the neutrino comoving momentum $y_D = m_\phi a_D/2$, with a_D the value of the scale factor at decay time. The peak shape can be modelled as a gaussian term, with an amplitude A such that

$$y^2 f_\nu(y) = y^2 \frac{1}{e^y + 1} + \frac{A}{\sqrt{2\pi\sigma^2}} \exp\left[-\frac{(y - y_D)^2}{2\sigma^2}\right]. \tag{4.96}$$

If these decays take place before the last photon-scattering epoch, the produced neutrino burst directly influences the CMB anisotropy spectrum, as well as large-scale structure formation; see Chapter 6. Interestingly, current cosmological observations of light-element abundances (and CMB and LSS data) are still compatible with very large deviations from the standard picture, i.e., high values of $y_D \lesssim 20$ and $A \lesssim 1$.

We conclude this section with a final remark about right-handed neutrino states. We have seen that a Dirac mass term could in principle be responsible for their production in the primordial plasma. This possibility is excluded by the smallness of neutrino masses, because the typical interaction rates are suppressed by the factor $(m_\nu/T)^2$; see the discussion at the end of Section 2.4.2. However, it is still possible to populate ν_R if there are right-handed currents mediated by W_R bosons, via processes mediated by the lagrangian density term $\bar{\nu}_R W_R \nu_R$ or analogous coupling with right-handed charged leptons. These are possible in some extensions of the standard electroweak model. If one assumes that the right-handed interaction has the same form as the left-handed one but with heavier intermediate bosons, one can obtain from BBN a lower limit on their mass on the order of $m_{W_R} \gtrsim 75\, m_W$ (Steigman *et al.*, 1979; Olive *et al.*, 2000; Barger *et al.*, 2003), which depends, however, on the exact particle spectrum of the physics beyond the SM up to $\sim 75\, m_W$.

4.4.4 Sterile neutrinos and Big Bang nucleosynthesis

We described in Section 4.2.5 the main aspects of active–sterile neutrino oscillations in the early universe, where we saw that they are effective before the onset of BBN for a wide range of mixing parameters. Therefore, it is an important issue to calculate the effect of active–sterile mixing on the primordial production of light elements. This has been considered in many papers, an incomplete list being Barbieri and Dolgov, 1990; Kainulainen, 1990; Barbieri and Dolgov, 1991; Cline, 1992; Enqvist *et al.*, 1992b; Shi *et al.*, 1993; Kirilova and Chizhov, 1998a; Lisi *et al.*, 1999; Kirilova and Chizhov, 2000; Abazajian, 2003; Dolgov and Villante, 2004; Cirelli *et al.*, 2005. The BBN phenomenology is as wide as the possible values of the active–sterile mixing parameters, but we can summarize the main effects as follows:

- Effective active–sterile oscillations lead to a partial or total population of ν_s with respect to the equilibrium neutrino density. The radiation energy density is enhanced, corresponding to $N_{\text{eff}} > 3$, affecting the expansion rate and the BBN outcome.
- The number density of electron neutrinos and antineutrinos can be depleted if sterile neutrino production is effective only after the decoupling of active neutrinos. This modifies the weak rates that fix the neutron-to-proton ratio and the abundance of ^{4}He. In some particular cases, such a depletion can be ν_e–$\bar{\nu}_e$ asymmetric, again changing the $n \leftrightarrow p$ weak rates.
- If a large lepton asymmetry is dynamically generated by active–sterile neutrino oscillations, its consequences for BBN could be important.

Note that the first two effects, an enhanced N_{eff} and a global depletion of ν_e's and $\bar{\nu}_e$'s, lead to a higher neutron-to-proton ratio and to a more abundant production of primordial D and ^{4}He. Therefore, we expect that at some level a BBN analysis can lead to constraints on the active–sterile mixing parameters in ranges which are not accessible to laboratory experiments, such as the case of very small mixing angles in vacuum.

Here we discuss only the case of one additional sterile neutrino, which in the majority of the analyses of active–sterile oscillations has been considered in the two-neutrino limit, either $\nu_e - \nu_s$ or $\nu_\mu - \nu_s$ (or equivalently $\nu_\tau - \nu_s$), characterized by the mixing parameters Δm^2 and θ but neglecting mixing among active neutrinos.

Although this assumption simplifies the calculations, note that this is necessarily an unphysical approximation of the real situation because of the experimental evidence for flavour oscillations among the three active species. When an additional neutrino mass eigenstate ν_4, with mass m_4, is added to the three-neutrino system, the evolution of the 4×4 system should be obtained from the kinetic equations in the density matrix formalism. Calculations in this general case are more complex because, depending on the mixing parameters, one can expect one or more resonances among active and sterile neutrinos. Only a few works, including Dolgov and Villante, 2004; Cirelli *et al.*, 2005; Mirizzi *et al.*, 2012 have taken into account the active mixing parameters as determined by oscillation experiments.

Let us summarize the main results of Dolgov and Villante, 2004, where a quite thorough analysis was performed, at least for the range $\Delta m^2 \lesssim 1 \text{ eV}^2$, and the two-neutrino limit was also computed for comparison. The quantum kinetic equations of the active–sterile system were numerically solved under certain approximations and the effects on BBN of an enhanced N_{eff} and a $\nu_e/\bar{\nu}_e$ density depletion were calculated as follows. We assume that electron neutrinos are in kinetic equilibrium and Boltzmann statistics are valid, neglecting the electron mass. Within this approximation, the neutron-to-proton interconversion rate Γ_{np} is proportional to $(1 + \kappa_e)/2$, where κ_e is the factor relating the actual ν_e spectrum to an equilibrium one: in comoving variables $\rho_{ee}(x, y) = \kappa_e(x) \exp(-y)$. This means that the neutron freeze-out takes place at a temperature which scales as

$$T_f \propto \left[\frac{g_*^{1/2}}{(1 + \kappa_e)/2} \right]^{1/3}, \qquad (4.97)$$

where $g_* = 10.75 + (7/4)\Delta N_{\text{eff}}$. Here ΔN_{eff} is the extra contribution of oscillating neutrinos to radiation. The global effect on BBN can be estimated as a variation in the effective number of neutrinos,

$$\Delta N^{\text{BBN}} = \frac{4}{7} \left[\frac{10.75 + (7/4)\Delta N_{\text{eff}}}{(1 + \kappa_e)^2/4} - 10.75 \right]. \qquad (4.98)$$

An alternative calculation of the effects on BBN can be found in (Cirelli *et al.*, 2005), where two different effective N_ν associated with the final abundance of ^{2}H and ^{4}He were defined.

The main results of Dolgov and Villante, 2004 in the two-neutrino limit are shown in Fig. 4.18, as ΔN^{BBN} isolines for the two oscillation cases of ν_e–ν_s or ν_μ–ν_s and either nonresonant ($\Delta m^2 > 0$) or resonant ($\Delta m^2 < 0$) production of ν_s, as described in Section 4.2.5. The dotted lines in each plot were calculated from analytical estimates of the extra contribution to radiation from active–sterile oscillations, reasonably accurate for large squared-mass differences (see Dolgov and Villante, 2004 for details). At large Δm^2, the main effect is the enhanced contribution to radiation, in particular for the ν_μ–ν_s case, and the results are very similar to the ΔN_{eff} contours in Fig. 4.6. Instead, for small Δm^2, significant deviations exist when the effect of $\nu_e/\bar{\nu}_e$ depletion becomes relevant, which is obviously more important for ν_e–ν_s mixing. This happens if sterile neutrino production takes place at low temperatures, when inverse annihilation processes are too slow with respect

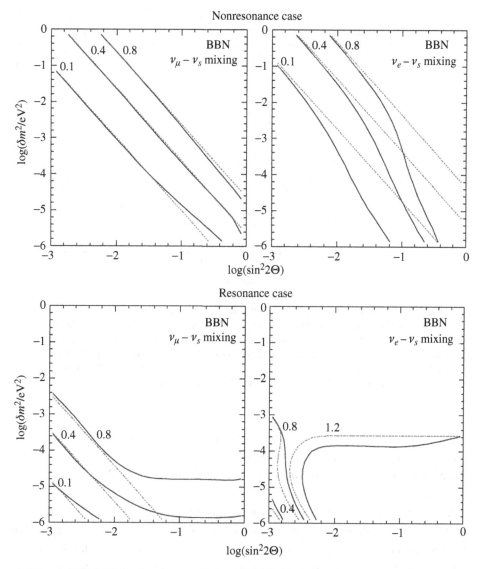

Figure 4.18 Effect of active–sterile neutrino oscillations on BBN in the approximation of two-neutrino mixing, parameterized in terms of ΔN^{BBN}. Each solid line corresponds to a fixed value of ΔN^{BBN} found according to Eq. (4.98). Dotted lines correspond to analytical estimates as discussed in Dolgov and Villante, 2004. (Adapted from Dolgov and Villante, 2004. Reprinted with permission from Elsevier.)

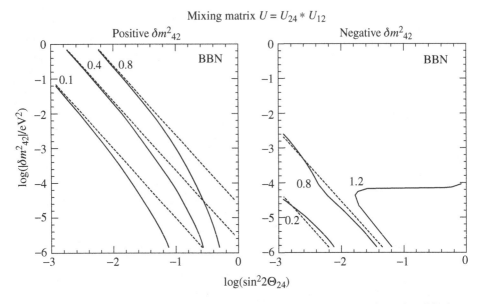

Figure 4.19 Effect of active–sterile neutrino oscillations on BBN in a simplified
model with four neutrinos (where the neutrino mixing matrix can be written as
$U = U_{24} \cdot U_{12}$), parameterized in terms of ΔN^{BBN}. Each solid line corresponds to
a fixed value of ΔN^{BBN} found according to Eq. (4.98). Dotted lines correspond to
analytical estimates as discussed in (Dolgov and Villante, 2004). (Adapted from
Dolgov and Villante, 2004. Reprinted with permission from Elsevier.)

to the expansion rate of the universe. This effect increases the BBN sensitivity
to active–sterile mixing, in particular to ν_e–ν_s oscillations with small Δm^2. As
expected from the results shown in Fig. 4.6, the influence on BBN is more relevant
in the resonant case.

A more realistic case of four mixed neutrinos is presented in Fig. 4.19, calculated
in Dolgov and Villante, 2004 with some simplifying assumptions regarding the
structure of the 4×4 mixing matrix, such as taking a small active–sterile neutrino
mixing. We have discussed the main aspects of active neutrino mixing in the
early universe when describing the small distortions of the neutrino spectra in
the noninstantaneous decoupling case. We saw that oscillations are effective at a
temperature above 1 MeV and lead to the existence of fast transitions between ν_e,
ν_μ and ν_τ. Clearly this means that the corresponding BBN bound on mixing with
sterile neutrinos will change with respect to the two-neutrino limit, because it is
expected that the effects related to $\nu_e/\bar{\nu}_e$ depletion will be more similar to the pure
ν_e–ν_s case, because electron neutrinos and antineutrinos are always involved in the
conversions. This expectation is confirmed with the results of the BBN effects for
four mixed neutrinos shown in Fig. 4.19.

The results for the case when active–sterile oscillations are effective after neutrino decoupling ($\Delta m^2 \lesssim 10^{-7}$ eV^2) complement what we have just discussed and have been presented in Kirilova and Chizhov, 1998a,b, 2000. In this case significant deviations of the active neutrino spectra from a pure Fermi–Dirac distribution could occur during the evolution, with the corresponding effect on BBN.

Finally, we again emphasize that the existence of an initial nonzero lepton asymmetry would modify the resonance condition and typically suppress active–sterile oscillations, reducing their impact on BBN (Foot and Volkas, 1995). This is especially interesting for the range of larger masses, where $m_s^2 \simeq \Delta m^2$, because the presence of a lepton asymmetry can lead to a relic abundance of keV neutrinos with the value required to account for the total DM density.[10]

---

[10] Massive sterile neutrinos in the MeV range must decay with short lifetimes (0.01–1 s) in order to agree with BBN observations, as shown in Dolgov *et al.*, 2000b,c; Ruchayskiy and Ivashko, 2012.

5

Neutrinos in the cosmic microwave background epoch

The statistical properties of CMB temperature and polarization anisotropy maps encode very precise information on the history and composition of our universe. They depend primarily on the behaviour of inhomogeneities in the photon and baryon medium until photon decoupling, which *feels* all other species in two ways: through their impact on the cosmological background evolution, and via their contribution to the local gravitational forces. This is why neutrinos play an indirect yet important role in the physics of CMB anisotropies, and why present (and future) data on these observables give us quite remarkable pieces of information on neutrino properties.

To understand this point quantitatively, we need to follow photon decoupling at a much more detailed level than in Section 2.4.1. This is the subject of Section 5.1, where we overview the main features of CMB physics, of cosmological perturbation equations, the different contributions to the spectrum of CMB temperature anisotropies, and the effect of each cosmological parameter on the CMB spectrum. Neutrinos will appear on stage in Section 5.2, where we focus on the evolution of their perturbations until photon decoupling, and in Section 5.3, where we infer the effect of neutrino abundance, masses and properties on CMB anisotropies. Finally, Section 5.4 is a brief summary of recent constraints on neutrino properties, exploiting CMB data alone.

In the following pages, physical quantities are written as the sums of their spatial averages, denoted by an overline, and their fluctuations in real or Fourier space. For instance, the density of a given species i will read $\rho_i(\eta, \vec{x}) = \overline{\rho}_i(\eta) + \delta\rho_i(\eta, \vec{x})$, where η stands for conformal time and $\vec{x}$ for comoving coordinate. To simplify the notation, we will use the same symbol for the perturbation in real space (e.g., $\delta\rho_i(\eta, \vec{x})$) and in Fourier space ($\delta\rho_i(\eta, \vec{k})$).

5.1 Cosmic microwave background anisotropies

5.1.1 Overview

CMB physics mainly depends on the behaviour of three species until recombination: baryons, electrons and, of course, photons.

Baryons and electrons remain strongly coupled throughout recombination via Coulomb scattering. Electric neutrality ensures that their density contrasts $\delta\rho_B/\overline{\rho}_B$ and $\delta\rho_e/\overline{\rho}_e$ are equal to each other, everywhere in the perturbed universe. They can be considered as a single fluid, whose energy density is dominated by baryons, because they are considerably more massive. For convenience, in the context of CMB physics, the tightly coupled baryon–electron medium is often referred as "baryons" only.

Photons and electrons interact through Compton scattering, which becomes inefficient around the time of recombination, as we saw in Section 2.4.1. At much higher redshift, the tightly coupled photon–baryon(–electron) plasma can be described as a single fluid. This approximation is accurate enough to understand the basics of acoustic oscillations, as we shall do in Section 5.1.5. As long as tight coupling holds, thermal equilibrium imposes a blackbody spectrum locally. Photons can then be described at each point by their temperature (plus some Stokes parameters specifying their polarization state). However, an observer at rest with respect to the coordinate frame but not to the fluid will see a Doppler shift. Hence, in a spherical harmonic expansion of the local temperature distribution, the perturbation at any given point is characterized by a monopole and a dipole.

When photons decouple, we could expect a priori that their distribution function becomes gradually nonthermally distorted, at least at the level of perturbations (we saw in Section 2.4.1 that the background distribution remains of the Bose–Einstein type). However, as long as photons interact only gravitationally, the equivalence principle implies that photons of different energy travelling along a given geodesic are redshifted by the same amount at each point, in such a way that the distribution of photons along this geodesic can experience a temperature shift, but no nonthermal distortions. Hence, after decoupling, photons can be described by a local value of temperature $\overline{T}(\eta) + \delta T(\eta, \vec{x}, \hat{n})$ (and of Stokes parameters for polarization), with extra dependence on the direction of propagation $\hat{n}$, unlike in thermal equilibrium. The same holds for standard decoupled relativistic neutrinos, described by $\overline{T}_\nu(\eta) + \delta T_\nu(\eta, \vec{x}, \hat{n})$. The same arguments cannot be extended to nonrelativistic neutrinos, for which gravitational interactions lead to some distortions in their momentum distribution. Such neutrinos will be described by a more complicated distribution function, depending also on momentum. We will review in Section 5.2.1 the evolution equation for these degrees of freedom.

The tightly coupled photon–baryon fluid experiences pressure forces due to photon pressure $P_\gamma = \rho_\gamma/3$. Any inhomogeneity propagates in the form of acoustic waves. Acoustic oscillations are affected by

- the baryon-to-photon ratio η_B (defined in Eq. (2.160)), which controls the effective pressure and sound speed of the photon–baryon fluid, and
- gravitational forces, caused by the self-gravity of the fluid and by gravitational interactions with other species.

The dispersion relation of these acoustic waves is such that a given perturbation propagates with only one wavefront, instead of multiple wavefronts as for waves on a water surface. This means that at a given time, a single correlation length appears in the spatial temperature distribution of the fluid. This correlation length is given simply by the distance travelled by a wavefront since some initial time, called the sound horizon. Because of the expansion of the universe, the value of the sound horizon at a given conformal time η is almost insensitive to the choice of initial time η_i, provided that $\eta_i \ll \eta$: the sound horizon does not depend on the early cosmological evolution close to the Big Bang, but rather on the late evolution around the recombination epoch.

After decoupling, free-streaming photons carry information on the local temperature and velocity of the fluid at recombination, near their last scattering point (with minor distortions acquired after decoupling and called secondary anisotropies). We will briefly comment on the fact that they also carry extra information encoded in their polarization state. Hence, to some extent we can reconstruct the map of temperature and polarization fluctuations on the last scattering surface, and extract the correlation length given by the sound horizon at decoupling, projected along a given angle. Indeed, the observed angular correlation function of CMB maps contains one characteristic feature, whose angular scale, shape and amplitude depend on the details of the photon–baryon plasma, the cosmological background evolution, and the presence of any other fluid interacting gravitationally with this plasma.

The correlation function can also be computed in multipole space after expanding CMB maps in spherical harmonics. The angular correlation function is usually noted $C(\theta)$, and its counterpart in harmonic space C_l is called the CMB (temperature or polarization) power spectrum. Strictly speaking, the angular correlation function $C(\theta)$ and the power spectrum C_l contain the same amount of information. However, the dependence of the C_l power spectrum on the underlying cosmological model is more transparent and easy to interpret. As in most of the literature, we will focus on this observable. The power spectrum contains several regularly spaced peaks, corresponding to the various harmonics of the single characteristic feature in $C(\theta)$. A large amount of cosmological information is encoded in the amplitude of each of these peaks.

5.1.2 Perturbation equations

The linearly perturbed universe can be described by a perturbed metric tensor $\bar{g}_{\mu\nu} + \delta g_{\mu\nu}$ and a set of dynamical variables for each species that we will review in this section. Perturbations can be described in several gauges, i.e., using several possible definitions of equal-time hypersurfaces on which spatial averaging is performed, all compatible with the assumption of a homogeneous background. We do not provide details here on the gauge issue, which is well described throughout the literature (see, e.g., Ma and Bertschinger, 1995). For simplicity, we restrict the following presentation to the Newtonian gauge in a spatially flat background spacetime, where the perturbed line element reads

$$ds^2 = a^2(\eta) \left[-(1 + 2\psi(\eta, \vec{x}))d\eta^2 + (1 - 2\phi(\eta, \vec{x}))d\vec{x}^2 \right] \tag{5.1}$$

and ψ plays the role of the Newtonian gravitational potential on scales much smaller than the Hubble radius. Any other gauge choice would be equally good, because truly observable quantities are always gauge-invariant. We will occasionally refer to the impact of a nonzero spatial curvature, without introducing the full equations of evolution in that case. The whole system of evolution equations for first-order cosmological perturbations in a spatially flat background spacetime is presented in the seminal paper by Ma and Bertschinger (1995). This section contains a pedagogical summary of this reference, sufficient for the purposes of this book. It can also be used to make first contact with cosmological perturbations, before investigating more technical aspects in the original papers.

Energy–momentum vector in the perturbed universe

Let us consider a particle of mass m, of physical momentum p^i (defined in 2.18), and of energy $E = \sqrt{p^2 + m^2}$. As in Chapter 2, we define $y = ap$, and similarly rescale the energy as

$$\epsilon \equiv a E = \sqrt{y^2 + a^2 m^2}. \tag{5.2}$$

We will use circumflexes to denote unit vectors, so that the proper momentum can be written as $\vec{p} = p\,\hat{n}$, with components $p^i = p_i = p\,n_i$. In the linearly perturbed universe and in the Newtonian gauge, the components of the comoving energy–momentum vector P_μ can be shown to be related to metric perturbations and to (y, ϵ) through

$$P_0 = -(1 + \psi)\epsilon, \qquad P_i = (1 - \phi)y\,n_i. \tag{5.3}$$

The geodesic equation can be written in terms of the above variables, and gives the time variation of y for a freely falling particle in the linearly perturbed universe,

$$\frac{dy}{d\eta} = -y\,\phi' - \epsilon\,\hat{n}\cdot\vec{\nabla}\psi, \tag{5.4}$$

with $'$ denoting the derivative with respect to conformal time. The geodesic equation also gives the variation of the direction of propagation $d\hat{n}/d\eta$, accounting for gravitational lensing effects. Both $dy/d\eta$ and $d\hat{n}/d\eta$ are of order one in perturbations, because in a homogeneous universe particles would propagate along straight lines with fixed comoving momentum. A crucial point is that for relativistic particles, we can use $\epsilon = y$ and rewrite the momentum evolution as

$$\frac{d\ln y}{d\eta} = -\phi' - \hat{n}\cdot\vec{\nabla}\psi. \tag{5.5}$$

This equation shows that when travelling across metric perturbations, relativistic decoupled particles experience a relative momentum shift which does not depend on the momentum itself. Hence, for any relativistic free-streaming species which attained an equilibrium distribution, the phase-space distribution keeps the functional form of a Fermi–Dirac or Bose–Einstein function, with the temperature now becoming a function not only of time and space, but also of direction. For these species, nonthermal distortions can be generated only through nongravitational couplings. On the other hand, for nonrelativistic decoupled particles, gravitational interactions do produce nonthermal distortions at the level of perturbations, because the momentum dependence cannot be eliminated from the r.h.s. of (5.4).

Perturbed stress–energy tensor

It is useful to classify the 10 degrees of freedom of the perturbed stress–energy tensor $\delta T_{\mu\nu}$ of each species as scalars, vectors and tensors under spatial rotations (Bardeen, 1980). In fact, in linear perturbation theory, the scalar, vector and tensor sectors are decoupled from each other. In this book, we will focus only on the scalar degrees of freedom. Indeed, the vectors describe vorticity, which quickly decays on all cosmologically interesting scales, at least in the standard picture. The tensors, briefly mentioned in Section 5.1.8, describe gravitational waves. These can be excited by inflation and lead to a signature in the CMB, but in our current understanding, they cannot be observed with a precision sufficient to probe any neutrino effect.

For every perturbed stress–energy tensor, there are four scalar degrees of freedom. They are usually defined as the relative density fluctuation $\delta \equiv \delta\rho/\bar\rho$, a function θ related to the bulk velocity divergence (or energy flux divergence), the pressure perturbations δP and a dimensionless potential σ associated to anisotropic stress (i.e., anisotropic pressure). These four degrees of freedom are related to δT_ν^μ through

$$\overline{\rho}\,\delta = -\delta T_0^0 \tag{5.6}$$

$$(\overline{\rho} + \overline{P})\theta = \sum_i \partial_i \delta T_i^0 \tag{5.7}$$

$$\delta P = \frac{1}{3}\sum_i \delta T_i^i \tag{5.8}$$

$$(\overline{\rho} + \overline{P})\nabla^2\sigma = -\sum_{i,j}\left(\partial_i\partial_j\frac{1}{3}\nabla^2\delta_{ij}\right)\delta T_j^i. \tag{5.9}$$

The total stress–energy tensor of the multicomponent universe is obtained by summing over quantitites on the l.h.s. in the above equations (for instance, the total δT_0^0 is given by $-\sum_i \overline{\rho}_i \delta_i$, and leads to a total density fluctuation $\delta_{\text{tot}} = \sum_i \overline{\rho}_i \delta_i / \overline{\rho}_{\text{tot}}$).

The conservation equation $\nabla_\mu T_\nu^\mu = 0$ leads to two equations of motion for the scalar sector, the energy conservation equation and the Euler equation. We see that two more relations would be needed to close the four-variable system. One of these relations is often provided by knowledge of the equation of state or, in the case of fluids with nonadiabatic perturbations, by knowledge of the sound speed. This makes it possible to replace δP as a function of δ. For a perfect fluid, microscopic interactions enforce isotropic pressure and the anisotropic stress vanishes, so that the system is closed. For some classes of imperfect fluids, the anisotropic stress can still be expressed as a function of other variables. In more general cases, and in particular for weakly coupled or collisionless species, the stress–energy tensor and its conservation equations are not sufficient for describing the perturbation evolution.

We will mainly be interested in species that can be characterized by their phase-space distribution function f obeying the (perturbed) Boltzmann equation. This distribution can be decomposed into a background term f_0 depending on y and occasionally on η (as already seen in Section 2.2), plus a perturbation $\delta f(\eta, \vec{x}, \vec{p})$ or $\delta f(\eta, \vec{x}, y, \hat{n})$, depending explicitly on time and on the full phase-space coordinates. For such systems, the stress–energy tensor and the scalar degrees of freedom (δ, θ, δp, σ) can be extracted from f_0 and δf using the perturbed version of Eq. (2.42),

$$T_\nu^\mu = g \int \frac{d^3 p}{(2\pi)^3}\frac{P^\mu P_\nu}{E} f(\eta, \vec{x}, \vec{p}). \tag{5.10}$$

This relation can be combined with the Boltzmann equation in order to derive the energy conservation and Euler equations in a different way than from $\nabla_\mu T_\nu^\mu = 0$.

Photons

Before decoupling, the dominant photon interaction is Compton scattering off electrons. CMB physics can be studied in the Thomson limit of Compton scattering, in which the photon energy is assumed to be much smaller than the electron rest energy. The evolution is described by the perturbed version of the Boltzmann equation (2.92). This means that the Liouville operator on the l.h.s. of (2.92) now includes partial derivatives with respect to each phase-space coordinate x^i and p^i. The collisional integral $\mathbf{C}(f_\gamma(\eta, \vec{x}, \vec{p}); f_i)$ on the r.h.s. should be chosen to describe Thomson scattering, and can be simplified by noticing that Pauli blocking and stimulated emission play a negligible role in this context (see Section 2.2.1).

We learned from Eq. (5.5) that even when photons decouple, their phase-space distribution keeps the functional form of a Bose–Einstein distribution, but with a direction-dependent temperature. We can then write the full photon phase-space distribution at any time (before, during and after recombination) as

$$f_\gamma(\eta, \vec{x}, \vec{p}) = \left[\exp\left(\frac{y}{a(\eta)\,\overline{T}(\eta)\left\{1 + \Theta_\gamma(\eta, \vec{x}, \hat{n})\right\}} \right) - 1 \right]^{-1}, \qquad (5.11)$$

where $\Theta_\gamma \equiv \delta T/\overline{T}$ stands for the relative photon temperature shift. We recall that after electron–positron annihilation and until today, the product $a(\eta)\,\overline{T}(\eta)$ remains constant in time (see Section 2.4.1). In the absence of temperature fluctuations, we recover the expected background distribution

$$f_{\gamma 0}(y) = \left[\exp\left(\frac{y}{a\overline{T}} \right) - 1 \right]^{-1} = \left[\exp\left(\frac{p}{\overline{T}} \right) - 1 \right]^{-1}. \qquad (5.12)$$

One can replace f_γ in the Boltzmann equation, expand it at first order in perturbations and turn it into a linear equation of evolution for Θ_γ. After the Liouville operator is expanded, $d\vec{x}/d\eta$ can be replaced by $\hat{n}$ and $dy/d\eta$ by its expression in Eq. (5.4). The expression for $d\hat{n}/d\eta$ is not needed because this vector is contracted with the spatial gradient of f_γ with respect to the direction $\hat{n}$ and the product vanishes at first order in perturbations. After some algebra and simplifications, the Boltzmann equation reduces to

$$\Theta_\gamma' + \hat{n} \cdot \vec{\nabla}\Theta_\gamma - \phi' + \hat{n} \cdot \vec{\nabla}\psi = an_e\sigma_{\mathrm{T}}(\Theta_{\gamma 0} - \Theta_\gamma + \hat{n} \cdot \vec{v}_{\mathrm{B}}), \qquad (5.13)$$

where n_e is the number density of free electrons and σ_{T} the Thomson cross section (both previously defined in Section 2.4.1). $\Theta_{\gamma 0}$ is the temperature perturbation monopole, i.e., the average of Θ_γ over all directions $\hat{n}$, whereas $\vec{v}_{\mathrm{B}}$ is the (common) bulk velocity of baryons and electrons, as these particles are tightly coupled by Coulomb interactions. Because we are studying scalar perturbations, we only need to consider the irrotational part of $\vec{v}_B$, which is fully given in terms of its divergence θ_B.

The interaction term on the r.h.s. of (5.13) is rather intuitive. In the tightly coupled limit, Thomson scattering forces Θ_γ to be equal to $\Theta_{\gamma 0} + \hat{n} \cdot \vec{v}_B$, i.e., to be independent of direction apart from a dipole term due to the Doppler effect, or in other words to the motion of the photon–baryon fluid with respect to an observer with fixed coordinates. Indeed, in a particular frame comoving with the baryons, $\vec{v}_B$ would vanish, and the role of Thomson scattering would be to enforce an isotropic temperature distribution such that $\Theta_\gamma = \Theta_{\gamma 0}$. This is what we expect in a fluid in thermal equilibrium.

The role of a CMB Boltzmann code such as CMBFAST (Seljak and Zaldarriaga, 1996), CAMB (Lewis *et al.*, 2000), CMBEASY (Doran, 2005) or CLASS (Blas *et al.*, 2011) is precisely to solve Eq. (5.13) coupled with other equations of evolution for baryons, dark matter or neutrinos (the latter two being coupled only gravitationally). This linear system of differential equations describes the probability evolution in the stochastic theory of cosmological perturbations (as we shall see in more detail in Section 5.1.4). The system can be more conveniently solved in Fourier space. It is also more practical to reduce the dimensionality of the problem by expanding $\Theta_\gamma(\eta, \vec{k}, \hat{n})$ in spherical harmonics. We note that in the Fourier transform of Eq. (5.13), $\hat{n}$ enters only through the combination $\mu \equiv \hat{n} \cdot \hat{k}$, i.e., the cosine of the angle between the direction of propagation and the wavevector. This rotational symmetry of the equation around the axis defined by $\hat{n}$ is a consequence of the isotropy of the Friedmann background. Moreover, initial conditions respect the same symmetry, because the temperature of tightly coupled photons can only depend on direction through a monopole and a dipole oriented along $\hat{n}$. Hence, we can consider Θ_γ as a function of $(\eta, \vec{k}, \mu)$ only and perform a one-dimensional Legendre transformation with respect to μ, instead of a two-dimensional transformation with respect to $\hat{n}$,

$$\Theta_\gamma(\eta, \vec{k}, \mu) = \sum_l (-i)^l (2l+1) \Theta_{\gamma l}(\eta, \vec{k}) P_l(\mu), \qquad (5.14)$$

where $\Theta_{\gamma 0}(\eta, \vec{k})$ is just the Fourier transform of the monopole which already appeared in Eq. (5.13). After these transformations, the Boltzmann equation leads to an infinite hierarchy of coupled equations for the multipole moments of Θ_γ:

$$\Theta_{\gamma 0}' = -k\Theta_{\gamma 1} + \phi'$$

$$\Theta_{\gamma 1}' = \frac{k}{3}\left(\Theta_{\gamma 0} - 2\Theta_{\gamma 2} + \psi\right) + an_e\sigma_T\left(\frac{\theta_B}{3k} - \Theta_{\gamma 1}\right) \qquad (5.15)$$

$$\Theta_{\gamma l}' = \frac{k}{(2l+1)}\left[l\Theta_{\gamma(l-1)} - (l+1)\Theta_{\gamma(l+1)}\right] - an_e\sigma_T\Theta_{\gamma l}, \qquad \forall l \geq 2.$$

One can show that the first three momenta are related to the scalar degrees of freedom introduced in the previous subsection through

$$\delta_\gamma = 4\Theta_{\gamma 0}, \qquad \theta_\gamma = 3k\Theta_{\gamma 1}, \qquad \sigma_\gamma = 2\Theta_{\gamma 2}. \qquad (5.16)$$

We notice that in the tightly coupled regime, $\Theta_\gamma(\eta, \vec{k}, \mu)$ is the sum of an isotropic term and a Doppler term. This means that all multipoles $\Theta_{\gamma l}(\eta, \vec{k})$ with $l \geq 2$ are negligible in this limit.

Baryons

Baryons (or, as we mentioned previously, the tightly coupled baryon–electron fluid) can be described by exactly the same Boltzmann equation as photons, maintaining their phase space density f_B, with a coupling term opposite to the one in the photon equation. As for photons, it reduces to an equation for the baryon temperature fluctuation Θ_B. However, further simplifications arise in the limit of nonrelativistic baryons, because all Legendre momenta Θ_{Bl} can be neglected apart from the monopole, related to the density fluctuation δ_B, and the dipole, which gives the bulk velocity divergence θ_B. After some algebra, the full Boltzmann equation eventually reduces to the energy conservation equation and the Euler equation sourced by Thomson scattering,

$$\delta_B' = -\theta_B + 3\phi'$$
$$\theta_B' = -\frac{a'}{a}\theta_B + k^2\psi + R^{-1} a n_e \sigma_T(\theta_\gamma - \theta_B), \qquad (5.17)$$

where we have defined the baryon-to-photon ratio R, rescaled by a factor of 3/4 to get simpler equations in the following:

$$R(\eta) \equiv \frac{3}{4}\frac{\overline{\rho}_B(\eta)}{\overline{\rho}_\gamma(\eta)}. \qquad (5.18)$$

The factor $\overline{\rho}_B$ appearing in the denominator in the baryon Boltzmann equation (5.17) should not be a surprise. The limit $\overline{\rho}_B \to \infty$ corresponds to an infinite average baryon mass, i.e., to the limit where no photon carries enough energy to transfer momentum to the baryon–electron fluid in each Thomson scattering process. In this case, the coupling term does not affect the bulk velocity of baryons. Note that in Eqs. (5.17), we neglected any pressure term (i.e., any particle velocity dispersion) for the baryons. This approximation is valid for the large cosmological scales in which we are interested in this book. It would be incorrect on scales smaller than the baryon Jeans length, which is today in the range of galactic scales, i.e., deep in the nonlinear regime.

Cold dark matter

Cold dark matter (CDM) is assumed to have decoupled from other species when already nonrelativistic and well before the CMB formation epoch. Its Boltzmann equation reduces to the same equations as for baryons but without the coupling term:

$$\delta'_C = -\theta_C + 3\phi'$$

$$\theta'_C = -\frac{a'}{a}\theta_C + k^2\psi. \tag{5.19}$$

Another way to derive these equations is to write the energy conservation and Euler equations for a species such that the pressure $\overline{P}$, pressure perturbation δP, and anisotropic pressure potential σ all vanish.

Neutrinos

Because the impact of neutrinos on the CMB is the main topic of this chapter, we will present and discuss neutrino equations separately in Section 5.2. In the current section, we limit our description of CMB physics to the case of a neutrinoless universe.

Einstein equations

The list of evolution equations introduced up to now would form a closed system if metric perturbations ϕ and ψ were known. These functions can be inferred from the total perturbed stress–energy tensor through Einstein equations. For scalar perturbations, Einstein equations provide four independent relations which, along with the previous equations of motion, would lead to a redundant system. The reason is that through Bianchi identities, Einstein equations imply two equations of conservation for the scalar part of the total stress–energy tensor. The same equations could be derived by combining the equations of motion of individual species. Hence, a practical way to solve the full system in the Newtonian gauge without redundancy is to use the equation of motion of each species, plus two Einstein equations playing the role of constraint equations, and providing ϕ and ψ at each step as a function of total matter perturbations. The first can be found from $\delta G^0_0 = 8\pi G \, \delta T^0_0$:

$$k^2\phi + 3\frac{a'}{a}\left(\phi' + \frac{a'}{a}\psi\right) = -4\pi G \, a^2 \sum_i \delta\rho_i. \tag{5.20}$$

Deep inside the Hubble length, this equation gives the Poisson equation, because for $k \gg \frac{a'}{a}$ one gets

$$-\frac{k^2}{a^2}\phi = 4\pi G \, \delta\rho_{\text{tot}}. \tag{5.21}$$

On the l.h.s., we recognize the Fourier transform of $a^{-2}\Delta\phi$, where Δ is the Laplacian defined with respect to comoving coordinates, such that $a^{-2}\Delta$ is the physical Laplacian. The r.h.s. is the usual source term in the Poisson equation (in the context of general relativity, the energy density fluctuation plays the role of the total mass density in the Newtonian version of the Poisson equation). To close the system, we need one more component of the Einstein equations, for instance, the one corresponding to the anisotropic stress component of the stress–energy tensor:

$$k^2(\phi - \psi) = 12\pi G a^2 \sum_i (\overline{\rho}_i + \overline{p}_i)\sigma_i. \tag{5.22}$$

We see that as long as the total anisotropic stress can be neglected, the two metric fluctuations coincide. Strictly speaking, it is the gradient of $\phi - \psi$ which vanishes, but $\phi = \psi$ is the only solution of $\Delta(\phi - \psi) = 0$ which is regular everywhere, does not diverge at infinity, and remains compatible with linear perturbation theory.

5.1.3 Adiabatic and isocurvature modes

The full system of coupled differential equations describing linear perturbations that we introduced in the last subsection includes two first-order equations for baryons, two for CDM, and an infinite hierarchy of equations for photons. However, we should keep in mind that as long as photons are tightly coupled, or for wavelengths greater than the Hubble scale, all multipoles above $l = 0$ (the monopole accounting for local density fluctuations) and $l = 1$ (the dipole due to the relative motion between the tightly coupled fluid and the coordinate frame) are negligible. We have seen in the last section how the coupling term in the Boltzmann equation enforces such a simplification, which is a natural consequence of the fact that in a strongly interacting fluid, the kinetics can be entirely described by a bulk velocity. This reduces the effective number of equations of evolution for photons to two, as for baryons. When computing the CMB or the large scale structure (LSS) power spectra, one always chooses initial conditions deep in the super-Hubble and tightly coupled regime, in which this simplification holds. In summary, on super-Hubble scales, all perturbations in a universe with three species (photons, baryons, CDM) can be described by six first-order equations. More generally, for N species, one would get $2N$ first-order equations.[1]

Such a system admits $2N$ independent initial conditions. Half of them can be shown to seed decaying modes that we cannot observe today, because the other half lead to much larger fluctuations. In studying possible mechanisms for the generation

---

[1] We will see in Section 5.2.2, where we introduce the so-called neutrino velocity isocurvature mode, that collisionless species such as neutrinos constitute a small exception to this rule.

of initial conditions (inflation, dynamics of spontaneous symmetry breaking, etc.), one task consists of identifying which combination of the N nondecaying solutions gets excited. Eventually, in complicated mechanisms in which the generation of initial conditions arises from a superposition of random processes, more than one combination can be excited, with or without statistical correlations between the various initial modes (e.g., in multiple-field inflation).

One particular combination has a particularly simple physical interpretation. In a perfectly homogeneous universe, the Friedmann law combined with our knowledge of particle physics and thermodynamics allows us to predict the evolution of homogeneous densities and pressures $\overline{\rho}_i(t)$ and $\overline{P}_i(t)$ for each species (here t stands for whatever definition of time we are adopting in the Friedmann universe). The simplest realization of an inhomogeneous universe that we can think of is the following: assume that some mechanism introduces a local time-shift (accounting, for instance, for inflationary fluctuations: in single-field inflation, the inflaton is the only clock in the quasi-De Sitter universe, and its fluctuations can be seen as local shifts with respect to average time). In such a hypothetical universe, densities and pressures would be described by functions

$$\rho_i(t, \vec{x}) = \overline{\rho}_i(t + \delta t(\vec{x})) \simeq \overline{\rho}_i(t) + \dot{\overline{\rho}}_i(t)\,\delta t(\vec{x})$$

$$P_i(t, \vec{x}) = \overline{P}_i(t + \delta t(\vec{x})) \simeq \overline{P}_i(t) + \dot{\overline{P}}_i(t)\,\delta t(\vec{x}), \qquad (5.23)$$

where the time-shift function $\delta t(\vec{x})$ (assumed to be of order one in perturbations) would be the same for all species. This assumption makes sense and is preserved by the time evolution, at least on super-Hubble wavelengths, for which different worldlines can be thought to have independent histories. This singles out a specific subclass of initial conditions subject to

$$\frac{\delta\rho_i}{\overline{\rho}_i + \overline{P}_i} = \frac{\dot{\overline{\rho}}_i(t)}{\overline{\rho}_i + \overline{P}_i}\delta t(\vec{x}) = -3\frac{\dot{a}(t)}{a(t)}\delta t(\vec{x}), \qquad (5.24)$$

which *does not* depend on the species i (for the last equality, we have used the energy conservation equation (2.48)). In such a perturbed universe, some remarkable properties would emerge. First, each species would have an adiabatic sound speed at least on super-Hubble scales; i.e., the ratio $\delta P_i/\delta\rho_i$ would be given by

$$\frac{\delta P_i(t, \vec{x})}{\delta\rho_i(t, \vec{x})} = \frac{\dot{\overline{P}}_i(t)}{\dot{\overline{\rho}}_i(t)} \equiv c_{a,i}^2(t). \qquad (5.25)$$

In addition, the total perturbations (summed over all species) would also be described by an effective sound speed:

$$\delta P(t, \vec{x}) = c_s^2(t)\delta\rho(t, \vec{x}), \qquad \text{with } c_s^2(t) \equiv \frac{\sum_i \dot{\overline{\rho}}_i(t)c_{a,i}^2(t)}{\sum_i \dot{\overline{\rho}}_i(t)}. \qquad (5.26)$$

This property is not true in the general case. Without assuming Eq. (5.23), we could write the total pressure perturbation only as a sum over N independent functions of $\vec{x}$, which could eventually be arranged as

$$\delta P(t, \vec{x}) = \sum_i c_{\mathrm{s},i}^2(t)\delta\rho_i(t, \vec{x})$$

$$= c_\mathrm{s}^2(t)\delta\rho(t, \vec{x}) + \sum_{i \neq j} d_{ij}(t)\left[\frac{\delta\rho_i}{\overline{\rho}_i + \overline{P}_i} - \frac{\delta\rho_j}{\overline{\rho}_j + \overline{P}_j}\right], \quad (5.27)$$

where the term between brackets stands for the entropy perturbation δS_{ij} of the fluid i compared to another fluid j (chosen arbitrarily as a reference), and the coefficients $d_{ij}(t)$ correspond to the partial derivatives of total pressure P with respect to S_{ij}. Hence, any set of perturbations satisfying Eq. (5.23) is such that the fluctuations of the total effective fluid have adiabatic properties. These solutions of the perturbation equations are called *isentropic* or *adiabatic* modes, whereas any other solution would feature entropy perturbations.

In order to define actual adiabatic initial conditions, one should impose Eq. (5.23) and solve the remaining system of differential equations, composed of only two independent equations. This leads to a basis of two independent initial conditions. When this basis is chosen appropriately, one of the solutions becomes quickly negligible with respect to the other one as time evolves. These two solutions (defined up to an arbitrary normalization constant) are called the decaying and growing *adiabatic modes*. Without Eq. (5.23) being imposed, the full system would have a basis of initial conditions including these two modes plus $2N - 2$ entropy modes. A customary way to fix the basis of nonadiabatic modes consists of picking up linear combinations such that all perturbations vanish, except for two fluids with opposite density fluctuations exactly cancelling each other in the expression of spatial curvature fluctuations (at least, in the asymptotic super-Hubble limit). These combinations are called *isocurvature modes*. Any general initial condition can be decomposed on the basis of solutions formed by one growing adiabatic mode, $(N - 1)$ growing isocurvature modes, and N decaying modes. Nonadiabatic modes include CDM isocurvature modes (with entropy perturbations between CDM and photons), baryon isocurvature modes, plus other modes if there are more species present in the universe (neutrino isocurvature modes will be mentioned in Section 5.2.2).

Although the simplest mechanisms for the generation of primordial perturbations can be formulated as a unique time-shift effect, and lead to adiabatic perturbations, isocurvature modes might be excited by more complicated mechanisms such as multiple-field inflation in the early universe, leading to a superposition of several independent time-shift functions. However, even when some isocurvature modes are excited at early times, the assumption that all species were once in thermal equilibrium implies that they share the same local fluctuations in number density, i.e., that any entropy perturbation has been erased. Hence, isocurvature modes should be taken into account only if they are attached to an always decoupled species. A known exception to this rule applies to particles with a sizeable chemical potential, as discussed by Malik and Wands, 2009. Even in thermal equilibrium, such particles can have a chemical potential $\mu(t, \vec{x})$, which could in principle fluctuate spatially, in such a way as to break the adiabaticity condition. This cannot be the case for the photons, baryons and CDM particles discussed here, but we will come back to this issue in the case of neutrinos.

In summary, isocurvature modes require some very specific assumptions and are not expected to be relevant in the simplest cosmological models. Of course, this conclusion is based on theoretical priors. Ultimately, the data should tell us whether isocurvature modes are present or not. So far, all CMB and LSS observations are compatible with purely adiabatic initial conditions, and put strong limits on the amplitude of isocurvature modes which, if any, should be small with respect to the dominant adiabatic contribution. Whenever more precise data become available, these limits will become stronger, or on the contrary we might discover a tiny isocurvature contribution. In the rest of this book, we will assume for simplicity that the universe can be described in terms of purely adiabatic initial conditions.

5.1.4 Power spectra and transfer functions

Power spectrum

The theory of cosmological perturbations is a stochastic theory, whose goal is to predict the statistical properties of perturbations at some arbitrary time η, given the statistical properties of perturbations at initial time η_{in} (inferred from quantum field theory in the case of inflationary cosmology). The simplest possible assumption compatible with all observations at the time of writing is that the early universe features gaussian fluctuations. As long as perturbations remain linear, gaussianity is preserved and all fluctuations can be described entirely by their two-point correlation function in real or in Fourier space, for instance, $\langle A(\eta, \vec{k})A^*(\eta, \vec{k}')\rangle$ for an arbitrary quantity A. For a stochastic gaussian field, different wavevectors are uncorrelated, and the Fourier two-point correlation function is proportional to the Dirac distribution $\delta^{(3)}(\vec{k} - \vec{k}')$. The coefficient of proportionality is called the power spectrum P_A. In a statistically isotropic universe such as the Friedmann universe, the power spectrum can be a function of only the wavenumber k, not of the direction $\hat{k}$. Finally, many authors use the notation $\mathcal{P}_A$ for the power spectrum rescaled by a factor $k^3/(2\pi^2)$:

$$\mathcal{P}_A(k) = \frac{k^3}{2\pi^2} P_A(k). \tag{5.28}$$

The technical reason for this redefinition is that $\mathcal{P}_A(k)$ represents the contribution of each logarithmic interval in wavenumber space to the two-point correlation function in real space.

The goal of the theory of linear cosmological perturbations is to predict the power spectrum of all quantities at any time, given some primordial power spectrum accounting for initial conditions, e.g., at the end of inflation.

Transfer functions

Like the power spectrum, the system of linear differential equations for cosmological perturbations does not depend on the wavevector direction $\hat{k}$ in the Friedmann universe. Hence this system can be solved only once for each wavenumber k, starting from an arbitrary initial condition. Let us assume, for instance, that we normalize the solution to $\Theta_{\gamma 0}(\eta_{\rm in}, \vec{k}) = 1$; the power spectrum of $\Theta_{\gamma l}$ at a given time would then be given by the actual power spectrum of $\Theta_{\gamma 0}$ at initial time, multiplied by the square of such a solution $\Theta_{\gamma l}(\eta, \vec{k})$.

By convention, in a universe with only adiabatic initial conditions, the initial normalization often refers to a quantity $\mathcal{R}$, called the comoving curvature perturbation. Indeed, in the comoving gauge (in which equal-time hypersurfaces are orthogonal at each point to the total velocity of the cosmic fluid), the dimensionless quantity $\mathcal{R}$ represents the local fluctuation of the spatial curvature in comoving units. In the Newtonian gauge, the curvature perturbation is given by

$$\mathcal{R} = \psi - \frac{1}{3} \frac{\delta\rho_{\rm tot}}{\overline{\rho}_{\rm tot} + \overline{P}_{\rm tot}}. \tag{5.29}$$

All equations can be solved starting from the arbitrary initial condition $\mathcal{R}(\eta, \vec{k}) = 1$. The power spectrum of a given quantity at time η will then be given by the square of the solution multiplied by the initial power spectrum of $\mathcal{R}$.

Equivalently, we could say that all equations of evolution can be divided by a normalizing quantity, namely $\mathcal{R}(\eta_{\rm in}, \vec{k})$ in the usual case. We can then solve the system and find the solution for

$$\Theta_{\gamma l}(\eta, k) \equiv [\Theta_{\gamma l}(\eta, \vec{k})/\mathcal{R}(\eta_{\rm in}, \vec{k})] \tag{5.30}$$

$$\delta_{\rm B}(\eta, k) \equiv [\delta_{\rm B}(\eta, \vec{k})/\mathcal{R}(\eta_{\rm in}, \vec{k})] \tag{5.31}$$

and similar rescaled variables for other perturbations. The quantities on the l.h.s., depending on k rather than $\vec{k}$, are called transfer functions. Their square multiplied by the initial curvature power spectrum gives the power spectrum of the corresponding quantity at time η. In the following, any perturbation of photons, baryons, CDM, etc., written as a function of a wavenumber instead of a wavevector, will stand for a transfer function normalized with respect to curvature: $f(\eta, k) \equiv [f(\eta, \vec{k}) / \mathcal{R}(\eta_{in}, \vec{k})]$.

Primordial spectrum

The primordial curvature spectrum $\mathcal{P}_{\mathcal{R}}(k)$ is not a function of time, because comoving curvature is conserved on super-Hubble scales (at least in the absence of isocurvature modes). According to the previous definitions, we have

$$\langle \mathcal{R}(\eta_{in}, \vec{k}) \, \mathcal{R}^*(\eta_{in}, \vec{k}') \rangle = \frac{2\pi^2}{k^3} \mathcal{P}_{\mathcal{R}}(k) \, \delta^{(3)}(\vec{k} - \vec{k}'). \tag{5.32}$$

This spectrum can be used to express all other spectra at any time; for instance,

$$\langle \Theta_{\gamma l}(\eta, \vec{k}) \, \Theta_{\gamma l}^*(\eta, \vec{k}') \rangle = \frac{2\pi^2}{k^3} \mathcal{P}_{\mathcal{R}}(k) \, [\Theta_{\gamma l}(\eta, k)]^2 \, \delta^{(3)}(\vec{k} - \vec{k}') \tag{5.33}$$

for all η.

Inflation predicts a nearly scale-invariant power spectrum (i.e., a nearly flat $\mathcal{P}_{\mathcal{R}}(k)$), as a consequence of the nearly constant energy density of the universe during the inflation stage. At first order, deviations from scale-invariance are accounted for by a tilt n_s close to unity,

$$\mathcal{P}_{\mathcal{R}}(k) = A_s (k/k_0)^{n_s - 1}, \tag{5.34}$$

where A_s stands for the primordial spectrum amplitude at the arbitrary pivot scale k_0. Beyond leading order, one could consider the running of the tilt with k, the running of the running, etc. A detailed study of quantum scalar perturbations during inflation shows that the tilt running and higher terms in the expansion are negligible (unless in very specific, nonminimal models), and makes it possible to relate A_s and n_s to the amplitude and shape of the inflaton potential $V(\phi)$ within a small range of field values $\Delta\phi$, called the observable window of inflation.

5.1.5 Acoustic oscillations

Sound speed

A precise solution of the system of differential equations describing cosmological perturbations can only be obtained numerically. Many analytic approximations have been discussed in the literature, at different levels of precision (see, e.g., Hu, 1995). Here, we will remain at the level of a qualitative discussion.

On times and scales for which the baryons and photons can be considered a single tightly coupled fluid, we can compute the sound speed of perturbations in this fluid as

$$c_s^2 = \frac{\delta P_\gamma + \delta P_B}{\delta\rho_\gamma + \delta\rho_B}. \tag{5.35}$$

Photons and baryons are then in thermal equilibrium with a temperature $T(\eta, \vec{x})$. Relativistic photons satisfy $\delta_\gamma = 4\,\delta T/T$ and $\delta P_\gamma = \delta\rho_\gamma/3$, whereas for nonrelativistic baryons we can use $\delta_b = 3\,\delta T/T$ and neglect δP_B. The sound speed is then equal to

$$c_s^2 = \frac{1}{3(1+R)}, \tag{5.36}$$

where we recall that $R \equiv 3\bar{\rho}_B/(4\bar{\rho}_\gamma)$. This ratio increases like the scale factor. It remains much smaller than one during radiation domination, and becomes of order one at the beginning of matter domination, i.e., near the recombination time. Hence, the sound speed in the fluid remains equal to $1/\sqrt{3}$ during radiation domination (or $c/\sqrt{3}$ in physical units), and then drops slowly down to zero. As long as the sound speed is nonzero, acoustic waves propagate in the fluid with a velocity c_s.

Sound horizon

If initial gravitational and pressure forces would compensate for each other exactly at each point, the tightly coupled fluid would be in equilibrium, with no propagation of acoustic waves. However, primordial perturbations (seeded by inflation, and/or eventually by some alternative mechanism) drive the system locally out of equilibrium at the initial time, for both adiabatic and nonadiabatic initial conditions. Starting from such perturbations, acoustic waves propagate causally, within a distance called the sound horizon. The comoving sound horizon (i.e., the comoving distance travelled by a wavefront since some arbitrary time deep inside the radiation-dominated regime) is given simply by

$$r_s(\eta) = \int_{t_{in}}^{t} \frac{c_s(t)dt}{a(t)} = \int_{\eta_{in}}^{\eta} c_s(\eta')d\eta'. \tag{5.37}$$

This quantity is indeed independent of η_{in} as long as it is much smaller than η. During radiation domination, r_s is equal to the comoving causal horizon η divided by $\sqrt{3}$.

A driven oscillator

Acoustic waves are density waves in the photon–baryon fluid. They can be represented by the variable $\Theta_{\gamma 0}(\eta, k)$, which remains equal to $\frac{1}{4}\delta_\gamma(t, k)$ and $\frac{1}{3}\delta_B(\eta, k)$ as a consequence of thermal equilibrium as long as the tight-coupling approximation holds. The most naive expectation would be that in Fourier space, this variable evolves like $\Theta_{\gamma 0}(\eta, k) \sim \cos[kr_s(\eta) + \varphi]$. In particular, for $kr_s(\eta) \ll 1$, i.e., for wavelengths much larger than the sound horizon, perturbations should still be frozen.

Such solutions indeed emerge from tight-coupling approximations to the full differential system, but not in such a trivial form. Indeed, the equation governing the evolution of $\Theta_{\gamma 0}(\eta, k)$ differs form that of a harmonic oscillator for essentially three reasons:

- The sound speed depends on time, so the oscillator equation for $\Theta_{\gamma 0}(\eta, k)$ has a time-dependent mass.
- The photon–baryon fluid feels gravitational forces, seeded by its own overdensities, and by gravitational interactions with other species (CDM, or neutrinos, as we will see in the next sections). It is also affected by other general relativity effects, such as local time dilation. These effects are accounted for by gradients and time derivatives of metric fluctuations, which act as a driving term in the oscillator equation for $\Theta_{\gamma 0}(\eta, k)$. Because baryons are nonrelativistic, this driving term also evolves when the baryon-to-photon ratio R increases.
- The increasing value of R also changes other properties of the baryon–photon fluid, such as its inertia.

We can track these different effects in the equation governing the evolution of $\Theta_{\gamma 0}(\eta, k)$ at leading order in the tight-coupling limit, which is easy to obtain from Eqs. (5.15) and (5.17). In the limit $\sigma_T \to \infty$, we have already seen that Eq. (5.15) implies $\theta_B = \theta_\gamma = 3k\Theta_{\gamma 1}$, and $\Theta_{\gamma l} = 0$ for $l \geq 2$. We can combine the second of Eqs. (5.15) with the second of Eqs. (5.17) to eliminate the interaction term. In the remaing equation, we use $\theta_B = 3k\Theta_{\gamma 1}$, $\Theta_{\gamma 2} = 0$, and we eliminate $\Theta_{\gamma 1}$ using the first of Eqs. (5.15). Finally, we can replace $\frac{a'}{a}R$ by R', and $3(1 + R)$ by c_s^{-2}. We are left with the second-order differential equation

$$\Theta_{\gamma 0}'' + \frac{R'}{1 + R}\Theta_{\gamma 0}' + k^2 c_s^2 \Theta_{\gamma 0} = -\frac{k^2}{3}\psi + \frac{R'}{1 + R}\phi' + \phi'', \qquad (5.38)$$

where we can clearly identify a baryon-induced damping term $\frac{R'}{1+R}\Theta_{\gamma 0}'$, a pressure term with time-dependent effective mass $k^2 c_s^2 \Theta_{\gamma 0}$, and a gravitational driving term on the r.h.s.

Diffusion damping

On top of these effects, we must take into account the fact that close to recombination, the tight-coupling approximation breaks down. Random scattering processes tend to erase perturbations below the photon diffusion length. In a first-order approximation, this length can be found by treating photon diffusion as a random walk, in which photons would pick up a completely random direction at each new interaction with an electron. In this limit, the comoving distance travelled by a photon between the early universe and some time η would follow

$$r_d^2(\eta) \sim \int_{\eta_{\rm in}}^{\eta} d\eta\, \Gamma_\gamma\, r_\gamma^2, \qquad (5.39)$$

where Γ_γ stands for the interaction rate computed with respect to conformal time, and r_γ for the photon mean free path in comoving space. In our case, we know that $\Gamma_\gamma = a n_e \sigma_{\rm T}$. The comoving mean free path (always in natural units) is then given by $r_\gamma = \delta x = \delta\eta = 1/\Gamma_\gamma = (a n_e \sigma_{\rm T})^{-1}$. Finally, we get the expression

$$r_d^2(\eta) \sim \int_{\eta_{\rm in}}^{\eta} \frac{d\eta}{a n_e \sigma_{\rm T}}, \qquad (5.40)$$

which does not depend on $\eta_{\rm in}$, provided that $\eta \gg \eta_{\rm in}$. A better approximation to r_d can be found for instance in Hu, 1995, but the preceding result is by far sufficient for understanding the effect of diffusion damping on the CMB spectrum. Photon diffusion will erase perturbations with a wavelength smaller than $\lambda_d \equiv a\, r_d$, i.e., with a wavenumber greater than $k_d \equiv 2\pi/r_d$.

All these driving and damping effects lead to an interesting phenomenology for acoustic oscillations, that we will summarize below. This discussion is essential for understanding the impact of cosmological parameters (and later of neutrinos) on the CMB spectrum. We will distinguish two stages: radiation domination, and the epoch between radiation/matter equality and decoupling.

Radiation domination: Constant acoustic oscillations

During radiation domination, approximate analytic solutions are easy to obtain. One can work in the limit where $R = 0$, $c_s = 1/\sqrt{3}$, baryons or cold dark matter density perturbations are negligible with respect to the photons' perturbations and the photon–baryon fluid is tightly coupled. A combination of the $0-0$ and $i-i$ components of the Einstein equation then leads to a simple second-order differential equation. Its two solutions correspond to the growing and decaying adiabatic modes. Indeed, as long as we neglect all fluids but one (the photons), we cannot find entropy modes. The growing solution, involving the Bessel function $J_{3/2}(k c_s \eta)$, has the following asymptotic behaviour:

- At wavelengths larger than the sound horizon, all transfer functions are constant over time,

$$4\Theta_{\gamma 0}(\eta, k) = \delta_\gamma(\eta, k) = \frac{4}{3}\delta_{\rm B}(\eta, k) = -2\phi(\eta, k) = -2\psi(\eta, k) = -\frac{4}{3} \quad (5.41)$$

(given our definition of transfer functions in Section 5.1.4, this means that the photon perturbations are related to the initial curvature perturbation through

$\Theta_{\gamma 0}(\eta, \vec{k}) = -\frac{1}{3}\mathcal{R}(\eta, \vec{k}))$. In this regime, acoustic waves propagation is negligible because the comoving sound horizon $r_s \simeq \eta/\sqrt{3}$ is very small with respect to the comoving wavelength $\frac{2\pi}{k}$, and the modes are still frozen to their initial value.

- Inside the sound horizon,

$$4\Theta_{\gamma 0}(\eta, k) = \delta_\gamma = \frac{4}{3}\delta_B(\eta, k) = 4\cos\left[kr_s(\eta)\right] \tag{5.42}$$

$$-2\phi(\eta, k) = -2\psi(\eta, k) = -\frac{6}{(k\eta)^2}\cos\left[kr_s(\eta)\right]. \tag{5.43}$$

Thus, after entering the sound horizon, the modes of photon density oscillate with a constant amplitude, whereas metric fluctuations decay with time. Gravitational forces and time dilation effects then become negligible with respect to pressure forces. Indeed, by inserting these solutions into Eq. (5.38), we clearly see that for $(k\eta) \gg 1$, the driving term on the r.h.s. can be neglected with respect to the term $k^2 c_s^2 \delta_\gamma$ induced by photon pressure. In this limit, Eq. (5.38) reduces to a simple harmonic oscillator equation $\delta_\gamma'' + \frac{k^2}{3}\delta_\gamma = 0$. One can also check that the Poisson equation (5.21) combined with the Friedmann equation gives the correct relation between the asymptotic solutions (5.42) and (5.43).

From equality to decoupling: Damped acoustic oscillations

Several phenomena make the evolution more complicated from the time of matter/radiation equality until that of decoupling:

- The growth of the baryon fraction R and the decrease of the sound speed c_s affect the amplitude of the acoustic oscillations.
- The growth of R also implies that the photon–baryon fluid couples more and more to gravity for the same amount of pressure, shifting the zero point of oscillations in such a way as to increase the fluid density in gravitational potential wells. We can check this from Eq. (5.38). Neglecting the time variation of ϕ, the zero point of temperature oscillations corresponds to $k^2 c_s^2 \Theta_{\gamma 0} = -\frac{k^2}{3}\psi$, i.e., to $\Theta_{\gamma 0} = -(1 + R)\psi$. Taking the real space counterpart of this relation, we see that for a gravitational potential well ($\psi < 0$), the value of the overdensity $\delta_\gamma = 4\Theta_{\gamma 0} > 0$ that corresponds to this equilibrium increases with R.
- The metric perturbations are now also influenced by the nonrelativistic matter components (baryons and potentially cold dark matter). So inside the Hubble radius, ϕ and ψ do not decay as quickly as during radiation domination. Thus the gravitational driving term in Eq. (5.38) affects the temperature evolution in a different way than during radiation domination.

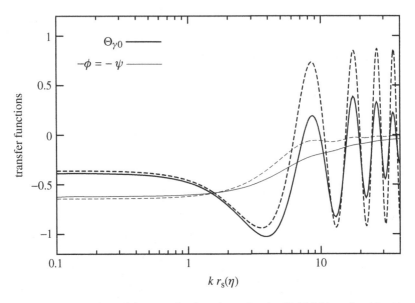

Figure 5.1 A snapshot of the transfer functions $\Theta_{\gamma 0}(\eta, k)$ (thick) and $-\phi(\eta, k) = -\psi(\eta, k)$ (thin) at the time of equality (dashed) and decoupling (solid), in a neutrinoless ΛCDM universe. The wavenumber axis has been rescaled in each case by the exact value of $r_s(\eta)$, to show that the phase of the oscillations is determined by the sound horizon. The negative signs of $\Theta_{\gamma 0}$ and $-\psi$ in the long-wavelength limit comes from the fact that transfer functions have all been normalized with respect to the curvature perturbation $\mathcal{R}$. A positive curvature perturbation corresponds to a gravitational potential hill ($\psi > 0$) and to an underdense region ($\delta_\gamma < 0$) with lower temperature ($\Theta_{\gamma 0} < 0$), and vice versa.

- The baryons and photons cannot be modeled as a single tightly coupled fluid at wavelengths below the diffusion length of the photon λ_d. In Fourier space, scattering processes introduce an exponential cutoff in $\Theta_{\gamma 0}(\eta, k)$ shaped like $\exp[-(k/k_d)^2]$.

To keep our discussion reasonably brief, we do not treat each individual effect here, but refer the reader to, e.g., Hu, 1995 for more details. The leading effects are, however, easy to understand. They are illustrated in Fig. 5.1, which shows the transfer functions $\Theta_{\gamma 0}(\eta, k)$ and $-\psi(\eta, k)$ at the times of equality (dashed curves) and of decoupling (solid curves). We recall that $\phi(\eta, k)$ is equal to $\psi(\eta, k)$ as long as there is no anisotropic stress contribution to the Einstein equations, which is the case in a neutrinoless universe before photon decoupling (because $\Theta_{\gamma 2} \simeq 0$). These transfer functions have been solved numerically with the CMB Boltzmann code CLASS (Blas *et al.*, 2011). The difference between the dashed and solid curves corresponds precisely to the few effects described earlier.

At equality, we see that the numerical results agree with the asymptotic solutions (5.41, 5.42, 5.43), with temperature oscillations roughly symmetric around $\Theta_{\gamma 0} = -\psi$ inside the sound horizon, and of constant amplitude in the large-k limit, in which metric fluctuations are negligible.

At decoupling, the amplitude of photon oscillations has been reduced on all subhorizon scales. The zero point of oscillations has been shifted down with respect to $\Theta_{\gamma 0} = -\psi$, leading to an enhancement of the absolute value of the first/third/fifth/... extremum in $\Theta_{\gamma 0}(\eta, k)$ with respect to the second/fourth/sixth/... extremum. Finally, we can see the exponential damping of oscillations in the large-k limit. These three crucial effects are controlled respectively by the duration of the transition stage between equality and decoupling, by the baryon fraction at decoupling and by the value of the diffusion wavenumber k_d.

A last feature can be noticed in Figure 5.1: even in the super-Hubble limit $k \to 0$, one can see a small evolution in the photon and metric transfer functions between equality and decoupling. The reason is that, although the curvature perturbation $\mathcal{R}$ is a conserved quantity on super-Hubble scales in a universe with adiabatic initial conditions, density contrasts and metric perturbations are not. At the time of equality, the equation of state of the universe changes, and all perturbations readjust to another set of constant values, slightly different from those during radiation domination. Using Einstein equations, one can show that for super-Hubble scales and during matter domination the transfer functions are subject to

$$4\Theta_{\gamma 0}(\eta, k) = \delta_\gamma(\eta, k) = \frac{4}{3}\delta_B(\eta, k) = -\frac{8}{3}\phi(\eta, k) = -\frac{8}{3}\psi(\eta, k) = -\frac{8}{5}. \quad (5.44)$$

The coefficients of the last three terms differ slightly from those in Eq. (5.41), valid during radiation domination. In Fig. 5.1, the value of each transfer function at equality and for $k \to 0$ is halfway between Eq. (5.41) and Eq. (5.44), whereas at decoupling it follows Eq. (5.44). This technical detail was worth mentioning in order to understand numerical factors in the Sachs–Wolfe formula that will be introduced in Section 5.1.6.

From decoupling to the current epoch: Gravitational clustering

The evolution of perturbations after decoupling will be discussed in detail in Chapter 6. To anticipate, we simply mention here that in the ideal case of a purely matter-dominated universe, $\phi = \psi$ is constant over time on all scales (super-Hubble and sub-Hubble). In the real universe, this is not true:

• Close to radiation domination, because we know that during radiation domination metric fluctuations quickly decay inside the sound horizon. So, at decoupling, there is still a residual decay of ϕ and ψ.

- During a cosmological constant or dark energy (DE)-dominated stage, which also leads to a decay of the potentials.
- On small scales if there are neutrinos, as we shall see in Chapter 6.

During these stages, nonrelativistic matter components are self-gravitating and their density contrast grows, leading to structure formation.

5.1.6 Temperature anisotropies

Temperature power spectrum

Temperature anisotropies on the last scattering surface can be expanded in spherical harmonics as

$$\frac{\delta T}{T}(\hat{n}) = \sum_{lm} a_{lm} Y_{lm}(\hat{n}). \tag{5.45}$$

The temperature anisotropy seen in a direction $\hat{n}$ is a property of photons traveling along the direction $-\hat{n}$. Hence $\frac{\delta T}{T}(\hat{n})$ coincides with the function $\Theta_\gamma(\eta, \vec{x}, -\hat{n})$ studied in Section 5.1.2, computed today ($\eta = \eta_0$) and at the position of the observer, which we can choose to be the origin for simplicity ($\vec{x} = \vec{o}$). Each a_{lm} can be extracted from the sky map,

$$a_{lm} = (-1)^l \int d\hat{n} \, Y_{lm}^*(\hat{n}) \Theta_\gamma(\eta_0, \vec{o}, \hat{n}), \tag{5.46}$$

where we performed a change of variable $\hat{n} \to -\hat{n}$ inside the integral and used $Y_{lm}(-\hat{n}) = (-1)^l Y_{lm}(\hat{n})$. The function $\Theta_\gamma(\eta, \vec{x}, \hat{n})$ can be expanded in Fourier space and in Legendre multipoles, as in Section 5.1.2. After some simple algebra, the expression for the a_{lm}'s reduces to

$$a_{lm} = (-i)^l \int \frac{d^3k}{2\pi^2} Y_{lm}(\hat{k}) \Theta_{\gamma l}(\eta_0, \vec{k}). \tag{5.47}$$

Given that there is a linear relation between multipoles a_{lm} and Fourier modes $\Theta_{\gamma l}(\eta_0, \vec{k})$, it is clear that

- as long as we assume that first-order cosmological perturbations are gaussian, the a_{lm}'s are also gaussian distributed, and their statistics is fully described by two-point correlation functions $\langle a_{lm} a_{l'm'}^* \rangle$;
- because different Fourier modes of a gaussian random field are uncorrelated, $\langle \Theta_{\gamma l}(\eta, \vec{k}) \Theta_{\gamma l}(\eta, \vec{k}')^* \rangle \propto \delta^{(3)}(\vec{k} - \vec{k}')$, different multipoles are also uncorrelated, $\langle a_{lm} a_{l'm'}^* \rangle \propto \delta_{ll'} \delta_{mm'}$;
- the isotropy of the universe, implying an isotropic power spectrum in Fourier space (depending on k but not on $\hat{k}$), also implies an isotropic harmonic spectrum

(depending on l but not m), denoted as

$$C_l = \langle a_{lm}a_{lm}^* \rangle, \qquad \forall m. \tag{5.48}$$

The harmonic power spectrum C_l is precisely the quantity that we want to compute for a given cosmological model and to compare with observations. As a matter of principle, the true harmonic power spectrum in our universe cannot be extracted from observations, because we observe only one realization of the underlying theory describing its evolution, in a finite fraction of our past light cone. However, using an assumption of ergodicity, we can build an estimator of the true power spectrum, taking advantage of the fact that all multipoles with a given l should have the same variance C_l. For an ideal all-sky experiment providing multipoles a_{lm}^{obs}, the best estimator reads

$$C_l^{\text{obs}} = \frac{1}{2l+1} \sum_{-l \leq m \leq l} |a_{lm}^{\text{obs}}|^2. \tag{5.49}$$

Note that for a realistic experiment affected by partial sky coverage, anisotropic foregrounds and instrumental noise, building optimal estimators becomes a non-trivial task.

This estimator is expected to deviate randomly from the unknown underlying spectrum. The average deviation for a given l can easily be computed for an ideal full-sky experiment. The estimator C_l^{obs} is obtained by averaging over $(2l+1)$ independent gaussian numbers centered at zero, each with variance C_l. Hence, it obeys a χ^2 distribution with $(2l+1)$ degrees of freedom, a mean equal to C_l and a variance $\sqrt{2/(2l+1)}C_l$ (note that this distribution is asymmetric around its peak, especially at low l). As expected, the variance decreases with increasing l, because for high multipoles we can average over more independent realizations of the same stochastic process. This random deviation, playing the role of a theoretical error, is called *cosmic variance*.

Brute-force calculation of the temperature anisotropy spectrum

A brute-force approach to computing the temperature anisotropy spectrum up to some multipole l_{max} consists of integrating all equations with at least l_{max} multipoles in the photon equations Eq. (5.15), between some initial time (at which all wavenumbers of interest are on super-Hubble scales) and today. Using Eq. (5.48), (5.47), (5.33) and the relation of orthogonality of spherical harmonics, we find that the temperature spectrum is given as a function of the photon transfer function (evaluated today) and of the curvature primordial spectrum by

$$C_l = \frac{1}{2\pi^2} \int \frac{dk}{k} [\Theta_{\gamma l}(\eta_0, k)]^2 \mathcal{P}_{\mathcal{R}}(k). \tag{5.50}$$

A technical issue comes from the fact that the hierarchy of coupled photon equations is infinite. Any given numerical algorithm can only integrate over a finite number of multipoles, and the truncation in l-space may cause a reflection of power down to lower multipoles. Let us briefly discuss how to avoid such problems.

As discussed in Sections 5.1.2 and 5.1.3, before photon decoupling, only the first two photon multipoles are important. After that time, higher and higher multipoles are populated. An approximate free-streaming solution (neglecting the gravitational coupling of photons, i.e., neglecting the sourcing of photon perturbations by metric perturbations) reads $\Theta_{\gamma l}(\eta, k) \propto j_l(k\eta)$, where $j_l(x)$ denote spherical Bessel functions. These functions peak near $x = l + \frac{1}{2}$, and feature damped oscillations for $x > l$. This gives a rough idea of the shape of the actual solution. In particular, it shows that at a given time η and as a function of l, $\Theta_{\gamma l}(\eta, k)$ peaks around $l \sim k\eta$, whereas multipoles with $l \gg k\eta$ are vanishingly small. This can be understood geometrically: at time η, the observer sees photons emitted on his own last scattering surface, at a comoving radius $r = (\eta - \eta_{\mathrm{LS}})$ away from him (assuming no spatial curvature). As soon as $\eta \gg \eta_{\mathrm{LS}}$, we can approximate r by η. Fourier modes of comoving wavelength $2\pi/k$ on this surface are seen by the observer under an angle $\theta = 2\pi/(kr)$ in the small-angle approximation. This means that the angular separation between a maximum and a minimum is $\theta = \pi/(kr)$, and that in multipole space the contribution peaks near $l \sim \pi/\theta \sim kr \sim k\eta$. Hence, for a brute-force calculation of the anisotropy spectrum today and until multipole $l_{\max}$, one should choose a maximum value of k on the order of $k_{\max} \sim l_{\max}/\eta_0$. A typical choice in numerical codes is $k_{\max} = 2l_{\max}/\eta_0$. This ensures that even today and for the largest wavenumbers, the transfer function $\Theta_{\gamma l}(\eta, k)$ has vanishingly small multipoles for $l > l_{\max}$, so that a truncation at $l_{\max}$ is harmless.

This brute-force approach is extremely time-consuming, because it relies on the integration of thousands of coupled equations for each wavenumber. In addition, it does not shed much light on the underlying physics. In the next section, we will introduce the alternative line-of-sight approach. This method provides a much more intuitive expression for the various contributions to each C_l, and speeds up Boltzmann codes in a spectacular way.

Line-of-sight integral

Let us consider photons travelling along a given geodesic, between one point on the last scattering surface and a CMB detector. This geodesic is not a straight line, because of gravitational lensing effects caused by intervening matter fluctuations. However, changes in the direction of propagation of photons are only important for second-order perturbation theory. Because in this chapter we limit ourselves to first-order perturbations, we can approximate each photon geodesic as a straight line. This means that a bunch of photons reaching us today from a direction $-\hat{n}$

has been traveling since decoupling in a constant direction $\hat{n}$. Their comoving coordinate at time η was $\vec{x} = -r\hat{n} = -(\eta_0 - \eta)\hat{n}$, because relativistic particles travel toward us with an evolution of the radial coordinate r given by $dr = -d\eta$ in natural units. Any function $\mathcal{F}(\eta, \vec{x}, \hat{n})$ evolves along such a trajectory according to the total derivative

$$\frac{d\mathcal{F}}{d\eta} = \mathcal{F}' + \frac{d\vec{x}}{d\eta} \cdot \vec{\nabla}\mathcal{F} = \mathcal{F}' + \hat{n} \cdot \vec{\nabla}\mathcal{F}, \tag{5.51}$$

where we used the straight-line approximation $d\hat{n}/d\eta = 0$. We wish to integrate the Boltzmann equation over the photon trajectory and relate the temperature fluctuation observed today to the one on the last scattering surface. For this purpose, it is necessary to consider the function $\mathcal{F}(\eta, \vec{x}, \hat{n}) \equiv \Theta_\gamma(\eta, \vec{x}, \hat{n}) + \psi(\eta, \vec{x})$, where Θ_γ is the photon temperature perturbation and ψ one of the two metric perturbations (both introduced in Section 5.1.2). The total derivative of this sum is

$$\frac{d}{d\eta}(\Theta_\gamma + \psi) = \Theta_\gamma' + \psi' + \hat{n} \cdot \vec{\nabla}(\Theta_\gamma + \psi). \tag{5.52}$$

We can use the Boltzmann equation (5.13) and write the r.h.s. as

$$\frac{d}{d\eta}(\Theta_\gamma + \psi) = a n_e \sigma_T (\Theta_{\gamma 0} - \Theta_\gamma + \hat{n} \cdot \vec{v}_B) + \psi' + \phi'. \tag{5.53}$$

We want to put all the terms involving $\Theta_\gamma(\eta, \vec{x}, \hat{n})$ inside a total derivative, in order to integrate it along the line of sight. This is possible after introducing the integral of the scattering rate along the line of sight,

$$\tau(\eta) \equiv \int_\eta^{\eta_0} d\eta \, a n_e \sigma_T, \tag{5.54}$$

called the optical depth. The visibility function $g(\eta) \equiv -\tau(\eta)' e^{-\tau(\eta)}$ can be interpreted as the probability of a photon reaching us today having experienced its last scattering at time η. The decoupling time η_{LS} can actually be defined as the maximum of this function. Note that n_e scales like a^{-3} between electron–positron annihilation and recombination, and remains very small after that time. As a consequence, $\tau(\eta)$ is huge for $\eta \ll \eta_{rec}$ and tiny for $\eta \gg \eta_{rec}$, expressing the transition from an opaque to a transparent universe taking place around recombination. We can now take Eq. (5.53), replace $a n_e \sigma_T = -\tau'$, multiply both sides by $e^{-\tau}$, and rearrange the terms as

$$\frac{d}{d\eta}\left[e^{-\tau}(\Theta_\gamma + \psi)\right] = g\left(\Theta_{\gamma 0} + \psi + \hat{n} \cdot \vec{v}_B\right) + e^{-\tau}(\phi' + \psi'). \tag{5.55}$$

We finally integrate this relation along the line of sight, beween some arbitrary initial time η_{in} chosen well before photon decoupling, and the present time η_0. We can use the fact that in the limit $\eta_{in} \ll \eta_{LS}$, the exponential $e^{-\tau(\eta_{in})}$ is vanishingly small, whereas $e^{-\tau(\eta_0)}$ equals one by definition. We obtain

$$\Theta_\gamma(\eta_0, \vec{o}, \hat{n}) = -\psi(\eta_0, \vec{o}) + \int_{\eta_{in}}^{\eta_0} d\eta \left\{ g \left(\Theta_{\gamma 0} + \psi + \hat{n} \cdot \vec{v}_B \right) + e^{-\tau}(\phi' + \psi') \right\}.$$

(5.56)

The quantity on the l.h.s. is the temperature fluctuation for photons crossing the origin today in a direction $\hat{n}$. Hence, it represents the temperature anisotropy seen by the observer in the direction $-\hat{n}$. The first term on the right is the metric fluctuation today at the observer's location, causing a local blueshifting of incoming photons if we live in a potential well. Because this term is isotropic, a CMB experiment could not distinguish between this contribution and a shift in the average photon temperature, typically of the order of $[\psi(\eta_0, \vec{o}) \times \overline{T}] \sim 10^{-5}\overline{T}$. This effect is tiny, so from now on we will neglect the term $\psi(\eta_0, \vec{o})$ in Eq. (5.56). The remaining terms in the integral show that in a given direction and for a given recombination history, the observed temperature anisotropy depends entirely on two quantities, namely

- the sum $(\Theta_{\gamma 0} + \psi + \hat{n} \cdot \vec{v}_B)$ along the portion of the line-of-sight where the visibility function is not negligible, i.e., around the time of decoupling (and also close to us, if the universe gets reionized at low redshift);
- the sum $(\phi' + \psi')$ along the portion of the line of sight where $e^{-\tau}$ is not negligible, i.e., between decoupling and today.

Because for $\eta \ll \eta_{LS}$, the two functions $g(\eta)$ and $e^{-\tau(\eta)}$ are negligible, the result of the integral in Eq. (5.56) does not depend on the value of η_{in}. This equation is therefore often written in the limit $\eta_{in} = 0$.

Equation (5.56) proves that the knowledge of all photon multipoles $\Theta_{\gamma l}(\eta, \vec{x}, \hat{n})$ with $l > 1$ is not directly needed for computing CMB anisotropies. This suggests that the whole CMB anisotropy spectrum could be obtained with a much more economical method than the brute–force approach described previously. Indeed, the approach that we summarized here in real space can be transposed into Fourier space, as pointed out by Zaldarriaga and Harari, 1995. Starting from the Fourier-expanded version of the Boltzmann equation (5.13), we can put all terms involving $\Theta_\gamma(\eta, \vec{k}, \mu)$ inside a total time derivative, integrate along the line of sight, perform an integration by parts in order to put all the μ-dependence in a factor $e^{ik\mu(\eta - \eta_0)}$, expand this factor in Legendre space using spherical Bessel functions, and obtain the following exact expression for the photon transfer function:

$$\Theta_{\gamma l}(\eta_0, k) = \int_{\eta_{in}}^{\eta_0} d\eta \left\{ g \left(\Theta_{\gamma 0} + \psi \right) + \left(gk^{-2}\theta_B \right)' + e^{-\tau}(\phi' + \psi') \right\} j_l[k(\eta_0 - \eta)],$$

(5.57)

where we omit the argument (η, k) of the four transfer functions $(\phi, \psi, \Theta_{\gamma 0}, \theta_b)$ standing on the r.h.s. The quantity between curly brackets is called the temperature source function, usually denoted by $S^{\mathrm{T}}(\eta, k)$. The temperature anisotropy spectrum then follows from Eq. (5.50).

Equations (5.57) and (5.50) show that to compute the temperature spectrum up to some arbitrary multipole $l_{\max}$, a Boltzmann code needs only to find the evolution of the four quantities $(\phi, \psi, \Theta_{\gamma 0}, \theta_B)$. Of course, to use Einstein equations, it is necessary to include a few more variables, namely the density, velocity and shear perturbation of each species. Including higher-temperature multipoles $\Theta_{\gamma l}$ with $l > 2$ in the differential equations remains necessary to avoid some artificial reflection of power caused by the truncation in l-space. But the truncation scheme should be designed only for accurately computing the first few multipoles. A nice scheme was proposed by Ma and Bertschinger, 1995 in the context of brute-force calculations. Because free-streaming photons approximately follow $\Theta_{\gamma l}(\eta, k) \propto j_l(k\eta)$ in the limit of negligible gravitational interactions, the recurrence relation statisfied by spherical Bessel functions can be used to extrapolate the behaviour of a multipole $\Theta_{\gamma(l+1)}$ given that of multipoles $\Theta_{\gamma l}$ and $\Theta_{\gamma(l-1)}$. This offers a way to smoothly truncate the hierarchy at a given l, limiting the reflection of power with respect to setting $\Theta_{\gamma(l+1)}$ brutally to zero. With such a scheme, a truncation of the Boltzmann hierarchy at an $l_{\max \gamma}$ of a few tens will introduce a very small error into the solution for multipoles $l = 0, 1, 2$. The power of the line-of-sight approach is summarized by the fact that a truncation at $l_{\max \gamma} = O(10)$ is sufficient for computing the C_l's up to $l_{\max} = O(10^3)$.

However, the value of $l_{\max}$ still determines the maximum wavenumber $k_{\max} \sim l_{\max}/\eta_0$ at which the source function needs to be evaluated, because this wavenumber carries information about perturbations on the last scattering surface seen today under an angle $\theta \sim \pi/l_{\max}$. The reduction of the number of differential equations takes place in multipole space, not in wavenumber space. But the factor $[l_{\max \gamma}/l_{\max}]$ makes it possible to gain a few orders of magnitude in computation time over the brute-force approach.

The line-of-sight approach was implemented for the first time in the Boltzmann code CMBFAST by Seljak and Zaldarriaga, 1996, and is used by all other modern codes.

Sachs–Wolfe, Doppler and integrated Sachs–Wolfe contributions

One can understand physically the few terms contributing to the observed CMB anisotropies in Eq. (5.56):

1. As first shown by Sachs and Wolfe, 1967, the most obvious contribution to the observed temperature fluctuation in one direction is given by the intrinsic

temperature fluctuation on the last scattering surface in the same direction, corrected by a gravitational shift (photons coming from a gravitational potential well [resp. hill] on the last scattering surface are redshifted [resp. blueshifted] when they leave this surface, and are seen with a lower [resp. higher] temperature). The *Sachs–Wolfe contribution* (SW) can be defined as the part of temperature anisotropies sourced by the term $g(\Theta_{\gamma 0} + \psi)$ in Eq. (5.56). The last scattering surface can really be thought of as a surface rather than a thick shell in the instantaneous decoupling limit, i.e., in the ideal situation in which the mean free path of photons would go from zero to infinity at $\eta = \eta_{LS}$. In this limit, one can replace the visibility function $g(\eta)$ with the Dirac delta function $\delta(\eta - \eta_{LS})$, which has the correct normalization. Indeed, one can easily check from the definition of g that $\int d\eta \, g(\eta) = 1$. One can then integrate the Sachs–Wolfe term in Eq. (5.56) and find a contribution

$$\Theta_{\gamma}^{SW}(\eta_0, \vec{o}, \hat{n}) \simeq \Theta_{\gamma 0}(\eta_{LS}, \vec{x}_{LS}, \hat{n}) + \psi(\eta_{LS}, \vec{x}_{LS}), \qquad (5.58)$$

where $\vec{x}_{LS} \equiv (\eta_{LS} - \eta_0)\hat{n}$ is the comoving coordinate at the time of decoupling of a photon hitting us today from the direction $-\hat{n}$. We know from Eq. (5.44) that for small values of k (corresponding to super-Hubble scales) and during matter domination, there is a relation $\Theta_{\gamma 0} = -2/3 \, \psi$. This means that in a CMB map smeared over small-scale fluctuations, the SW contribution to the temperature anisotropy in one direction is given by

$$\Theta_{\gamma}^{SW, smoothed}(\eta_0, \vec{o}, \hat{n}) \simeq \frac{1}{3}\psi(\eta_{LS}, \vec{x}_{LS}) \simeq -\frac{1}{2}\Theta_{\gamma 0}(\eta_{LS}, \vec{x}_{LS}, \hat{n}), \qquad (5.59)$$

i.e., that hot regions in the observed anisotropy map correspond to cold regions on the last scattering surface, and vice versa. The reason is that photons leaving a hot, overdense region lose a lot of energy when climbing out of the gravitational potential well. Stated differently, the gravitational redshift effect wins against the intrinsic temperature contribution.

2. Photons are emitted from a tightly coupled baryon–electron fluid, with a different peculiar velocity at each point on the last scattering surface. This velocity, projected along the line of sight, induces a Doppler shift in the photon wavelenght. The *Doppler contribution* is the one sourced by the term $g\hat{n} \cdot \vec{v}_B$ in Eq. (5.56). In the instantaneous decoupling limit, this term simply gives the familiar Doppler formula

$$\Theta_{\gamma}^{Doppler}(\eta_0, \vec{o}, \hat{n}) \simeq \hat{n} \cdot \vec{v}_B(\eta_{LS}, \vec{x}_{LS}). \qquad (5.60)$$

3. Along the line of sight, photons are continuously redshifted or blueshifted by metric fluctuations. The Boltzmann equation itself shows that the temperature changes because of gradients in the gravitational potential ψ, and because of

time variations in the metric fluctuation ϕ. Physically, the first effect accounts for the gain or loss of energy of photons feeling gravitational forces, and the second effect represents a local correction to the average time dilation, responsible for cosmological redshift in an expanding universe. If the gravitational potential were static, the first effect would be conservative; i.e., the integral of the gradient over the line of sight would give the difference between ψ at the point of last scattering and at the observer's location. We already discussed this contribution: the value of ψ at last scattering is included in the Sachs–Wolfe term, and the value at the observer's location appears in Eq. (5.56) and has been shown to be negligible. However, for a time-varying gravitational potential, the gradient does not coincide any more with the total derivative along the line of sight, and photons pick up a cumulative temperature shift accounted for by the integral of ψ' along the line of sight. Similarly, the cumulative effect of local time dilation is accounted for by the integral of ϕ'. These two effects form the so-called *integrated Sachs–Wolfe contribution* (ISW), encoded in the term $e^{-\tau}(\phi' + \psi')$ in Eq. (5.56). In the instantaneous decoupling limit, we can replace $e^{-\tau}$ by the Heaviside function $H(\eta - \eta_{LS})$ and obtain

$$\Theta_\gamma^{ISW}(\eta_0, \vec{o}, \hat{n}) \simeq \int_{\eta_{LS}}^{\eta_0} d\eta(\phi' + \psi'). \tag{5.61}$$

The features of the cosmic microwave background spectrum: Sachs–Wolfe plateau, acoustic peaks and damping tail

We discussed the various terms appearing in Eq. (5.56), which give the contribution to the temperature anisotropy observed in a given direction. We should now use this decomposition to understand the shape of the temperature power spectrum C_l in multipole space.

The temperature spectrum is given by Eqs. (5.57) and (5.50), i.e., by the line-of-sight integral of the source function $S^T(\eta, k)$, followed by a convolution with the primordial power spectrum. The source function contains the three generic terms identified in the previous subsection: a Sachs–Wolfe contribution $g(\Theta_{\gamma 0} + \psi)$, a Doppler contribution $(gk^{-2}\theta_B)'$, and an integrated Sachs–Wolfe contribution $e^{-\tau}(\phi' + \psi')$. Figure 5.2 shows the contribution of each of these three terms to the full temperature spectrum.

Using the facts that the visibility function is very peaked around recombination, that the spherical Bessel function $j_l(x)$ is peaked near $x \sim l$ and that the primordial spectrum $\mathcal{P}_\mathcal{R}$ is nearly scale-independent, it is easy to derive mathematically a result that can easily be understood geometrically, namely that the SW contribution to the temperature spectrum C_l multiplied by l^2 is qualitatively similar to the square

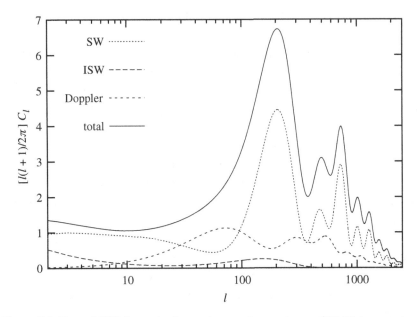

Figure 5.2 For a ΛCDM neutrinoless universe, the spectrum of CMB temperature anisotropies obtained numerically from the full temperature source function, or from individual contributions explained in the text: Sachs–Wolfe term (SW), integrated Sachs–Wolfe term (ISW) and Doppler term. Note that the full spectrum is not exactly given by the sum of the other three because of correlations between the terms.

of the SW transfer function in Fourier space,

$$l^2 C_l \propto \left[\Theta_\gamma(\eta_{LS}, k) + \psi(\eta_{LS}, k) \right]^2_{k=l/(\eta_0-\eta_{LS})}, \qquad (5.62)$$

where in the relation between k and l, we assume a flat universe for simplicity. The reason is that anisotropy multipoles at a given l are created mainly by Fourier modes of wavelength $\lambda \sim 2\pi a(\eta_{LS})/k$ on the last scattering surface seen today under an angle $\theta = \lambda/d_A(\eta_{LS}) \sim 2\pi/l$ (note the factor of 2: for a given multipole, π/l gives the angle between a maximum and a minimum, corresponding to half a wavelength of the perturbation on the surface). In a spatially flat universe, we have $d_A(\eta_{LS}) = a(\eta_{LS})(\eta_0 - \eta_{LS})$, whereas in a spatially nonflat universe, this expression should be changed accordingly, using (2.83). This gives $2\pi/l \sim 2\pi/[k(\eta_0 - \eta_{LS})]$.

This is only an approximate result, because in reality a given wavenumber contributes to an ensemble of multipoles, with the maximum contribution corresponding to the preceding relation. However, this picture gives a rather good understanding of the shape of the SW contribution in Fig. 5.2, which is indeed very similar to the square of the sum of $\Theta_\gamma(\eta_{LS}, k)$ and $\psi(\eta_{LS}, k)$ plotted in Figure 5.1. On small l's, the flat asymptote of the SW spectrum, called the Sachs–Wolfe plateau, corresponds to large wavelengths still above the sound horizon at decoupling, which are frozen. On larger l's, we recognize the acoustic peaks already discussed in Section 5.1.5. The first is given by the correlation length on the last scattering surface, corresponding to the sound horizon at decoupling. Secondary peaks represent harmonics of the same feature, and are modulated by all the effects due to baryons, gravitational effects and diffusion damping mentioned in Section 5.1.5. We see clearly in Figure 5.2 the enhancement of odd peaks with respect to even peaks, which one would also see by taking the square of $[\Theta_{\gamma 0} + \psi]$ in Fig. 5.1, explained by the shift in the zero point of oscillations given by $\Theta_{\gamma 0} = -(1 + R)\psi$. Finally, we see how diffusion damping affects secondary peaks: the suppression, accounted for by a factor of $\exp[-(k/k_\mathrm{D})^2]$ in Fourier space, is well represented by a factor $\exp[-(l/l_\mathrm{D})^2]$ in multipole space, with $l_d \sim k_d(\eta_0 - \eta_\mathrm{LS})$. Diffusion damping of CMB fluctuations is often called Silk damping, although the famous work of Silk (1968) refers primarily to the damping of baryon density perturbations, occurring roughly at the same epoch.

We can carry on a similar discussion for the Doppler term, despite its slightly more complicated geometrical structure, because this term comes from the projection of a gradient along the line of sight. This explains the derivative in the Doppler source function $(g\theta_\mathrm{B})'/k^2$. Still, this term is sourced by θ_B and θ_B' mutiplied by functions very peaked near $\eta = \eta_\mathrm{LS}$, and the contribution of the Doppler term to the C_l's is set by the value of the transfer function $\theta_\mathrm{B}(\eta, k)$ and its time derivative at the time of decoupling. This contribution is negligible on the scale of the Sachs–Wolfe plateau, because outside the sound horizon perturbations are frozen, and the photon–baryon fluid velocity remains very low. Inside the sound horizon, θ_B and θ_B' exhibit the same oscillatory patterns as $\Theta_{\gamma 0}$, but with a phase shifted by respectively $\pi/2$ and π, as for any oscillator. This explains the shape of the Doppler contribution in Fig. 5.2.

Finally, the ISW contribution would vanish if between decoupling and today, the universe was a perfect matter-dominated universe. In this case, as mentioned in Section 5.1.5 and to be proved in Section 6.1.1, the metric perturbations would be static everywhere and at any time, $\phi' = \psi'$. The ISW contribution picks up nonzero contributions:

- First, because at the time of decoupling, the universe is still near the beginning of matter domination, and metric fluctuations, which were quickly decaying inside the Hubble radius during radiation domination, did not completely freeze out. The residual time variation of ϕ and ψ generates an early integrated Sachs–Wolfe effect (EISW).

- Second, because at late times, when the cosmological constant Λ (or more generally dark energy) changes the universe's expansion-rate behaviour, metric fluctuations start decaying again. This late integrated Sachs–Wolfe (LISW) effect could be considered a secondary rather than a primary anisotropy, in the sense that it is not related to cosmological perturbations near the last scattering surface, but to gravitational interactions between free-streaming CMB photons and neighbouring galaxy clusters.

These two distinct contributions can easily be identified in the ISW curve of Fig. 5.2. The EISW effect cannot affect modes which are still outside the sound horizon at decoupling, for which no time evolution can occur, so that the EISW is negligible for $l \ll 100$. For modes inside the sound horizon, the EISW contribution to $l^2 C_l$ tends to decrease as a function of l, as a consequence of the factor k^{-2} in the Doppler term of the source function. Therefore, in Fig. 5.2, the EISW contribution is the one peaking around $l \sim 200$, enhancing the first acoustic peak in the total C_l's. The LISW effect is present on all scales, because dark energy domination produces a decay of metric fluctuations at all wavelengths. Because this effect also decreases with l for the same reason, it peaks at $l = 2$, and is only visible in Fig. 5.2 for $l \leq 30$. This term tilts the Sachs–Wolfe plateau in the total spectrum.

Shape and parameter dependence of the temperature spectrum

To conclude this section, it is worthwhile to summarize the dependence of the CMB temperature spectrum on the various effects discussed so far, and to relate them to the free cosmological parameters of a neutrinoless minimal ΛCDM model.

Let us first recall the relevant effects affecting the CMB temperature spectrum for each multipole range:

- For $l \ll 100$, the spectrum receives a SW contribution from modes which are still above the sound horizon at decoupling, leading to a Sachs–Wolfe plateau that depends on the primordial spectrum amplitude and tilt n_s. It also depends upon a LISW contribution which is related to the duration of Λ or dark energy domination and tends to tilt the Sachs–Wolfe plateau.
- For $l \geq 100$, the spectrum exhibits a series of acoustic peaks corrected by sub-dominant Doppler peaks, corresponding to the fundamental mode and harmonic decomposition of the correlation length $d_s(\eta_{LS})$ on the last scattering surface. These peaks are modulated by various effects: their amplitude is globally suppressed during the transition era between equality and decoupling; odd peaks are enhanced with respect to even peaks when the baryon content of the universe increases; the first peak is further enhanced by the EISW effect; finally, diffusion damping causes an exponential suppression at high l.

There is one last effect to be taken into account in the minimal ΛCDM model, which we did not discuss previously because it is caused by astrophysical phenomena after photon decoupling. When the first stars form (typically around redshift 10), the universe is believed to be partially reionized by starlight. A small fraction of CMB photons can then be rescattered by free electrons. This effect is negligible for modes entering into the Hubble scale very recently (at $z \ll 10$), but for all other modes it results in a scale-independent suppression of the CMB spectrum, accounted for by a factor $\exp[-\tau_{\text{reion}}]$, where τ_{reion} is the optical depth of reionization, measured to be on the order of ~ 0.1. Hence, reionization produces a steplike supression in the C_l's, with a step location around $l_{\text{step}} \sim 40$ (corresponding to modes crossing the Hubble radius near $z \sim 10$). For $l \gg l_{\text{step}}$, the effect of reionization is completely parameterized by τ_{reion}, whereas around l_{step} the shape of the step could depend on the details of the reionization history, which is poorly constrained by current data.

We can now relate these effects more explicitly to the free parameters of a neutrinoless minimal ΛCDM model, whereas neutrino effects will be introduced later in Section 5.3. This model has six free parameters:

- The primordial spectrum amplitude A_s and tilt n_s.
- The baryon density $\omega_B = \Omega_B h^2$.
- The total nonrelativistic matter density $\omega_M = (\Omega_B + \Omega_C)h^2$.
- Either the cosmological constant density fraction Ω_Λ or the Hubble parameter today, $H_0 = 100 \, h \, \text{km s}^{-1} \text{Mpc}^{-1}$. Because we are assuming the universe to be spatially flat, for fixed ω_M, there is a one-to-one correspondence between values of Ω_Λ and of $h = \sqrt{\omega_M/(1 - \Omega_\Lambda)}$.
- The optical depth to reionization τ_{reion}.

We consider $\omega_\gamma = \Omega_\gamma h^2$ as a fixed parameter, because the CMB temperature today is accurately measured. In this model, the redshift of equality between radiation and matter is set by ω_M, because the radiation density $\omega_R = \Omega_R h^2$ is fixed by the measurement of the CMB temperature today. The redshift of equality between matter and Λ depends only on Ω_Λ.

The degrees of freedom controlling the shape of the CMB temperature spectrum are as follows:

(C1) The peak location, which depends on the angle $\theta = d_s(\eta_{\text{LS}})/d_A(\eta_{\text{LS}})$. The sound horizon at decoupling, $d_s(\eta_{\text{LS}})$, depends on the expansion history and sound speed in the photon–baryon fluid until decoupling. It is affected by changes in ω_B (setting the baryon-to-photon ratio in the sound speed expression) and ω_M (setting the redshift of equality between matter and radiation). The angular diameter distance at decoupling, $d_A(\eta_{\text{LS}})$, depends on the expansion history after decoupling, i.e., on Ω_Λ or h (controlling the time of equality between matter and Λ and the critical density today).

(C2) The contrast between odd and even peaks depends on (ω_B/ω_γ) (with ω_γ fixed by the CMB temperature today), i.e., on the balance between gravity and pressure in the tightly coupled photon–baryon fluid before decoupling.

(C3) The amplitude of all peaks further depends on the amount of expansion between equality and decoupling. Because the redshift of decoupling is almost fixed by the thermodynamics, this amount mainly depends on the redshift of equality, i.e., on (ω_M/ω_R). A larger ω_M means that equality took place earlier. All peaks are then smaller, because the amplitude of acoustic oscillations decreased during a longer stage. The first peak is additionally supressed by a smaller EISW effect: at decoupling, metric fluctuations were closer to their constant asymptotic value. Conversely, a smaller ω_M leads to higher peaks, especially the first one.

(C4) The envelope of the secondary peaks in the large-l limit depends on the angle $\theta = \pi/l_d = \lambda_d(\eta_{LS})/d_A(\eta_{LS})$. The diffusion length λ_d is controlled by the expansion history and recombination history before decoupling. Because the integrand in Eq. (5.40) is increasing very rapidly with time before recombination (when n_e drops to nearly zero), λ_d depends essentially on the free electron density n_e at decoupling, which is fixed by ω_B in the minimal ΛCDM model, and on the conformal time at decoupling η_{LS}: the integral in Eq. (5.40) has a negligible dependence on the expansion and electron fraction at the time of equality and before. Therefore, λ_d depends essentially on ω_B and ω_M, whereas we know that $d_A(\eta_{LS})$ depends on ω_M and Ω_Λ or h.

(C5) The global amplitude of the C_l's is proportional to the amplitude A_s of the primordial spectrum, as can be checked from Eq. (5.50).

(C6) The global tilt of the C_l's depends trivially on the tilt n_s of the primordial spectrum, through the convolution in Eq. (5.50) (in the approximation of Eq. (5.62) in which a given k contributes essentially to a given l, this is even more obvious).

(C7) The slope of the Sachs–Wolfe plateau is controlled not only by the tilt n_s, but also by the LISW effect, which enhances the first few multipoles. The amplitude of the LISW depends on the duration of the Λ-dominated stage, i.e., on the time of equality between matter and Λ, fixed by the ratio $(\Omega_\Lambda/\Omega_M)$ (in terms of the parameter basis discussed above, this ratio is given by $\Omega_\Lambda/(1 - \Omega_\Lambda)$). A larger Ω_Λ implies a longer Λ domination and an enhanced LISW effect.

(C8) The global amplitude of the spectrum at $l \gg 40$ relative to the one at $l < 40$ depends on τ_{reion}. This effect is not degenerate with that of the time of equality, which also suppresses the peaks, but starting from a higher l and not in a constant way; it is not degenerate either with the effect of A_s, provided that the spectrum is observed also for $l \leq 40$.

For simplicity, we have been assuming that the recombination redshift was fixed. In fact, the free electron density $n_e(z)$ and the redshift of recombination z_{LS} have a small dependence at least on the baryon density and on the primordial helium fraction Y_p. Hence these parameters can slightly affect the sound horizon at decoupling, the duration of the transition era and the photon diffusion length, with some additional impact on the CMB through the effects (C1), (C3) and (C4). However, the CMB is much more sensitive to ω_B through the relative amplitude of the peaks than through its small impact on z_{LS}, and the effect of Y_p is negligible when this parameter is varied within the range indicated by measurements of primordial element abundances or by standard BBN. Therefore, for most purposes, the recombination history and redshift of decoupling can indeed be considered as fixed.

This enumeration shows eight distinct ways to alter the shape of the temperature spectrum, controlled by only six parameters. So in principle, the measurement of the CMB temperature spectrum alone should give independent constraints on each parameter of the minimal ΛCDM model. To some extent, this is already the case with WMAP temperature results (Komatsu *et al.*, 2011). In practice, however, cosmic variance and instrumental noise lead to large error bars on the C_l's for the smallest and largest multipole values, leading to partial parameter degeneracies and to a degradation of some of the constraints. For instance, in the minimal ΛCDM case, Ω_Λ and τ_{reion} are poorly constrained by the CMB temperature alone, because effects (C7) and (C8) can be constrained only through measurements at low l, plagued by cosmic variance. Degeneracies can be reduced by combining CMB temperature data with polarization information, or with other types of cosmological observations, or by performing more accurate measurements: this is the goal of the Planck satellite, which will make very accurate observations up to scales where primary anisotropies are anyway suppressed by diffusion damping ($l \sim 2500$).

5.1.7 Polarization anisotropies

The cross section of Thomson scattering depends on photon polarization: more precisely, on the orientation of the polarization vector relative to the scattering plane. However, in the tight-coupling regime, the photon polarization remains isotropically distributed at any point. Indeed, in the frame comoving with the tightly coupled photon–baryon(–electron) fluid, all quantities can only be isotropic because of the high interaction rate.

Around the time of recombination, such an isotropy disappears. The photons experience their last interactions in regions of space where anisotropies are growing. The quadrupolar component $\Theta_{\gamma 2}(\eta, \vec{x})$ of these anisotropies is responsible for a net polarization of scattered photons. Hence, polarization patterns appear on the last scattering surface and are strongly correlated with temperature patterns.

The observable polarization map is a vector field on a sphere, rather than a scalar field like temperature. Hence, it can be decomposed into two modes: a gradient field and a curl field, or by analogy with electromagnetism, an E-polarization and a B-polarization component. It is possible to define the harmonic power spectrum of the E and B modes, as well as three cross-correlation spectra for the products TE, TB and EB. Parity invariance implies that only the first cross-correlation term is nonzero after last scattering. The TB and EB cross-correlation spectra can be generated only at the level of secondary anisotropies, in particular, through the weak lensing of the last scattering surface. Moreover, as long as the total stress–energy tensor in the universe contains only scalar perturbations (following the definition of Section 5.1.2), no B modes can be excited.

The calculation of the spectra C_l^{EE} and C_l^{TE} in a universe containing only scalar perturbations can be carried on along the same line as for temperature anisotropies. The photon polarization is described by a new degree of freedom, whose evolution is described by the Boltzmann equation. In the previous section, we neglected polarization in the scattering term and obtained a simplified equation governing the evolution of temperature anisotropies $\Theta_\gamma(\eta, \hat{n}, \vec{x})$. The full Boltzmann equation leads to two hierarchies of equations for temperature and polarization multipoles. The two sets of equations are coupled, but the impact of polarization terms in the evolution of temperature anisotropies is rather small, so that our previous discussion of temperature anisotropies remains valid to a very good approximation. As for temperature, one can define a source function for E-mode polarization and express the polarization anisotropy as the integral of this source along the line of sight.

For brevity, we do not provide explicit expressions for polarization anisotropies here, because the neutrino effect on polarization is not qualitatively different from that on temperature (discussed in Section 5.3). Measuring the CMB polarization spectrum and TE cross spectrum helps in removing parameter degeneracies in the ΛCDM model, because some physical effects have intrinsically different consequences in the polarization sector. For instance, reionization imprints a distinct effect on C_l^{EE}. Because the neutrino effect is qualitatively similar for temperature and polarization, and because current CMB experiments measure the temperature spectrum with much higher accuracy, we will not make any other references to polarization anisotropies in the rest of this book.

5.1.8 Tensor perturbations

As mentioned in Section 5.1.2, Bardeen (1980) classified metric and stress–energy tensor perturbations according to their behaviour under spatial rotations: scalar, vector and tensor modes. At first order in perturbation theory, these sectors are decoupled and can be studied independently.

Until this point our discussion has addressed only scalar perturbations. Vector perturbations accounting for vorticity tend to decay after their possible excitation by nonlinear phenomena such as phase transitions or topological defects. In standard cosmology, they do not make any sizable contribution to CMB anisotropies. Tensor perturbations are fundamentally different from scalars and vectors, because they describe two propagating degrees of freedom (called the polarization states of the graviton in the context of quantum gravity, or the polarization states of gravitational waves at the classical level). For scalar and vector modes, metric perturbations are responding only to excitations in the stress–energy tensor of matter, whereas for tensor modes, the two polarization states can be excited by quantum effects during inflation. Starobinsky (1979) found that very generically, after a stage of inflation, primordial gravitational waves are generated with a nearly scale-invariant power spectrum, whose amplitude depends on the square of the Hubble parameter when observable wavelengths cross the Hubble scale during inflation.

Gravitational waves are coupled to all species having non-negligible tensor degrees of freedom in their perturbed stress–energy tensor $\delta T_{\mu\nu}$. These degrees of freedom are contained in the nondiagonal part of the spatial stress tensor δT_{ij}. They vanish to a very good approximation for cold dark matter because of its very small velocity dispersion, as well as for baryons and tightly coupled photons, because interactions enforce a diagonal stress tensor accounting for isotropic pressure. Hence, the only species efficiently coupled with gravitational waves are photons after decoupling and collisionless species before their nonrelativistic transition (namely, in the minimal ΛCDM scenario, neutrinos).

Extra gravitational waves can be generated after inflation by nonlinear phenomena such as phase transitions and topological defects. In the minimal cosmological scenario, it is assumed that the former can impact only small scales, comparable to the Hubble radius during the transition and cannot affect CMB anisotropies, whereas topological defects (if any) do not contribute significantly to total large-scale perturbations. Hence we only need to consider primordial gravitational waves from inflation, which are known to contribute to temperature anisotropies on large angular scales (typically $l < 150$) and to E-type and B-type polarization anisotropies on all scales.

The influence of neutrinos on tensor CMB anisotropies was studied analytically by Weinberg, 2004 and implemented by Lewis in the Boltzmann code CAMB.[2] We will see in Section 5.3 that the influence of neutrinos on scalar CMB anisotropies can only be significant for modes crossing the horizon during radiation domination or at the beginning of matter domination, i.e., on small scales. This is also true for their influence on tensor anisotropies.

---

[2] See http://cosmologist.info/notes/CAMB.pdf, Section XVI.

Even if we assume that inflation took place at high energy and produced a detectable amount of primordial gravitational waves, small-scale temperature and E-polarization anisotropies are largely dominated by scalar contributions. For B-polarization, there is no scalar contamination to primordial anisotropies, but on small scales gravitational lensing effects generate a leak from E-type to B-type polarization, in such a way that the contribution of tensor perturbations to the spectrum C_l^{BB} is also subdominant.

Hence tensor modes could only be detected in the CMB on large scales, on which the impact of neutrinos must be negligible. For this reason, we choose not to discuss tensor perturbations in the following.

5.2 Neutrino perturbations

Having presented in Section 5.1 the basics of linear cosmological perturbation theory in the absence of neutrinos, we now focus on the specific evolution of neutrino perturbations.

5.2.1 Perturbation equations

In this book, we are mainly interested in the description of the three standard active neutrinos. We present here the formalism describing the cosmological perturbations of such neutrinos in the massless limit, as well as in the more realistic massive case. Massless neutrino equations can safely be employed for collisionless particles which are still relativistic today, i.e., in the case of ordinary neutrinos, when their mass m_ν is smaller than $T_{\nu,0} \simeq 1.68 \times 10^{-4}$ eV, which might be the case for only one neutrino mass state given experimental results on flavour oscillations (see Chapter 1). We will also occasionally consider more exotic cases such as neutrinos with high chemical potentials, large nonthermal distortions, light or heavy sterile neutrinos. We will see that all these cases can still be described by one of the two sets of equations introduced hereafter. Only neutrinos with nonstandard interactions require a different set of equations, which would depend very much on the type of interactions. This exotic assumption will only be mentioned briefly at the end of this chapter.

Massless neutrinos and other relativistic relics

When studying the impact of massless neutrinos on the CMB, we need only to consider fully decoupled neutrinos. Indeed, the details of neutrino decoupling would only impact cosmological perturbations which are already inside the Hubble length when $T \sim$ MeV. Such scales are not observable. In the CMB spectrum, they have been suppressed by diffusion damping and are masked anyway by foreground contamination in real data sets. In the large-scale structure spectrum, a strong nonlinear evolution has erased all memory of the early linear evolution on such small scales.

We have seen in Sections 2.4.2 and 4.1 that standard active neutrinos have a frozen background distribution function $f_{\nu 0}(y)$ of the Fermi–Dirac type (with negligible mass and chemical potential),

$$f_{\nu 0}(y) = \left(e^{\frac{y}{a\bar{T}_\nu}} + 1 \right)^{-1}, \tag{5.63}$$

with a constant product $a\,\bar{T}_\nu$ after neutrino decoupling. Nonthermal distortions of this distribution due to entropy release at the $e^{\pm}$ annihilation stage are very small, on the level of a few percent at most, and can be completely neglected in our discussion because all present (and forthcoming) data are completely blind to such a small effect. This means that the neutrino equation of evolution can be derived following the same steps as for photons, with just a sign difference in f accounting for Fermi–Dirac statistics (but this makes no difference in the final equations), and of course assuming no interaction. Hence, we could write an equation of evolution for the neutrino temperature perturbation Θ_ν which would be strictly identical to Eqs. (5.13) or (5.15) for Θ_γ in the collisionless limit $\sigma_T = 0$.

We can remain at a more general level and assume that we simply deal with any kind of decoupled relativistic particles, with an arbitrary background distribution function $f_{\nu 0}(y)$. The absence of an explicit time dependence in this distribution is appropriate if the species is indeed decoupled. This covers the case of massless neutrinos with chemical potentials or a nonthermal distribution, as well as any other decoupled relativistic relic. The way to simplify the Boltzmann equation in this case is not to parameterize f_ν explicitly in terms of temperature fluctuations, as in Eq. (5.11), but to introduce the momentum average of the perturbation of the distribution function relative to the background value

$$F_\nu(\vec{x}, \hat{n}, \eta) = \frac{\int y^2 dy\, y[f_\nu(\vec{x}, y, \hat{n}, \eta) - f_{\nu 0}(y)]}{\int y^2 dy\, y f_{\nu 0}(y)}. \tag{5.64}$$

By plugging into the above formula the phase space distribution of thermal relics with a negligible chemical potential, one can easily show that this quantity exactly coincides with the temperature perturbation Θ_ν multiplied by four. If we take the Boltzmann equation (2.92) in the collisionless limit (sometimes called the Vlasov equation), integrate the Liouville operator over momentum and simplify it using the same steps as for photons in Section 5.1.2 (always at order one in perturbations), we obtain an equation of motion for F_ν,

$$F_\nu' + \hat{n} \cdot \vec{\nabla} F_\nu - 4\phi' + 4\hat{n} \cdot \vec{\nabla}\psi = 0. \tag{5.65}$$

We can expand this equation in Fourier space, and as for photons, notice that the isotropy of the background implies that the equation is axially symmetric around $\hat{n}$. Moreover, F_ν follows the same symmetry at initial time. In fact, before neutrino decoupling, neutrino perturbations can only depend on direction through a Doppler-induced dipole along the axis defined by $\hat{n}$, and after decoupling, this remains true for all modes above the Hubble radius, for causality reasons. So the symmetry is preserved throughout the evolution, and the Fourier transform of $F_\nu(\vec{x}, \hat{n}, \eta)$ depends only on the arguments $F_\nu(\vec{k}, \mu, \eta)$, with $\mu = \hat{k} \cdot \hat{n}$. This function follows

$$F_\nu' + ik\mu F_\nu - 4\phi' - ik\mu 4\psi = 0. \tag{5.66}$$

Expanding this equation in Legendre polynomials,

$$F_\nu(\vec{k}, \mu, \eta) = \sum_l (-i)^l (2l + 1) F_{\nu l}(\vec{k}, \eta) P_l(\mu), \tag{5.67}$$

we get an infinite hierarchy of equations:

$$F_{\nu 0}' = -k F_{\nu 1} + 4\phi'$$

$$F_{\nu 1}' = \frac{k}{3}(F_{\nu 0} - 2F_{\nu 2} + 4\psi) \tag{5.68}$$

$$F_{\nu l}' = \frac{k}{(2l+1)}\left[l F_{\nu(l-1)} - (l+1)F_{\nu(l+1)}\right], \qquad \forall l \geq 2.$$

The relation between the first multipoles of F_ν and the perturbations $\delta_\nu, \theta_\nu, \sigma_\nu$ can be derived from Eq. (5.10):

$$\delta_\nu = F_{\nu 0}, \qquad \theta_\nu = \frac{3}{4}k F_{\nu 1}, \qquad \sigma_\nu = \frac{1}{2}F_{\nu 2}. \tag{5.69}$$

One can check that the first two equations in (5.68) coincide with the general continuity and Euler equations for a species with equation of state $\overline{P}/\overline{\rho} = 1/3$, a sound speed $c_s^2 = \delta P/\delta\rho = 1/3$, and some unspecified anisotropic pressure.

Massive neutrinos

Massive neutrinos should be described by a set of equations interpolating from CDM equations in the large mass limit to massless neutrino equations in the small mass limit. We can assume that at the time at which we impose initial conditions, neutrinos are already decoupled but still relativistic, with a given background distribution $f_{\nu 0}(y)$. We know that for standard active neutrinos $f_{\nu 0}(y)$ is given by Eq. (5.63), but because we are also interested in sterile neutrinos and nonstandard active neutrinos, we keep the discussion at a more general level and assume only that $f_{\nu 0}(y)$ has no explicit time dependence after neutrino decoupling.

The difference between massive and massless neutrinos appears when the terms $d\vec{x}/d\eta$ and $dy/d\eta$ are replaced in the Liouville operator. For nonrelativistic particles, these terms are given respectively by $y\hat{n}/\epsilon$ and by Eq. (5.4), and depend explicitly on the mass through the energy ϵ. In that case, the collisionless Boltzmann equation cannot be simplified by integrating over momentum. The physical explanation is that for nonrelativistic particles, gravitational interactions induce a relative momentum shift depending on the momentum itself, i.e., some nonthermal distortions of the perturbed phase-space distribution. The Boltzmann equation can still be simplified to some extent by introducing the relative fluctuation of the phase-space distribution,

$$\Psi_\nu(\eta, \vec{x}, y, \hat{n}) = \frac{f_\nu(\eta, \vec{x}, y, \hat{n})}{f_{\nu 0}(\eta, y)} - 1, \tag{5.70}$$

expressed at first order in perturbations.

As long as neutrinos remain relativistic, the quantity $\Psi_\nu(\eta, \vec{x}, y, \hat{n})$ can be related to the variables introduced previously for massless neutrinos. In the case of standard neutrinos with a Fermi–Dirac distribution, we can write the perturbed distribution in the particular form $f_\nu(\eta, \vec{x}, y, \hat{n}) = f_{\nu 0}(y/[a\overline{T}_\nu\{1 + \Theta_\nu(\eta, \vec{x}, \hat{n})\}])$ with constant $a\overline{T}_\nu$ and make a Taylor expansion at first order in Θ_ν to find

$$\Psi_\nu(\eta, \vec{x}, y, \hat{n}) = -\Theta_\nu(\eta, \vec{x}, \hat{n})\frac{d \ln f_{\nu 0}(y)}{d \ln y}. \tag{5.71}$$

In the general case, Ψ_ν can be identified in the relativistic limit with

$$\Psi_\nu(\eta, \vec{x}, y, \hat{n}) = -\frac{1}{4}F_\nu(\eta, \vec{x}, \hat{n})\frac{d \ln f_{\nu 0}(y)}{d \ln y}. \tag{5.72}$$

Equations (5.71) and (5.72) are no longer valid when the particles become nonrelativistic and gravitational interactions introduce nonthermal distortions.

If we take the Boltzmann equation (2.92) in the collisionless limit, replace $f_\nu(\eta, \vec{x}, y, \hat{n})$ as a function of $\Psi_\nu(\eta, \vec{x}, y, \hat{n})$ and simplify the Liouville operator using the same steps as for photons and massless neutrinos (always at order one in perturbations), we obtain an equation of motion for Ψ_ν:

$$\Psi_\nu' + \frac{y}{\epsilon}\hat{n} \cdot \vec{\nabla}\Psi_\nu + \frac{d \ln f_{\nu 0}}{d \ln y}\left[\phi' - \frac{\epsilon}{y}\hat{n} \cdot \vec{\nabla}\psi\right] = 0. \tag{5.73}$$

In the relativistic limit, the ratio (y/ϵ) is equal to unity, and using Eq. (5.72), we immediately recover the massless neutrino Boltzmann equation. As for photons and massless neutrinos, one can transform Eq. (5.73) to Fourier space, make use of the axial symmetry around $\hat{n}$, expand Ψ_ν in Legendre polynomials and obtain an infinite hierarchy of equations:

$$\Psi_{\nu 0}' = -\frac{yk}{\epsilon}\Psi_{\nu 1} - \phi'\frac{d\ln f_{\nu 0}}{d\ln y}$$

$$\Psi_{\nu 1}' = \frac{yk}{3\epsilon}(\Psi_{\nu 0} - 2\Psi_{\nu 2}) - \frac{\epsilon k}{3y}\psi\frac{d\ln f_{\nu 0}}{d\ln y} \tag{5.74}$$

$$\Psi_{\nu l}' = \frac{yk}{(2l+1)\epsilon}\left[l\Psi_{\nu(l-1)} - (l+1)\Psi_{\nu(l+1)}\right], \qquad \forall l \geq 2.$$

Finally, using Eq. (5.10), one can show that the usual four scalar degrees of freedom of the perturbed stress–energy tensor can be obtained by integrating $\Psi_{\nu\, l=0,1,2}$ over momentum:

$$\delta\rho_\nu = \overline{\rho}_\nu\delta_\nu = 4\pi a^{-4}\int y^2 dy\,\epsilon\, f_{\nu 0}(y)\Psi_0 \tag{5.75}$$

$$\delta P_\nu = \frac{4\pi}{3}a^{-4}\int y^2 dy\,\frac{y^2}{\epsilon}\, f_{\nu 0}(y)\Psi_0 \tag{5.76}$$

$$(\overline{\rho}_\nu + \overline{P}_\nu)\theta_\nu = 4\pi a^{-4}\int y^2 dy\,\epsilon\, f_{\nu 0}(y)\Psi_1 \tag{5.77}$$

$$(\overline{\rho}_\nu + \overline{P}_\nu)\sigma_\nu = 4\pi a^{-4}\int y^2 dy\,\frac{y^2}{\epsilon}\, f_{\nu 0}(y)\Psi_2. \tag{5.78}$$

By integrating the first two equations (5.74) over momentum y, one can recover the general continuity and Euler equations, this time with no exact analytic expression for $\overline{P}_\nu/\overline{\rho}_\nu$, and no exact relation between $\delta\rho_\nu$, δP_ν and σ_ν. For details on this procedure, see Shoji and Komatsu, 2010; Lesgourgues and Tram, 2011.

Deep in the nonrelativistic limit, the ratio (y/ϵ) goes asymptotically to zero. In that case, the integrand of δP_ν is suppressed by a factor of $(y/\epsilon)^2$ with respect to that of $\delta\rho_\nu$; i.e., $\delta P_\nu/\delta\rho_\nu$ is of order $(\overline{T}_\nu/m_\nu)^2$ and pressure perturbations can be neglected. The same is true for the ratios $\overline{P}_\nu/\overline{\rho}_\nu$ and σ_ν/δ_ν. Thus, the continuity and Euler equations for nonrelativistic neutrinos become identical to those for cold dark matter.

5.2.2 Neutrino isocurvature modes

The number of independent initial conditions for the whole system of cosmological perturbations is equal to the number of first-order equations describing their evolution of super-Hubble wavelengths. We know that there are two first-order equations for baryons, two for CDM, and also two for photons, despite the infinite hierarchy of Legendre momenta in the Boltzmann equation for photons. As explained in Section 5.1.3, the reason is that as long as photons are tightly coupled, all multipoles above $l = 0$ (the monopole acounting for local density fluctuations) and $l = 1$ (the dipole due to the relative motion between the tightly coupled fluid and the coordinate frame) are negligible, as clearly seen from the interaction terms in Eq. (5.15), in the limit of an infinite interaction rate.

A similar simplification holds for each distinct neutrino species at leading order in the long-wavelength approximation. As long as neutrinos are in thermal equilibrium, they have vanishing multipoles for $l > 1$, for the same reasons as for photons. After decoupling occurs, these multipoles remain suppressed by a factor $(k\eta)^l$ on super-Hubble wavelengths, as can be checked from Eq. (5.68). Because $F'_{\nu l}$ is sourced by $k F_{\nu(l-1)}$, higher multipoles are suppressed by increasing powers of $(k\eta)$. Physically, this is a simple consequence of causality. It takes some amount of time for high multipoles to grow from zero to a significant value. For a given wavelength, this happens after Hubble crossing, i.e., after entering the causal horizon (computed starting from some initial time after the end of inflation).

Thus, as long as we neglect multipoles $l > 1$ when fixing initial conditions on super-Hubble scales, we can say that all perturbations in a universe with N species can be described by $2N$ equations, leading to $2N$ independent solutions: one growing adiabatic mode, $(N - 1)$ isocurvature growing modes and N decaying modes. Moreover, if we choose this initial time so that all neutrino species are relativistic, they can all be described by the same pair of equations (the first two of Eqs. (5.74) in the limit $F_{\nu 2} = 0$), and they count as only one species. We see that in the minimal ΛCDM model, including the usual three neutrino species, one has $N = 4$, because the counting runs over photons, baryons, CDM and generic relativistic neutrinos.

This picture has been refined by the analysis of Bucher *et al.*, 2000, who pointed out that if we search for the solutions of a system with one more equation (that for the evolution of the neutrino anisotropic stress $\sigma_\nu = F_{\nu 2}/2$), then there is one more isocurvature growing mode, corresponding to an initial excitation of the neutrino flux divergence θ_ν, compensated for by an opposite excitation of θ_γ in order to keep a diagonal stress–energy tensor in the limit $(k\eta) \to 0$. In other words, if we allow an initial excitation of θ_ν, then all multipoles $F_{\nu l}$ with $l \geq 1$ are larger by one power of $(k\eta)$, and one should keep the $l = 2$ equation in order to derive a consistent solution at leading order. This new solution has been called the *neutrino isocurvature velocity* mode (NIV). The previously found neutrino isocurvature mode was then renamed the *neutrino isocurvature density* mode (NID). However, no plausible mechanism leading to an excitation of the NIV mode has been proposed so far.

This logic cannot be pushed to higher order. The next step would consist of searching for a solution with an initial excitation of the neutrino anisotropic stress, compensated for by an opposite excitation of the photon anisotropic stress.

However, σ_γ must remain very small due to Thomson scattering, as well as σ_ν until neutrino decoupling.

We already mentioned that even if some isocurvature modes are excited in the primoridal universe, thermal equilibrium forces species to share the same local fluctuations in number density, and erases any entropy perturbation. Hence, isocurvature modes should be taken into account only if they are attached to an ever-decoupled species, whereas neutrinos are coupled until $T \sim 1$ MeV. This argument could in principle be avoided in two cases: if the universe reheats at very low temperature and neutrinos never reach thermal equilibrium, or if they are thermal relics with high chemical potentials, corresponding to a large lepton asymmetry. The chemical potentials $\mu_{\nu_i}(t, \vec{x})$ could in principle fluctuate spatially in such a way as to violate the adiabaticity condition. One can show that under such assumptions, the NID mode can be excited (Malik and Wands, 2009).

In summary, the NIV mode sounds very unlikely to be produced in the early universe, and the NID mode would require a complicated mechanism implying large lepton asymmetries. Moreover, none of these modes is favoured by current data (Castro *et al.*, 2009). For these reasons, we will not consider isocurvature modes in the following.

5.2.3 Adiabatic mode in the presence of neutrinos

We explained in Section 5.1.3 that adiabatic initial conditions have a simple meaning on super-Hubble scales. They correspond to a perturbed universe described at each point by the background evolution, with a local time-shift function $t \longmapsto t + \delta t(\vec{x})$ (or the same using conformal time η). We showed that this leads to a universal relation (5.24) giving for the neutrino density transfer function

$$\delta_\nu(\eta, k) = \delta_\gamma(\eta, k). \qquad (5.79)$$

On super-Hubble scales, the Einstein 0–0 equation (5.20) relates ψ to the total density perturbation in the universe, coming from photons and neutrinos during radiation domination. Because photons and neutrinos share the same density contrast, the relation $\delta_\gamma = -2\psi$ found in the neutrinoless case is still valid in the presence of neutrinos. Hence the curvature perturbation $\mathcal{R}$, given by Eq. (5.29), is also related to δ_γ in the same way. All these quantities are still constant over time for adiabatic initial conditions. Higher neutrino momenta θ_ν/k and σ_ν vanish in the $\eta \to 0$ limit, because they contribute to the nondiagonal part of the stress–energy tensor, whereas the background stress–energy tensor is diagonal. The leading contribution to θ_ν/k (resp. σ_ν) in the Newtonian gauge is of order one (resp. two) in a $(k\eta)$ expansion (Ma and Bertschinger, 1995). The coefficients of these terms and of higher-order terms can be computed analytically by solving the full system of equations in the super-Hubble limit. Higher-order terms are generally included in the initial condition equations implemented in Boltzmann codes.

However, the relation $\phi = \psi$ is not valid any more. One can check from Eq. (5.22) (using the Friedmann equation) that a total anisotropic stress proportional to η^2 implies a constant offset between the two metric perturbations. The growing adiabatic solution of the full system of equations is such that

$$\phi - \psi = \frac{2}{5} R_\nu \psi \tag{5.80}$$

with

$$R_\nu \equiv \frac{\overline{\rho}_\nu}{\overline{\rho}_\gamma + \overline{\rho}_\nu} = \frac{\frac{7}{8} \left(\frac{4}{11}\right)^{4/3} N_{\mathrm{eff}}}{1 + \frac{7}{8} \left(\frac{4}{11}\right)^{4/3} N_{\mathrm{eff}}} = \frac{0.2271 \, N_{\mathrm{eff}}}{1 + 0.2271 \, N_{\mathrm{eff}}}, \tag{5.81}$$

where we use the definition of the effective neutrino number in (2.198). In summary, in the presence of neutrinos, the relation (5.41) between density and metric perturbations during radiation domination and on super-Hubble scales becomes

$$4\Theta_{\gamma 0}(\eta, k) = \delta_\gamma(\eta, k) = \delta_\nu(\eta, k) = \frac{4}{3}\delta_{\mathrm{B}}(\eta, k)$$

$$= -2\left(1 + \frac{2}{5} R_\nu\right)\phi(\eta, k) = -2\psi(\eta, k) = -\frac{4}{3}. \tag{5.82}$$

The presence of the factor R_ν in this relation gives the confusing impression that the leading-order solution for density perturbations on super-Hubble scales is affected by the neutrino anisotropic stress, responsible for the constant offset between ϕ and ψ. This would contradict the general interpretation of the adiabatic mode: if adiabatic perturbations are equivalent to a local time-shift of background quantities, the neutrino anisotropic stress should be irrelevant on super-Hubble scales, because it contributes only to the vanishing nondiagonal part of the stress–energy tensor. In fact, there is no contradiction, because the shift of ϕ by a constant amount $\frac{2}{5} R_\nu \psi$ has no observable consequences. All equations of evolution depend on the metric through the time-dilation term ϕ' and the gravitational force $k^2 \psi$. The derivative ϕ' is not affected by the offset, and its leading contribution is of order $(k\eta)$. We conclude that the neutrino anisotropic stress (and more generally the presence of neutrinos) changes the evolution of all densities only at next-to-leading order in a $(k\eta)$ expansion. The fact that metric perturbations themselves are affected at leading order is a kind of gauge artifact. In other gauges, it is clear that R_ν plays no role in metric perturbations at leading order.

This technical discussion will become important when the effect of neutrinos on the CMB and large-scale structure power spectra is discussed: we will use the fact that the presence of neutrinos has no impact on other species on super-Hubble scales. This result was illustrated here during radiation domination, but it remains true during matter domination.

5.2.4 Free-streaming length

After their decoupling, neutrinos free-stream, or in other words, evolve like freely falling particles. Because the universe is expanding, free-streaming is not relevant on all scales at a given time. We can introduce two quantities describing the scale above which neutrino free-streaming can be ignored:

1. a dynamical quantity, the *free-streaming scale* λ_{fs} (or k_{fs} in comoving Fourier space), showing on which scales free-streaming can be neglected in the equations of evolution at any given time;
2. an integrated quantity, the *free-streaming horizon* d_{fs} (or r_{fs} in comoving space), giving the average distance travelled by neutrinos between the early universe and a given time, and hence showing which scales could be affected at all by neutrino free-streaming at this time.

The first quantity could be defined as the product of the average neutrino velocity c_ν by the Hubble time $t_{\text{H}} \equiv 1/H$, but most authors prefer to define the free-streaming length by analogy with the Jeans length:

$$\lambda_{\text{fs}}(\eta) = a(\eta)\frac{2\pi}{k_{\text{fs}}} \equiv 2\pi \sqrt{\frac{2}{3}\frac{c_\nu(\eta)}{H(\eta)}}. \tag{5.83}$$

In the Jeans case, the quantity c_ν should be replaced by the sound speed c_{s} of a given fluid. The justification of the prefactor $2\pi\sqrt{2/3}$ for the Jeans length is that for a fluid i with constant sound speed dominating the expansion of the universe, the continuity, Euler, Friedmann and Poisson equations can be combined into the following equation of evolution on sub-Hubble scales:

$$\delta_i'' + \frac{a'}{a}\delta_i' + \left(k^2 - \frac{3a^2H^2}{2c_{\text{s}}^2}\right)c_{\text{s}}^2\delta_i = 0. \tag{5.84}$$

The effective mass $(k^2 - k_{\text{J}}^2)c_{\text{s}}^2$ changes sign when k is equal to the Jeans wavenumber. Hence the Jeans length represents the scale below which pressure forbids gravitational collapse in the fluid. By analogy, the free-streaming length is the scale below which collisionless particles cannot remain confined in gravitational potential wells, because of their velocity dispersion.

The second quantity is defined like any other comoving horizon scale:

$$d_{\text{fs}}(\eta) = a(\eta)\,r_{\text{fs}}(\eta) \equiv a(\eta)\int_{\eta_{\text{in}}}^{\eta} c_\nu(\eta)\,d\eta. \tag{5.85}$$

We recall that neutrinos travel along geodesics with on average $dx = c_v \, dt/a = c_v \, d\eta$, and provided that η_{in} is chosen after the end of inflation (which makes perfect sense, because neutrinos were produced by reheating after that time), $d_{fs}(\eta)$ is independent of η_{in} as long as $\eta_{in} \ll \eta$.

Relativistic neutrinos

If neutrinos are still relativistic today, their velocity is given at all times by $c_v = c = 1$, and we find

$$\lambda_{fs} = 2\pi \sqrt{\frac{2}{3}} H^{-1}, \qquad d_{fs} = a\eta; \tag{5.86}$$

i.e., the free-streaming scale is given by the Hubble radius times a numerical factor, and the free-streaming horizon is equal to the particle horizon. Both quantities are very close to each other. During radiation domination $a\eta = H^{-1}$, and during matter domination $a\eta = 2H^{-1}$.

Neutrinos becoming nonrelativistic during matter domination

Neutrinos become nonrelativistic when their average momentum $\langle p \rangle$ falls below their mass m_ν. For a relativistic Fermi–Dirac distribution with negligible chemical potential, the average momentum is given as a function of the temperature by $\langle p \rangle = 3.15 \, T_\nu$. We will denote the temperature of ordinary active neutrinos in the instantaneous decoupling limit as

$$T_\nu^a \equiv (4/11)^{1/3} T. \tag{5.87}$$

The active neutrino temperature today, $T_{\nu 0}^a$, can easily be inferred from the known value of the CMB temperature, T_0. Ordinary neutrinos with $T_\nu \simeq T_\nu^a$ become nonrelativistic during matter or Λ domination if their mass is in the range from

$$3.15 \, (4/11)^{1/3} T_0 = 5.28 \times 10^{-4} \, \text{eV} \tag{5.88}$$

to approximately

$$(1 + z_{eq}) \, 5.28 \times 10^{-4} \, \text{eV} \simeq 1.5 \, \text{eV} \tag{5.89}$$

(because current observations indicate that the redshift of equality is close to $z_{eq} = 2900$). More generally, the redshift of the transition is given by

$$z_{nr} = \left(\frac{m_\nu}{5.28 \times 10^{-4} \, \text{eV}} \right) \left(\frac{T_\nu^a}{T_\nu} \right) - 1. \tag{5.90}$$

After the nonrelativistic transition, the velocity dispersion (also called thermal velocity)

$$c_\nu = \frac{\langle p \rangle}{m_\nu} = 3.15 \frac{T_\nu}{m_\nu} = 158 \,(1 + z) \left(\frac{T_\nu^a}{T_\nu} \right) \left(\frac{1 \text{ eV}}{m_\nu} \right) \text{km s}^{-1} \quad (5.91)$$

scales like $c_\nu \propto a^{-1} \propto \eta^{-2}$. Hence, during matter domination, the free-streaming length of nonrelativistic neutrinos increases like η while the comoving free-streaming length decreases like η^{-1}. The free-streaming length and wavenumber are given by

$$\lambda_{\text{fs}} = 8.10 \,(1 + z) \frac{H_0}{H(z)} \left(\frac{T_\nu}{T_\nu^a} \right) \left(\frac{1 \text{ eV}}{m_\nu} \right) h^{-1} \text{Mpc} \quad (5.92)$$

$$k_{\text{fs}} = 0.776 \,(1 + z)^{-2} \frac{H(z)}{H_0} \left(\frac{T_\nu}{T_\nu^a} \right) \left(\frac{m_\nu}{1 \text{ eV}} \right) h \text{ Mpc}^{-1} \quad (5.93)$$

with $H(z)^2 = H_0^2[\Omega_\Lambda + \Omega_k(1 + z)^2 + \Omega_M(1 + z)^3]$ during matter and Λ domination; see (2.61). In the second equation and in the rest of this section we also assume that the scale factor today is fixed to one.

At the time of the transition, the free-streaming wavenumber passes through a minimum usually denoted k_{nr}. An approximation to k_{nr} is found by simply plugging Eq. (5.90) into Eq. (5.93),

$$k_{\text{nr}} \equiv k_{\text{fs}}(\eta_{\text{nr}}) \simeq 0.0178 \, \Omega_M^{1/2} \left(\frac{T_\nu^a}{T_\nu} \right)^{1/2} \left(\frac{m_\nu}{1 \text{ eV}} \right)^{1/2} h \text{ Mpc}^{-1}, \quad (5.94)$$

assuming that the transition takes place during matter domination, when $H(z)/H_0 \simeq \Omega_M^{1/2}(1 + z)^{3/2}$. Note that sometimes k_{nr} is simply defined as the wavenumber such that $k_{\text{nr}} = aH$ at the time of the nonrelativistic transition. With respect to the definition adopted previously, there is only a factor $\sqrt{3/2}$ of difference.

The comoving free-streaming horizon is equal to η in the relativistic regime. After that time, in the approximation in which c_ν would switch from 1 to $(\eta_{\text{nr}}/\eta)^2$ at $\eta = \eta_{\text{nr}}$, it is given by

$$r_{\text{fs}}(\eta > \eta_{\text{nr}}) \simeq \eta_{\text{nr}} \,(2 - \eta_{\text{nr}}/\eta) \simeq \sqrt{\frac{3}{2}} \frac{4}{k_{\text{nr}}} \left(1 - \frac{1}{2} \left[\frac{1 + z}{1 + z_{\text{nr}}} \right]^{1/2} \right), \quad (5.95)$$

where we used the expression $\eta = 2/(aH) \propto a^{1/2}$ valid during matter domination. Taking a recent Λ or curvature-dominated stage into account would change the result by a very small amount at late times. For $z \ll z_{\text{nr}}$, the comoving free-streaming horizon r_{fs} remains very close to the minimum comoving free-streaming scale $(2\pi/k_{\text{nr}})$ (they differ at most by 28%). Strictly speaking, the comoving free-streaming horizon is the right quantity to be computed if one wishes to know above which comoving scale free-streaming can be neglected. However, in most of the literature this role is attributed to $(2\pi/k_{nr})$, which makes no difference in practice, at least for particles becoming nonrelativistic after the time of equality between matter and radiation.

Neutrinos becoming nonrelativistic during radiation domination

Most of this book is focused on ordinary active neutrinos with sub-eV mass. Occasionally we will refer to heavy sterile neutrinos with mass in the keV range. In this case the redshift of the nonrelativistic transition, still given by Eq. (5.90) in the case of a thermal momentum distribution, falls within the radiation-dominated era. After that time, the neutrino velocity c_ν (still subject to Eq. (5.91)) evolves like $c_\nu \propto a^{-1} \propto \eta^{-1}$. At the same time $H = 1/a\eta \propto \eta^{-2}$. Hence, during radiation domination, the free-streaming length of nonrelativistic neutrinos increases like $\lambda_{fs} \propto \eta$, while the comoving free-streaming length remains constant. After the time of equality between radiation and matter, we have just seen that the free-streaming length decreases. Hence, it reaches its maximum between η_{nr} and η_{eq}. The free-streaming length and wavenumber are still given by (5.92) and (5.93) with now $H(z)^2 = H_0^2[\Omega_\Lambda + \Omega_k(1 + z)^2 + \Omega_M(1 + z)^3 + \Omega_R(1 + z)^4]$. Hence the minimum value of k_{fs} is given by

$$k_{nr} \equiv k_{fs}(\eta_{nr}) \simeq 0.776\, \Omega_R^{1/2} \left(\frac{T_\nu^a}{T_\nu}\right) \left(\frac{m_\nu}{1\,\text{eV}}\right) h\,\text{Mpc}^{-1} \qquad (5.96)$$

and differs from its counterpart (5.94) for neutrinos with sub-eV mass.

In the approximation in which c_ν would switch from 1 to (η_{nr}/η) at $\eta = \eta_{nr}$, and to $(\eta_{nr}\eta_{eq}/\eta^2)$ at $\eta = \eta_{eq}$, the comoving free-streaming horizon reads

$$r_{fs}(\eta > \eta_{eq}) \simeq \eta_{nr}\left(2 + \ln(\eta_{eq}/\eta_{nr}) - \eta_{eq}/\eta\right) \qquad (5.97)$$

$$\simeq \sqrt{\frac{3}{2}}\frac{2}{k_{nr}}\left(1 + \frac{1}{2}\log\left(\frac{1 + z_{nr}}{1 + z_{eq}}\right) - \frac{1}{2}\left(\frac{1 + z}{1 + z_{eq}}\right)^2\right),$$

where we used the expression $\eta = 1/(aH) \propto a$ during radiation domination and $\eta = 2/(aH) \propto a^{1/2}$ during matter domination (taking a recent Λ or curvature-dominated stage into account would change the result by a very small amount at late times). The last term is usually negligible, but the second term, with the logarithm, can be important for heavy particles becoming nonrelativistic at high redshift. In this case, r_{fs} can be significantly larger than $(2\pi/k_{nr})$, and the latter does not give a good approximation of the scale below which free-streaming can be neglected. We see that the difference between the two comes from the stage between η_{nr} and η_{eq}, during which the comoving free-streaming scale is constant while the comoving free-streaming horizon grows logarithmically (Boyarsky *et al.*, 2009d).

5.2.5 *Linear evolution of neutrino perturbations*

Relativistic regime

If relativistic neutrinos were not coupled gravitationally, the equation of motion for perturbations in Fourier space (5.66) would be solved by simple plane waves, $F_\nu(\eta, k, \mu) \propto \exp(-ik\mu\eta)$. Because the Legendre coefficients of plane waves are spherical Bessel functions, the solution in multipole space would read $F_{\nu l}(\eta, k) \propto j_l(k\eta)$. This explains why the last of Eqs. (5.68) coincides exactly with the recurrence relation of spherical Bessel functions. In the super-Hubble limit, we see that $F_{\nu 0} = \delta_\nu$ would be constant whereas other multipoles would grow like $(k\eta)^l$: this is indeed the correct behaviour expected for the growing adiabatic mode in the Newtonian gauge.

In the actual equation of evolution, the first few multipoles $F_{\nu l}$ are sourced by metric perturbations. More precisely, in the Newtonian gauge, the evolution of the monopole is sourced by a time-dilation term involving ϕ', whereas the evolution of the dipole is driven by the gradient of the gravitational potential ψ. The full solution is found to be qualitatively similar to that of the homogenous equation, although the amplitude and phase of the multipoles are slightly modified by the coupling with metric perturbations.

Spherical Bessel functions $j_l(x)$ grow as a function of x until $x \simeq l + \frac{1}{2}$, and then feature damped oscillations. For neutrinos, this means that as time evolves, power is transferred from small to large multipoles. In the super-Hubble limit, all the power is in $F_{\nu 0} = \delta_\nu$. At Hubble crossing, it is transferred to $l = 1$, then to $l = 2$, etc. A given multipole reaches its maximum close to $\eta \sim l/k$. This can be understood geometrically. At time η, the observer sees neutrinos emitted on his own neutrino last scattering surface, i.e., on a sphere centered on him and of comoving radius $r = (\eta - \eta_{\nu LS})$, where $\eta_{\nu LS}$ is the neutrino decoupling time. At typical times of interest (close to photon decoupling), $\eta \gg \eta_{\nu LS}$ and the radius of the sphere can be approximated as $r = \eta$. Fourier modes of comoving wavelength $2\pi/k$ on the sphere are seen by the observer under an angle $\theta = 2\pi/(kr)$ in the small-angle approximation. This means that in multipole space the contribution of a given wavenumber peaks near $l \sim 2\pi/\theta \sim kr \sim k\eta$.

Nonrelativistic regime

When neutrinos become nonrelativistic, the stress part of their stress–energy tensor becomes negligible. We saw explicitly in Eq. (2.47) that in the limit $\langle E \rangle \sim m \gg$

$\langle p \rangle$, the pressure P becomes much smaller than the energy density ρ. As shown at the end of Section 5.2.1, the ratios $\delta p_\nu / \delta \rho_\nu$ and $\sigma_\nu / \delta \rho_\nu$ become tiny for the same reasons. Massive neutrinos can then be followed by exploiting the first two of Eqs. (5.74) only. By integrating these equations over momentum in the limit $y \ll \epsilon$, one could show that they are *asymptotically* equivalent to the CDM equations of evolution (5.19).

5.2.6 Practical implementation and approximations

As for photons, a technical issue comes from the fact that the hierarchy of coupled neutrino equations is infinite. A given numerical code can integrate only a finite number of equations, but a truncation of the hierarchy at some multipole $l_{\max \nu}$ may cause a reflection of power from $l_{\max \nu}$ down to lower multipoles. Let us briefly discuss how to avoid such problems.

Brute-force integration

Because at a given time and for a given wavenumber, massless neutrino perturbations are populated until $l \sim k\eta$, the most secure way to avoid any reflection of power for a given mode k would be to truncate the neutrino equations well above $l = k\eta_0$, where η_0 is the conformal age of the universe. For instance, a sharp truncation at $l_{\max \nu} = 2k\eta_0$ gives accurate results. In the case of massive neutrinos, all multipoles with $l \geq 2$ decay after η_{nr}, so the truncation only needs to be performed at $l_{\max \nu} > k\eta_{\mathrm{nr}}$.

These schemes are accurate but heavy. We have seen in Section 5.1.6 that for computing the CMB spectrum until $l_{\max}$, we must compute the evolution of perturbations up to a wavenumber $k_{\max} \sim l_{\max}/\eta_0$, even when using the line-of-sight approach. So, for massless neutrinos, the truncation should be implemented at $l_{\max \nu} \sim 2l_{\max}$ at least for the largest wavenumbers. This implies that thousands of equations must be integrated for many wavenumbers. For massive neutrinos, the number of multipoles can be reduced by a factor of $(\eta_{\mathrm{nr}}/\eta_0)$, but on the other hand the Boltzmann equation Eq. (5.74) must be integrated for a discrete set of momenta y. In both cases the computation time is prohibitive.

Usual truncation scheme

In reality, we need to get accurate solutions for only the first three multipoles $l = 0, 1, 2$, i.e., for $(\delta_\nu, \theta_\nu, \sigma_\nu)$, because these are the only quantities appearing in the linearized Einstein equations. So the goal of a given truncation scheme is to avoid the reflection of power at $l_{\max \nu}$ contaminating the multipoles $l = 0, 1, 2$ significantly. Hence the situation for neutrinos is identical to that for photons in the line-of-sight approach.

The truncation formula proposed by (Ma and Bertschinger, 1995) for photons and mentioned in Section 5.1.6 can be applied to neutrinos too, because it only assumes that high multipoles satisfy the same recurrence relation as the spherical Bessel function $j_l(k\eta)$. For massless neutrinos this gives

$$F_{\nu\,[l_{\max\nu}+1]} = \frac{(2l_{\max\nu}+1)}{k\eta} F_{\nu\,l_{\max\nu}} - F_{\nu\,[l_{\max\nu}-1]}, \tag{5.98}$$

whereas for massive neutrinos

$$\Psi_{\nu\,[l_{\max\nu}+1]} = \frac{(2l_{\max\nu}+1)\epsilon}{yk\eta} \Psi_{\nu\,l_{\max\nu}} - \Psi_{\nu\,[l_{\max\nu}-1]}. \tag{5.99}$$

With such a scheme and for both massless and massive neutrinos, $l_{\max\nu}$ can be reduced typically to 20 or less, depending on the required precision. This approach is the one most commonly implemented in Boltzmann codes and leads to a reasonable computational time. Even with such a gain, neutrinos are still the species described by the largest number of equations in a Boltzmann code. For this reason, several approximation schemes have been devised.

Implicit solutions

The truncation formula of Ma and Bertschinger, 1995 would be exact if gravitational source terms in the Boltzmann equation could be neglected. By applying Green's method, it is possible to derive an exact implicit solution for Eq. (5.66) including these terms (Shoji and Komatsu, 2010; Blas *et al.*, 2011). For massless neutrinos, the solution for the transfer function $F_\nu(k, \mu, \eta)$ is

$$F_\nu(k, \mu, \eta) = F_\nu(k, \mu, \eta_{\rm in})e^{-ik\mu\eta} + \int_{\eta_{\rm in}}^{\eta} e^{-ik\mu(\eta-\tilde{\eta})} 4[\phi'(k, \tilde{\eta}) + ik\mu\,\psi(k, \tilde{\eta})]d\tilde{\eta}.$$

$$\tag{5.100}$$

For adiabatic modes, the initial condition is given by $F_\nu(k, \mu, \eta_{\rm in}) = \delta_\nu(k, \eta_{\rm in}) = \delta_\gamma(k, \eta_{\rm in})$, plus higher-order terms in $(k\eta)$. To write this solution in Legendre space, it is necessary to integrate the term $[ik\mu\,\psi]$ by parts so that the integral depends on μ only through the factor $e^{-ik\mu(\eta-\tilde{\eta})}$, and then to expand the exponential in spherical Bessel functions (see Blas *et al.*, 2011 for a synchronous gauge version of this calculation). The same steps are described by Shoji and Komatsu, 2010 for the massive neutrino equation.

These formal solutions are not immediately useful, because metric fluctuations cannot be considered an external source term. In other words, their evolution depends on the neutrino fluctuation evolution and cannot be computed in advance before integrating Eq. (5.100). However, these solutions can be useful for deriving or justifying several approximation schemes.

Imperfect fluid approximations

Usually, imperfect fluid equations describe weakly coupled species, such that the mean free path of particles exceeds the typical scale on which macroscopic quantities (defined by coarse grain) experience spatial fluctuations. One might think that the case of decoupled neutrinos is very different, because they are collisionless and cannot be considered a fluid at all. However, in a cosmological context, an imperfect fluid approach could be applied to neutrino perturbations expanded in Fourier space. Indeed, for wavelengths larger than the Hubble radius (or more precisely than the free-streaming scale), the propagation of neutrinos can be neglected. In this limit, the neutrino perturbation equations are equivalent to perfect fluid equations. We have already seen that with adiabatic initial conditions, neutrino density fluctuations are identical to photon density fluctuations on super-Hubble scales, although neutrinos are collisionless and photons are tightly coupled at high redshift. Around the time of Hubble crossing, the mean free path (which can be identified with the free-streaming horizon) is still not much bigger than the characteristic length in the Friedmann universe, namely the Hubble radius: thus, an imperfect fluid approximation may in principle give good results. Later, for modes well inside the horizon, it is obvious that the free-streaming horizon cannot be considered small anymore, and so there is no reason for an imperfect fluid approximation to work. However, in this limit, the gravitational back-reaction of neutrinos on photons, baryons and CDM is very small (as we shall see in Sections 5.3.3 and 6.1.4). So we do not need neutrino perturbation approximations to be very precise, unless we want to understand the clustering properties of the neutrinos themselves.

In general, a theory of imperfect fluids involves equations for shear viscosity, bulk viscosity, heat conduction, entropy flux, etc. These possibilities have not all been investigated in detail in the context of neutrino cosmology, but several people have proposed an approximate evolution equation for shear viscosity, i.e., for the anisotropic stress σ_ν, on top of the continuity and Euler equations. In other words, the Boltzmann hierarchy (5.68) or (5.74) could be closed at $l = 2$ (and integrated over momentum in the case of massive neutrinos), using some particular approximation. This can be very efficient because it reduces the number of neutrino equations to three for each wavenumber.

An imperfect fluid approximation could be used at any time or could be substituted into the full equations only inside the Hubble radius, when $k\eta$ exceeds some threshold value $[k\eta]_{\text{fluid}}$ (Blas *et al.*, 2011; Lesgourgues and Tram, 2011). The advantage of the second case is that an error in the neutrino sector would have a moderate impact on other quantities, because the gravitational coupling between neutrinos and other species is relevant mainly around Hubble crossing. This approach is also efficient because at the time when the approximation is switched, only modes up to $l \sim [k\eta]_{\text{fluid}}$ are populated. So before using the approximation, we can truncate the hierarchy with Eq. (5.98) or (5.99) at a lower $l_{\text{max}\,\nu} \sim [k\eta]_{\text{fluid}}$.

The first concrete fluid approximation was proposed by Hu, 1998. The idea is to use a shear equation of the form

$$\sigma_\nu' = -3\frac{a'}{a}\frac{c_{\text{eff}}^2}{w}\sigma_\nu + \frac{8}{3}\frac{c_{\text{vis}}^2}{1+w}\theta_\nu \tag{5.101}$$

with two given coefficients c_{eff} (the effective sound speed) and c_{vis} (the effective viscosity speed). Hu proposes to replace the first coefficient with the adiabatic sound speed ($c_{\text{eff}}^2 = \bar{P}'/\bar{\rho}'$), and c_{vis}^2 with the equation of state parameter $w = \bar{P}/\bar{\rho}$. For relativistic neutrinos, these different factors reduce to $c_{\text{eff}}^2 = c_{\text{vis}}^2 = w = 1/3$. Also, during radiation domination, a'/a can be replaced with $1/\eta$. In these two limits, the above equation becomes equivalent to the shear equation in Eq. (5.68) when the usual truncation scheme is applied at $l_{\text{max}\,\nu} = 2$ (so F_3 is replaced as a function of σ_ν and θ_ν using Eq. (5.98)).

More accurate alternatives have been discussed in the literature. Shoji and Komatsu (2010) start from the implicit exact solution for massive neutrinos, and assume a sharp truncation at $l = 3$ ($\Psi_3 = 0$). Blas *et al.* (2011) start from the implicit exact solution for massless neutrinos, and infer a shear evolution equation by keeping only the leading terms inside the Hubble radius, when $(k\eta) \gg 1$. Lesgourgues and Tram (2011) generalize the latter approach to massive neutrinos. The merits of these different schemes are discussed in these works. It appears that the accuracy of imperfect fluid approximations can be largely sufficient for fitting the CMB and LSS spectrum in future experiments.

Other approximations

The most computationally expensive part in a Boltzmann code is usually the integration of massive neutrino equations, because the Legendre momenta Ψ_l of the perturbed phase space distribution must be followed for several discrete values of momentum y (and even if the number of momenta can be kept small thanks to a quadrature approach; see Lesgourgues and Tram, 2011). We have seen in Eq. (5.72) that as long as neutrinos are relativistic, the y dependence of Ψ_l is known and given by $[d \ln f_{\nu 0}(y)/d \ln y]$. Deviations from this form can be captured by a Taylor expansion in the parameter

$$\frac{\sqrt{\epsilon^2 - y^2}}{y} = \frac{am}{y}, \tag{5.102}$$

which grows from zero to one in the relativistic regime, and from one to infinity in the nonrelativistic one. Then the transfer function $\Psi_l(\eta, k, y)$ can be written in the form

$$\Psi_l(\eta, k, y) = -\frac{1}{4} \frac{d \ln f_{\nu 0}(y)}{d \ln y} \left[\sum_i \left(\frac{am}{y} \right)^i \tilde{\Psi}_l(\eta, k)^{(i)} \right]. \tag{5.103}$$

Howlett *et al.* (2012) used the exact equation of motion for Ψ_l to show that

- $\tilde{\Psi}_l^{(0)}$ reduces to F_ν, as expected from the relativistic limit of Eq. (5.72);
- $\tilde{\Psi}_l^{(1)}$ vanishes;
- $\tilde{\Psi}_l^{(2)}$ is subject to a hierarchy of equations, like that for F_ν but with different coefficients;
- higher-order terms can be neglected for the purpose of computing CMB anisotropies, even if good precision is required.

In this scheme, the full massive neutrino equations are replaced with those for massless neutrinos, plus a duplicate set of equations for $\tilde{\Psi}_l^{(2)}$. This approach is not as fast as fluid approximations, but it is simpler, robust, and easy to generalize to higher order.

5.3 Effects of neutrinos on primary cosmic microwave background anisotropies

5.3.1 How can decoupled species affect the cosmic microwave background?

It is always difficult to discuss the impact of a given species (neutrinos, CDM, early dark energy, etc.) on the CMB, because several effects mix with each other, and it is sometimes impossible to separate them clearly. However, we can make a general classification of these effects, which will help us to understand the effect of massless or massive neutrinos in the next subsections.

Species coupled only gravitationally with the photon–baryon(–electron) fluid can only impact the CMB through the Friedmann equation, or through the recombination history, or finally, through the linearized Einstein equations via metric perturbations. Hence, it is useful to adopt a classification in terms of background and perturbation effects.

(a) *Background effects* are encoded in a modified evolution of the scale factor $a(\eta)$, or possibly of the free electron fraction $X_e(\eta)$. We can notice already that neutrinos have no impact on the recombination history, and only affect $a(\eta)$ (during radiation domination in the case of massless neutrinos, and also slightly during matter and Λ domination in the case of massive neutrinos). In Section 5.1, we showed that the spectrum of CMB anisotropies in the minimal ΛCDM model depends on a small number of effects or "degrees of freedom", each of them being related to characteristic scales (sound horizon, diffusion length, angular diameter distance), to characteristic times (equality between radiation and matter, equality between matter and Λ), to the primordial spectrum and to density ratios (e.g., the balance between pressure and gravity in the tightly coupled photon–baryon fluid relates to ω_B, or more precisely to ω_B/ω_γ, with ω_γ fixed by the measurement of the CMB temperature today). The list of these effects is presented at the end of Section 5.1.6 with labels (C1)–(C8). In the following sections, we will relate the impact of neutrinos to variations of these characteristic scales, times and ratios. We will also see that the background effects of a given species depend on which particular set of other parameters is kept fixed when the density of this species is increased; sometimes these background effects can be completely absorbed by tuning other parameters at the same time. For this reason, background effects can be thought to be "indirect" and difficult (or sometimes even impossible) to probe with observations. We will provide a detailed discussion of these issues in the next subsections.

(b)   *Perturbation effects* can instead be considered as "direct". The presence of a decoupled species can modify the evolution of metric fluctuations, and back-reacts on the perturbations of the photon–baryon fluid. In the next subsections, we will do our best to cancel the background effect of neutrinos (by tuning other parameters) in order to single out their direct perturbation effect. Let us discuss separately perturbation effects occuring before and after photon decoupling:

(b.1) *Before decoupling*, a key observation is that the impact of a species coupled only gravitationally to photons and baryons can only be important around the time of sound horizon crossing for that species. Indeed, on super-Hubble scales, the growing adiabatic mode is subject to universal relations between δ_γ, δ_B and the curvature perturbation $\mathcal{R}$, not affected by the presence of other species: we checked this explicitly for neutrinos in Section 5.2.3. In the opposite limit, well inside the sound horizon, we have seen in Section 5.1.5 that metric fluctuations have decayed. Once $|\phi|$ and $|\psi|$ are much smaller than $|\Theta_\gamma|$, there can be no gravitational feedback on photon fluctuations in Eq. (5.38), and neutrinos or other decoupled species cannot be relevant. Hence, the only region in which the effect of neutrinos on primary CMB anisotropies should be discussed is that of sound horizon crossing before photon decoupling.

(b.2) *Soon after decoupling*, a given species can impact the CMB via the early ISW effect. We mentioned earlier that the EISW is negligible on small scales, and peaks for wavenumbers crossing the Hubble scale roughly around the time of decoupling. In the space (k, η), this region is close to the one described in (b.1) and extends it to higher η and slightly smaller k. The EISW depends on the time variation of $(\phi + \psi)$ at a given wavenumber after decoupling. As briefly mentioned at the very end of Section 5.1.5 and shown in Chapter 6, this variation would vanish in a fully matter-dominated universe. Changing the neutrino abundance or mass implies that we play with the fraction of total matter which does not cluster like CDM (because neither relativistic neutrinos nor nonrelativistic neutrinos cluster below their free-streaming scale). Hence, neutrinos can play a crucial role in the determination of the time needed for $(\phi + \psi)$ to reach a nearly constant asymptotic value after decoupling, and of the amplitude of the EISW. This effect is distinct from the global enhancement of the EISW caused by a delay in the time of equality between radiation and matter, which we would classify as a background effect.

(b.3) *At low redshift*, massless neutrinos cannot play a role in the late ISW effect, because their density is completely negligible, but massive neutrinos can, because below their free-streaming scale they do not cluster like CDM and introduce a very small but nonzero derivative $(\phi' + \psi')$. This effect is distinct from that of a change in the time of equality between matter and Λ, which would be classified as a background effect. Because it is related to secondary anisotropies imprinted in the recent universe, we will discuss it in Section 6.3.1 of the next chapter (and find that it is almost negligible).

5.3.2 Effects of massless neutrinos

A single free parameter, the effective neutrino number

We know that all relativistic collisionless species are described by the same equations (5.68) integrated over momentum, including all neutrinos and relics such that $m \ll \langle p \rangle$ (at least until today), with or without chemical potentials and/or non-thermal distortions. All these cases are encoded in one single parameter, the total density of such species today (which can be scaled back to the past, because for relativistic species $\rho_R \propto a^{-4}$). This density can be specified, e.g., by giving the total radiation density ω_R (including photons and extra relics), or the fractional radiation density Ω_R, or the effective neutrino number N_{eff} defined in (2.198). In this section we choose to use N_{eff} as a free parameter. The neutrinoless ΛCDM model discussed in section 5.1 corresponds to $N_{eff} = 0$, whereas the actual minimal ΛCDM model including the usual three families of active neutrinos (with masses assumed to be negligible) assumes $N_{eff} = 3.046$, as explained in Section 4.1. Here, N_{eff} will be considered as a seventh free parameter on top of the six ΛCDM parameters $(A_s, n_s, \omega_B, \omega_M, \Omega_\Lambda$ or $h = \sqrt{\omega_M/(1 - \Omega_\Lambda)}, \tau_{reion})$ defined in Section 5.1.6.

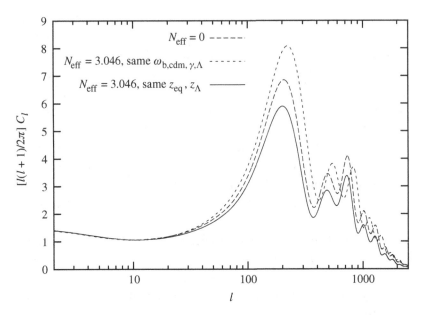

Figure 5.3 Spectrum of CMB temperature anisotropies for a neutrinoless model (middle curve) a model with $N_{\text{eff}} = 3.046$ and identical parameters ($A_s, n_s, \omega_B, \omega_M, \Omega_\Lambda, \tau_{\text{reion}}$) (top curve) and a model with $N_{\text{eff}} = 3.046$ and the same redshift of equality between radiation and matter and between matter and Λ (lower curve; see the text for details). The difference between the middle and lower curves can be attributed entirely to neutrino perturbation effects, plus a shift in the diffusion damping scale.

Background effects

We start by comparing two models with $N_{\text{eff}} = 0$ and $N_{\text{eff}} = 3.046$, sharing identical values of ($A_s, n_s, \omega_B, \omega_M, \Omega_\Lambda, \tau_{\text{reion}}$). We will see later that the choice to keep these parameters fixed is not very illuminating, and we will take a second approach in the next paragraphs.

The middle and top curves in Fig. 5.3 show the temperature spectrum of these two models. The difference can be explained by the sum of some perturbation effects (that we will described later separately), plus the following background effects:

1. A different redshift of equality (described as effect (C3) at the end of Section 5.1.6). Indeed, in the presence of relativistic relics, this redshift is given by

$$z_{\text{eq}} = \frac{\omega_M}{\omega_\gamma \left[1 + \frac{7}{8}\left(\frac{4}{11}\right)^{4/3} N_{\text{eff}}\right]} = \frac{\omega_M}{\omega_\gamma \left[1 + 0.2271 N_{\text{eff}}\right]}. \tag{5.104}$$

Because ω_γ is fixed by the measurement of the CMB temperature today, known up to four digits, this formula has two free parameters, ω_M and N_{eff}. Increasing N_{eff} with ω_M fixed implies a shorter stage between equality and decoupling, leading to higher peaks (especially the first one), as explained in Section 5.1.6.

2. A different peak scale (effect (C1), related to the ratio $d_s(\eta_{LS})/d_A(\eta_{LS})$ or $r_s(\eta_{LS})/r_A(\eta_{LS})$). The comoving sound horizon at decoupling depends on the expansion history before η_{LS}, and in particular on the time of equality. A later equality implies a smaller sound horizon at decoupling, so all peaks are shifted to higher l's. The comoving angular diameter distance $r_A(\eta_{LS})$, depending only on the expansion history after η_{LS}, is the same in these two models.

3. A tiny change in the diffusion scale (effect (C4), related to the ratio $d_d(\eta_{LS})/d_A(\eta_{LS})$ or $r_d(\eta_{LS})/r_A(\eta_{LS})$). Both $r_s(\eta_{LS})$ and $r_d(\eta_{LS})$ are given by integrals running from $\eta = 0$ to η_{LS}, respectively Eq. (5.37) and Eq. (5.40). However, the integrand of Eq. (5.40) increases more quickly with time, roughly as a^2. Hence $r_D(\eta_{LS})$ depends only on the expansion and thermal history immediately before decoupling and is affected by a change in z_{eq} only by a negligible amount. This explains why in Figure 5.3, the secondary peaks of the middle and upper curves share roughly the same envelope $\exp[-(l/l_D)^2]$. However, the third and higher peaks of the model with $N_{eff} = 3.046$ moved to larger l in such a way as to fall inside the diffusion damping region: hence their amplitude is suppressed, even if effect (C3) alone would have the opposite effect.

Other effects (C2), (C5), (C6), (C7), (C8) are controlled by the parameters $(A_s, n_s, \omega_B, \Omega_\Lambda, \tau_{reion})$ and are the same in the two models.

In the previous comparison, the direct perturbation effect of neutrinos was masked by the large impact of the change in the time of equality. To make a more useful comparison, we should vary N_{eff} not with other cosmological parameters fixed, but with the quantities governing the effects (C1)–(C8) unchanged, if this is possible. This approach was taken previously by (Bashinsky and Seljak, 2004) and summarized by (Hou *et al.*, 2011).

In increasing the effective neutrino number from zero to N_{eff}, we enhance the radiation density by a factor

$$\alpha \equiv (1 + 0.2271 \, N_{eff}). \tag{5.105}$$

If we simultaneously increase the matter density ω_M by the same amount, the redshift of equality between radiation and matter will not vary, and effect (C3) will be the same for the two models. If this is done by enhancing the CDM density with fixed ω_B, effect (C2) will also remain constant. We should keep Ω_Λ fixed in order to maintain the same redshift of equality between matter and Λ, and leave effect (C7) unaffected. Finally, A_s, n_s, τ_{reion} should also be left invariant, in order to preserve (C5), (C6), (C8). We must now check if the remaining effects (C1) and (C4) are affected by such a transformation or not. In summary:

- We increase ω_R and ω_M by the same amount α, while keeping Ω_Λ fixed.
- Hence the reduced Hubble parameter $h = \sqrt{\omega_M/(1 - \Omega_\Lambda)}$ increases by $\sqrt{\alpha}$.
- This implies that ω_Λ also increases by α, so the three densities $\rho_R(z)$, $\rho_M(z)$ and $\rho_\Lambda(z)$ are rescaled by the same amount. Indeed, the proper way to keep the two redshifts of equality fixed is to multiply all densities by the same factor. The Friedmann equation shows that the Hubble parameter as a function of redshift, $H(z)$, is rescaled by $\sqrt{\alpha}$.
- The functions $\eta(z)$, $r_s(z)$, $r_A(z)$, obtained by integrating over $d\eta$ (or equivalently, after a change of variable, over dz/H), are all rescaled by $\alpha^{-1/2}$.
- The square of the function $r_D(z)$ is obtained by integrating over $(d\eta/an_e)$, or equivalently over $(zdz/n_e H)$. Because the reionization history controlled by the parameter $X_e(z)$ is unaffected by the transformation, $r_D(z)^2$ is also rescaled by $\alpha^{-1/2}$.

We conclude that effect (C1) is unaffected by this transformation: both $r_s(\eta_{LS})$ and $r_A(\eta_{LS})$ have been rescaled by the same factor, with a constant ratio.[3] Instead, effect (C4) has been changed, because r_D decreases by $\alpha^{-1/4}$, whereas the ratio $r_D(\eta_{LS})/r_A(\eta_{LS})$ increases by $\alpha^{1/4}$. So in the second model, diffusion damping occurs at larger angles, i.e., at smaller l's.

We conclude that the difference between the original model and the rescaled model should only be attributed to diffusion damping and to neutrino perturbation effects. This is illustrated in Fig. 5.3. The lower curve has been obtained by rescaling all densities as described. The peaks of the $N_{eff} = 0$ and $N_{eff} = 3.046$ models now have roughly the same location and shape, with, however, an overall amplitude suppression (resulting from neutrino free-streaming, as developed in the next subsection) and a smaller damping tail.

Our goal was to increase N_{eff} without affecting effects (C1)–(C8), in order to eliminate background effects and to isolate the direct gravitational effect of neutrinos. This was not fully achieved by the previous transformation, because effect (C8) varied. However, in the framework of the minimal ΛCDM model with free N_{eff}, (C1) and (C8) cannot be kept fixed simultaneously. Because ω_M is determined by the redshift of equality and ω_B by the baryon-to-photon ratio, we

---

[3] One could go further and show that in the equation of evolution for the tightly coupled photon–baryon fluid, a rescaling of $H(z)$ with fixed $R(z)$ can be completely eliminated. At the same time as $H(z)$ is rescaled by $\sqrt{\alpha}$, all wavenumbers should be rescaled by the same factor. But because d_A is rescaled by the inverse of $\sqrt{\alpha}$, after a projection in harmonic space, the spectrum C_l is left invariant. Hence, as long as tight coupling holds, the transformation discussed here leaves the CMB invariant. Differences can show up only beyond the tight-coupling approximation, i.e., at the level of diffusion damping, or through the gravitational back-reaction of other species mediated by metric fluctuations.

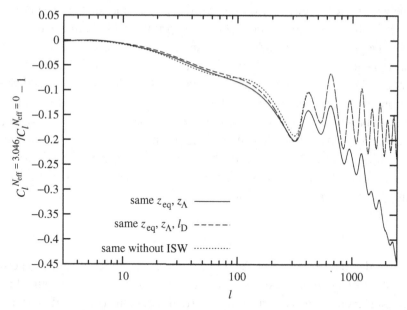

Figure 5.4 CMB temperature spectrum for models with $N_{eff} = 3.046$ divided by the spectrum of a model with $N_{eff} = 0$. The lower curve corresponds to the case in which the two models share the same redshift of equality between radiation and matter and between matter and Λ (so it is the ratio of the lower and middle curves in Figure 5.3). The upper curve is obtained by additional tuning of the primordial helium fraction Y_p, in such a way as to keep the diffusion damping scale r_d/r_A constant. Hence the upper curve illustrates the effect of neutrinos at the level of perturbations only.

can adjust effects (C1) and (C8) only by playing with h. But r_s is proportional to h^{-1} whereas r_d is proportional to $h^{-1/2}$, so both effects cannot be compensated for at the same time. However, we could try to change the recombination history. As noticed by Bashinsky and Seljak, 2004, this is a useful way to gain further insight into the effect of relativistic neutrinos.

In Fig. 5.4, the lower curve shows the spectrum of the previous model with $N_{eff} = 3.046$ divided by that of the neutrinoless model. This ratio can be described as a steplike suppression with superimposed oscillations, plus a rapid decrease for $l > 700$. We suspect that this rapid decrease is due to diffusion damping, i.e., to the shifting of the envelope $\exp[-(l/l_D)^2]$ to a smaller l_D in the $N_{eff} = 3.046$ model. This can be checked by decreasing the primordial helium fraction Y_p defined in (2.157) in the model with neutrinos. Indeed, changing this parameter leads to a rescaling of n_e at the redshifts of interest, leaving all other characteristic times and scales fixed, apart from a very small change in the decoupling time η_{LS} which turns out to be negligible. In order to increase r_D and keep the same ratio $r_d(\eta_{LS})/r_A(\eta_{LS})$

as in the neutrinoless model, we need to decrease Y_p to unrealistically small values, around $Y_p = 0.02$. In a real data analysis, the primordial helium fraction is given either by the prediction of standard Big Bang nucleosynthesis (to about 0.25, with a weak dependence on ω_B), or by direct measurements of primordial element abundances. Values significantly smaller than 0.24 would conflict with both BBN predictions and direct observations, but the exercise performed here should be regarded as purely formal. The spectrum of the model with $N_{\text{eff}} = 3.046$ and low helium fraction, divided by that of the original neutrinoless model, is shown in Fig. 5.4 (upper curve). As expected, the exponential suppression at $l > 700$ has now been removed. The remaining features can be attributed to the direct effect of neutrinos at the level of perturbations.

Perturbation effects

As discussed at the beginning of Section 5.3, the direct effect of neutrino perturbations on the CMB can be important when a given mode crosses the sound horizon, and acoustic oscillations are driven by metric fluctuations (i.e., by the term on the r.h.s. of Eq. (5.38)). Metric fluctuations quickly decay inside the sound horizon, and their time variation tends to boost temperature fluctuations until ϕ and ψ become negligible with respect to Θ_γ (more details on the effect of the driving term can be found, e.g., in Hu, 1995). After that stage, we have seen that oscillations go on with a constant amplitude during the radiation-dominated stage, or with a decreasing amplitude after the time of equality, caused by the increasing baryon fraction and by diffusion damping on small scales.

In the presence of free-streaming neutrinos, we expect metric fluctuations to be smaller for wavelengths inside the free-streaming scale (which coincides with the Hubble radius for relativistic neutrinos), because neutrinos cannot cluster on those scales. During radiation domination, neutrinos account for a large fraction of total matter, so they can reduce metric fluctuations by a significant amount. Hence, neutrinos reduce the boosting of temperature fluctuations during the driven oscillation stage. At the end, we observe smaller CMB anisotropies for all scales entering the sound horizon before decoupling, and especially during radiation domination.

An analytic approximation of the impact of neutrinos on the driving term was derived by Hu and Sugiyama, 1996, leading to the conclusion that the oscillation amplitude inside the sound horizon is reduced by $(1 + 4/15R_\nu)^{-1}$ with respect to a neutrinoless model. So we expect the CMB peaks to be reduced by

$$\frac{\Delta C_l}{C_l} = \left(1 + \frac{4}{15}R_\nu\right)^{-2} = \left(1 + \frac{4}{15} \times \left[\frac{0.2271 N_{\text{eff}}}{1 + 0.2271 N_{\text{eff}}}\right]\right)^{-2}. \qquad (5.106)$$

For the two models that we compared in Figs. 5.3 and 5.4, this formula predicts 19% suppression between $N_{\text{eff}} = 0$ and $N_{\text{eff}} = 3.046$, in rather good agreement with the average value of the upper curve in 5.4 for the largest values of l. Notice that for intermediate values corresponding to the first two peaks, the effect is only partial because the modes entered the sound horizon at the beginning of matter domination. For a small variation of N_{eff} around three, Eq. (5.106) corresponds to a variation of the C_l's in the region of acoustic oscillations given by

$$\frac{\Delta C_l}{C_l} = -0.072 \, \Delta N_{\text{eff}}. \tag{5.107}$$

A more detailed analysis was presented later by (Bashinsky and Seljak, 2004). This work concludes with a reduction of the amplitude with respect to the neutrinoless case by

$$\frac{\Delta C_l}{C_l} = \left(1 - 0.2683 R_\nu + O(R_\nu^2)\right)^2, \tag{5.108}$$

found to be in very good agreement with Eq. (5.106). The same reference also shows that in real space, neutrinos (traveling at the speed of light) tend to pull temperature perturbations (propagating at a lower velocity $c_s \sim c/\sqrt{3}$) out of gravitational potential wells. In Fourier space, this "neutrino drag" effect tends to shift the phase of oscillations in such a way that acoustic peaks are seen on larger scales, i.e., for smaller values of l. The analytic approximation of Bashinsky and Seljak, 2004 predicts a shift with respect to the neutrinoless case by

$$\Delta l_{\text{peak}} = -\frac{r_A(\eta_{\text{LS}})}{r_s(\eta_{\text{LS}})} \left(0.1912\pi R_\nu + O(R_\nu^2)\right). \tag{5.109}$$

This approximation does not work as well as that for the amplitude of the peaks. For instance, it would correspond to a shift $\Delta l_{\text{peak}} \simeq 20$ between the middle and lower curves of Fig 5.3, whereas the actual shift is closer to 10. The phase shift also explains the oscillatory pattern of the curves in Fig. 5.4.

Note that in the two models that we are comparing, ω_C is different. Like neutrinos, CDM couples gravitationally with the photon–baryon fluid. So one could argue that the effects observed here are a superposition of gravitational effects due to both neutrinos and CDM. However, during radiation domination, the CDM density is negligible. So the perturbation effects of an enhanced CDM component are subleading for modes entering the sound horizon during radiation domination. This explains why the analytic prediction for the neutrino effect in Eq. (5.106) is a very good approximation to the full numerical result shown in Fig. 5.4.

So far, we have discussed only effects occuring before decoupling, through the driving term in the photon–baryon oscillator equation. As mentioned in the previous subsection, neutrinos can also induce a difference in the EISW effect. However, this effect is subleading, as shown by the dashed curve in Fig. 5.4, in which we removed the whole ISW contribution to each spectrum before taking the ratio. For the sake of brevity, we will not discuss the details of this tiny correction.

5.3.3 *Effects of massive neutrinos*

A mode-dependent situation

Whereas a single parameter N_{eff} catches all possible effects of massless neutrinos (or more generally collisionless ultrarelativistic relics), massive neutrino effects are described by more parameters, especially if one goes beyond the minimal picture. The two parameters N_{eff} (describing the neutrino abundance in the early universe, when they are still ultrarelativistic) and ω_ν (the total density of neutrinos today, dominated by the contribution of at least two nonrelativistic mass eigenstates) are certainly playing a key role. However, in the perturbation equation for neutrinos, it is not possible to integrate quantities in momentum space in order to obtain reduced equations as for massless neutrinos, and it is not possible to factor out the total neutrino mass. In principle, individual masses could play a role, as well as any modification of the phase-space distribution function due, for instance, to chemical potentials or nonthermal distortions. In the following, we will assume for simplicity that all neutrinos have the same mass (as in the degenerate limit of the normal and inverted hierarchy scenarios) and share the same Fermi–Dirac distribution function (which is almost true in the minimal scenario, because nonthermal distortions imprinted around the time of electron–positron annihilation are extremely small, as discussed in Section 4.2.3). At the end of this section we will briefly discuss the effect of different individual masses and see that for standard active neutrinos, only the total mass M_ν is detectable in practice.

We saw in Section 5.2.4 that active neutrinos with a mass $m_\nu < 1.5\,\text{eV}$ become nonrelativistic after the time of equality between matter and radiation. We can safely restrict our discussion to this case. We will see later that heavier active neutrinos would strongly contradict current cosmological bounds.

Background effects

We have seen in Section 5.3.2 that the most relevant way to study the impact of a given parameter on the CMB spectra is to vary it while keeping fixed at the same time all characteristic times, scales and density ratios, which control the physical effects (C1)–(C8) described at the end of Section 5.1.6. A particularly important quantity is the redshift of equality between matter and radiation. Because we assumed that $m_\nu < 1.5\,\text{eV}$, we should count neutrinos as radiation at that time, so

z_{eq} is given by

$$z_{eq} = \frac{\omega_B + \omega_C}{\omega_\gamma \left[1 + 0.2271 N_{eff}\right]}, \tag{5.110}$$

where the neutrinos are counted in the N_{eff} factor. The comparison between different masses should be performed at fixed $(\omega_B + \omega_C)$, or better at fixed ω_B and ω_C in order to keep a fixed ratio (ω_B/ω_γ) (and not affect (C2) or (C3)). This means that the total nonrelativistic matter density today, $\omega_M = \omega_B + \omega_C + \omega_\nu$, should be increased at the same time as the neutrino mass.

Two models with different neutrino masses but sharing the same values of $(A_s, n_s, \omega_B, \omega_C, \tau_{reion})$ are strictly equivalent until the time of the nonrelativistic transition at redshift $z_{nr} \propto m_\nu$. After that time, neutrinos dilute like $\rho_\nu \propto a^{-3}$ instead of a^{-4} and play the role of extra nonrelativistic matter. Therefore, the mass has an impact on the expansion history $H(z)$ at $z < z_{nr}$, on the comoving angular diameter distance to recombination $r_A(\eta_{LS})$, and on the redshift of equality between matter and the cosmological constant z_Λ. Only if neutrinos are nonrelativistic at decoupling, i.e., roughly for $m_\nu \geq 0.6$ eV (see Eq. (5.90)) does their mass also impact the comoving sound horizon $r_s(\eta_{LS})$ and the damping scale $r_D(\eta_{LS})$ at decoupling.

By tuning h and Λ, still related through $h = \sqrt{\omega_M/(1 - \Omega_\Lambda)}$, we have the possibility of keeping constant one of the two quantities $d_A(\eta_{LS})$ or z_Λ, but not both simultaneously. Because the scales of the peak and of the damping tail (controlled by $r_A(\eta_{LS})$ through effects (C1) and (C7)) are much better constrained by the data than the slope of the Sachs–Wolfe plateau (controlled by z_Λ through (C7)), the most interesting comparison is achieved by keeping $d_A(\eta_{LS})$ fixed. For masses much smaller than 0.6 eV, this will lead to the same peak scale and damping tail. For masses close to 0.6 eV or higher, $r_s(\eta_{LS})$ and $r_d(\eta_{LS})$ are also affected, but we can always tune h to keep the same peak scale, at the expense of a small change in the diffusion damping scale.

In summary, we can increase m_ν with a constant redshift of equality and peak scale. The effects (C1)–(C8) then remain unaffected with the exception of

1. the LISW effect (C7), due to a shift in z_Λ;
2. eventually the diffusion damping effect (C4), but only for neutrinos already nonrelativistic at decoupling, i.e., with $m_\nu \geq 0.6$ eV.

Differences in the CMB spectra beyond those two effects should be attributed to the direct gravitational impact of massive neutrinos at the perturbation level.

We compare in Fig. 5.5 three spectra obtained with the same values of $(A_s, n_s, \omega_B, \omega_C, \tau_{reion})$ and three active neutrinos with a common mass $m_\nu = M_\nu/3$, with either $M_\nu = 0$ eV, 3×0.3 eV or 3×0.6 eV. The two massive models

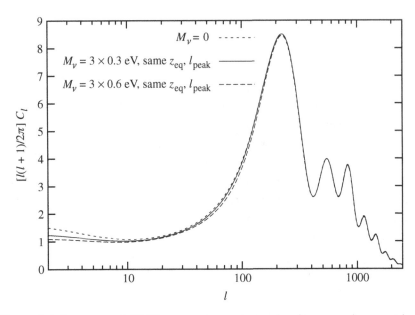

Figure 5.5 Spectrum of CMB temperature anisotropies for a massless neutrino model (highest curve for $l \leq 100$), a model with three degenerate neutrinos of total mass $M_\nu = 3 \times 0.3 \, \text{eV}$ (middle curve for $l \leq 100$), and one with $M_\nu = 3 \times 0.6 \, \text{eV}$ (lowest curve for $l \leq 100$). In all cases the redshift of equality between radiation and matter and the scale of the peaks has been kept constant.

correspond to a nonrelativistic transition at respectively $z_{nr} = 570$ or $z_{nr} = 1100$. When increasing the mass, we decrease h in order to keep the same peak scale. The ratios of the two massive cases divided by the massless case are shown in Fig. 5.6.

The difference due to the LISW is the most obvious one, leading to different slopes for $l < 20$ in Fig. 5.5, and to the sharp decrease for the same multipoles in Fig. 5.6. Models with a higher mass have a smaller h, a smaller Ω_Λ, and hence a smaller LISW effect. The small shift in the damping scale is seen better in Fig. 5.6 in the form of a rise at $l > 500$, especially for the model with nonrelativistic neutrinos at decoupling, as expected. Remaining differences in the range $20 < l < 500$ should be attributed to direct perturbation effects induced by neutrino masses.

Perturbation effects

We expect perturbation effects to be caused either by the gravitational driving of the photon–baryon oscillator equation before decoupling (effect described as (b.1)) at the beginning of Section 5.3), or by the neutrino mass impact on the evolution of $(\phi + \psi)$ after decoupling, i.e., by the EISW effect (described as (b.2)). To separate these two effects, we show in Fig. 5.6 the difference between the ratios obtained

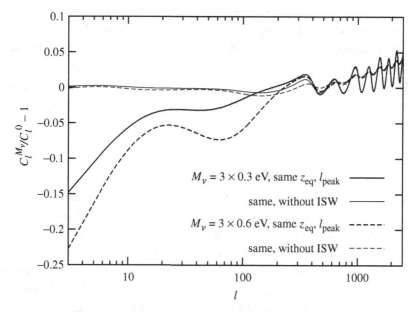

Figure 5.6 CMB temperature spectrum for models with either $M_\nu = 3 \times 0.3$ eV or $M_\nu = 3 \times 0.6$ eV divided by the spectrum of a model with three massless neutrinos sharing the same (standard) temperature. From left to right, the differences can be attributed to the LISW effect, EISW effect, gravitationally driven oscillations effect and diffusion damping, as explained in the text. To separate better the impact of the LISW plus EISW effect, the thin lines show the result of the comparison when the full ISW contribution is removed from each spectrum.

with the full temperature spectra (solid curves) and with the spectra without the ISW contribution (dashed curves).

The largest effect in the range $20 < l < 500$ is caused by the EISW: it leads to a depletion of the spectrum for $20 < l < 200$. This depletion is also visible in Fig. 5.5. Its amplitude is roughly given by

$$\frac{\Delta C_l}{C_l} \simeq -\left(\frac{m_\nu}{10 \text{ eV}}\right), \tag{5.111}$$

i.e., 3% for $m_\nu = M_\nu/3 = 0.3$ eV. Its location is also related to the neutrino mass, because this effect is caused by the fact that for wavenumbers $k < k_{\text{nr}}$ neutrinos behave as a clustering component, whereas in the absence of neutrino mass they would behave as a free-streaming component like photons. Hence $(\phi + \psi)$ experiences less decay for these modes: we are closer to the limit of a fully matter-dominated universe for which $(\phi + \psi)$ would remain constant at horizon crossing. This implies a smaller EISW effect for $k < k_{\text{nr}}$, visible in the CMB spectra above a given angle. The magnitude of this effect scales with the ratio $(\overline{\rho}_R/\overline{\rho}_M)$ at the time when $k = k_{\text{nr}}$ crosses the horizon, so it decreases with the neutrino mass.

A smaller effect related to the evolution prior to recombination is visible in the curves without ISW in the range $30 < l < 500$. For the model with relativistic neutrinos at decoupling, $M_\nu = 3 \times 0.3$ eV, this effect remains below the percent level. For the model with $M_\nu = 3 \times 0.6$ eV, we see a 3% depletion of the spectrum around $l \sim 100$, caused by the fact that the last wavenumbers approaching the sound horizon just before decoupling do not see free-streaming neutrinos, and do not experience the full boosting effect already mentioned in the case of massless neutrinos.

Effect of mass splitting

In principle, the depletion in the CMB power spectrum caused by the EISW effect depends on the time at which neutrinos become nonrelativistic, i.e., on individual masses. However, in practice, the mass-splitting effect is too small to be detectable. The difference between the temperature spectrum for the minimal-inverted-hierarchy scenario and for a normal-hierarchy scenario with the same mass (see Section 1.4.1) remains below the level of one per mill for all multipoles, i.e., beyond the sensitivity level of Planck and even of a next generation of CMB satellite. We will see in Section 6.1.4 that the effect of mass splitting on large-scale structure is significantly greater, but still very difficult to observe.

5.3.4 Effects of interacting neutrinos

If neutrinos are not collisionless at temperature $T \ll 1$ MeV because of some kind of nonstandard interaction, all the "direct perturbation effects" described earlier can be modified. For any given interaction between neutrino and other species, or any self-interaction term, the CMB spectrum can be computed with a modified Boltzmann code featuring a coupling term in the hierarchy of Boltzmann equations for neutrinos. This has been studied, for instance, in the case of neutrino interactions with putative extra light degrees of freedom or with dark matter or in the mass-varying neutrino (MaVaN) scenario, in which neutrinos are assumed to be coupled with a scalar field accounting for dark energy. This case will be reported in Section 6.5.3 of the next chapter, because this scenario affects LSS observables but not the CMB (such neutrinos would still be relativistic and have negligible interactions at the epoch of decoupling).

In another simple limiting case, neutrinos would reduce to a perfect or imperfect fluid because of a nonstandard self-interaction term. In this situation, they could be subject to fluid equations with some unknown sound speed c_s relating pressure to density perturbations, and eventually (in the case of an imperfect fluid with shear viscosity) with some unknown viscosity parameter c_{vis} relating the neutrino anisotropic stress σ_ν to the bulk velocity θ_ν. In such models, neutrinos would not free-stream inside the Hubble radius, and their perturbation evolution would be radically different. The picture presented in Section 5.3.2 and Fig. 5.4 would not hold any more: such neutrinos would not reduce and shift the CMB peaks like ordinary neutrinos. We will not discuss this case in further detail because recent analyses (De Bernardis *et al.*, 2008 or Archidiacono *et al.*, 2011) show that ordinary collisionless neutrinos provide a better fit to the data than such a self-interacting neutrino fluid. More detail on this type of analysis is also presented in Section 6.5.3.

5.4 Bounds on neutrinos from primary cosmic-microwave-background anisotropies

5.4.1 Cosmic microwave background and homogeneous cosmology data sets

The best current constraints on the CMB temperature spectrum for $l \leq 1000$ come from the seven years of observations with the Wilkinson Microwave Anisotropy Probe satellite (WMAP7, Dunkley *et al.*, 2009). Other experiments probe smaller angular scales, with the most recent constraints provided by the 2008 observation campaign of the Atacama Cosmology Telescope (ACT-2008, Dunkley *et al.*, 2011) and by the South Pole Telescope (SPT, Keisler *et al.*, 2011).

The accuracy of parameter extraction from CMB data is usually bounded by instrumental errors at high l's, and by cosmic variance at low l's. These limitations lead to a poor determination of parameter combinations along directions of degeneracies. For instance, we have seen in Section 5.1.6 that the effect (C7), namely the late ISW effect, is difficult to measure accurately because of cosmic variance. Hence, the value of the cosmological constant of the minimal ΛCDM model can only be probed through other effects such as (C1) and (C4) for which Ω_Λ impacts the data in combination with other parameters such as ω_M. In the parameter basis $(\omega_M, \omega_B, h, \tau_{\text{reion}}, A_s, n_s)$ with $\Omega_\Lambda = 1 - \omega_M/h^2$, this leads to a degeneracy in the space of reconstructed (ω_M, h) values, and to a poor determination of these two parameters from CMB data alone.

Therefore, it is useful to combine CMB data with an external data set. We leave the discussion of large scale structure observations for the next chapter. We will only discuss in this section the combination of CMB data with probes of homogeneous cosmology, i.e., of the expansion history and geometry of the universe. Currently, the most stringent constraints are derived from

- Direct measurements of the Hubble parameter today, which can be obtained from various techniques. One of them is based on measuring the slope of the luminosity distance–redshift relation of nearby Type Ia supernovae (SNIa) at $z < 0.1$. This technique is identical to the one used to probe the universe's acceleration, but applied here to objects at smaller redshift. Indeed, the luminosity distance–redshift relation of nearby supernovae depends mainly on H_0, whereas at high redshift it probes also the spatial curvature and acceleration of the universe, as we discussed in Chapter 2. For some of these supernovae, the relation between the absolute magnitude and the time of extinction can be accurately calibrated by resolving cepheids in the same galaxy and measuring the distance independently. Using this method, Riess *et al.* (2009) obtain $H_0 = 74.2 \pm 3.6\,\mathrm{km\,s^{-1}Mpc^{-1}}$ (68% confidence level). This measurement will be referred to hereafter as "H_0".

- Measurements of the angular scale of baryon acoustic oscillations. The origin of baryonic oscillations in the matter power spectrum will be explained in the next chapter, in Section 6.1.2. By measuring only the characteristic angular scale of these oscillations, and comparing them to the scale of acoustic peaks in the CMB, one essentially probes the angular diameter–redshift relation in the redshift range where the matter power spectrum is measured (and not the clustering of matter on large scales, which would fall in the category of LSS data). We will refer to recent constraints on the angular diameter distance at redshifts $z = 0.2$ and $z = 0.35$ by Percival *et al.* (2010) as "BAO". These measurements are useful for constraining the late-time cosmology and resolving degeneracies, e.g., between H_0 and ω_M (they are also sensitive to curvature in non-spatially flat models).
- Measurements of the luminosity distance-redshift relation with distant SNIa, probing simultaneously the Hubble radius and the parameters ruling the expansion law at small redshifts, such as $\Omega_M = 1 - \Omega_\Lambda$ in a flat universe, or the cosmological constant and curvature in a more general case. Recent data analyses include Hicken *et al.*, 2009; Kessler, 2009; Conley *et al.*, 2011.

The level of systematic errors in SNIa data sets is still under investigation, although considerable progress is being made in this area. At the moment of writing, CMB+SNIa constraints are very interesting but still treated with care, whereas H_0 and BAO constraints are considered very robust. We will report below several constraints based on the combination of CMB with BAO and H_0 data.

Other techniques are making fast progress. For instance, Moresco *et al.* (2012a) recently used a determination of the Hubble parameter $H(z)$ as a function of redshift in the range $0.09 < z < 1.75$, based on the observation of the evolution of early-type galaxies, treated as "cosmic chronometers" (Moresco *et al.*, 2012b). We will refer to this data set as "OHD" (standing for Observational Hubble Parameter).

5.4.2 Neutrino abundance

We explained in Section 5.3.2 how the abundance of relativistic relics, given in terms of the effective neutrino number N_{eff}, impacts the CMB temperature spectrum. We saw that varying N_{eff} with fixed ω_M shifts the time of equality between radiation and matter, but because this time is very well constrained by the CMB data through effect (C3) of Section 5.1.6, it is more relevant to study the impact of N_{eff} with fixed z_{eq} rather than fixed ω_M. We showed that N_{eff} can be varied together with ω_M and h in such a way as to keep a constant redshift of equality between radiation and matter, and between matter and Λ. This transformation results in a small variation of the CMB for $l \leq 1000$, due to direct perturbation effects, and in a more radical variation for $l > 1000$, due to a change in the diffusion damping scale. This discussion was illustrated by Fig. 5.4, which presents a rather extreme comparison between two models with $N_{\text{eff}} = 0$ and $N_{\text{eff}} = 3.046$. Small changes on the order of $\Delta N_{\text{eff}} \sim 1$ around the standard value $N_{\text{eff}} \simeq 3$ lead to much smaller variations than on this figure.

Because the small variation of the temperature spectrum for $l \leq 1000$ can easily be mimicked by a change in the primordial spectrum tilt and amplitude, a CMB experiment precise only in the range $2 < l < 900$ such as WMAP cannot measure N_{eff} accurately alone. Dunkley *et al.* (2009) reported a bound[4] $N_{\text{eff}} > 2.7$ at the 95% confidence level (C.L.) using WMAP 5-year data. Interestingly, WMAP alone brings indirect evidence for the existence of the cosmological neutrino background: a model with $N_{\text{eff}} = 0$ degrades the effective χ^2 of the data by 8.2 with respect to a model with $N_{\text{eff}} = 3$. Of course, it is still possible that the neutrino background predicted by the standard cosmological model is not present, whereas other relativistic relics would account for this nonzero N_{eff}. The only way to disprove this assumption would be through a direct measurement of cosmological neutrinos, which is far beyond current technology, as explained in Chapter 7. However, this eventuality appears as unlikely as unnecessarily complicated.

Because varying N_{eff} with a constant z_{eq} affects the diffusion damping scale and the high-l tail of the CMB temperature spectrum, we expect that the inclusion of more CMB data on small angular scales could resolve the degeneracy and tighten bounds on N_{eff}. This has been confirmed by data from the ACT and SPT experiments: WMAP7+ ACT-2008 give $N_{\text{eff}} = 5.3 \pm 1.3$ (68% C.L.) (Dunkley *et al.*, 2011; Hamann, 2011), whereas WMAP7+SPT give $N_{\text{eff}} = 3.85 \pm 0.62$ (68% C.L.) (Keisler *et al.*, 2011). The standard value $N_{\text{eff}} = 3.046$ is well inside the 95% C.L. interval.

---

[4] All the bounds reported in this book are Bayesian credible intervals. They stand for the preferred interval for a given parameter, at a given confidence level, within a given model, with given priors on model parameters (unless otherwise stated, flat priors), after marginalization over all other parameters. Some different bounds inspired by the frequentist approach to data fitting are sometimes discussed in the literature. As a rule of thumb, one should keep in mind that all different definitions of parameter bounds tend to match each other when a parameter is really measured by a given data set, whereas they may differ significantly for parameters that are not *required* by the data and can only be bounded from above or below by the data. In that case the bounds are only valid under some precise assumptions and should be regarded as indicative only.

Because N_{eff} participates in parameter degeneracy together with ω_{M} and h (along which z_{eq} remains constant, as explained in Section 5.3.2), any external data set bringing information on these other two parameters makes it possible to tighten the bounds on N_{eff}. Komatsu *et al.* (2011) found that the combination of WMAP7+BAO+H_0 leads to $N_{\mathrm{eff}} = 4.34^{+0.86}_{-0.88}$ (68% C.L.), a result well compatible with the standard value.

Finally, the combination of WMAP7+ACT-2008+BAO+H_0 gives $N_{\mathrm{eff}} = 4.56 \pm 0.75$ (68% C.L.), that of WMAP7+SPT+BAO+H_0 gives $N_{\mathrm{eff}} = 3.86 \pm 0.42$ (68% C.L.), and WMAP7+SPT+H_0+OHD gives $N_{\mathrm{eff}} = 3.5 \pm 0.3$ (68% C.L.). Currently, these are the most stringent bounds on N_{eff} not involving LSS data (and we shall see in Section 6.5.2 that current LSS data do not bring significant improvements for the determination of this particular parameter). They show a marginal preference for extra relativistic degrees of freedom, because the standard value 3.046 sits roughly at the lower end of the 95% C.L. credible interval. However, no robust conclusion can be drawn at the moment. Indeed, this marginal preference for $N_{\mathrm{eff}} > 3$ could be due to

- Yet unknown systematic errors in the high-l CMB data sets.
- Underestimated systematic errors in the BAO+H_0 data sets, shifting the results along the valley of degeneracy in $(N_{\mathrm{eff}}, \omega_{\mathrm{M}}, h)$ space. For instance, if future determinations of H_0 prefer lower values than today (closer to $70\,\mathrm{km\,s^{-1}Mpc^{-1}}$ or slightly below), the marginal evidence for an excess of radiation will disappear.
- The fact that we are not fitting the correct model to the data, and we are missing the impact of extra physics and parameters. For instance, when the primordial helium fraction is left as a free parameter, there is a degeneracy between N_{eff} and Y_p such that the WMAP7+SPT bound enlarges to $N_{\mathrm{eff}} = 3.4 \pm 1.0$ (68% C.L.) (Keisler *et al.*, 2011). However, if the WMAP7+SPT data are not affected by systematics, lowering N_{eff} to 3.046 requires a high helium fraction in slight tension with direct measurement of the helium abundance (Nollett and Holder, 2011), or in strong tension with standard BBN predictions for Y_p, given the allowed range for ω_{B} indicated by CMB data. Bounds on N_{eff} also become weaker when other degrees of freedom are introduced into the fitted model, such as free neutrino masses, a dark energy component with an arbitrary equation of state instead of a cosmological constant, a spatial curvature, or a significant amount of primordial gravitational waves.

In conclusion, CMB data combined with homogeneous cosmology data are able to establish the existence of a relativistic relic background, with a density compatible to standard predictions for the neutrino background at the 95% C.L. Despite some marginal evidence, there is no reason at the moment to claim an excess which, if confirmed, could be explained in several different ways, related to neutrino physics (light sterile neutrinos, leptonic asymmetry, nonthermal distorsions due to new particles decaying into neutrinos) or not (other light or massless relics, early dark energy, energy density of a gravitational wave background, modifications to the Friedmann equation due to extensions of Einstein gravity, etc.).

Future CMB experiments will allow a more accurate determination of N_{eff}. Using data from the Planck satellite alone (Perotto *et al.*, 2006), the 1σ error bar (equivalent to half of the 68% C.L. allowed range) is expected to shrink down to $\sigma(N_{\text{eff}}) \sim 0.46$, without using CMB lensing extraction techniques mentioned in the next chapter. Major progress will be achieved by combining future CMB data with other probes of large-scale structure, as we shall see in Section 6.5.2.

5.4.3 Neutrino masses

We have seen in Section 5.3.3 that CMB experiments can hardly resolve the splitting of the total neutrino mass $M_\nu = \sum_i m_{\nu_i}$ between different families. Most constraints discussed so far in the literature refer to three degenerate neutrinos with mass $m_\nu = M_\nu/3$. The corresponding bounds on M_ν apply in fact to all scenarios with three light neutrinos. We will discuss the case of extra light species separately, together with LSS bounds, in Section 6.5.2. We do not address in this chapter bounds on heavy sterile neutrinos, with masses in the keV range, falling into the category of warm dark matter (WDM). Such relics have no impact on CMB physics, and can only be probed with LSS experiments, as explained in Chapter 6.

For three light degenerate neutrinos, we explained in Section 5.3.3 that if the neutrino mass is increased while (ω_B, ω_C) are kept fixed and h is decreased to keep the peak scale constant, then the CMB temperature spectrum varies by a very small amount, with the most pronounced effects being on angular scales $l < 20$ (LISW effect) and $20 < l < 500$ (EISW effect). This discussion was illustrated by Fig. 5.6.

WMAP data alone can constrain the total neutrino mass, with an upper bound $M_\nu < 1.3$ eV (95% C.L.) for the minimal ΛCDM model with massive neutrinos (seven free parameters). Because the leading effects appear for $l < 500$, adding small-scale CMB data from ACT-2008 or SPT does not bring about significant improvements.

However, this transformation (corresponding to a direction of degeneracy in parameter space) requires h to decrease when M_ν increases, in order to keep a constant peak scale. Any additional constraint on h does help reduce this degeneracy. Indeed, the bound from WMAP7+BAO+H$_0$ is significantly stronger: $M_\nu < 0.58$ eV

(95% C.L.), whereas the combination WMAP7+OHD+H_0 gives[5] $M_\nu < 0.48$ eV (95% C.L.).

These mass bounds appear to be very robust from the point of view of the data being used. However, as usual, they can be relaxed by introducing other physical ingredients and free parameters into the model. One parameter known to be slightly degenerate with M_ν is the number of light/massless species N_{eff}. Hence mass bounds must be investigated separately when allowing for the presence of extra massless or light relics. The same is true with other ingredients such as an arbitrary dark energy equation of state, a spatial curvature, or a significant amount of primordial gravitational waves. All these cases have been considered, but in combination with LSS data, so we will report the corresponding mass bounds in Chapter 6.

The expected sensitivity of the Planck satellite alone to neutrino masses is $\sigma(M_\nu) \sim 0.2$ eV for a minimal model with seven parameters, not using CMB lensing extraction techniques mentioned in the next chapter. Much better bounds can be obtained using CMB lensing extraction and other LSS observations, as we shall see in Section 6.5.1.

---

[5] Here we assume that for this particular data set, the 95% C.L. bound is equal to twice the 68% C.L. bound, which seems to be the case, judging from Fig. 8 of Moresco *et al.*, 2012a.

6

Recent times: neutrinos and structure formation

There are several ways to observe the large-scale structure of our universe on different scales and redshifts. This structure is even more sensitive to the neutrino abundance, mass spectrum and properties than CMB anisotropies, and offers several opportunities to measure neutrino parameters. If the universe can be described by general relativity and does not feature significant dark energy perturbations, all observables related to large-scale structure can be inferred from the matter power spectrum $P(\eta, k)$ defined through

$$\langle \delta_{M}(\eta, \vec{k}) \delta_{M}^{*}(\eta, \vec{k}') \rangle = P(\eta, k)\, \delta^{(3)}(\vec{k} - \vec{k}'). \tag{6.1}$$

Here δ_M is the relative density perturbation of nonrelativistic matter components. In a ΛCDM universe with CDM, baryons and nonrelativistic neutrinos, δ_M can be decomposed as

$$\delta_{M} = \delta \rho_{M}/\bar{\rho}_{M} = \left(\sum_{i=B,C,\nu} \bar{\rho}_i \delta_i \right) \bigg/ \left(\sum_{i=B,C,\nu} \bar{\rho}_i \right). \tag{6.2}$$

In this chapter, for simplicity, we will often use the same letter to denote a function of conformal time η, or of the corresponding redshift z, or finally of the corresponding scale factor a. For instance, we will write indifferently $P(\eta, k)$, $P(z, k)$ or $P(a, k)$.

As we shall see in Section 6.4, galaxy or cluster redshift surveys probe $P(\eta, k)$ modulo a light-to-mass bias factor (or function). Instead, observations of weak lensing and of the late integrated Sachs–Wolfe contribution to the CMB probe the power spectrum of metric fluctuations, on scales smaller than the Hubble radius. The latter can be related to $P(\eta, k)$ using the Poisson equation

$$k^2 \phi = k^2 \psi = -4\pi G a^2 \bar{\rho}_M \delta_M \tag{6.3}$$

(during matter and Λ domination, we can ignore the photon and neutrino anisotropic stresses, implying $\phi = \psi$, and in the gravitational source term $\sum_i \bar{\rho}_i \delta_i$ we can ignore the subdominant contribution of relativistic species). Finally, Lyman-α observations probe the one-dimensional matter power spectrum, related to the three-dimensional one by a simple convolution.

Because the matter power spectrum is so crucial, we will devote the first two sections of this chapter to the computation and parameter dependence of $P(\eta, k)$. Fluctuations on a sufficiently large scale and/or at sufficiently large redshift can be described by linear theory. Observations of $P(\eta, k)$ in the linear regime store a maximum amount of information concerning cosmological parameters, and are less affected by theoretical errors, bias and non-gaussianity issues. Section 6.1 will be entirely devoted to the properties of the linear matter power spectrum, and in particular to the impact of neutrinos on this quantity. However, it is also important to compute neutrino effects on the nonlinear power spectrum, either because some observational methods probe only the mildly nonlinear regime, or in order to increase the amount of information extracted from a given data set covering both linear and nonlinear scales, or finally because some specific neutrino effects show up only in the nonlinear region (for instance, for heavy sterile neutrinos). We will review nonlinear issues in Section 6.2. Then we will come back to CMB fluctuations, and explain in Section 6.3 how secondary anisotropies are influenced by neutrino properties at small redshift. We will give a brief summary of current and future techniques for measuring the matter power spectrum in Section 6.4. Finally, we will present observational constraints on neutrino parameters in Section 6.5.

6.1 Linear matter power spectrum

The computation of the linear matter power spectrum is as subtle as that of the CMB temperature spectrum. In this section, we will review this topic in a simple and qualitative way, insisting on the physical issues that are important for understanding the possible impact of neutrinos on $P(\eta, k)$.

Because the matter power spectrum of a ΛCDM model (with eventually relativistic or nonrelativistic neutrinos) depends on the evolution of the three density fluctuations δ_C, δ_B, δ_ν, the relative abundances of the underlying components are important. They can be parameterized by the CDM, baryon and neutrino fractions

$$f_C \equiv \frac{\bar{\rho}_C}{\bar{\rho}_C + \bar{\rho}_B + \bar{\rho}_\nu}, \quad f_B \equiv \frac{\bar{\rho}_B}{\bar{\rho}_C + \bar{\rho}_B + \bar{\rho}_\nu}, \quad f_\nu \equiv \frac{\bar{\rho}_\nu}{\bar{\rho}_C + \bar{\rho}_B + \bar{\rho}_\nu} \quad (6.4)$$

subject to $f_C + f_B + f_\nu = 1$. Because CDM and baryons are nonrelativistic during the CMB and structure formation epochs, f_B/f_C can always be regarded as constant. The neutrino fraction f_ν becomes asymptotically constant after the nonrelativistic transition of neutrino species. If one neutrino species is much heavier than the others, f_ν becomes roughly constant as soon as this species becomes nonrelativistic, because lighter species represent a negligible fraction of f_ν. If several species have masses of the same order of magnitude, f_ν can have a nontrivial evolution with various steps before reaching its constant asymptote.

To introduce technical difficulties one after another, we will first summarize in Section 6.1.1 the properties of the matter power spectrum in a universe containing no neutrinos ($f_\nu = 0$) and a negligible baryon fraction ($f_B \ll 1$). In this limit, CDM does not experience gravitational feedback from baryons and neutrinos, and simple solutions can be derived. We will then discuss the impact of baryons in Section 6.1.2, of relativistic relics such as massless neutrinos in Section 6.1.3, of hot dark matter such as light neutrinos in Section 6.1.4, and of warm dark matter such as heavy sterile neutrinos in Section 6.1.5.

6.1.1 Neutrinoless universe with cold dark matter

Equation of evolution for cold dark matter

Baryons always play a crucial role in the dynamics of the tightly coupled photon–baryon fluid, but if $f_B \ll f_C$, they play a negligible role with respect to CDM for linear structure formation. In a neutrinoless universe, the only significant contribution to the matter power spectrum then comes from CDM perturbations, so we only need to follow the evolution of $\delta_C(\eta, k)$. On super-Hubble scales and for adiabatic initial conditions, we know from Eqs. (5.24), (5.41) and (5.44) that δ_C is constant and related to metric fluctuations through

$$\delta_C(\eta, k) = -\frac{3}{2}\phi(\eta, k) = -\frac{3}{2}\psi(\eta, k) = -1 \quad (\eta < \eta_{eq}) \tag{6.5}$$

$$\delta_C(\eta, k) = -2\phi(\eta, k) = -2\psi(\eta, k) = -\frac{6}{5} \quad (\eta > \eta_{eq}) \tag{6.6}$$

(these relations apply to transfer functions normalized to $\mathcal{R}(\eta_{ini}, \vec{k}) = 1$, as explained in Section 5.1.4). On sub-Hubble scales, we expect δ_C to grow because of gravitational collapse. The growth rate could potentially depend on

- The expansion rate: faster expansion implies less efficient gravitational interactions, because all physical distances between two bodies increase with the scale factor. In typical equations of evolution, this effect is accounted by a "Hubble friction" term.
- Gravitational interactions with other species (in the present case, only with photons, because we assumed baryons to be negligible in terms of density and gravitational back-reaction, and neutrinos to be absent).

The equation of evolution for δ_C is obtained by combining the continuity and Euler equations (5.19) into

$$\delta_C'' + \frac{a'}{a}\delta_C' = -k^2\psi + 3\phi'' + 3\frac{a'}{a}\phi'. \tag{6.7}$$

The second term on the left-hand side accounts for Hubble friction. The first term on the right-hand side represents the gravitational attraction force (generated by CDM itself and all other species if relevant), whereas the next two terms are related to dilation, i.e., to local variations of the expansion rate.

Modes crossing the Hubble scale during matter domination

Equation (6.7) is easy to solve after the time of equality between matter and radiation. First, we notice that only collisionless relativistic particles (only decoupled photons in a neutrinoless universe) can contribute to the total anisotropic stress σ. Because the density of photons is subdominant after radiation-to-matter equality, anisotropic stress can be neglected, and the Einstein equation (5.22) implies $\phi = \psi$. Next, because cold dark matter is a pressureless self-gravitating fluid in this regime, it will not experience acoustic oscillations, and both δ_C and ϕ must be smooth over a Hubble time scale, rather than oscillating with a pulsation of order k. So the time derivatives of ϕ on the right-hand side should be negligible with respect to $k^2\psi = k^2\phi$. The latter quantity is given inside the Hubble radius by the Poisson equation (5.21)

$$k^2\phi = k^2\psi = -4\pi G a^2 \bar\rho_C \delta_C. \tag{6.8}$$

Finally, the equation of evolution for CDM inside the Hubble radius and after equality can be approximated as

$$\delta_C'' + \frac{a'}{a}\delta_C' - 4\pi G a^2 \bar\rho_C \delta_C = 0. \tag{6.9}$$

During matter domination, the scale factor evolves like $a \propto \eta^2$ and the Friedmann equation gives

$$4\pi G a^2 \bar\rho_C \simeq 4\pi G a^2 \bar\rho_{\text{tot}} = \frac{3}{2}\left(\frac{a'}{a}\right)^2. \tag{6.10}$$

Then equation (6.9) becomes

$$\delta_C'' + \frac{2}{\eta}\delta_C' - \frac{6}{\eta^2}\delta_C = 0 \tag{6.11}$$

and has two trivial solutions. The growing mode, $\delta_C \propto \eta^2 \propto a$, gives the famous growth rate of structures during matter domination in the linear regime. During Λ domination, one should replace (a'/a) using the Friedmann law in the presence of both $\bar\rho_C$ and $\bar\rho_\Lambda$. The solution is a hypergeometric function and is the same for each k modulo an arbitrary constant of integration, because the equation of evolution is k-independent. We will not write this solution explicitly but will incorporate it into a function $g(a; \Omega_M)$ defined as

$$g(a; \Omega_M) = \frac{a_M \, \delta_C(a, k)}{a \, \delta_C(a_M, k)}, \tag{6.12}$$

where a_M is the scale factor at some arbitrary time deep inside the matter-dominated regime, when $\delta_C \propto a$. This definition is convenient because it allows us to write the growing mode for δ_C as

$$\delta_C(a, k) \propto a \, g(a; \Omega_M) \tag{6.13}$$

during both matter and Λ domination, with $g(a; \Omega_M) = 1$ deep inside the matter-dominated regime. Hence g stands for the correction to the linear growth factor of δ_C induced by the cosmological constant. By enhancing the universe's expansion and the Hubble friction term, Λ tends to slow down structure formation; hence g decreases monotonically from one to zero. A simple approximation for g today (Kofman *et al.*, 1993) reads

$$g(a_0; \Omega_M) \simeq (\Omega_M)^{0.2} / \left[1 + 0.003(\Omega_\Lambda / \Omega_M)^{4/3} \right] \tag{6.14}$$

with $\Omega_\Lambda = 1 - \Omega_M$, because we restrict the discussion to a spatially flat universe.

For modes crossing the Hubble scale during matter domination, it is easy to match the super-Hubble solution of Eq. (6.6) to this sub-Hubble solution. This can be done either by solving Eqs. (5.20) and (6.7) simultaneously during matter domination, or even more simply by using the Einstein equation $\delta G^i_i = \delta T^i_i$:

$$\left(2\frac{a''}{a} - \left(\frac{a'}{a}\right)^2 \right) \psi + \frac{k^2}{3}(\phi - \psi) + \phi'' + \frac{a'}{a}(\psi' + 2\phi') = 4\pi G a^2 \, \delta p. \tag{6.15}$$

Deep inside the matter-dominated era, this equation simplifies considerably, because

- With a negligible contribution of relativistic species and of Λ to the Friedmann equation, the scale factor evolves as $a \propto \eta^2$, and the factor in front of ψ vanishes.
- Because the density of photons is subdominant, anisotropic stress can be neglected during matter domination, and the Einstein equation (5.22) implies $\phi = \psi$. Hence the second term in Eq. (6.15) can be neglected.
- Similarly, the total pressure term on the right-hand side of Eq. (6.15) receives only contributions from relativistic particles (photons in a neutrinoless universe), which are subdominant during the matter era: at leading order we can neglect this term also.

Hence, to a first approximation, the equation governing the evolution of $\phi = \psi$ during the matter era is

$$\phi'' + \frac{6}{\eta}\phi' = 0. \tag{6.16}$$

The solutions are combinations of a constant mode and of a decaying mode $\phi \propto \eta^{-5}$. After equality, the decaying mode quickly becomes negligible, and metric fluctuations are constant in time on all scales. It is always surprising to see that when structures form, the gravitational potential is constant even inside the Hubble radius. This results from a cancellation between the effects of gravitational clustering and of cosmological expansion during matter domination.

We can now come back to Eq. (6.7) with a constant source term on the right-hand side, and write the most general solution,

$$\delta_C = D_C^g + \frac{D_C^d}{\eta} - \frac{k^2\eta^2}{6}\psi, \tag{6.17}$$

where D_C^g and D_C^d are two constants of integration depending on $\vec{k}$. These constants can be found explicitly by matching with the super-Hubble limit given in Eq. (6.6). Neglecting the decaying mode proportional to D_C^d, we find that the result for the transfer function of CDM perturbations reads

$$\delta_C(\eta, k) = -\frac{6}{5} - \frac{k^2\eta^2}{10}. \tag{6.18}$$

If we want an expression depending on a rather than η, we can use the two relations

$$k^2\eta^2 = \frac{4k^2}{a^2 H^2} \quad \text{and} \quad \frac{H^2}{H_0^2} = \Omega_M \left(\frac{a}{a_0}\right)^{-3}, \tag{6.19}$$

both valid during matter domination, even if they involve Ω_M, the fractional matter density today, and H_0, the Hubble rate today. In combination with Eq. (6.18), these relations give

$$\delta_C(a, k) = -\frac{6}{5} - \frac{2}{5}\Omega_M^{-1}\frac{k^2}{a_0^2 H_0^2} \cdot \frac{a}{a_0} \tag{6.20}$$

during matter domination. Finally, a matching with the solution $a\, g(a; \Omega_M)$ valid during both matter and Λ domination shows that the CDM transfer function reads

$$\delta_C(a, k) = -\frac{6}{5} - \frac{2}{5}\Omega_M^{-1}\frac{k^2}{a_0^2 H_0^2} \cdot \frac{a\, g(a; \Omega_M)}{a_0} \tag{6.21}$$

at any time between equality and today, and for any wavenumber k crossing the Hubble scale during matter domination. The square of this transfer function multiplied by the primordial spectrum (defined in Section 5.1.4) gives the matter power spectrum for all modes $k < k_{eq}$ with

$$k_{eq} \equiv a_{eq} H_{eq}. \tag{6.22}$$

Moreover, we can only measure the matter power spectrum on scales that are sub-Hubble today, so we can assume $k \gg a_0 H_0$ and neglect the constant term in Eq. (6.21). Finally, the large-wavelength branch of the power spectrum is

$$P(a, k) = \delta_C(a, k)^2 P_{\mathcal{R}}(k) \tag{6.23}$$

$$= \frac{2\pi^2}{k^3} \left(\frac{2}{5} \Omega_M^{-1} \frac{k^2}{a_0^2 H_0^2} \cdot \frac{a \, g(a; \Omega_M)}{a_0} \right)^2 P_{\mathcal{R}}(k), \qquad (a_0 H_0 < k < k_{eq}).$$

Modes crossing the Hubble scale during radiation domination

We have achieved only half of the task of this section: we have not yet discussed the evolution of scales crossing the Hubble radius during radiation domination. This involves subtle issues related to the way to treat Eq. (6.7) for $\eta \leq \eta_{eq}$. One could assume naively that because CDM does not dominate the expansion during radiation domination, the Poisson equation

$$k^2 \psi = -4\pi G a^2 \left(\bar{\rho}_\gamma \delta_\gamma + \bar{\rho}_C \delta_C \right) \tag{6.24}$$

can be simplified by neglecting the CDM contribution. In that case, ψ could be inferred from the oscillatory evolution of the self-gravitating photon fluid, and could be plugged into the right-hand side of Eq. (6.7) as an external source term. This approach is, however, incorrect. CDM fluctuations can be neglected in the Poisson equation only as long as

$$\left| \frac{\delta_C}{\delta_\gamma} \right| < \frac{\bar{\rho}_\gamma}{\bar{\rho}_C} = \frac{a_{eq}}{a}. \tag{6.25}$$

The term on the right-hand side of this inequality is greater than one during radiation domination but decreases with expansion. The term on the left-hand side starts from $(3/4)$ on super-Hubble scales (see Eq. (5.24)), and grows on sub-Hubble scales, because CDM is pressureless and experiences gravitational clustering (we will see later at what rate this growth takes place). Hence, inevitably, the CDM contribution to the Poisson equation becomes dominant well inside the Hubble radius. In this regime, it is the CDM component rather than the relativistic one that behaves like a self-gravitating fluid. Tightly coupled photons actually decouple from gravity in this limit: we have already seen in Section 5.1.5 that pressure forces take over from gravitational forces in the photon equation of evolution (5.38). Hence, deep inside the Hubble scale and during radiation domination, CDM is already governed by the same equation as during matter domination, namely Eq. (6.9).

The most subtle issue consists of solving the CDM equation of motion just after Hubble crossing, when both photons and CDM are contributing to the Poisson equation. The literature offers various analytic approaches for studying this regime. It is possible to solve the coupled equations of motion simultaneously for tightly coupled photons and CDM perturbations. After a few approximations, one is led to analytic solutions of a rather complicated form (involving the cosine integral). A much simpler approach has been used since the work of Mészáros (1974), although Weinberg (2002) was the first to establish it on firm mathematical ground. Weinberg pointed out that the solution of the full system describing the evolution of $(\delta_\gamma, \delta_C, \phi, \psi)$ can be decomposed into fast modes (oscillating with a pulsation on the order of k or kc_s) and slow modes (only evolving over a Hubble time scale). Power counting arguments show that for fast modes, δ_C is always subdominant in the Poisson equation, whereas for slow modes, it is δ_γ that is subdominant.[1] This nontrivial result implies that fast modes can be inferred from the photon equation of motion and Einstein equations with δ_C set to zero, and vice-versa for slow modes. In other words, the gravitational back reaction of one species on the other species can be neglected. At the end of the calculation, the full solution for metric fluctuations is given by the sum of all modes, i.e., by the superposition of damped acoustic oscillations with zero mean value (for fast modes) and a smooth function shifting the zero-point of the oscillations (for slow modes), accounting for CDM gravitational clustering.

As long as we are interested in the CDM evolution, we need to bother only about slow modes, and we can employ the very same equation as in other regimes, namely Eq. (6.9). This equation can be combined with the Friedmann equation sourced by the sum of $\bar{\rho}_R = \bar{\rho}_\gamma$ and $\bar{\rho}_M = \bar{\rho}_C$,

$$\left(\frac{a'}{a}\right)^2 = \frac{8\pi G}{3} a^2 (\bar{\rho}_M + \bar{\rho}_R). \tag{6.26}$$

After introducing the parameterization

$$\bar{\rho}_M = \bar{\rho}_{eq}(a/a_{eq})^{-3}, \qquad \bar{\rho}_R = \bar{\rho}_{eq}(a/a_{eq})^{-4}, \qquad y \equiv a/a_{eq} \tag{6.27}$$

and changing variables from η to y, one obtains a simple second-order equation for δ_C involving only y, called the Mészáros equation,

$$y(1+y)\frac{d^2\delta_C}{dy^2} + \left(1 + \frac{3y}{2}\right)\frac{d\delta_C}{dy} - \frac{3}{2}\delta_C = 0, \tag{6.28}$$

---

[1] The proof presented in Weinberg (2002) applies to equations in the synchronous gauge, but on sub-Hubble scales the synchronous and Newtonian gauges lead to the same equations and solutions.

with two exact analytical solutions first derived by (Mészáros, 1974; Groth and Peebles, 1975; see also Weinberg, 2008), which we do not write here for conciseness. It is sufficient for the rest of the discussion to admit that the fastest-growing solution is proportional to $\ln(y)$ deep inside radiation domination, and to y deep inside matter domination. During radiation domination, the transfer function $\delta_C(a, k)$ properly normalized to adiabatic initial conditions has a sub-Hubble limit

$$\delta_C(a, k) \longrightarrow \alpha + \beta \ln\left(\frac{k}{aH}\right) \qquad (\eta \ll \eta_{eq}, k \gg aH), \qquad (6.29)$$

where α and β are two numerical coefficients independent of k. We see that CDM clusters much more slowly during radiation domination than during matter domination (because δ_C grows logarithmically rather than linearly with the scale factor), as a consequence of the different expansion law. This is a key result for understanding structure formation. In total, for a qualitative understanding of the matter power spectrum, it is sufficient to approximate the growth of CDM perturbations during radiation domination with Eq. (6.29) evaluated at the time of equality, i.e., when $aH = a_{eq}H_{eq} \equiv k_{eq}$, and to perform a crude matching with the sub-Hubble solution during matter and Λ domination, i.e., with the second term in Eq. (6.21). This means that during the matter- and Λ-dominated stages, the CDM transfer function is

$$\delta_C(a, k) = \left(\alpha + \beta \ln(k/k_{eq})\right) \frac{a\, g(a; \Omega_M)}{a_{eq}}. \qquad (6.30)$$

To homogenize this result with its counterpart of Eq. (6.23) for modes $k < k_{eq}$, we notice that at equality

$$\frac{H_{eq}^2}{H_0^2} = \Omega_R(a_0/a_{eq})^4 + \Omega_M(a_0/a_{eq})^3 = 2\Omega_M(a_0/a_{eq})^3, \qquad (6.31)$$

from which we infer

$$\frac{a}{a_{eq}} = \frac{a\, a_{eq}^2 H_{eq}^2}{2\Omega_M a_0^3 H_0^2} = \frac{a\, k_{eq}^2}{2\Omega_M a_0^3 H_0^2}. \qquad (6.32)$$

Finally, following the same steps as in Eq. (6.23), the matter power spectrum of modes crossing the Hubble scale during radiation domination is

$$P(a, k) = \frac{2\pi^2}{k^3} \left(\frac{1}{2}\Omega_M^{-1} \frac{k_{eq}^2}{a_0^2 H_0^2} \cdot \frac{a\, g(a; \Omega_M)}{a_0} \left[\alpha + \beta \ln\left(\frac{k}{k_{eq}}\right)\right]\right)^2 \mathcal{P}_\mathcal{R}(k). \qquad (6.33)$$

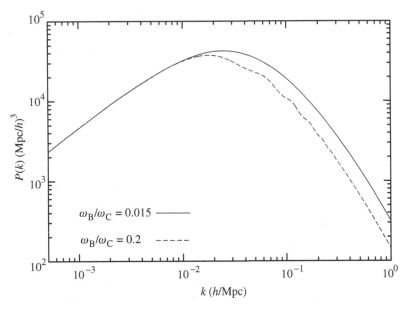

Figure 6.1 Matter power spectrum for two neutrinoless ΛCDM models with the same primordial spectrum and Ω_M, but different baryon-to-CDM ratios ω_B/ω_C. The difference between the two curves clearly shows the baryon effects described in the text (suppression of power for $k \geq k_{eq}$ and imprint of small acoustic oscillations).

Matter power spectrum in practical units

In summary, the two asymptotes of the matter power spectrum (6.23, 6.33) read

$$P(a, k) = \left(\frac{a\, g(a; \Omega_M)}{a_0}\right)^2 \frac{k\, \mathcal{P}_\mathcal{R}(k)}{\left(\Omega_M a_0^2 H_0^2\right)^2} \times \begin{cases} \frac{8\pi^2}{25} & (a_0 H_0 < k < k_{eq}) \\[2mm] \frac{k_{eq}^4}{2k^4}\left(\alpha + \beta \ln\left(\frac{k}{k_{eq}}\right)\right)^2 & (k > k_{eq}). \end{cases}$$

(6.34)

For a scale-invariant power spectrum (a constant $\mathcal{P}_\mathcal{R}$), the matter power spectrum scales like k on large scales, and like $k^{-3} \ln(k)^2$ on small scales. This behavior is indeed well reproduced by numerical results (see for instance Fig. 6.1, where the upper curve corresponds to a neutrinoless ΛCDM model with negligible baryon fraction). An analytic approximation to the matter power spectrum in the intermediate range $k \sim k_{eq}$ can be obtained from the full solution of the Mészáros equation (see, e.g., Weinberg, 2008). Accurate fits to the numerical results have also been derived by Eisenstein and Hu, 1998. The fact that the spectrum of a ΛCDM model with negligible baryon fraction is smooth instead of keeping track of acoustic oscillations is counterintuitive; it is a consequence of the absence of efficient gravitational back reaction of photons over CDM during radiation domination, as already explained after Eq. (6.25).

The matter power spectrum is usually expressed as a function of k in units of h/Mpc. It is worth coming back on this choice. Although comoving wavenumbers do not represent physical quantities, the ratio a_0/k is a physical distance that can be expressed in units of megaparsecs (because $2\pi a_0/k$ is the current wavelength of the comoving Fourier mode k). However, the evolution of cosmological perturbations inside the Hubble radius depends primarily on how far inside the Hubble radius is a given Fourier mode, i.e., on the dimensionless ratio $k/(aH)$. Hence it is useful to plot the power spectrum as a function of the dimensionless number $k/(a_0 H_0)$. A different choice for the value of H_0 will then leave the overall shape of the power spectrum $P(a, k/(a_0 H_0))$ invariant (if all other relevant characteristic times or scales do not change simultaneously). However, by definition of the reduced Hubble parameter and in units such that $c = 1$, we have $H_0 = h/3000 \text{ Mpc}^{-1}$. Hence

$$\frac{k}{a_0 H_0} = 3000 \left[\frac{k/a_0}{1\, h\,\text{Mpc}^{-1}} \right]. \tag{6.35}$$

Thus it is equivalent to plot the spectrum as a function of $k/(a_0 H_0)$ or of

$$\tilde{k} \equiv \left[\frac{k/a_0}{1\, h\,\text{Mpc}^{-1}} \right]. \tag{6.36}$$

Because astrophysicists often use the implicit assumption that $a_0 = 1$ in plots involving comoving wavenumbers, the matter power spectrum is usually displayed as a function of k in units of h/Mpc along the horizontal axis, i.e., as a function of $\tilde{k}$. We must reexpress our results for P as a function of $\tilde{k}$ if we want to comply with standard conventions and remove a trivial dependency on H_0.

The same discussion holds for the units of the power spectrum itself. In Eq. (6.34), the quantities k^{-1} and $(a_0 H_0)$ are comoving scales. Other quantities such as Ω_M, (a/a_0), $g(a; \Omega_M)$ or $\mathcal{P}_{\mathcal{R}}$ are dimensionless. So P represents a cube comoving scale, and $a_0^3 P$ has the dimension of the cube of physical length. The quantity left invariant by a change of H_0 is then the ratio of $a_0^3 P$ to the cube of the Hubble radius today, i.e., the dimensionless number $a_0^3 H_0^3 P$. Because

$$a_0^3 H_0^3 P = 3000^{-3} \left[\frac{a_0^3 P}{1\, h^{-3}\text{Mpc}^3} \right], \tag{6.37}$$

plotting $a_0^3 H_0^3 P$ is equivalent to plotting

$$\tilde{P} \equiv \left[\frac{a_0^3 P}{1\, h^{-3}\text{Mpc}^3} \right]. \tag{6.38}$$

This explains why plots usually show P in units of $h^{-3}\mathrm{Mpc}^3$, with an implicit assumption that $a_0 = 1$. The expression for $\tilde{P}$ as a function of $\tilde{k}$ can easily be obtained from Eq. (6.34):

$$\tilde{P}(a, \tilde{k}) = \mathcal{N} \left(\frac{a\, g(a; \Omega_{\mathrm{M}})}{a_0\, \Omega_{\mathrm{M}}} \right)^2 \tilde{k}\, \mathcal{P}_{\mathcal{R}}(\tilde{k}) \times \begin{cases} \frac{8\pi^2}{25} & (\tilde{k} < \tilde{k}_{\mathrm{eq}}) \\[2ex] \frac{\tilde{k}_{\mathrm{eq}}^4}{2\tilde{k}^4} \left(\alpha + \beta \ln\left(\frac{\tilde{k}}{\tilde{k}_{\mathrm{eq}}} \right) \right)^2 & (\tilde{k} > \tilde{k}_{\mathrm{eq}}) \end{cases}$$

$$\tag{6.39}$$

with

$$\mathcal{N} \equiv \left(\frac{1\, h\, \mathrm{Mpc}^{-1}}{H_0} \right)^4 = 3000^4. \tag{6.40}$$

As expected, the rescaled power spectrum does not depend explicitly on the value of H_0. Apart from the primordial power spectrum, it actually depends on only two free parameters: Ω_{M} and $\tilde{k}_{\mathrm{eq}}$ (we recall that α and β are uniquely fixed by the solution of the Mészáros equation with adiabatic initial conditions). Indeed, Ω_{M} (equal to $(1 - \Omega_{\Lambda})$) governs the time of equality between matter and Λ, and the amount of perturbation damping during Λ domination, encoded in $g(a; \Omega_{\mathrm{M}})$, whereas $\tilde{k}_{\mathrm{eq}}$ defines the scale of the transition between the two asymptotes, or in other words the scale of the peak in the matter power spectrum. The Ω_{M} factor in the denominator just comes from the fact that we expressed the power spectrum as a function of a/a_0 rather than conformal time, and the coefficient of proportionality between a/a_0 and η^2 during matter domination depends on the matter density in the universe, i.e., on $\Omega_{\mathrm{M}} H_0^2$ (as can be checked from Eq. (6.19)).

6.1.2 Neutrinoless universe with cold dark matter and baryons

Because current observations favour a value of the baryon fraction close to $f_{\mathrm{B}} \sim 0.2$, the results of the previous section do not provide a realistic description of linear structures in the universe. Let us follow the same steps as in Section 6.1.1 and identify the changes brought about by the presence of baryons.

Baryon drag epoch

An important characteristic time for structure formation is that of baryon decoupling, which does not coincide exactly with that of photon decoupling. In Section 5.1.6, we defined the photon decoupling time η_{LS} as the maximum of the photon visibility function, i.e., the most likely time for the last scattering of photons. To understand structure formation, we should compute the time at which baryons are effectively released by the photons and start to cluster like free-falling nonrelativistic particles. The baryon interaction term on the r.h.s. of Eq. (5.17) (often called the baryon drag term, because it describes how baryons tend to drag photons towards gravitational potential wells) corresponds to an interaction rate $R^{-1}an_e\sigma_T$, from which we can define the drag depth (see, e.g., Hu and Sugiyama, 1996),

$$\tau_{dr}(\eta) \equiv \int_{\eta}^{\eta_0} d\eta \, R^{-1}an_e\sigma_T. \tag{6.41}$$

The drag depth is similar to the photon optical depth of Eq. (5.54) with an additional R^{-1} factor. It goes from infinity in the tightly coupled limit to zero in the decoupled limit. The characteristic time η_{dr} at which baryon drag stops being efficient, defined through $\tau_{dr}(\eta_{dr}) = 1$, is usually called the baryon drag time (a slightly misleading terminology, because it marks the end of drag effects). Because in realistic models recombination takes place at the beginning of matter domination with $R = 3\bar{\rho}_B/4\bar{\rho}_\gamma < 1$, the release of baryons takes place soon after photon decoupling (in terms of redshifts, $z_{dr} < z_{LS}$). We mentioned in Section 5.1.6 that the value z_{LS}, fixed by the recombination history, has a very mild dependence on the few parameters affecting the free electron fraction evolution $n_e(z)$, such as ω_B, the primordial helium fraction Y_P, and eventually N_{eff}. For realistic parameter values, z_{LS} varies by such small amounts that it can usually be considered a fixed number. Instead z_{dr} has a strong dependence on ω_B due to the factor R appearing in the definition of τ_{dr}. Baryons are released earlier if the baryon density is increased.

We have seen in Section 5.1.5 that in the tight-coupling regime $\eta \ll \eta_{dr}$, baryons track photons with $\delta_B = \frac{3}{4}\delta_\gamma$. At recombination and until η_{dr}, baryon fluctuations on scales smaller than the photon and baryon mean free paths are erased by diffusion effects. This mechanism is called Silk damping (Silk, 1968). After the baryon drag epoch, baryon fluctuations evolve according to the same equation as any other colisionless species, identical to Eq. (6.7) for CDM:

$$\delta_B'' + \frac{a'}{a}\delta_B' = -k^2\psi + 3\phi'' + 3\frac{a'}{a}\phi'. \tag{6.42}$$

Modes crossing the Hubble scale after the baryon drag epoch

As in Section 6.1.1, we first discuss the evolution of perturbations on large scales – more precisely, on comoving scales crossing the Hubble radius after the baryon drag epoch. For such modes, the CDM and baryon density perturbations are equal to each other at any time, because they start from the same initial conditions (5.41), (5.44) and they are subject to the same equation of evolution (6.7), (6.42) during and after Hubble crossing. The total matter perturbation is then equal to

$$\delta_M = (1 - f_B)\delta_C + f_B\delta_B = \delta_C = \delta_B. \tag{6.43}$$

The three quantitites δ_C, δ_B and δ_M evolve exactly like δ_C in a universe with $f_B \ll 1$, and the power spectrum is still given by the large-scale asymptote of Eq. (6.39), independent of f_B.

Modes crossing the Hubble scale before the baryon drag epoch

Before the baryon drag time, the behavior of baryon fluctuations $\delta_B = \frac{3}{4}\delta_\gamma = 3\Theta_{\gamma 0}$ can be inferred from that of photon fluctuations, already studied in Section 5.1.5. Baryons experience acoustic oscillations, with a boost after horizon crossing due to gravitational driving forces, a constant amplitude during radiation domination deep inside the Hubble length, a decreasing amplitude with a shift of the zero point of oscillations at the beginning of matter domination, and finally, an exponential suppression on small scales during recombination due to Silk damping.

To understand the evolution of CDM, the arguments of Weinberg, 2002 concerning fast and slow modes can still be applied. Soon after Hubble crossing and for slow modes, δ_B and δ_γ are negligible with respect to δ_C. For fast modes, the contrary is true. Hence, the evolution of δ_C is still given by Eq. (6.9). Because $\bar{\rho}_C = \bar{\rho}_M - \bar{\rho}_B$, this equation can be written as

$$\delta_C'' + \frac{a'}{a}\delta_C' - 4\pi G\bar{\rho}_M(1 - f_B)\delta_C = 0 \tag{6.44}$$

and can be put in the Mészáros form following the same steps as in (6.26) and (6.27), but with now an extra factor $(1 - f_B)$,

$$y(1 + y)\frac{d^2\delta_C}{dy^2} + \left(1 + \frac{3y}{2}\right)\frac{d\delta_C}{dy} - \frac{3}{2}(1 - f_B)\delta_C = 0. \tag{6.45}$$

We see that the impact of baryons consists of reducing the gravitational force term, for a fixed Hubble friction term. By shifting the balance between gravitational attraction and background expansion, baryons reduce the growth rate of CDM. This is confirmed by the solution of the equation, during both radiation and matter domination. For instance, using the Friedmann equation, it is easy to show that deep inside matter domination, the fastest-growing solution of Eq. (6.45) is

$$\delta_C \propto a^{1-\frac{3}{5}f_B}. \tag{6.46}$$

In summary, CDM perturbations reach the baryon drag epoch with a reduced amplitude with respect to the $f_B = 0$ limit. Modes entering earlier in the Hubble radius are subject to a reduced growth rate for a longer time, so $\delta_C(\eta_{dr}, k)$ is negatively tilted with respect to the $f_B = 0$ limit, with less and less power at large k. The suppression of $\delta_C(\eta_{dr}, k)$ tends to saturate in the large k limit, because in the early radiation-dominated regime, the impact of baryons is very small.

At the baryon drag time, it is obvious that $|\delta_B| \ll |\delta_C|$ for wavenumbers well inside the Hubble radius. Indeed, baryon fluctuations have been oscillating together with photons and have been further suppressed by Silk damping. Meanwhile, CDM fluctuations have been growing because of gravitational clustering.

The evolution of baryons after the drag time is governed by Eq. (6.42), which has a simple solution during matter domination in the limit of constant metric fluctuations,

$$\delta_B = D_B^g + \frac{D_B^d}{\eta} - \frac{k^2\eta^2}{6}\psi. \tag{6.47}$$

We see that baryon perturbations reach asymptotically the same value as those of CDM, given by $-(k^2\eta^2\psi)/6$. This is not surprising, because both species are nonrelativistic and fall into the same potential wells. Because they start from a much smaller value, the matching of this solution with that for $\eta \leq \eta_{dr}$ leads to coefficients D_B^g and D_B^d such that $D_B^g + D_B^d/\eta_{dr}$ is negative. The time needed for baryon fluctuations to reach the same value as CDM fluctuations is greater for smaller scales, because in the large-k limit the ratio $|\delta_B/\delta_C|$ is smaller at η_{dr}. However, on all cosmological scales of interest in this book (i.e., scales which are still in the linear or midly nonlinear regime today), baryon and CDM fluctuations become equal to each other at a redshift varying between $z_{dr} \sim 1000$ for large scales and $z \sim 100$ for small scales. After that time, we can follow a single variable $\delta_M = f_B\delta_B + (1 - f_B)\delta_C$, whose evolution follows

$$\delta_M'' + \frac{a'}{a}\delta_M' - 4\pi G\bar{\rho}_M\delta_M = 0. \tag{6.48}$$

This equation is again identical to the one governing the CDM evolution in a universe with $f_B \simeq 0$. In fact, it describes the exact evolution of δ_M at any time after baryon drag, even when $|\delta_B| < |\delta_C|$ (as can be shown using Eqs. (5.21), (6.7) and (6.42)). Therefore, instead of following separately δ_B and δ_C between baryon drag and the time at which $\delta_B = \delta_C$, it is sufficient to perform a matching at $\eta = \eta_{dr}$ between the solutions $\delta_B(\eta, k)$, $\delta_C(\eta, k)$ for $\eta \leq \eta_{dr}$ and the solution of the above equation for $\eta \geq \eta_{dr}$. A priori, the matching process should give the coefficient of the growing mode of δ_M (growing linearly with a during matter domination) as a linear combination of the four numbers $\delta_C(\eta_{dr}, k)$, $\delta_C(\eta_{dr}, k)'$, $\delta_B(\eta_{dr}, k)'$ and $\delta_B(\eta_{dr}, k)'$.

This matching has been studied in detail by Hu and Sugiyama, 1996 and Eisenstein and Hu, 1998. The leading term turns out to be the first one, and also the last one if f_B is not too small. The fact that the contribution of $\delta_C(\eta_{dr}, k)'$ is negligible comes from the fact that $\delta_C(\eta, k)$ is slowly varying before the matching. Instead, baryon fluctuations are quickly evolving before the baryon drag time, and their time derivative at η_{dr} plays a crucial role in the subsequent evolution of baryon fluctuations.

We know that the transfer function $\delta_C(\eta_{dr}, k)$ is smoothly increasing with k (see Eq. (6.30)), whereas $\delta_B(\eta_{dr}, k)$ is oscillating and quickly decreasing with k (because of combined effects of acoustic oscillations and Silk damping). The derivative $\delta_B(\eta_{dr}, k)'$ is also oscillating and decreasing with k, with a phase shifted by $\pi/2$. Hence in the large k limit, the contribution of baryons to the growing mode of δ_M is negligible, and the result of the matching is as simple as

$$\delta_M(\eta_{dr}, k) \simeq \left[\frac{\bar{\rho}_C}{\bar{\rho}_C + \bar{\rho}_B} \right] \delta_C(\eta_{dr}, k). \tag{6.49}$$

The factor between brackets is simply equal to $f_C = 1 - f_B$ in the present case, but more generally, if other species such as massive neutrinos were present, it would be given by $(1 + \omega_B/\omega_C)^{-1}$.

In the large-k limit, baryon fluctuations do not contribute directly to the final matter power spectrum, but the amplitude of total matter fluctuations is suppressed by a factor f_C with respect to that of CDM fluctuations at $\eta = \eta_{dr}$. This accounts for the reduced growth rate of δ_C (see Eq. (6.46)) between baryon drag and the time at which $\delta_B = \delta_C$.

In the limit $f_B \ll 1$, the CDM transfer function during matter and Λ domination (for modes crossing the Hubble radius during radiation domination) was given by Eq. (6.30). The previous discussion suggests that baryon effects can be accounted for by modifying this result in the following way:

$$\delta_C(a, k) = \left[\alpha\left(\tfrac{\omega_B}{\omega_C}\right) + \beta\left(\tfrac{\omega_B}{\omega_C}\right) \ln\left(\frac{k}{k_{eq}}\right) + \gamma(k) \sin[kr_s(\eta_{dr})] \right] \frac{a\, g(a; \Omega_M)}{a_{eq}}.$$

$$\tag{6.50}$$

The coefficients α and β are now functions of the baryon-to-CDM fraction, where this dependence accounts for the reduced growth of CDM perturbations during radiation and matter domination, until the time at which $\delta_B = \delta_C$. The extra contribution from baryon perturbations, significant only on intermediate scales crossing the Hubble radius at the end of radiation domination, oscillates like $\sin[kr_s(\eta_{dr})]$ because it comes from the derivative of δ_B or $\Theta_{\gamma,0}$, which oscillate like $\cos[kr_s(\eta_{dr})]$. The envelope of these oscillations is given by a function $\gamma(k)$ decreasing with k. We did not write explicitly the dependence of γ on parameters such as the baryon density and the Silk damping scale.

In summary, the presence of a significant baryon fraction affects the shape of the matter power spectrum on small scales. The expression (6.39) for $\tilde{P}(a, \tilde{k})$ should be replaced in the $\tilde{k} \gg \tilde{k}_{eq}$ limit by

$$\tilde{P}(a, \tilde{k}) = \mathcal{N} \left(\frac{a \, g(a; \Omega_M)}{a_0 \Omega_M^0} \right)^2 \tilde{k} \, \mathcal{P}_{\mathcal{R}}(\tilde{k}) \qquad (6.51)$$

$$\times \frac{\tilde{k}_{eq}^4}{4\tilde{k}^4} \left(\alpha \left(\frac{\omega_B}{\omega_C} \right) + \beta \left(\frac{\omega_B}{\omega_C} \right) \ln \left(\frac{k}{k_{eq}} \right) + \gamma(k) \sin[k r_s(\eta_{dr})] \right)^2 .$$

The baryon impact is illustrated in Fig. 6.1.

Spectrum shape and evolution as a function of cosmological parameters

We can now review the effect of cosmological parameters on the matter power spectrum in a neutrinoless ΛCDM universe, in the same way as in Section 5.1.6 for the CMB temperature spectrum. We introduced in Eq. (6.38) the quantity $\tilde{P}(a, \tilde{k})$ referring to the matter power spectrum expressed in units of $h^{-3}\text{Mpc}^3$, as a function of the scale factor and of Fourier wavenumbers expressed in units of $h\text{Mpc}^{-1}$. When evaluated today, $\tilde{P}(a_0, \tilde{k})$ depends on

(P1) The scale $\tilde{k}_{eq}$, which determines the location of the maximum in the matter power spectrum. Following equations (6.35, 6.36), this scale is given by the dimensionless ratio $[a_{eq} H_{eq}]/[a_0 H_0]$ divided by a factor of 3000. Using Eq. (6.31), is easy to show that this ratio is equal to $[2\Omega_M(1 + z_{eq})]^{1/2}$.

(P2) The baryon-to-CDM fraction (ω_B/ω_C), which alters the shape of the large-scale asymptote. When this fraction increases, the spectrum is suppressed for $\tilde{k} \geq \tilde{k}_{eq}$, accounting for a reduction in the growth rate of CDM perturbations as long as $|\delta_B| \ll |\delta_C|$, but on intermediate scales some small baryon acoustic oscillations (BAO's) inherited from the baryon density fluctuations prior to the baryon drag epoch are imprinted.

(P3) The phase of the BAO's depends on the sound horizon at the drag epoch $r_s(\eta_{dr})$, and the decrease of the oscillation amplitude with k (more difficult to observe precisely) depends on the Silk damping scale $r_d(\eta_{dr})$. Both parameters depend strongly on ω_B. Note that the scale of BAO oscillations differs from that of CMB oscillations not only due to the difference between $r_s(\eta_{LS})$ and $r_s(\eta_{dr})$, but also due to a phase shift by $\pi/2$.

(P4) The overall amplitude of the matter spectrum depends on both Ω_M (given by $(1 - \Omega_\Lambda)$, because we are dealing with spatially flat ΛCDM models) and on the primordial spectrum amplitude A_s (defined in Eq. (5.34)).

(P5) The overall tilt of the matter spectrum depends on that of the primodial spectrum n_s (also defined in Eq. (5.34)).

As a function of time, the matter power spectrum of a neutrinoless universe has a fixed shape (as long as we do not reach such high redshifts that on small scales

$\delta_B \neq \delta_C$). It evolves with the scale factor like $[a\, g(a; \Omega_M)]^2$, or as a function of redshift like

$$\tilde{P}(z, \tilde{k}) = (1 + z)^{-2} g(z; \Omega_M)^{-2}. \tag{6.52}$$

Prior to Λ domination, when $g(z; \Omega_M) \simeq 1$, the product $(1 + z)^2 \tilde{P}(z, \tilde{k})$ is independent of redshift.

6.1.3 Impact of massless neutrinos

We have seen in Section 5.3.2 that the impact of massless neutrinos or any relativistic relics on the evolution of linear cosmological perturbations can be parameterized by a single degree of freedom, the effective neutrino number N_{eff} (defined in Eq. (2.198)), irrespective of the details of the phase-space distribution function. Following the same logic as in Section 5.3.1, we notice that a variation in N_{eff} can potentially affect the matter power spectrum through

(a) *Background effects*: Enhancing the radiation density can change some characteristic times and scales that control the matter power spectrum. According to the results of the previous section, for a fixed primordial spectrum $\mathcal{P}_{\mathcal{R}}(\tilde{k})$, the matter power spectrum is affected by four effects (P1)–(P4) depending mainly on $(\Omega_M, z_{eq}, \omega_B/\omega_C, \omega_B)$. The variation of these four parameters with N_{eff} depends on which quantities are chosen to be fixed when N_{eff} varies, as we shall see later.

(b) *Perturbation effects*: Neutrino perturbations can have a direct impact on matter perturbations due to the gravitational coupling between neutrinos, CDM and baryons. As explained in Section 5.3.1, this coupling can only be important during radiation domination and immediately after Hubble crossing; otherwise relativistic neutrino fluctuations are too small to back-react on matter perturbations.[2]

We can already guess that direct perturbation effects are small. Indeed, we know from Section 5.2.5 that the perturbations $\delta_\nu(\eta, k) = F_{\nu 0}(\eta, k)$ of free-streaming relativistic particles oscillate after Hubble crossing with a pulsation k. Hence, in Weinberg's decomposition of the full solution in fast modes and slow modes (see Section 6.1.1), neutrinos couple only with fast modes. Therefore, before the baryon drag epoch, neutrino perturbations cannot have a significant back-reaction on CDM perturbations, which are described entirely in terms of slow modes. The Mészáros equation (6.28) (or (6.45) if the baryon fraction is not negligible) still applies to scenarios with relativistic relics, with no further modification.

---

[2] This category of effect was called (b.1) in Section 5.3.1. Effects called (b.2) and (b.3) are specific to photons and are irrelevant for the matter power spectrum calculation.

Fast modes (describing photon and baryon oscillations) are instead affected by the presence of neutrino perturbations. We actually summarized the corresponding effects in Section 5.3.2. The gravitational feedback of free-streaming neutrinos just after horizon crossing reduces the amplitude of acoustic oscillations in the coupled photon–baryon fluid and slightly shifts the oscillation phase (Hu, 1995; Bashinsky and Seljak, 2004). These effects are imprinted in baryon fluctuations $\delta_B(\eta, k)$ prior to the baryon drag time, and in BAO features in the total matter power spectrum after that time.

In summary, we expect direct effects of neutrino perturbations to show up only in the amplitude and phase of baryon acoustic oscillations.

To check this with numerical results, we would like to vary N_{eff} and cancel all background effects, in order to isolate direct neutrino perturbation effects. This is impossible, because background effects depend on $(\Omega_M, z_{eq}, \omega_B/\omega_C, \omega_B)$: if we keep these four quantities fixed, then N_{eff} is also fixed through Eq. (5.104) (assuming of course that ω_γ is also fixed by the measurement of the CMB temperature):

$$z_{eq} = \frac{\omega_M}{\omega_\gamma \left[1 + 0.2271 N_{eff}\right]} \quad \Longrightarrow \quad [1 + 0.2271 N_{eff}] = \frac{\omega_B(1 + \omega_C/\omega_B)}{z_{eq}\, \omega_\gamma}.$$

(6.53)

However, if we vary N_{eff} while keeping only $(\Omega_M, z_{eq}, \omega_B/\omega_C)$ fixed, we expect the fraction of the matter power spectrum inherited from CDM perturbations (i.e., from slow modes during radiation domination) to be unaffected. In the following subsections, we will see how the spectrum changes if we vary N_{eff} while keeping fixed either $(\Omega_M, z_{eq}, \omega_B/\omega_C)$ or $(\Omega_M, z_{eq}, \omega_B)$. The second option is also relevant, because LSS data are usually used in combination with CMB data, which tend to fix accurately ω_B (in fact, the ratio ω_B/ω_γ) rather than ω_B/ω_C, as explained in Section 5.1.6.

Varying N_{eff} with fixed redshifts of equality and baryon-to-cold dark matter ratio

Fixing $(\Omega_M, z_{eq}, \omega_B/\omega_C)$, i.e., the two redshifts of equality (between radiation and matter, and between matter and Λ) and the baryon-to-CDM ratio, can be achieved by multiplying the radiation density, matter density and squared Hubble parameter by the same factor. For instance, to compare a neutrinoless model with a model of given N_{eff}, we can perform the transformation

$$(\omega_C, \omega_B, h) \longrightarrow (\alpha\omega_C, \alpha\omega_B, \sqrt{\alpha}h) \tag{6.54}$$

with $\alpha = [1 + 0.2271 N_{eff}]$. This leaves the three quantities $(\Omega_M, z_{eq}, \omega_B/\omega_C)$ invariant. We have done a similar transformation in Section 5.3.2 in order to identify

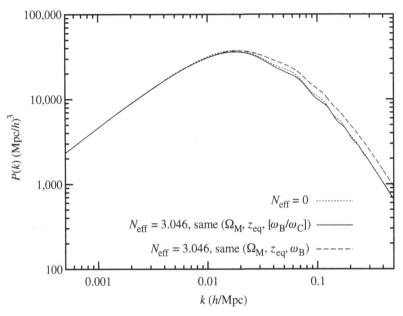

Figure 6.2 Matter power spectrum for a neutrinoless model (dotted line), a model with $N_{eff} = 3.046$ and the same redshifts of equality and baryon-to-CDM ratio (solid line) and a model with $N_{eff} = 3.046$ and the same redshifts of equality and baryon density (dashed line). The difference between the middle and lower curves (a shift in BAO scales and amplitude) is due to a combination of neutrino perturbation effects and a different baryon density ω_B, affecting the baryon drag time and hence the sound horizon at that time.

neutrino perturbation effects on the CMB. The difference was that in Section 5.3.2, the matter density $\omega_M = \omega_C + \omega_B$ was increased with a fixed baryon density. In the present case, our goal is to leave effects (P1), (P2), (P4) of Section 6.1.2 unaffected (as well as (P5), because the primordial spectrum is unchanged). The only expected modifications to the matter power spectrum come from (P3), because ω_B has changed, and from direct neutrino perturbation effects. However, these two effects can alter only the phase and amplitude of BAO's. We show in Fig. 6.2 that this is indeed the case. The BAO phase shift is mainly due to the fact that the model with a higher N_{eff} has a high baryon density ω_B, so baryons are released earlier with a low value of the sound horizon $r_s(\eta_{dr})$. It follows that the BAO peaks are shifted to slightly smaller scales (larger wavenumbers). The phase shift due to neutrino drag is of opposite sign, but of smaller amplitude.

Varying N_{eff} with fixed redshifts of equality and baryon density

Fixing ω_B/ω_C was useful for illustrative purposes, but in practice the CMB tends to fix ω_B with very good precision, and LSS data are used to resolve remaining

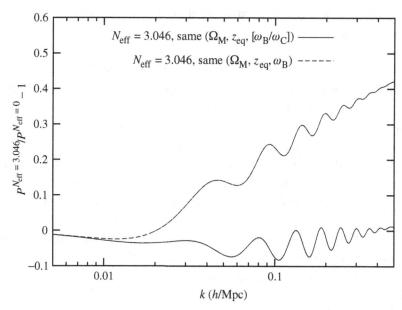

Figure 6.3 Matter power spectrum for models with $N_{eff} = 3.046$ divided by the spectrum of a model with $N_{eff} = 0$. The lower curve corresponds to the case in which the two models share the same redshifts of equality and baryon-to-CDM ratio, whereas in the upper one they have the same redshifts of equality and baryon density.

parameter degeneracies. If we now perform the transformation

$$(\omega_C, \omega_B, h) \longrightarrow ([\alpha\omega_C + (\alpha - 1)\omega_B], \omega_B, \sqrt{\alpha}h), \tag{6.55}$$

still with $\alpha = [1 + 0.2271 N_{eff}]$, the two redshifts of equality are still left invariant, but the baryon-to-CDM ratio decreases. This is exactly the same transformation as in Section 5.3.2. We now expect the matter power spectrum to be modified by effect (P2) and by small neutrino perturbation effects. This is confirmed by Fig. 6.2: the low baryon-to-CDM ratio results in a high amplitude on small scales; the BAO amplitude is damped both by the smaller baryon fraction and by neutrino perturbation effects; the small phase shift in BAOs due to neutrino drag is hardly visible in Fig. 6.2 but appears in Fig. 6.3 when we take the ratio of the two spectra.

6.1.4 Impact of hot dark matter

We will now scrutinize the effect of neutrino masses on the linear matter power spectrum. As we will see later, cosmological bounds are such that we can restrict the discussion to the case of neutrinos with $m_\nu \leq 1.5$ eV. Hence we assume that neutrinos become nonrelativistic after the time of equality between radiation and matter (see Eq. (5.90)).

Evolution of neutrino perturbations

The effect of neutrinos on the matter power spectrum presents many similarities to that of baryons (studied in Section 6.1.2), with an analogy between the baryon drag time and the nonrelativistic transition time. Before the nonrelativistic transition, neutrinos free-stream below the wavelength λ_{fs} defined in Section 5.2.4, i.e., roughly on all sub-Hubble scales. Their density $\delta_\nu(\eta, k)$ is strongly suppressed with respect to that of nonrelativistic species, because pressure forces are much stronger than gravitational forces and prevent gravitational collapse. After the nonrelativistic transition, we have seen at the end of Section 5.2 that the equation of state $w = \bar{P}_\nu/\bar{\rho}_\nu$, the sound speed $\delta p_\nu/\delta \rho_\nu$ and the anistropic-stress-to-density ratio δ_ν/σ_ν decay proportionally to a^{-2}. Their order of magnitude is given roughly by c_ν^2, i.e., by $(\bar{T}_\nu/m_\nu)^2$; (see Eq. (5.91)).

If all these ratios vanished, neutrinos would be described by the same equation as CDM and decoupled baryons:

$$\delta_\nu'' + \frac{a'}{a}\delta_\nu' = -k^2\psi + 3\phi'' + 3\frac{a'}{a}\phi'. \tag{6.56}$$

However, pressure perturbations and shear viscosity cannot be completely neglected and continue to play a role in the equation of motion on sub-Hubble scales. Like decoupled baryons, neutrinos fall in the same gravitational potential wells as any other nonrelativistic species, and tend to an equilibrium solution, but because of pressure and shear, this solution differs from that for baryons (Ringwald and Wong, 2004):

$$\delta_\nu \sim (k_{fs}/k)^2\delta_B = (k_{fs}/k)^2\delta_C. \tag{6.57}$$

However, the growth rate of $\delta_\nu(\eta, k)$ between the nonrelativistic transition and the time at which equilibrium is reached is much lower than for baryons, because of pressure forces and viscosity. A detailed study of this evolution shows that today, $\delta_\nu = \delta_C$ holds only for wavenumbers $k \leq k_{nr}$. For larger wavenumbers the ratio $\delta_\nu(\eta_0, k)/\delta_C(\eta_0, k)$ decreases slowly as a function of k, as illustrated in Figure 6.4.

Impact on the total matter power spectrum

This has important consequences for the total matter power spectrum:

- For modes $k < k_{nr}$, the matter power spectrum of a ΛCDM model with massive neutrinos is the same as that of a massless model (if the two models share the

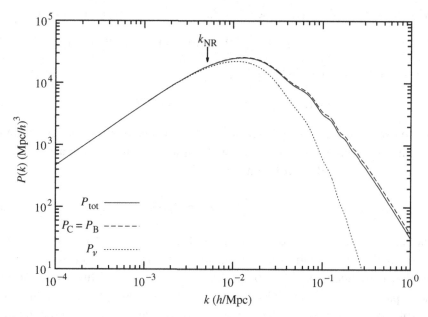

Figure 6.4 Matter power spectrum at redshift zero for a ΛCDM model with three degenerate massive neutrino species ($m_\nu = 0.3$ eV), compared to the individual power spectrum of CDM, baryon and neutrino density perturbations. In this model k_{nr} is equal to $5.1 \times 10^{-3} h$/Mpc (see Eq. (5.94)). For wavenumbers $k > k_{nr}$, neutrino perturbations remain smaller than CDM and baryon perturbations, because of their low growth rate after the nonrelativistic transition.

same Ω_M and primordial power spectrum). Indeed, before the Hubble radius is crossed, all perturbations are subject to the usual universal relations given by Eq. (5.24) for adiabatic initial conditions. After Hubble crossing, if $k < k_{nr}$, neutrino free-streaming can be neglected: massive neutrinos share the same evolution as CDM and fall into the same potential wells, with δ_ν quickly reaching the asymptotic value of Eq. (6.57). Hence all quantitites evolve exactly as described in Section 5.24, with neutrinos being counted as part of the cold dark matter component. Because the matter power spectrum depends only on Ω_M and $\mathcal{P}_\mathcal{R}(k)$ for wavenumbers $k < k_{eq}$, and because $k_{nr} < k_{eq}$, two models with different neutrino masses but the same total matter fraction and primordial spectrum are indistinguishable on those scales.

- for $k \gg k_{nr}$, we can use the fact that at low redshift and for the cosmological scales of interest in this book, $|\delta_\nu| \ll |\delta_C| = |\delta_B|$. If we expand the total matter fluctuation as

$$\delta_M = f_C \delta_C + f_B \delta_B + f_\nu \delta_\nu \qquad (6.58)$$

(with $f_C + f_B + f_\nu = 1$), we see that the matter power spectrum is given by

$$P(\eta, k) \simeq (f_C + f_B)^2 P_C(\eta, k) = (1 - f_\nu)^2 P_C(\eta, k), \qquad (6.59)$$

where $P_C(\eta, k)$ is the CDM power spectrum

$$\langle \delta_C(\eta, \vec{k}) \delta_C^*(\eta, \vec{k}') \rangle = P_C(\eta, k) \delta^{(3)}(\vec{k} - \vec{k}'). \qquad (6.60)$$

So we can understand the impact of neutrino masses on those scales by simply studying the evolution of δ_C. We know that after Hubble crossing and before the baryon drag epoch, this evolution is described by Eq. (6.9):

$$\delta_C'' + \frac{a'}{a} \delta_C' = 4\pi G \bar{\rho}_C \delta_C. \qquad (6.61)$$

The term on the right-hand side is derived from the Poisson equation. It does not involve photon, relativistic neutrino and baryon perturbations for the reasons developed in Sections 6.1.1, 6.1.2 and 6.1.3: we are dealing with the slow mode part of the solution, whereas other perturbations are either negligible or contributing to fast modes. The right-hand side cannot involve nonrelativistic neutrino perturbations either, because after the nonrelativistic transition $|\delta_\nu| \ll |\delta_C|$ for wavenumbers $k \gg k_{nr}$.

After the baryon drag epoch, we have seen in Section 6.1.2 that there is a simple equation of evolution (6.48) for the weighted average between δ_C and δ_B. We cannot denote this average δ_M any more because total matter now also includes neutrinos. We should rather denote it as $\delta_{CB} \equiv (f_C \delta_C + f_B \delta_B)/(f_C + f_B)$, subject to the equation of evolution

$$\delta_{CB}'' + \frac{a'}{a} \delta_{CB}' - 4\pi G(\bar{\rho}_C + \bar{\rho}_B)\delta_{CB} = 0. \qquad (6.62)$$

Again, neutrino perturbations do not appear in this equation, for the same reasons as in Eq. (6.9). The conclusion is that neutrino masses can affect the evolution of δ_C and δ_{CB} only through the magnitude of the expansion rate (a'/a) relative to the density of clustering species $\bar{\rho}_C$ or $(\bar{\rho}_C + \bar{\rho}_B)$. The background density of massive neutrinos $\bar{\rho}_\nu$ contributes to the expansion rate through the Friedmann equation, but not to the density of clustering species. Hence massive neutrinos enhance the Hubble friction term relative to the self-clustering term in the preceding equations. This leads to a reduction of the growth rate of δ_C and δ_{CB}, which will be discussed below in more detail.

- In the intermediate region (k slightly larger than k_{nr}), neutrino perturbations, although smaller than CDM perturbations, are not completely negligible, at least at small redshift. Hence there is a smooth transition between the region where neutrino masses have no effect, and that in which they have a maximal effect.

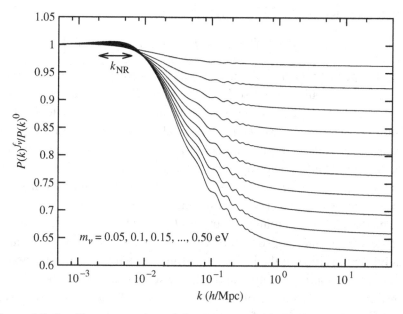

Figure 6.5 Steplike suppression of the matter power spectrum due to neutrino mass. The power spectrum of a ΛCDM model with two massless and one massive species has been divided by that of a massless model, for several values of m_ν between 0.05 eV and 0.50 eV, spaced by 0.05 eV. All spectra have the same primordial power spectrum and the same parameters (Ω_M, ω_M, ω_B).

In summary, neutrino masses produce a smooth steplike suppression of the matter power spectrum on scales $k > k_{nr}$. This step is shown in Fig. 6.5 for various masses. In the next subsection, we show how to estimate the suppression factor analytically as a function of neutrino masses in the small-scale limit.

Suppression factor for $k \gg k_{nr}$

Several approaches to estimating the neutrino mass impact on small scales analytically or semianalytically have been discussed in the literature. A very accurate (but also very technical) discussion has been presented in Hu and Eisenstein, 1998 (see also Holtzman, 1989; Pogosian and Starobinsky, 1995; Ma, 1996; Novosyadlyj *et al.*, 1998). For conciseness, we prefer to follow here the simple approach of Lesgourgues and Pastor, 2006 (although with a more precise and elaborate discussion of matching issues). An even simpler discussion was presented in Tegmark (2005), at a very sketchy level.

Our goal is to estimate the ratio of the matter power spectrum with neutrino masses (with a given neutrino fraction f_ν) to that with massless neutrinos ($f_\nu = 0$), in the large-wavenumber limit $k \gg k_{nr}$. Equation (6.59) shows that this ratio can be formulated in terms of CDM transfer functions

$$P(\eta, k)^{f_\nu} / P(\eta, k)^0 = (1 - f_\nu)^2 \left[\delta_C^{f_\nu}(\eta, k) / \delta_C^0(\eta, k) \right]^2. \tag{6.63}$$

We need to specify which parameters should be kept fixed in the comparison. The two models should of course have the same Ω_M and the same primordial spectrum, in order to have the same large-scale limit. Following the logic of similar discussions in Section 5.3, we would like ideally to cancel any difference in the background evolution of the two models by playing with other parameters than neutrino masses. This is obviously impossible, because the background density of each massive species $\bar{\rho}_{\nu_i}(a)$ has a nontrivial evolution, switching from $\bar{\rho}_{\nu_i} \propto a^{-4}$ to $\bar{\rho}_{\nu_i} \propto a^{-3}$ around $a = a_{nr}(m_{\nu_i})$. However, if we fix ω_M to the same value in the two models, the background evolution $H(a)$ is identical at least for $a \gg a_{nr}$ (more precisely, in the case of neutrinos with nondegenerate masses, after the nonrelativistic transition of the heaviest neutrinos, when the total neutrino density $\bar{\rho}_\nu$ scales at least approximately like a^{-3}). A fixed ω_M can easily be achieved by compensating for the increase of ω_ν by a decrease of ω_B, ω_C or both. We will discuss different choices in the following.

In summary, we wish to compare a reference massless ΛCDM model that we call **R**, with parameters $(\omega_C^R, \omega_B^R, \omega_M^R)$ such that

$$\omega_M^R = \omega_C^R + \omega_B^R, \tag{6.64}$$

to a massive ΛCDM model that we call **M**, with parameters $(\omega_C^M, \omega_B^M, \omega_\nu^M, \omega_M^M)$ such that

$$\omega_M^M = \omega_C^M + \omega_B^M + \omega_\nu^M$$
$$\omega_M^M = \omega_M^R$$
$$\omega_\nu^M = f_\nu \, \omega_M^R$$
$$\omega_C^M + \omega_B^M = (1 - f_\nu) \, \omega_M^R. \tag{6.65}$$

Our approach consists of approximating the background evolution of the massive model, in order to use known results for the massless case. Figure 6.6 shows an example of background evolution for two models $\mathbf{M}_1$ and $\mathbf{M}_3$ with the same f_ν and two different mass splittings, compared with their common reference model **R** with the same ω_M. As a first approximation, we could try to compute the evolution of $\delta_C(\eta, k)$ in a given model **M**, assuming the background expansion of model **R**. This background approximation would lead to the correct density in the relativistic and nonrelativistic limits, but it would be very crude and would make the transition much smoother than it is in reality, with an overestimate of the expansion rate close to the transition, as can be seen in Fig. 6.6. Moreover, the error would increase

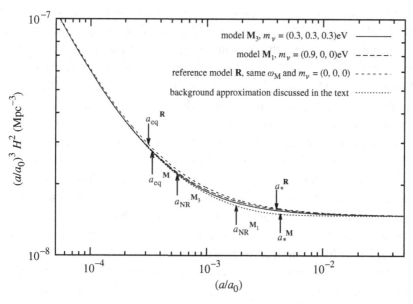

Figure 6.6 Background evolution for two models $\mathbf{M}_1$ and $\mathbf{M}_3$ with the same f_ν and two different mass splittings, compared with their common reference model $\mathbf{R}$ with the same ω_M. To see the difference between the various models clearly, we plot the squared expansion rate H^2 multiplied by $(a/a_0)^3$: this ratio is constant deep inside the matter-dominated regime and after the nonrelativistic transition of neutrinos. For each model, we show the scale factor at equality and, if relevant, at the time of the nonrelativistic transition for massive neutrinos. Finally, we show another background evolution discussed in the text and used for deriving an approximate analytic formula for the small-scale suppression factor, as well as the scale factor a_* at which we perform a matching between two classes of solutions in this approximation scheme.

in the small-f_ν limit, which is the most interesting one given current bounds on neutrino masses. Indeed, in this limit, the expansion rate in model $\mathbf{M}$ remains lower than that in model $\mathbf{R}$ for a longer time.

We will discuss a different approximation based on the fact that the neutrinos we are considering become nonrelativistic after the time of equality. Hence, before the nonrelativistic transition, the massive model is practically equivalent to a massless model $\tilde{\mathbf{M}}$ with the same CDM and baryon density parameters $(\omega_\mathrm{C}^\mathbf{M}, \omega_\mathrm{B}^\mathbf{M})$. The times of equality in models $\mathbf{R}$ and $\mathbf{M}$ or $\tilde{\mathbf{M}}$ differ by a factor

$$\frac{\omega_\mathrm{C}^\mathbf{M} + \omega_\mathrm{B}^\mathbf{M}}{\omega_\mathrm{C}^\mathbf{R} + \omega_\mathrm{B}^\mathbf{R}} = (1 - f_\nu), \qquad (6.66)$$

with equality taking place later in the model $\mathbf{M}$ or $\tilde{\mathbf{M}}$. When approaching the neutrino nonrelativistic transition, the background evolution of $\mathbf{M}$ and $\tilde{\mathbf{M}}$ starts to differ, but we can choose to go on approximating the massive model with model $\tilde{\mathbf{M}}$ until the total density in $\tilde{\mathbf{M}}$ reaches the true density of $\mathbf{M}$ during matter domination. After that time, we neglect the photon density and consider that the total matter density scales like a^{-3} until Λ domination. In other words, we approximate the true background evolution of $\mathbf{M}$ with that of the massless model $\tilde{\mathbf{M}}$, matched at a given time to a radiationless model with the right matter and Λ densities. Such an approximation tends to underestimate the expansion rate at times close to the nonrelativistic transition, as can be seen from the lower curve in Fig. 6.6. However, the decisive advantage of this scheme is that it becomes exact in the small-f_ν limit. Also, we will see that it leads to simple analytical predictions. The error made on the expansion rate is larger when at least one neutrino species becomes nonrelativistic close to equality, i.e., in the large-f_ν limit, and for a fixed f_ν, when the mass is concentrated in a single species. This can be seen in Fig. 6.6 by comparing the $\mathbf{M}_1$ and $\mathbf{M}_3$ curves with the curve corresponding to our approximation: the $\mathbf{M}_1$ curve is further away from the approximation. In this plot, in order to enhance the difference between the various curves, we choose a large total mass (0.9 eV), already in tension with current data. For smaller masses the approximation becomes very good.

To ensure the continuity of $H(a)$, the matching between perturbations in the $\tilde{\mathbf{M}}$ model and in the radiationless model should be performed at a scale factor $a_*^{\mathbf{M}}$ such that[3]

$$\omega_{\mathrm{M}}^{\tilde{\mathbf{M}}}\left(\frac{a_*^{\mathbf{M}}}{a_0}\right)^{-3} + \omega_{\mathrm{R}}^{\tilde{\mathbf{M}}}\left(\frac{a_*^{\mathbf{M}}}{a_0}\right)^{-4} = \omega_{\mathrm{M}}^{\mathbf{R}}\left(\frac{a_*^{\mathbf{M}}}{a_0}\right)^{-3}, \tag{6.67}$$

i.e., for

$$\frac{a_0}{a_*^{\mathbf{M}}} = \frac{f_\nu\,\omega_{\mathrm{M}}^{\mathbf{R}}}{\omega_\gamma\,(1 + 0.2271 N_{\mathrm{eff}})}. \tag{6.68}$$

The Mészáros equation (modified in the presence of baryons; see Eq. (6.45)) gives the evolution of δ_{C} as a function of $y = a/a_{\mathrm{eq}}$, and depends only on $\bar{\rho}_{\mathrm{C}}/(\bar{\rho}_{\mathrm{C}} + \bar{\rho}_{\mathrm{B}})$. If the baryon-to-CDM density ratio is the same in models $\tilde{\mathbf{M}}$ and $\mathbf{R}$, i.e., if

$$\omega_{\mathrm{C}}^{\mathbf{M}} = (1 - f_\nu)\,\omega_{\mathrm{C}}^{\mathbf{R}} \tag{6.69}$$

$$\omega_{\mathrm{B}}^{\mathbf{M}} = (1 - f_\nu)\,\omega_{\mathrm{B}}^{\mathbf{R}}, \tag{6.70}$$

then all perturbations are identical in the two models $\tilde{\mathbf{M}}$ and $\mathbf{R}$ for $a \leq a_*^{\mathbf{M}}$, modulo a shift in time

$$\delta_{\mathrm{C}}^{\tilde{\mathbf{M}}}(a, k) = \delta_{\mathrm{C}}^{\mathbf{R}}((1 - f_\nu)a, k). \tag{6.71}$$

---

[3] The calculation of the suppression factor presented in Lesgourgues and Pastor (2006) is identical to the present one, except for a different (semiempirical) choice for the matching time, $a_*^{\mathbf{M}} = a_{\mathrm{nr}}/\sqrt{1 - f_\nu}$, not compatible with a continuous $H(a)$.

After $a = a_*^M$, because we assume that the total background density scales exactly like a^{-3} until Λ domination, the relevant equations of evolution (Eq. (6.9) before baryon drag or (6.62) after baryon drag) can be approximated during the rest of the matter-dominated stage as

$$\delta_C'' + \frac{2}{\eta}\delta_C' - \frac{6}{\eta^2}(1 - f_\nu)\delta_C = 0 \quad \text{for } \eta \leq \eta_{dr} \tag{6.72}$$

$$\delta_{CB}'' + \frac{2}{\eta}\delta_{CB}' - \frac{6}{\eta^2}(1 - f_\nu)\delta_{CB} = 0 \quad \text{for } \eta \geq \eta_{dr}. \tag{6.73}$$

These equations have two exact solutions proportional to $\eta^{2p_\pm} \propto a^{p_\pm}$ with

$$p_\pm = \frac{-1 + \sqrt{1 + 24(1 - f_\nu)}}{4}. \tag{6.74}$$

For $f_\nu \ll 1$, these exponents read

$$p_+ \simeq 1 - (3/5)f_\nu + O(f_\nu^2), \tag{6.75}$$

$$p_- \simeq -(3/2) + (3/5)f_\nu + O(f_\nu^2). \tag{6.76}$$

Current bounds on neutrino masses are such that terms of order f_ν^2 can safely be neglected, so from now on we will use these first-order expressions for $p_\pm$. We see that they have opposite signs and correspond respectively to growing and decaying modes. When the cosmological constant starts to play a role, we know that the coefficient of Eqs. (6.72), (6.73) change, and that in the absence of massive neutrinos the growing mode would be given by $[a\, g(a; \Omega_M)]$ (with the function g defined in Section 6.1.1). Numerical simulations confirm that in a very good approximation, the solutions with massive neutrinos can be approximated as $[a\, g(a; \Omega_M)]^{p_\pm}$.

We should now perform a matching between the solution $\delta_C^{\tilde{M}}(a, k)$ for $a \leq a_*^M$ and a linear combination of the two solutions $[a\, g(a; \Omega_M)]^{p_\pm}$ for $a \geq a_*^M$. We will assume for simplicity that the matching selects a pure growing mode, such that the density perturbation just after the matching reads

$$\delta_C^M(a, k) = \left[\frac{a}{a_*^M}\right]^{p_+} \delta_C^{\tilde{M}}\left(a_*^M, k\right). \tag{6.77}$$

At late times, taking into account the effect of baryon drag and of Λ domination, the perturbations read

$$\delta_{CB}^M(a, k) = \delta_C^M(a, k) = F(\omega_B/\omega_C)\left[\frac{a\, g(a; \Omega_M)}{a_*^M}\right]^{p_+} \delta_C^{\tilde{M}}\left(a_*^M, k\right), \tag{6.78}$$

where the function F is equal to one if the matching at a_*^M takes place after the baryon drag epoch, or to $(1 + \omega_B/\omega_C)^{-1}$ in the opposite situation (see Eq. (6.49)).

Using the equivalence between the models $\tilde{\mathbf{M}}$ and $\mathbf{R}$ summarized by Eq. (6.71), we can rewrite this equality as

$$\delta_{\mathrm{CB}}^{\mathbf{M}}(a, k) = F(\omega_{\mathrm{B}}/\omega_{\mathrm{C}}) \left[\frac{a\, g(a; \Omega_{\mathrm{M}})}{a_*^{\mathbf{M}}} \right]^{p_+} \delta_{\mathrm{C}}^{\mathbf{R}}\left((1 - f_\nu)a_*^{\mathbf{M}}, k\right). \tag{6.79}$$

To find the suppression factor, we should finally relate the CDM density perturbation at $a_*^{\mathbf{R}} \equiv (1 - f_\nu)a_*^{\mathbf{M}}$ to the matter density perturbation today in the reference model. We can follow the same steps as in the massive model, i.e., match the solution at $a \leq a_*^{\mathbf{R}}$ with a linear combination of the two solutions $[a\, g(a; \Omega_{\mathrm{M}})]^{p_\pm}$ for $a \geq a_*^{\mathbf{R}}$, with now $p_+ = 3/5$ and $p_- = -3/2$ in the massless case. Also, we need to be consistent with previous assumptions concerning the derivative of the solution at the time of matching. By selecting the growing mode in the previous matching, we explicitly assumed that in the model $\tilde{\mathbf{M}}$,

$$\frac{d}{da}\delta_{\mathrm{C}}^{\tilde{\mathbf{M}}}(a, k) = [1 - (3/5)f_\nu]\, \delta_{\mathrm{C}}^{\tilde{\mathbf{M}}}(a, k) \tag{6.80}$$

at $a = a_*^{\mathbf{M}}$. Then the exact correspondence between models $\tilde{\mathbf{M}}$ and $\mathbf{R}$, summarized by Eq. (6.71), implies that

$$\frac{d}{da}\delta_{\mathrm{C}}^{\mathbf{R}}(a, k) = [1 - (3/5)f_\nu]\, \delta_{\mathrm{C}}^{\mathbf{R}}(a, k) \tag{6.81}$$

at $a = a_*^{\mathbf{R}}$. A matching with the two solutions $a^{3/5}, a^{-3/2}$ in the subsequent matter-dominated stage gives the coefficients of the growing and decaying modes:

$$\delta_{\mathrm{C}}^{\mathbf{R}}(a, k) = \left[\left(1 - \frac{6}{25}f_\nu \right) \left(\frac{a}{a_*^{\mathbf{R}}} \right) + \frac{6}{25}f_\nu \left(\frac{a}{a_*^{\mathbf{R}}} \right)^{-3/2} \right] \delta_{\mathrm{C}}^{\mathbf{R}}\left(a_*^{\mathbf{R}}, k\right). \tag{6.82}$$

At late times, we should keep only the growing mode and take into account baryon drag and Λ domination,

$$\delta_{\mathrm{M}}^{\mathbf{R}}(a, k) = F(\omega_{\mathrm{B}}/\omega_{\mathrm{C}}) \left(1 - \frac{6}{25}f_\nu \right) \left[\frac{a\, g(a; \Omega_{\mathrm{M}})}{a_*^{\mathbf{R}}} \right] \delta_{\mathrm{C}}^{\mathbf{R}}\left(a_*^{\mathbf{R}}, k\right), \tag{6.83}$$

where $F(\omega_{\mathrm{B}}/\omega_{\mathrm{C}})$ is the same as in Eq. (6.79), because the two models $\mathbf{M}$ and $\mathbf{R}$ share the same baryon-to-CDM fraction. We now take the ratio of Eq. (6.79) and Eq. (6.83) and find the relation

$$\delta_{\mathrm{CB}}^{\mathbf{M}}(a, k) = \left(1 - \frac{6}{25}f_\nu \right)^{-1} \frac{a_*^{\mathbf{R}}}{a_*^{\mathbf{M}}} \left[\frac{a\, g(a; \Omega_{\mathrm{M}})}{a_*^{\mathbf{M}}} \right]^{-(3/5)f_\nu} \delta_{\mathrm{M}}^{\mathbf{R}}(a, k). \tag{6.84}$$

We can use $(a_*^R/a_*^M) = (1 - f_\nu)$ and take the square of this equality to obtain a relation between the matter power spectra of the two models (using Eq. (6.63)),

$$P^{f_\nu}(a, k) = \frac{(1 - f_\nu)^4}{(1 - (6/25)f_\nu)^2} \left[\frac{a\, g(a; \Omega_M)}{a_*^M} \right]^{-(6/5)f_\nu} P^0(a, k). \tag{6.85}$$

We can finally replace a_*^M by its explicit expression (6.68) and write our final result as

$$\frac{P(a, k)^{f_\nu}}{P(a, k)^0} = \frac{(1 - f_\nu)^4}{(1 - (6/25)f_\nu)^2} \left[\frac{a\, g(a; \Omega_M)}{a_0} \cdot \frac{f_\nu\, \omega_M}{\omega_\gamma\, (1 + 0.2271 N_{\text{eff}})} \right]^{-(6/5)f_\nu}. \tag{6.86}$$

We recall that in this equation N_{eff} should be summed over all neutrinos (massive and massless). This result could also be expressed as a function of the scale factor at equality, in either the massless or the massive model. For instance, if a_{eq}^0 denotes the redshift of equality in the massless model,

$$\frac{P(a, k)^{f_\nu}}{P(a, k)^0} = \frac{(1 - f_\nu)^4}{(1 - (6/25)f_\nu)^2} \left[\frac{a\, g(a; \Omega_M)}{a_{\text{eq}}^0}\, f_\nu \right]^{-(6/5)f_\nu}. \tag{6.87}$$

In the derivation of this result, we made four approximations:

- We used $p_+ \simeq 1 - (3/5)f_\nu$: this approximation leads to negligible errors for realistic models with $f_\nu \leq 0.1$.
- We assumed that solutions for $\delta_C^M(\eta, k)$ during Λ domination are given by $(a\, g(a; \Omega_M))^{p_\pm}$ instead of solving the full equation of evolution. However, this approximation is excellent for the growing mode, and the solution for the decaying mode is anyway irrelevant at late times.
- In the massive model, we approximated the background evolution in such a way that (a'/a) near the transition was underestimated, especially when the highest neutrino mass was large and the transition took place close to equality. More expansion implies a slower growth of perturbations in the massive model, i.e., a reduction of the suppression factor.
- In the massive model **M**, we assumed in the matching process that only the growing mode proportional to $a^{1-(3/5)f_\nu}$ was excited. If we assume that a small fraction of δ_c^M goes to the decaying mode, both δ_c^M and δ_c^R are affected, but the total effect is a reduction of the suppression factor.

We see that the first two approximations have a negligible impact, whereas the last two go in the same direction, and we expect Eq. (6.86) to slightly overestimate the suppression factor. It can be shown that the worst approximation is actually the third one; this is important for understanding the difference between the degenerate

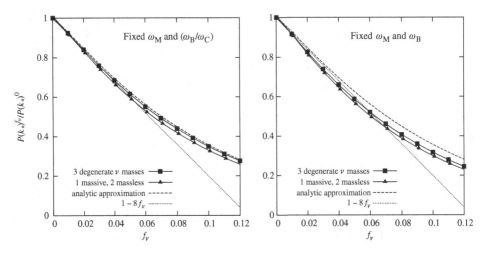

Figure 6.7 Suppression factor $P(\eta, k)^{f_v}/P(\eta, k)^0$ computed today for $k = 10\,h$/Mpc, as a function of the total neutrino fraction f_v. The result depends on the mass splitting but is always contained between the two displayed curves corresponding to a single massive neutrino or three degenerate neutrinos, i.e., to the maximal and minimal values of the heaviest mass that can be involved for each value of f_v. We show for comparison the analytic approximation of Eq. (6.86) as well as the well-known linear approximation $(1 - 8 f_v)$. On the left, the comparison was performed with fixed values of ω_M and (ω_B/ω_C), as explicitly assumed when deriving the analytical formula. On the right, we kept ω_M and ω_B fixed (with the increase in ω_v compensated for by a decrease in ω_C), and we compare with the same approximations.

and single-mass scenarios, and more generally for the discussion of mass splitting effects in the next subsection.

These expectations are confirmed by Fig. 6.7 (left plot). For the case of degenerate masses, the analytic prediction is accurate by better than 1%, whereas for a single massive species the error can reach 3% (at least for $f_v \leq 0.12$, i.e., throughout the range compatible with current data). Indeed, we have seen earlier that our approximation for the background evolution of the massive model gets worse when the time of nonrelativistic transition of at least one species gets closer to the time of equality. So, in the single mass case, the underestimation of the suppression factor is greater.

For a given neutrino fraction f_v, i.e., a total mass M_v, there are an infinity of models corresponding to different ways to split M_v between several species. However, the heaviest individual mass is always inside the range defined by the previous two cases, i.e., between the degenerate case ($m_v = M_v/N_v$) and the single-massive-neutrino case ($m_v = M_v$). Because it is mainly the heaviest mass that determines the time at which our background approximation becomes not so good and produces an error, the two models displayed in Fig. 6.7 can be seen as limiting cases, and the discrepancy between the analytic formula and the numerical results cannot be larger than for the single-mass model. We will come back to mass splitting issues in more detail in the next subsection.

In Fig. 6.7 (left plot), we also show the famous linear approximation of (Hu *et al.*, 1998) to the suppression factor,

$$\frac{P(a, k)^{f_\nu}}{P(a, k)^0} \simeq 1 - 8 f_\nu. \tag{6.88}$$

This fit is accurate at the 2% level for $f_\nu \leq 0.05$, but departs significantly from the numerical solution for a larger neutrino fraction.

In Fig. 6.7, the numerical estimate of the suppression factor was done at the scale $k = 10 \, h/\text{Mpc}$. Of course, on such a small scale, real matter perturbations are strongly nonlinear today. However, on this scale, the ratio $P(a, k)^{f_\nu}/P(a, k)^0$ has reached its asymptotic value up to very good accuracy: we checked that the factor inferred from $k = 100 \, h/\text{Mpc}$ agrees to better than 1% with that computed at $k = 10 \, h/\text{Mpc}$. This is not the case on larger scales: the factor computed at $k = 1 \, h/\text{Mpc}$ is as much as 2% larger (but, coincidentally, it is in even better agreement with the analytical prediction of Eq. (6.86)).

We performed all this exercise assuming that the baryon-to-CDM fraction is the same in the massive model and in the reference model. In practice, however, it is useful to consider the effect of neutrino masses when the baryon fraction ω_B is fixed, because this parameter is accurately determined by CMB observations. If we go along the same lines of reasoning fixing $\omega_B^R = \omega_B^M$, then we should assume that the ratio ω_B/ω_C is larger in model $\mathbf{M}$ than in model $\mathbf{R}$. Hence the matter power spectrum of $\mathbf{M}$ will be suppressed both by neutrino masses as described previously, and (to a smaller extent) by an increase in the baryon fraction, with the consequences described in Section 6.1.2. In this case, the suppression factor is smaller, as can be checked in the right plot of Fig. 6.7. The discrepancy between the analytic prediction of Eq. (6.86) and numerical results increases up to about 6% (for realistic values of ω_B and ω_C). In that case, the linear fit of Eq. (6.88) remains very accurate for $f_\nu \leq 0.01$, but in the range $f_\nu \in [0.01 \ 0.05]$ the error can reach 3%. A linear fit $(1 - 8.5 f_\nu)$ would be better in that case.

Mass splitting

The effect of neutrino masses on the matter power spectrum does not depend only on the total mass $M_\nu = \sum_i m_{\nu_i}$ for essentially two reasons:

1. The scale at which a given neutrino mass m_{ν_i} starts to have an impact on the matter power spectrum depends on $k_{nr} \propto m_{\nu_i}^{1/2}$ (see Eq. (5.94)). If there is a single

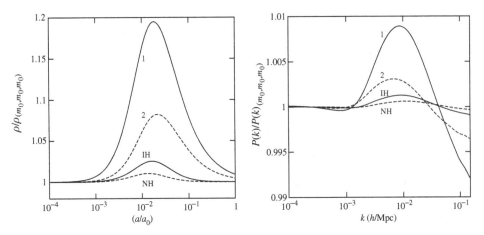

Figure 6.8 (Left) Evolution of the total neutrino energy density as a function of the scale factor for models where the same total mass $M_\nu = 0.12$ eV is distributed among the three species in various ways: only one massive eigenstate (1), two degenerate massive states (2), normal hierarchy (NH), inverted hierarchy (IH). All curves are normalized to the case with three degenerate massive states. (Right) Comparison of the matter power spectrum obtained for the same models, divided each time by that with three degenerate massive states. Differences in the mass splitting affect the position and amplitude of the break in the power spectrum. (Reprinted from Lesgourgues and Pastor, 2006, with permission from Elsevier.)

mass in the problem, as in scenarios with N massive neutrinos and $(N_\nu - N)$ massless ones, the steplike suppression starts at a smaller wavenumber k if the mass is smaller, i.e., if M_ν is distributed equally among all species. If there are several masses, the location and shape of the step are rather model-dependent: we will illustrate this below and compare the most relevant cases, namely the normal hierarchy (NH) and inverted hierarchy (IH) scenarios preferred by neutrino oscillation experiments.

2. We have seen in the previous subsection that the amplitude of the step is controlled mainly by f_ν, i.e., by M_ν, but with an additional dependence on the detailed background evolution close to the nonrelativistic transition(s). The case of degenerate masses is very close to the analytic approximation of the last section, because neutrinos remain relativistic for a maximum amount of time for a given M_ν, because no mass is larger than M_ν/N_ν. For any other mass splitting, the background density is enhanced for a limited amount of time with respect to the degenerate scenario. This was illustrated in Fig. 6.6 for the single-mass scenario, and more examples are given in Fig. 6.8 (left plot). Any temporary enhancement of the background density leads to slower growth of CDM and baryon fluctuations for a little while, and to a stronger steplike suppression.

These differences are of course very small for a fixed total mass. The maximum discrepancy can be observed in the comparison of the degenerate and single-mass scenarios, on small scales where the neutrino mass effect is maximum, and at $z = 0$. We see in Fig. 6.7 that even in this case the effect is only on the order of 3%. The discrepancy between realistic scenarios (NH and IH), on scales close enough to the linear regime to enable precise observations with an ideal instrument (i.e., k not too large and/or z not too small), is much smaller and typically below the percent level. It is therefore unlikely that these effects can be probed at a good significance level in the future, but because the most ambitious forecasts suggest that discrimination is marginally possible (Pritchard and Pierpaoli, 2008; Jimenez *et al.*, 2010), we will describe them in more detail.

In Fig. 6.8, we compare the total neutrino density evolution $\rho_\nu(a)$ and the current matter power spectrum $P(a_0, k)$ of five models with the same total mass $M_\nu = 0.12$ eV: one, two, three degenerate massive species (with two, one, zero massless species); three massive species with normal hierarchy; and three massive species with inverted hierarchy. All curves are normalized to the case of three degenerate species. For the NH and IH scenario, squared mass differences were chosen in the range probed by experimental data, leading to $(m_1, m_2, m_3) = (3.03, 3.15, 5.82) \times 10^{-2}$eV in the NH case, and $(m_1, m_2, m_3) = (5.11, 5.18, 1.71) \times 10^{-2}$eV in the IH case.

The model with a single massive species features the highest mass and the earliest nonrelativistic transition. In this model, the background density is significantly enhanced with respect to the degenerate model when the scale factor is in the range between $a_{nr}(M_\nu)$ and $a_{nr}(M_\nu/N_\nu)$. Hence the step starts at a larger wavenumber $k_{nr}(M_\nu)$, but goes deeper. So the spectra of the two models must cross each other at some point, and their ratio has the bumpy shape seen in Fig. 6.8 (right plot). The same arguments explain the difference between the two-masses scenario and the degenerate case.

The NH and IH scenarios are more similar to the degenerate case, because they feature three masses. The heaviest mass is roughly the same in the two cases, 0.0582 eV versus 0.0518 eV in our example. In the IH scenario, however, two neutrinos become nonrelativistic almost at the same time. There is a short period during which the NH scenario has one nonrelativistic neutrino with $m_\nu = 0.0582$ eV, whereas the IH scenario has two of them with a summed mass of $0.0511 + 0.0518 = 0.1029$ eV. Hence the total density is enhanced for a short amount of time in the IH case, with slower growth of CDM perturbations: this explains why the amplitude of the spectrum on scales $k \gg k_{nr}$ is a bit smaller in the IH scenario. For intermediate wavenumbers crossing the Hubble scale when the universe contains two relativistic species in the NH case, or one relativistic species

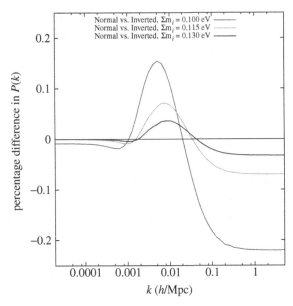

Figure 6.9 Ratio of matter power spectra for pairs of models with three massive neutrinos, subject to either the normal or the inverted hierarchy scenario, but with a common total mass for each pair: from bottom to top on the right, $M_\nu = 0.100$, 0.115 or 0.130 eV. Squared mass differences were fixed to $\delta m^2 = 7.6 \times 10^{-5}$eV and $\Delta m^2 = \pm 2.4 \times 10^{-3}$eV. The first total mass is very close to the minimum allowed value for the inverted hierarchy given these differences, $M_\nu = 0.0994$ eV. (Reprinted from Lesgourgues and Tram, 2011, with permission from IOP.)

in the IH case, however, the IH spectrum is less affected by neutrino free-streaming and appears to be enhanced.

The previous comparison is useful for pedagogical purposes, but in practice it is more useful to look at differences between the NH and IH cases for given values of the total mass. These differences are shown in Fig. 6.9. When the total mass increases, the two models become closer to the degenerate case and more similar to each other. For smaller masses, the effects discussed in the previous paragraph are clearly identified. Also, the IH model with the smallest total mass has its large-scale amplitude reduced by 0.01%. This effect, too small to be observable, is related to the presence of a very light neutrino in this model, just finishing its nonrelativistic transition today. Its nonnegligible pressure slightly affects metric perturbations at large wavelengths.

Because the power spectrum of realistic neutrino mass scenarios depart from the degenerate model by only a fraction of percent for a given f_ν, the analytic formula (6.86) is very accurate for these scenarios (to better than 1%, at least for $f_\nu \leq 0.12$).

Scale-dependent growth factor

We saw in Eq. (6.52) that in the absence of neutrino masses, the power spectrum evolves as a function of redshift (or of the scale factor) as $[g(z; \Omega_M)/(1 + z)]^2$ (or $[a \, g(a; \Omega_M)]^2$) on all scales. The linear growth factor is defined (up to a normalization factor) as the time-dependent part of the square root of the matter power spectrum. This means that in absence of neutrino masses, the linear growth factor is independent of k and proportional to $[g(z, \Omega_M)/(1 + z)]$.

We know that the power spectrum on large scales $k < k_{nr}$ is unaffected by neutrino masses (as long as Ω_M is kept fixed). On small scales $k \gg k_{nr}$, we must instead take into account a correction factor very well approximated by Eq. (6.86). This approximation is valid at least in the redshift range in which we can observe large-scale structure (in this range, f_ν can be considered as independent of time). Taking this correction into account, the small-scale matter power spectrum grows proportionally to $[g(z; \Omega_M)/(1 + z)]^{2-(6/5)f_\nu}$; i.e., the linear growth factor evolves like

$$[g(z; \Omega_M)/(1 + z)]^{1-(3/5)f_\nu}, \qquad (k \gg k_{nr}). \tag{6.89}$$

On intermediate scales such that k is not much bigger than k_{nr}, the linear growth factor interpolates smoothly from $[g(z; \Omega_M)/(1 + z)]$ to Eq. (6.89). An analytic approximation for intermediate scales can be found (e.g., in Hu and Eisenstein, 1998). Hence a crucial effect of neutrino masses is to render the linear growth factor scale-dependent. This can be formulated differently by saying that for a given total neutrino mass M_ν, the steplike suppression of the matter power spectrum is less pronounced at high redshift, as illustrated in Fig. 6.10. This observation is important for the detection of neutrino masses with LSS observations: by comparing the matter power spectrum at different redshifts, it is possible in principle to observe the redshift dependence of neutrino mass effects, on top of their wavenumber dependence.

Impact of the phase-space distribution function

We have seen in Section 4.1 that ordinary active neutrinos are expected to acquire small flavour-dependent nonthermal corrections at decoupling. These are, however, too small to be detectable with cosmological observables. At the level of precision reachable by current and future experiments, the phase-space distribution of ordinary active neutrinos can be approximated as a Fermi–Dirac distribution, with a temperature slightly enhanced with respect to $T_\nu^a \equiv (4/11)^{1/3} T_\gamma$ in such way that the three active species lead to an effective neutrino number equal to $N_{eff} = 3.046$ instead of 3 (as can be inferred from Eq. (4.21) and Table 4.1).

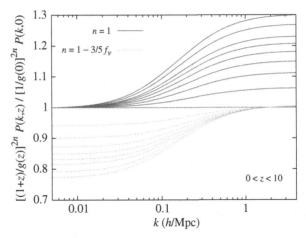

Figure 6.10 Shape distortion of the matter power spectrum $P(k, z)$ at different redshifts. *Upper solid curves:* The power spectra are divided by $[a\,g(a; \Omega_M)]^2 = [g(z)/(1 + z)]^2$, which accounts for the evolution on large scales. From top to bottom, the curves correspond to $z = 10, 8, 6, 5, 4, 3, 2, 1, 0$. *Lower dashed curves:* The power spectra are divided instead by $[a\,g(a; \Omega_M)]^{2-6/5\,f_\nu} = [g(z)/(1 + z)]^{2-6/5\,f_\nu}$, which approximates the power spectrum evolution on small scales. From top to bottom, the curves correspond to $z = 0, 1, 2, 3, 4, 5, 6, 8, 10$. All cosmological parameters are kept fixed and the neutrino density fraction is $f_\nu = 0.1$. (Reprinted from Lesgourgues and Pastor, 2006, with permission from Elsevier.)

In some extensions of the minimal cosmological model, it is necessary to compute the matter power spectrum in the presence of hot relics with a different phase-space distribution. First, active neutrinos may feature a large leptonic asymmetry described by chemical potentials, as discussed in Section 4.2.4. Under some even more radical assumptions, active neutrinos could also acquire large nonthermal distortions, e.g., through the decay of an unstable particle into neutrinos after neutrino decoupling (Cuoco *et al.*, 2005); see Section 4.4.3 . It has also been suggested that under very unusual assumptions, neutrinos could follow Bose–Einstein statistics (Dolgov *et al.*, 2005). Additional hot relics could also have a temperature different from that of active neutrinos, a chemical potential, a Bose–Einstein distribution or a nonthermal distribution.

Among all these possibilities, we can distinguish two situations with different consequences for cosmological perturbations. Hot relics may have

(a) A Fermi–Dirac distribution with arbitrary temperature, or a rescaled Fermi–Dirac distribution (e.g., for sterile neutrinos being populated by nonresonant oscillations with active neutrinos). These relics lead to the same signature as ordinary active neutrinos, with simply a different correspondence between the parameters describing the relic properties (mass, temperature, rescaling factor) and those describing effects in the matter power spectrum (k_{nr} and $f_\nu = \omega_\nu/\omega_M$).

(b) Another distribution. In that case they lead to the same signature as ordinary active neutrinos in the small- and large-scale limits, but to a different step shape on intermediate scales (for k slightly larger than k_{nr}).

Indeed, for case (a), let us consider some hot relics called "x" with a Fermi–Dirac distribution, possibly rescaled by a number χ, an arbitrary temperature T_x and two internal degrees of freedom (as for one pair (ν_i, $\bar{\nu}_i$)). After a change of variable from y to $\tilde{y} = (T_\nu^a/T_x)y$ and $\tilde{m}_x = (T_\nu^a/T_x)m_x$, the perturbation equations (5.74–5.78) for x are exactly identical to those of thermal relics with temperature T_ν^a, except for a global factor $\chi(T_x/T_\nu^a)^4$ in the expression for density and pressure. Hence a rescaling of the phase-space distribution is mathematically equivalent to a change of temperature. These two effects cannot be distinguished at the level of linear perturbations. Before their nonrelativistic transition, such particles would contribute to the radiation density as

$$\Delta N_{\text{eff}} = \chi(T_x/T_\nu^a)^4, \tag{6.90}$$

whereas today their density would be given by

$$\omega_x = \chi \left(\frac{T_x}{T_\nu^a}\right)^3 \frac{m_x}{94.1\text{eV}}. \tag{6.91}$$

Their average velocity would be subject to Eq. (5.91) with $T_\nu = T_x$, leading to

$$k_{nr} \equiv k_{fs}(\eta_{nr}) \simeq 0.0178\,\Omega_M^{1/2} \left(\frac{T_\nu^a}{T_x}\right)^{1/2} \left(\frac{m_x}{1\text{ eV}}\right)^{1/2} h\,\text{Mpc}^{-1}. \tag{6.92}$$

These particles would affect the matter power spectrum both through a change in the radiation density, parameterized by ΔN_{eff} and described in Section 6.1.3, and through a steplike suppression, parameterized by $f_x = \omega_x/\omega_M$ and k_{nr} and described previously in this subsection.

For case (b), the effects of chemical potentials, of different statistics or of nonthermal distortions should be studied case by case. In the relativistic regime, any nonthermal distortion is equivalent to a change in N_{eff}, as mentioned in Section 5.3.2. But in the nonrelativistic regime, all details in the distribution can be important, because the equation of motion for neutrinos cannot be integrated over momentum. We know that hot relics do not affect the largest scales, and that in the small-scale limit the matter power spectrum depends only on the CDM and baryon evolution, with no impact of the gravitational back reaction of neutrinos even in the radiation-dominated stage (see Section 6.1.3). So the phase-space distribution of hot dark matter is irrelevant in this limit for a given value of the HDM density fraction. But on intermediate scales (k slightly larger than k_{nr}), the matter power spectrum receives contributions from each nonrelativistic component: CDM, baryons and neutrinos. In this range, nontrivial phase-space distributions can leave a distinct signature, through modification of the shape of the neutrino power spectrum. The steplike suppression in the total matter power spectrum can then be sharper or smoother. This is illustrated in Hannestad *et al.*, 2005, for the case of a Bose–Einstein versus Fermi–Dirac statistics, and in Lesgourgues and Pastor, 1999, for active neutrinos with chemical potentials.

Instead of studying each subcase of (b) separately, one could try to constrain the phase-space distribution with a model-independent approach. For instance, an arbitrary distribution can be described by the infinite series of its statistical moments. Observations can be used to provide model-independent constraints at least on the first few moments. The analysis of Cuoco *et al.*, 2005 shows that due to parameter degeneracies (between neutrino and cosmological parameters), it is difficult to constrain more than the mass and the first two moments of a given species, related respectively to their density-to-mass ratio ω_ν/m_ν and their contribution to N_{eff}. However, this conclusion is based on the use of linear observables only (CMB and linear matter power spectrum). The nonlinear evolution could potentially enhance the impact of differences in higher momenta.

6.1.5 Impact of warm dark matter

Qualitative behavior of warm dark matter perturbations

If their mass and velocity dispersion fall into an appropriate range, dark matter particles may become nonrelativistic deep inside the radiation-dominated epoch, and preserve the shape of the CDM matter power spectrum down to cluster or galaxy scales, but not further. Such a dark matter component is called warm dark matter (WDM). For example, heavy sterile neutrinos could play the role of WDM. As for hot dark matter, the impact of WDM on the matter spectrum depends on its phase-space distribution function. There is a wide variety of equally plausible scenarios for the production of WDM particles, leading to either thermal or nonthermal distributions. For instance, the case of resonantly or nonresonantly produced sterile neutrinos was already mentioned in Section 4.2.5.

Because WDM particles become nonrelativistic during radiation domination, their free-streaming scale remains almost constant between the time of their nonrelativistic transition and of radiation-to-matter equality, as explained in Section 5.2.4. For wavenumbers much smaller than k_{nr} (defined in Eq. (5.96)), the evolution of cosmological perturbations is unaffected by the warm nature of dark matter particles, and equivalent to that in a scenario with an equivalent amount of CDM. The explanation is the same as for HDM in Section 6.1.4: on large scales, before horizon crossing, nonrelativistic matter perturbations obey a universal solution for adiabatic initial conditions, whereas after horizon crossing, the WDM sound speed and anisotropic stress can be neglected, and density fluctuations are subject to the same equation (6.9) as CDM.

For smaller scales, during the relativistic regime, the WDM mass is negligible and perturbations are subject to the same equations as ultrarelativistic relics. We know from Section 5.2.5 that in this regime, density fluctuations are strongly suppressed on sub-Hubble scales. After the time of the nonrelativistic transition, they start to grow under the effect of gravitational collapse, but at a rate reduced with respect to CDM, due to the fact that pressure and anisotropic stress decay slowly and still play a role in the equations. Hence the WDM power spectrum is suppressed with respect to that of CDM for all small scales with $k > k_{nr}$. The shape of the WDM power spectrum can be studied semianalytically (Boyanovsky and Wu, 2011) or numerically with Boltzmann codes.

We discussed in Section 5.2.4 the difference between the maximum comoving free-streaming scale $[2\pi/k_{nr}]$ and the comoving free-streaming horizon r_{fs}. We showed that for HDM particles becoming nonrelativistic after radiation domination, these two quantitites are very close to each other, whereas for WDM particles becoming relativistic during radiation domination, they represent two distinct scales, given by different combinations of parameters (see Eqs. (5.96) and (5.97)). We saw that the difference comes from the fact that between η_{nr} and η_{eq}: the former remains constant, whereas the latter grows logarithmically. Analytical and numerical approaches confirm that the characteristic scale below which the WDM spectrum differs from the CDM one is given by the free-streaming horizon. This matches physical expectations: the suppression of the WDM spectrum with respect to the CDM one is caused by diffusion damping in a medium of collisionless particles, and the diffusion length of an average WDM particle between the early universe and a given time is indeed given by the free-streaming horizon.

Pure warm dark matter

We first consider the case in which all dark matter in the universe is warm. In particular, the ΛWDM scenario shares the same cosmological parameters and physical ingredients as the ΛCDM scenario, except that cold particles are replaced by warm particles with a nonnegligible velocity dispersion. The power spectrum of this model can be parameterized as

$$P_{\Lambda WDM}(\eta, k) = P_{\Lambda CDM}(\eta, k) T(\eta, k)^2, \tag{6.93}$$

where $P_{\Lambda CDM}$ is the spectrum of a model with an equivalent amount of CDM, and $T(\eta, k)$ is a transfer function accounting for the effect of WDM velocities. The previous discussion suggests that $T(\eta, k) = 1$ for $k r_{fs}(\eta) < 1$, whereas $T(\eta, k) < 1$ for $k r_{fs}(\eta) > 1$.

The shape of the transfer function depends on the WDM phase-space distribution. The two most studied cases in the literature are those of

- a thermal WDM relic with a Fermi–Dirac distribution and an unknown temperature T_x,
- or a nonthermal relic with a rescaled Fermi–Dirac distribution, the same temperature $T_x = T_\nu$ as active neutrinos, and some unknown rescaling factor χ.

The latter case is motivated by the possibility that WDM consists of sterile neutrinos, populated by nonresonant interactions with active neutrinos. This model is often called DW after the work of Dodelson and Widrow (1994). We already discussed such distribution functions for light HDM particles in the previous section. Following the same lines, it is easy to show through a change of variable that a DW model with mass, temperature and rescaling factor (m_x, T_ν, χ) is described by the same equations as a thermal model with mass and temperature

$$(m_{th}, T_{th}) = (\chi^{1/4} m_x, \chi^{1/4} T_\nu). \tag{6.94}$$

Following Eq. (6.91), the common WDM density of these two equivalent models is given by

$$\omega_x = \chi \left(\frac{T_\nu}{T_\nu^a} \right)^3 \frac{m_x}{94.1 \text{eV}} = \left(\frac{T_{th}}{T_\nu^a} \right)^3 \frac{m_{th}}{94.1 \text{eV}}. \tag{6.95}$$

The first equality can be used to eliminate χ from Eq. (6.94) and find another useful relation between the equivalent mass of the DW and thermal models:

$$m_x = 4.43 \text{ keV} \left(\frac{T_\nu}{T_\nu^a} \right) \left(\frac{0.25(0.7)^2}{\omega_x} \right)^{1/3} \left(\frac{m_{th}}{1 \text{ keV}} \right)^{4/3}. \tag{6.96}$$

The numerical study of these two degenerate cases by Bode *et al.*, 2001 (updated by Viel *et al.*, 2005) shows that the transfer function $T(k, \eta)$ is nearly independent of time within the redshift range where the matter power spectrum can be observed, and decays asymptotically like $T(k) \propto k^{-10}$ in the large-k limit. The transition between the two asymptotes is well accounted for by the expression

$$T(k) = \left[1 + (k/k_{break})^p \right]^{-10/p} \tag{6.97}$$

with $p = 2.24$. The value of k_{break} was obtained as a function of the WDM mass-to-temperature ratio and of other cosmological parameters by fitting Eq. (6.97) to numerical results in a range of plausible parameter values:

$$k_{break} = \frac{1}{0.24} X^{0.83} \left(\frac{\omega_x}{0.25.(0.7)^2} \right)^{0.16} \text{Mpc}^{-1} \quad \text{with } X \equiv \frac{m_x / T_x}{1 \text{ keV} / T_\nu^a}. \tag{6.98}$$

This relation can be applied indifferently to DW and thermal WDM models, because two equivalent models share the same ratio $m_{th}/T_{th} = m_x/T_\nu$. Physically, this ratio represents the WDM velocity dispersion up to a numerical factor (we are dealing here with Fermi–Dirac-shaped distributions, with $c_x = 3.15\, T_x/m_x$, as in Eq. (5.91)). The reason for which k_{break} depends on both the velocity dispersion and the warm dark matter density is precisely that this scale is determined by the free-streaming horizon, rather than the maximum free-streaming scale. Equation (5.97) only gives a crude analytic approximation for r_{fs}, based on a simplification of the expansion law $a(t)$ and of the velocity evolution $c_x(a)$. Still, this approximation is sufficient to explain the parametric dependence of Eq. (6.98). Indeed, if we combine Eq. (5.97) (with the last term neglected for $z \ll z_{eq}$) with Eq. (5.96), and replace $(1 + z_{eq})$ with $(\omega_x + \omega_B)/\omega_R$, we get

$$r_{fs} = \frac{3.16 \times 10^{-3}}{\omega_R X} \left[1 + \frac{1}{2} \ln \left(\frac{\omega_R X}{5.28 \times 10^{-7}(\omega_x + \omega_B)} \right) \right] \text{ Mpc.} \qquad (6.99)$$

Replacing ω_R with its fixed value and ω_B with its observed value, one can check that this expression is well fitted by

$$r_{fs} \propto X^{-1}(X/\omega_x)^{0.16} = X^{0.84}\omega_x^{-0.16}, \qquad (6.100)$$

at least within the range of phenomenologically interesting values of X. This explains the parametric dependence of the fitting formula (6.98) for k_{break}. The WDM transfer function can in fact be written as

$$T(k) = \left[1 + \beta(kr_{fs})^p \right]^{-10/p}, \qquad (6.101)$$

where β is a numerical factor. An analytic study of the WDM perturbation evolution by Boyanovsky and Wu, 2011 leads to an approximate expression for $T(k)$ very close to the fitting formula (6.97).

For other nonthermal distributions, results can be derived case by case. By computing the velocity dispersion c_x and the density ω_x of a nonthermal WDM species and applying the previous formulas blindly, one may obtain totally inaccurate predictions. For instance, typical WDM distributions for resonantly produced sterile neutrinos (Shi and Fuller, 1999; Laine and Shaposhnikov, 2008) lead to a very different power spectrum than in the thermal or DW case. The reason is that a significant fraction of such WDM particles have a very low velocity and fall into the CDM category. Hence the impact of such a model on the matter power spectrum is closer to that of a mixture of CDM and thermal WDM, briefly reviewed in the next subsection, as illustrated in Boyarsky et al., 2009c. Other examples are discussed by Boyanovsky and Wu, 2011.

Mixture of warm dark matter and cold dark matter

Mixed cold plus warm dark matter models are relevant not only to scenarios where two different particles with very different mass and velocity dispersion contribute to the matter density in the universe but also as an approximation to the case of a unique dark matter particle with a nonthermal distribution (enhanced in the small momentum limit with respect to a thermal distribution), as mentioned in the previous paragraph.

This case is qualitatively similar to the mixed hot plus cold dark matter one, i.e., to a ΛCDM model with massive neutrinos. For scales such that $kr_{fs} \lesssim 1$, the warm nature of some of the dark matter particles does not play a role, and the model is equivalent to a ΛCDM model with the same total dark matter density. For smaller scales, WDM density perturbations are suppressed with respect to δ_C for the reasons explained at the beginning of Section 6.1.5. One can show that for $kr_{fs} > 1$ one still has $|\delta_w| < |\delta_C|$ today (as is the case for the density perturbation of massive neutrinos for $k > k_{nr}$). In the small-wavelength limit, the matter power spectrum of a mixed model with a warm fraction $f_w \equiv \omega_w/\omega_M$ is suppressed for two reasons:

- the total matter power spectrum comes from the CDM component only and is reduced by a factor of $(1 - f_w)^2$, in the same way as with HDM (see Eq. (6.59));
- CDM perturbations δ_C grow at a lower rate because in the usual equation of evolution (6.7), WDM contributes to the Hubble friction term, again as is the case with HDM.

Hence the mixed ΛCWDM power spectrum is steplike suppressed with respect to the ΛCDM one, with just a different scale, shape and amplitude than for light massive neutrinos. The suppression factor in the small-scale limit can be estimated as a function of f_w following the same lines as in Section 6.1.4. The only difference is that WDM particles are already nonrelativistic at the approach of radiation-to-matter equality. Boyarsky *et al.* (2009d) provide an approximation for this factor,

$$\frac{P_{\Lambda \mathrm{WDM}}(\eta_0, k)}{P_{\Lambda \mathrm{CDM}}(\eta_0, k)} \longrightarrow (1 - f_w)^2 \left(\frac{a_0\, g(a_0)}{a_{eq}} \right)^{-\frac{3}{2} f_w} \simeq 1 - 14 f_w + O(f_w^2), \quad (6.102)$$

which is clearly different from its counterpart for mixed ΛCHDM models in Eqs. (6.87) and (6.88). The full numerical ΛCWDM power spectrum is presented in the same reference.

We do not discuss here the case of mixed ΛWHDM or ΛCWHDM models, which are similar to the previously discussed cases but with an additional steplike suppression because of the hot relics.

6.2 Nonlinear matter power spectrum

The theory of linear cosmological perturbations presented in Sections 5.1, 5.2 and 6.1 cannot describe matter fluctuations in our universe at small redshift and on small scales. A given comoving Fourier mode k enters into the nonlinear regime when the variance of matter fluctuations integrated in spheres of comoving radius $2\pi/k$ becomes of the same order as the average matter density. The power spectrum of the ΛCDM model leads to hierarchical structure formation, with smaller scales entering earlier into the nonlinear regime.

Thanks to galaxy, cluster and weak lensing surveys, we can observe matter fluctuations on scales significantly larger than the nonlinear scale, but we can compare only observed power spectra with linear predictions up to that scale. The amount of information extracted from a given data set would increase if the comparison could be pushed to smaller scales, under the condition that theoretical errors remain smaller than instrumental errors. This provides a strong motivation for computing power spectra on mildly or even strongly nonlinear scales.

For instance, in measurements of the three-dimensional matter power spectrum $P(k)$, the number of independent modes in the survey scales as the cube of k_{max}, the maximum wavenumber at which we trust theoretical predictions and that we decide to include in theoretical fits of the data. Naively, this seems to imply that the error bar on cosmological parameters decreases as $k_{max}^{3/2}$. In fact, the amount of information contained in the matter power spectrum tends to saturate at large k_{max}, because deep inside the nonlinear regime, the nonlinear spectrum $P_{nl}(z, k)$ loses sensitivity to early linear perturbations and to the underlying cosmology. However, it remains true that modelling the nonlinear spectrum is a key ingredient for improving the sensitivity of LSS observations to neutrino parameters.

6.2.1 N-body simulations

Nonlinear structure formation can be simulated with N-body codes. In this approach, cold dark matter (as well as eventually baryons and other dark matter species) is represented as a set of N-body particles, with typical mass on the order of a million solar masses or more, not to be confused with fundamental particles. These particles are initially distributed inside a box in such a way that the power spectrum of the smoothed distribution coincides with the linear power spectrum $P(z_{ini}, k)$ of a given cosmological model. After this step, they need to be evolved only under gravitational interactions. In current N-body codes, the expansion of the universe is taken into account through a Hubble friction term in the particle equation of motion, but gravitational interactions are computed in the nonrelativistic limit, using the Poisson equation. Indeed, during matter and Λ

domination, simulated particles are nonrelativistic, implying metric fluctuations much smaller than one (because $\phi \sim (v/c)^2$). N-body simulations are performed in boxes in comoving space with periodic boundary conditions. Their comoving spatial resolution scales like the volume of the box divided by the number of particles N.

N-body algorithms

Because gravity is a long-range interaction, the motion of one particle depends on the force exercised by all other particles. A brute-force calculation would imply the computation of $N(N-1)$ forces between pairs of particles, which becomes impossible for millions of particles. Various clever approximations have been discussed and tested in N-body codes.

First, the so-called tree method consists of replacing the force exercised by many particles located in a given region with the force exercised by a unique particle, located at the center of this region and of equivalent mass. Such an approximation does not introduce significant errors provided that this region is far enough from the point at which we want to evaluate the force, and not too dense. The simulation box is divided into large cells, each of them split into smaller subcells, and so on. When evaluating the force on a given particle, a sophisticated algorithm chooses the most appropriate cell size to be used in different regions, with larger cells in remote underdense regions, and smaller cells in nearby overdense regions. Instead of evaluating $(N-1)$ forces from all other particles, the code typically needs to evaluate $\log N$ forces from variable-size cells. Hence, with the tree method, the simulation time scales like $N \log N$.

The particle-mesh (PM) method consists of replacing the true particle distribution at each new time step with a collection of other pointlike particles located at the nodes of a regular grid. The value of the mass placed at each node is obtained by interpolating the underlying particle distribution, in such a way that the two distributions smoothed over a distance larger than the grid size are identical. The new particle distribution can be represented as a mass distribution function $\rho(\vec{x}_i)$ in discrete space, which is easy to expand in Fourier space. The gravitational potential $\psi(\vec{k})$ is then trivially inferred from the Poisson equation and transformed back to real space. Its gradient gives the gravitational force at each point and is used to evolve the set of true particles over one time step. This approach is very fast even with respect to the tree method, but it mistreats gravitational interactions on scales comparable to the grid size or smaller.

The TreePM method is a hybrid approach in which long-range forces are calculated with the PM method, and short-range forces with the tree method. The two approaches are matched smoothly around a scale corresponding to a few PM grid cells. Some state-of-the-art codes such as GADGET (Springel, 2005) are based on the TreePM approach. There are other powerful methods and codes on the market, but we do not mention them here, because the following discussion of neutrino implementations in N-body codes refers mainly to TreePM methods.

Hot dark matter particles

The first difference between cold and hot particles in an N-body simulation appears at the level of the initial spatial distribution, because these species are subject to different linear power spectra featuring either a cutoff or a steplike suppression on small scales. When two different species coexist, distinct sets of N-body particles must be used for each component, and the initial spatial distribution of each species must reflect its own linear power spectrum. For instance, when simulating a ΛCDM model with massive neutrinos, it is necessary to introduce cold particles with a steplike suppressed spectrum and hot particles with a spectrum with a different slope on small scales, as illustrated in Fig. 6.4. Moreover, it is crucial to compute the linear power spectrum at the correct initial redshift with a linear Boltzmann code, because for mixed models the linear spectra have a nontrivial redshift dependence (and a scale-dependent growth factor). For instance, we have seen in Section 6.1.4 that for fixed neutrino masses, the steplike suppression in the CDM power spectrum is smaller at high redshift than today. Note that the distributions of the two species must be statistically correlated: in a universe with adiabatic initial conditions, overdensities in one species coincide with overdensities in another species, even if the power spectra are different.

This first step raises no particular difficulties. There exist various algorithms for drawing initial particle distributions under the constraint of a given arbitrary power spectrum. These codes are based either on the Zel'dovich approximation (ZA), consistent with first-order perturbation theory, or on second-order Lagrangian perturbation theory (2LPT) for more precise settings. They are not difficult to generalize to the case of several correlated species with distinct spectra.

Second, the initial velocity distribution of N-body particles should reflect the phase-space distribution of fundamental particles in a given HDM model. For cold particles, previously mentioned algorithms (ZA or 2LPT) assign initially to each particle a peculiar velocity consistent with the local gravitational flow, whose divergence is related to the scalar degree of freedom θ_C of linear perturbation theory. This peculiar velocity is often called the Zel'dovich velocity. For hot particles, after assigning Zel'dovich velocities (corresponding to the bulk motion inside the gravitational potential), it is necessary to add random velocities reflecting the phase-space distribution $f_0(y)$. Because N-body codes rely on Newtonian gravity, the starting redshift should be chosen in such way that particles are initially nonrelativistic, so the momentum appearing in the phase-space distribution can be replaced by $p = mv$ (or $y = amv$). Hence, for particles with a thermal distribution, typical velocities are of the order of $v \sim [\langle p \rangle / m] \sim [T/m]$. Even in the nonrelativistic regime, typical thermal velocities can be significantly higher than Zel'dovich velocities at the initial time. Later, they decrease as the inverse of the scale factor because of the universe's expansion, which is consistently taken into account by the simulation.

The presence of large thermal velocities raises computational difficulties for two reasons. A priori, the integration time step should scale like the inverse of typical velocities in order to resolve particle trajectories with good accuracy. If thermal velocities are higher than Zel'dovich velocities by one or two orders of magnitude, the simulation will be slowed down in the same proportions with respect to a CDM simulation. Second, HDM particles should fill a large volume in phase space. The phase-space volume of cold particles is null, because they are not given any velocity dispersion: initially, they have a unique possible velocity at each point, corresponding to the Zel'dovich velocity. Instead, in the ideal case, hot particles with a thermal velocity higher than the Zel'dovich velocity should have a nearly isotropic velocity distribution in each point, like the true distribution of underlying fundamental particles. In practice, one needs to introduce more hot particles than cold particles in the simulation to obtain consistent results down to a given resolution scale; otherwise the sampling of phase space is not sufficient, and shot noise quickly dominates the evolution, leading to incorrect predictions for the nonlinear power spectrum.

Despite the large CPU time and memory cost of such simulations, consistent results have been obtained by various groups for mixed ΛCHDM models (Brandbyge *et al.*, 2008; Viel *et al.*, 2010; Bird *et al.*, 2012). These works show that because neutrinos experience less clustering than CDM on small scales, their short-range gravitational interactions can be neglected. Hence, in the TreePM code GADGET, the tree part can be evaluated only for CDM particles, whereas the mesh part includes both species. It has also been shown that to follow CDM clustering correctly, it is sufficient to set the integration time step to the same value as in a pure CDM simulation. The main limitation of these simulations comes from memory requirements, in that many neutrino particles are needed to keep shot noise at an acceptable level. Because shot noise is directly related to the phase-space volume of the particles, i.e., to the magnitude of thermal velocities (compared to Zel'dovich velocities), simulations become more difficult for faster particles, i.e., lighter neutrinos. The starting redshift z_{ini} should decrease with the neutrino mass, in order to have low enough thermal velocities when initial conditions are set. But for small masses, z_{ini} may have to be pushed to values at which CDM particles are already in the nonlinear regime, so their initial conditions cannot be set correctly. This problem can be alleviated by making use of the 2LPT algorithm. So far, simulations with cold plus hot particles have been carried successfully with three degenerate neutrino species of total mass $M_\nu = 3m_\nu \geq 0.15$ eV. These simulations converge to a maximum wavenumber on the order of $k \sim 10\,h/\mathrm{Mpc}$.

Alternative approaches for hot dark matter

To study the small-scale nonlinear clustering of a given HDM species, it is unavoidable to model this species with N-body particles, and run heavy simulations. However, to study the nonlinear clustering of CDM and compute the total nonlinear matter power spectrum in mixed ΛCHDM models, it is possible to use several types of approximation. Mistreating neutrinos on small scales is not necessarily an issue, because we expect $\delta_\nu \ll \delta_\mathrm{C}$ on those scales. For instance, instead of introducing hot particles from the beginning, one can describe neutrinos as a fluid discretized on a grid, following linear equations solved by an external Boltzmann code (Brandbyge and Hannestad, 2009), until some late time at which they are converted into particles, when their thermal velocities are sufficiently small (Brandbyge and Hannestad, 2010). Alternatively, one can postulate some approximate fluid equations for the neutrino component (similar in spirit to those discussed in Section 5.2.5) and treat neutrinos with the same numerical machinery as the baryonic component, i.e., with a smoothed-particle hydrodynamics (SPH) approach, less time-consuming than a pure N-body approach (Hannestad *et al.*, 2012b). The accuracy of these approximate methods is still being discussed and investigated (Bird *et al.*, 2012).

Results for ΛCDM models with massive neutrinos

A consistent picture is emerging from all recent ΛCHDM simulations. We know from Section 6.1.4 that at the linear level, the massive-to-massless power spectrum ratio $P^{f_\nu}(z, k)/P^0(z, k)$ has the shape of a smooth step, departing from one above the wavenumber k_nr, and reaching a plateau roughly between $k = 1\,h/\mathrm{Mpc}$ and $10\,h/\mathrm{Mpc}$, depending on neutrino masses. The amplitude of the step for fixed $(\Omega_\mathrm{M}, \omega_\mathrm{M}, \omega_\mathrm{B}/\omega_\mathrm{C})$ is well approximated by Eq. (6.86) or Eq. (6.87), and remains close to $(1 - 8 f_\nu)$ for $f_\nu \leq 0.05$ and at $z = 0$.

The same ratio computed from the nonlinear results of N-body simulations has the shape of a spoon (Brandbyge *et al.*, 2008; Viel *et al.*, 2010; Bird *et al.*, 2012; Hannestad *et al.*, 2012b). The results of Bird *et al.*, 2012 for three massive species of total mass $M_\nu = 0.6$ eV are shown in Fig. 6.11. At $z = 0$, the ratio reproduces linear results up to a scale on the order of $k \sim 0.1\,h/\mathrm{Mpc}$; then it keeps decreasing until reaching a dip, and finally it goes up at least to the scale at which simulation errors blow up. At redshift zero, the scale of the dip is on the order of $k = 1\,h/\mathrm{Mpc}$, and the maximal suppression at this scale is close to $(1 - 10 f_\nu)$. At redshift $z > 0$, the same features are observed, but shifted to smaller scales.

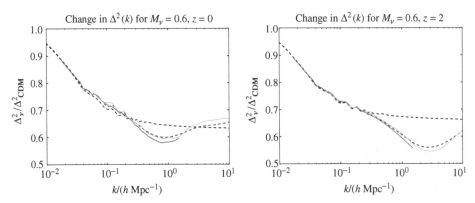

Figure 6.11 Spoon-shaped suppression of the matter power spectrum due to neutrino masses. The solid lines show the power spectrum of a ΛCDM model with three massive species of total mass $M_\nu = 0.6$ eV, divided by that of a massless model with the same parameters (Ω_M, ω_M). Baryons have been neglected in these simulations. In each plot, the two solid lines correspond to two different N-body simulations with different resolution scales. The step-shaped dashed line shows the linear predictions, studied in Section 6.1.4, whereas the spoon-shaped dashed curve shows the outcome of the fitting formula HALOFIT, recalibrated in order to incorporate massive neutrino effects. (Reprinted from Bird *et al.*, 2012, with permission from Elsevier.)

The spoon shape is easy to understand qualitatively. In the presence of massive neutrinos and for a fixed total matter density ω_M, CDM perturbations grow more slowly on scales $k \gg k_{nr}$, at least in the linear theory. Hence, they enter a bit later into the nonlinear regime. The enhancement of the linear power spectrum due to nonlinear clustering is then a bit smaller in the massive case, and this effect adds up to the suppression already observed at the linear level. This explains why the ratio falls to $(1 - 10f_\nu)$ instead of $(1 - 8f_\nu)$ at redshift zero. On smaller scales, another effect takes over. It is well known that deep inside the nonlinear regime, the matter power spectrum gradually loses memory of initial conditions and becomes insensitive, e.g., to the details of the growth rate of CDM perturbations in the linear regime. So, in the large k limit, the ratio between the power spectrum of a massive and a massless model should reach one asymptotically. At redshift $z > 0$, the same features can be identified at larger wavenumbers, because the scale of nonlinearity is smaller.

Bird *et al.* (2012) have used the results of N-body simulations to re-calibrate the nonlinear power spectrum fitting formula HALOFIT (Smith *et al.*, 2003) in the case of ΛCHDM models with three degenerate neutrinos of arbitrary mass. Brandbyge *et al.* (2010) focused on the impact of massive neutrinos on smaller scales and computed the neutrino density profile in galactic halos, as well as the halo mass function (number of halos as a function of their mass) in the presence of massive neutrinos.

Warm dark matter particles

In principle, all previous remarks concerning the implementation of HDM particles in N-body simulations may apply to WDM particles. However, the latter case is simpler, because for phenomenologically interesting WDM masses and phase-space distributions, the velocity dispersion of particles at the beginning of a typical N-body simulation (i.e., for $10 < z < 100$) are lower than the Zel'dovitch velocities (Boyarsky *et al.*, 2009d). This means that thermal velocities can be safely neglected in most cases, and that WDM particles are equivalent to CDM particles with an initially different power spectrum, featuring a cutoff for $kr_{fs} > 1$, as explained in Section 6.1.5. However, to resolve the same scales, it is necessary to introduce a larger number of particles in the WDM case (see, e.g., Wang and White, 2007; Lovell *et al.*, 2012). Also, simulations of mixed ΛCWDM models can be carried out with a single type of particles, initially following the linear power spectrum of the cold plus warm components.

Several recent simulations of ΛWDM and ΛCWDM have been carried out in different contexts, for instance, on mildly nonlinear scales, in order to fit data from Lyman-α absorption in quasar spectra, or on strongly nonlinear scales for studying galaxy halo profiles and galaxy satellite abundances. See, e.g., Macciò and Fontanot, 2009; Dunstan *et al.*, 2011; Polisensky and Ricotti, 2011; Schneider *et al.*, 2011; Lovell *et al.*, 2012; Macciò *et al.*, 2012a,b; Viel *et al.*, 2012.

6.2.2 Analytic approaches

Several analytic approaches have been proposed to deal with the computation of the nonlinear power spectrum on mildly nonlinear scales, with the goal of avoiding tedious numerical simulations. Many (but not all) of these methods try to solve the nonlinear continuity and Euler equations of dust (i.e., of cold particles),

$$\delta' + \vec{\nabla} \cdot \left[(1 + \delta)\vec{v} \right] = 0$$
$$\vec{v}' + \mathcal{H}\vec{v} + (\vec{v} \cdot \vec{\nabla})\vec{v} = -\vec{\nabla}\phi, \tag{6.103}$$

where the gravitational potential ϕ is given by the Poisson equation. These equations are valid in the single-flow approximation: they neglect shell crossing between various flows of CDM particles, which inevitably appear when modes enter in the deeply nonlinear regime. Such techniques include one-loop calculations, renormalized perturbation theory, and renormalization group methods. Some of these methods have been extended in order to account for a neutrino component: Wong

(2008) and Saito *et al.* (2009) studied one-loop perturbation theory, whereas Lesgourgues *et al.* (2009) used the time renormalization group method. These works treat neutrinos as a linearly evolving species, coupled gravitationally to CDM and baryons, that are subject to the preceding nonlinear equations. The results correctly reproduce the departure from linear theory, although in a very limited range of scales (roughly until $k \simeq 0.2\,h/\text{Mpc}$ for $z > 2$, or until $k \simeq 0.15\,h/\text{Mpc}$ at $z = 1$). In the future, new approaches may lead to significant progress in this field.

6.3 Impact of neutrinos on secondary cosmic microwave background anisotropies

We have studied in Chapter 5 the impact of neutrinos on primary CMB anisotropy, defined as the contribution to CMB maps of physical phenomena taking place until the epoch of recombination. Secondary anisotropy, caused by the rescattering or the deflection of CMB photons in the recent universe, can also be sensitive to neutrinos – interestingly, when those are already nonrelativistic. In Chapter 5, we mentioned two sources of secondary anisotropy: the late integrated Sachs–Wolfe (LISW) effect, and the reionization of the universe due to star formation. Without giving a list of all sources of secondary anisotropy, we will briefly review in this section the two secondary effects currently identified as the most relevant for constraining neutrinos: the LISW effect, and CMB weak lensing.

6.3.1 Late integrated Sachs–Wolfe effect

We have seen in Section 5.1.6 that CMB temperature anisotropies pick up an integrated Sachs–Wolfe (ISW) contribution,

$$\frac{\delta T}{T}(\hat{n}) = \int_{\eta_{\text{LS}}}^{\eta_0} d\eta \, (\phi' + \psi'), \tag{6.104}$$

where the integral runs along each line of sight. We know from Section 6.1.1 that in the minimal ΛCDM model (without massive neutrinos) and deep inside the matter-dominated regime, the two Newtonian gauge metric fluctuations ϕ and ψ are equal to each other and independent of time. Hence the ISW contribution can be separated into an early and a late term. The second one reads

$$\left.\frac{\delta T}{T}(\hat{n})\right|_{\text{LISW}} = 2 \int_{\eta_{\text{M}}}^{\eta_0} d\eta \, \phi', \tag{6.105}$$

where η_{M} is some arbitrary time deep inside the matter-dominated regime. Inside the Hubble radius, the metric fluctuation ϕ can be related to the total density fluctuation using the Friedmann and Poisson equations:

$$\phi(\eta, k) = -\frac{3}{2}\frac{a^2 H^2}{k^2}\delta_{\mathrm{M}}(\eta, k). \tag{6.106}$$

Under the matter-dominated regime, the product $[a^2 H^2]$ evolves like η^{-2}, whereas matter perturbation grows linearly with the scale factor, i.e., like η^2. We see that indeed ϕ is constant, at least in a first approximation. When we approach Λ domination, $\bar{\rho}_{\Lambda}$ starts to play a role in the Friedmann equation, and both $[a^2 H^2]$ and δ_{M} start to evolve differently. Then a net integrated Sachs–Wolfe effect accumulates along each line of sight.

In the presence of massive neutrinos, we have seen that the solution for $\delta_{\mathrm{M}}(\eta, k)$ and hence for $\phi'(\eta, k)$ is not affected at large scales such that $k \ll k_{\mathrm{nr}}$. At smaller scales, $\delta_{\mathrm{M}}(\eta, k)$ grows at a lower rate, as $(a\, g(a; \Omega_{\mathrm{M}}))^{1-(3/5)f_\nu}$ instead of $a\, g(a; \Omega_{\mathrm{M}})$ (see Eq. (6.76)). Hence, ϕ experiences some extra damping during all stages: it decays like $a^{-(3/5)f_\nu}$ during matter domination, and like a more complicated function during Λ domination. For realistic neutrino masses, f_ν is much smaller than one, and this extra damping is very small, but because this effect accumulates over a long duration, it can be important at small redshift.

A subtle point is that neutrino mass has two opposite effects on the LISW contribution. On one hand, in presence of neutrino mass, metric fluctuations are slightly erased on small scales, because of neutrino free-streaming, and the LISW effect tends to be suppressed. On the other hand, ϕ decreases a bit more quickly because of neutrino masses, so in absolute value ϕ' is enhanced. The competition between these two effects has been described with numerical solutions and analytic approximations by Lesgourgues *et al.*, 2008.

The LISW contribution is difficult to observe, because CMB maps mix primary and secondary anisotropy. A change in the LISW contribution to the first few multipoles $l \leq 10$ can affect the slope of the plateau in the CMB spectrum significantly, but these are the scales on which cosmic variance is very large. For higher multipoles the LISW signal is overseeded by primary anisotropies. As explained earlier, the neutrino mass effect is expected to be maximal on small scales. Hence, in practice, the CMB temperature spectrum is insensitive to LISW-induced neutrino mass effects.

However, the LISW contribution can be disentangled from the primary anisotropies by cross-correlating CMB and large-scale structure maps. Various techniques offer the possibility of reconstructing the matter distribution in shells centered on us. For instance, by observing galaxies in all directions and in a given redshift bin (centered on a value z_i), one can infer the map of matter fluctuations in a shell of radius $r_i(z_i)$, modulo a bias factor. With such a map $\delta_{\mathrm{G}_i}(\hat{n})$, one can compute a harmonic power spectrum $C_l^{\mathrm{G}_i \mathrm{G}_i}$, using the same expansion and definitions as

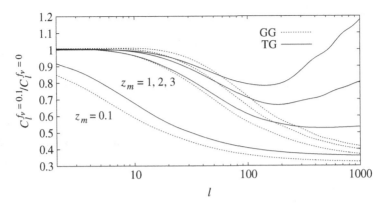

Figure 6.12 Matter autocorrelation spectrum $C_l^{G_iG_i}$ and matter–temperature cross-correlation spectrum $C_l^{TG_i}$ for a model with massive neutrinos, divided by the same quantitites for a model with massless neutrinos. Matter fluctuations have been evaluated in four shells centered on redshift $z_i \in \{0.1, 1, 2, 3\}$. The massive model has a HDM density fraction $f_\nu = 0.1$ due to three degenerate neutrino species with mass $m_\nu = 0.41$ eV each. As in Fig. 6.5, the massless and massive models share the same values of $(\Omega_M, \omega_M, \omega_B)$. (Reprinted with permission from Lesgourgues *et al.*, 2008. Copyright 2008 by the American Physical Society.)

for CMB maps (see Section 5.1.6). If the redshift bin is narrow, the map represents matter fluctuations in a thin shell, and the spectrum $C_l^{G_iG_i}$ is directly related to the matter power spectrum $P(z_i, k)$ with a one-to-one correspondence

$$k = \frac{a(\eta_i) l}{d_A(\eta_i)} \qquad (6.107)$$

in the small-angle limit.

The cross-correlation spectrum $C_l^{TG_i}$ of the temperature and density maps would vanish in the absence of secondary anisotropy; however, the LISW contribution depends on the same structures and gravitational potential distribution as the density map. Therefore, the cross correlation makes it possible to extract the LISW signal in a given redshift bin, given by Eq. (6.105), with the integral running only inside the bin. Because the LISW part depends on ϕ', which can be related to δ'_M through the Poisson equation, the cross-correlation spectrum $C_l^{TG_i}$ depends on both $P(z_i, k)$ and its derivative with respect to time or redshift.

Figure 6.12 illustrates the impact of neutrino masses on the autocorrelation and cross-correlation spectra $C_l^{G_iG_i}$ and $C_l^{TG_i}$ in four redshift bins centered on $z_i \in \{0.1, 1, 2, 3\}$. The autocorrelation spectrum is suppressed by neutrino masses at small angles (large l's) in the same way as the matter power spectrum. The scale of the step is constant in Fourier space (we know that it is given by k_{nr}). But in angular or harmonic space, this scale is seen under a larger angle (smaller l) at smaller redshift.

Because of these two opposite effects, the effect of neutrino masses on the cross-correlation spectrum is not so trivial. For $z_i = 0.1$, the cross-correlation spectrum follows almost the same steplike suppression as the autocorrelation spectrum. The step is, however, slightly balanced by the fact that in the massive case, time derivatives of metric fluctuations decrease faster. This second effect actually takes over on small angular scales and at high redshift. In this limit, the LISW effect is negligible with massless neutrinos (because metric fluctuations are static) but not with massive neutrinos (because metric fluctuations decay like $\phi \propto a^{-(3/5)f_\nu}$): neutrino masses enhance the cross-correlation spectrum. Lesgourgues *et al.* (2008) showed that for realistic experiments, these effects could be detectable in the future, but with such error bars that this method does not offer such a sensitive probe of neutrino masses as, e.g., galaxy or CMB weak lensing.

6.3.2 Cosmic microwave background lensing

Another source of secondary anisotropies comes from the weak gravitational lensing of the last scattering surface caused by large-scale structure (reviewed by Lewis and Challinor, 2006). The trajectories of CMB photons are slightly deflected by matter fluctuations localized at redshifts $z \leq 3$. At leading order in perturbations, CMB lensing can be described entirely in terms of a two-dimensional deflection field $\hat{d}(\hat{n})$. The deflection field represents the difference between the direction $\hat{n}$ in which photons have been emitted from the last scattering surface, and the direction $\hat{n} + \hat{d}(\hat{n})$ in which they are actually observed. It is given by the gradient of a lensing potential φ, related to the Newtonian metric perturbations ϕ and ψ through a convolution along the line of sight

$$\varphi(\hat{n}) = -\int_{\eta_{\mathrm{LS}}}^{\eta_0} d\eta \, \frac{\chi(\eta_{\mathrm{LS}}) - \chi(\eta)}{\chi(\eta)\chi(\eta_{\mathrm{LS}})} (\phi + \psi)_{(\eta, \vec{x}=r(\eta)\hat{n})}, \qquad (6.108)$$

where $\chi(\eta)$ is the comoving distance defined in Eq. (2.67) (in a flat universe $\chi(\eta) = r(\eta) = (\eta_0 - \eta)$). During matter domination, we can relate $\phi = \psi$ to matter density fluctuations through the Friedmann and Poisson equation (6.106). Therefore, the harmonic power spectrum $C_l^{\varphi\varphi}$ of a given map $\varphi(\hat{n})$ can be inferred from the matter power spectrum $P(z, k)$, convolved in redshift space with a given kernel. As expected, in the presence of neutrino masses, the lensing spectrum features steplike suppression on small scales, illustrated in Fig. 6.13.

It is not obvious how to infer the lensing power spectrum from observations, because we can observe CMB anisotropies only after lensing effects take place. The deflection field can, however, be extracted statistically, by studying the

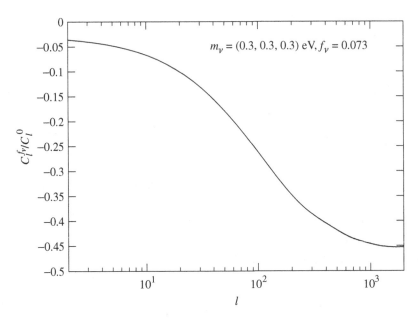

Figure 6.13 CMB lensing spectrum $C_l^{\varphi\varphi}$ for a model with massive neutrinos, divided by the same quantity for a model with massless neutrinos. The massive model has a hot dark matter density fraction $f_\nu = 0.073$ due to three degenerate neutrino species with mass $m_\nu = 0.3$ eV each. As in Fig. 6.5, the massless and massive models share the same values of $(\Omega_{\mathrm{M}}, \omega_{\mathrm{M}}, \omega_{\mathrm{B}})$.

nongaussianity and nontrivial correlations induced by weak lensing in temperature and polarization maps. Okamoto and Hu (2003) and Hirata and Seljak (2003) have discussed various estimators of the lensing power spectrum, which can be seen as nonlinear transformations of CMB maps that make it possible to estimate $C_l^{\varphi\varphi}$ with an error bar given by the spectrum of primary anisotropy and of instrumental noise. Other estimators can be used to reconstruct the lensing deflection map $\varphi(\hat{n})$ or the delensed CMB anisotropy maps. Current and future CMB experiments are able to detect the lensing power spectrum with a good signal-to-noise ratio. Thanks to this technique, the CMB offers an opportunity to probe the effect of neutrino masses on the growth of structure at small redshift $z \leq 3$, with a better sensitivity than through the previously discussed LISW effect.

6.4 Observing the large-scale structure

The matter power spectrum can be probed with various techniques on different scales and at different redshifts. In this section, we give a very brief summary of the methods used (or expected to be used in the future) for constraining neutrino parameters, without entering into any technical detail.

6.4.1 Galaxy and cluster power spectrum

Galaxy maps (or, similarly, cluster maps) can be smoothed over small scales and Fourier transformed in order to provide a power spectrum. Relating such a spectrum to the total matter power spectrum discussed in Sections 6.1 and 6.2 is a tricky exercise for several reasons:

- Galaxies are not perfect tracers of the total dark matter: there is no reason for the relative density perturbation in observed galaxies $\delta_G(\eta, \vec{x})$ to be equal to the total matter density perturbation $\delta_M(\eta, \vec{x})$. In Fourier space, the relation between galaxy clustering and total matter clustering can be parameterized with a light-to-mass bias function $b(\eta, k) \equiv \delta_G(\eta, k)/\delta_M(\eta, k)$. Fortunately, several arguments based on numerical simulations, analytical modeling of galaxy formation or comparisons between various data sets show that on linear scales, the bias is independent of k, so that we can relate the power spectrum of galaxies to that of total matter:

$$P_G(\eta, k) = b^2(\eta) P(\eta, k). \tag{6.109}$$

  However, the numerical value of the bias factor depends both on time and on the type of galaxies selected in a given data set, because this parameter is related to the way in which galaxies form inside gravitational potential wells. Different galaxy surveys may select different types of galaxies, with different median redshifts, corresponding to specific average bias values. When theoretical predictions are fitted to a given galaxy power spectrum $P_G(\eta, k)$, it is customary to leave the bias as a free parameter and to marginalize the posterior parameter probability over it. Hence, data on the galaxy power spectrum probe the shape of the matter power spectrum, but not its overall amplitude. Eventually, some prior on the bias can be inferred from higher-order statistics or from simulations.
- For each object, a given survey measures the sky coordinates (i.e., two angles (θ, ϕ)) and the redshift. Hence, actual galaxy maps are obtained in redshift space (θ, ϕ, z), rather than in spherical comoving coordinates (θ, ϕ, r). To measure the three-dimensional Fourier spectrum $P(k)$, it is necessary to convert redshifts into comoving distances. Assuming a fiducial cosmology, one can perform such a conversion using Eqs. (2.8), (2.67). These relations take into account the average Hubble flow, but not the peculiar velocities of individual objects, induced by coherent motions of galaxies inside potential wells, that affect each redshift through the Doppler effect. Hence, this operation introduces an error on the power spectrum (or in other words, a k-dependent bias) known as the redshift-space distortion. This distortion remains negligible for scales in the linear regime,

corresponding to sufficiently small Fourier modes of the velocity field, and becomes important on nonlinear scales.

- We have seen in Section 6.2 that on small scales, it is not a trivial exercise to model nonlinear corrections to the linear power spectrum with sufficient accuracy.

In summary, fitting a theoretical linear power spectrum to some real data set is not too difficult on large (linear) scales provided that the linear bias is marginalized over, whereas on small (nonlinear) scales it is necessary to model nonlinear biasing, redshift-space distortions and nonlinear corrections. Moreover, on small scales, the nonlinear evolution alters the (a priori gaussian) statistics of primordial perturbations. In principle, this should also be taken into account in the expression for the data likelihood.

These complications can be avoided by limiting theoretical fits to very large scales, but then a lot of information contained in the data is not used. It is a major challenge to push the comparison to smaller scales without introducing systematic errors larger than the statistical errors of the data set. Swanson *et al.* (2010) illustrate how uncertainties in the modeling of nonlinear corrections, nonlinear biasing and redshift space distortions affect the sensitivity of galaxy surveys to neutrino mass measurements.

In the following, we will show bounds on the shape of $P(z, k)$ derived from three recent data sets: the halo power spectrum of the Large Red Galaxy (denoted later as Gal-LRG), measured by Reid *et al.*, 2010a, the spectrum of the MegaZ catalogue (Gal-MegaZ), used by Thomas *et al.*, 2010 and the WiggleZ Dark Energy Survey (Gal-WiggleZ), used by Riemer-Sørensen *et al.*, 2012. The first two data sets are actually extracted from the same big survey, the Sloan Digital Sky Survey (SDSS). Technical details on the different methods and underlying assumptions can be found in these references.

For sufficiently deep galaxy surveys, it is possible to separate galaxies into redshift bins and compute different correlation functions at different redshifts. This technique, called tomography, can be very useful for constraining the scale-dependent growth factor induced by neutrino masses. In that case, the data can be reduced to a set of two-dimension power spectra in different shells, instead of a three-dimensional power spectrum with a given median redshift. For a given shell, corresponding to a given redshift bin, the density in each direction $\delta_G(\hat{n})$ can be expanded in spherical harmonics. It is then possible to estimate the harmonic power spectrum C_l^{GG} in each redshift bin, using the same definition as for the CMB power spectrum (see Section 5.1.6). We already saw such a quantity in Section 6.3.1. Recently, such a tomographic analysis was used by Xia *et al.*, 2012 for constraining neutrinos, using galaxies from the Canada–France–Hawaii-Telescope Legacy Survey (CFHTLS), split into three redshift bins covering the ranges

0.5 < z < 0.6, 0.6 < z < 0.8 and 0.8 < z < 1.0 (this data set will be denoted as Gal-CFHTLS).

Several galaxy surveys with better sensitivity and larger volume are about to release data or have been planned over the next decades, including the Baryon Oscillation Spectroscopic Survey[4] (BOSS), the Dark Energy Survey[5] (DES), the Large Synoptic Survey Telescope[6] (LSST) and the Euclid satellite.[7] Also, Wang *et al.*, 2005 pointed out that in the future, accurate measurements could be inferred from cluster surveys. Because clusters are more luminous than galaxies, they can be mapped up to higher redshift. One advantage of gathering high-redshift data is that they can be used to probe the same comoving scale at a time when perturbations are closer to the linear regime.

6.4.2 Cluster mass function

Instead of probing the matter power spectrum $P(z, k)$ directly from the spatial distribution of objects, it is possible to constrain integrated quantities of the type $\int dk\, P(k)W(k)$, where $W(k)$ stands for a given window function. One such quantity is related to the histogram of cluster masses. If the mass of a significant number of galaxy clusters within a given redshift bin is known, this histogram gives an estimate of the so-called *cluster mass function*, $dn(M, z)/dM$, with dn being the number of clusters of redshift approximately equal to z, and with a mass in the range $[M, M + dM]$. This function is related to $\sigma^2(M, z)$, the variance of the density in spheres enclosing a mass M, itself derived from the convolution of the power spectrum $P(z, k)$ with an appropriate window function. Although one should in principle compute $dn(M, z)/dM$ for each theoretical model and compare it to the raw data, it is customary to reduce observational constraints to a single bound on a combination of the two cosmological parameters to which this method is mostly sensitive. The first parameter, called σ_8, is related to the amplitude of the linear power spectrum today on typical cluster scales. Its square gives the variance of the density in spheres of radius $R = 8\,h^{-1}\mathrm{Mpc}$. This parameter is obtained by convolving $P(z, k)$ with a given window function at $z = 0$. The second parameter is Ω_M: it specifies both the fraction of the total energy density in the universe in the form of clustering matter, and the linear growth factor $g(z; \Omega_M)$ (neglecting massive neutrino effects). Approximate constraints from cluster abundances take the form of a measurement of $\sigma_8 \Omega_M^\alpha$, where α is an exponent depending on the details of the catalogue and of the method (in particular, on how the clusters have been selected).

---

[4] cosmology.lbl.gov/BOSS/.   [5] www.darkenergysurvey.org/.
[6] www.lsst.org/.   [7] sci.esa.int/euclid.

In the following we will refer to bounds derived from cluster abundances probed by X-ray observations from the ROSAT survey, presented by Mantz *et al.*, 2010 (denoted later as Clus-ROSAT), and by optical observations from the MaxBCG catalogue, presented by Reid *et al.*, 2010b (denoted later as Clus-MaxBCG). Cluster abundances are also probed by weak lensing observations, but current bounds on the total neutrino mass from this method do not compete with the ones mentioned (Kristiansen *et al.*, 2007).

6.4.3 Galaxy weak lensing

The images of observed galaxies are distorted by gravitational lensing effects, caused by density fluctuations along the line of sight. One of these effects is called cosmic shear. It corresponds to the squeezing of an image in one direction in the sky, and its stretching in the orthogonal direction. Because such distortions are coherent over the angular size of the lensing potential wells responsible for lensing, they tend to align the apparent major axis of galaxies slightly in a given patch of the sky. Hence the average cosmic shear in a given direction can be estimated statistically by averaging over major axis orientations.

Let us consider a catalogue of galaxy images, and assume that the number density of galaxies per redshift $g(z)$ is roughly isotropic. Assuming a fiducial cosmology, one can infer from $g(z)$ the density of galaxies with respect to comoving radius, $g(r)$. The galaxy catalogue can be divided into small solid angles or pixels. Because the intrinsic orientation of apparent galaxy major axes is randomly distributed (at least to a good approximation), the average orientation in each pixel gives an estimate of the average cosmic shear in the direction of the pixel, up to a shot noise term which decreases when the number of galaxies per pixel increases.

The distortion of a source galaxy sitting at coordinates (θ, ϕ, r_s), corresponding to the direction of observation $\hat{n} = (\theta, \phi)$, is related to the metric fluctuations ϕ and ψ integrated along the line of sight and weighted by a given kernel,

$$\varphi(\hat{n}; r_s) = - \int_{\eta_s}^{\eta_0} d\eta \, \frac{\chi(\eta_s) - \chi(\eta)}{\chi(\eta)\chi(\eta_s)} (\phi + \psi)_{(\eta, \vec{x}=r(\eta)\hat{n})}, \qquad (6.110)$$

where $\chi(\eta)$ and $r(\eta)$ are respectively the comoving distance and radial coordinate of an object seen at time η. This expression for the lensing potential φ is identical to the one presented in Section 6.3.2 for CMB lensing, although the source is now located at an arbitrary point rather than on the last scattering surface. Because observations probe the metric fluctuations at sub-Hubble distances, we can identify ϕ and ψ with the gravitational potential. For an ensemble of galaxies distributed according to the selection function $g(r_s)$, the average observed cosmic shear is given by the average lensing potential

$$\varphi(\hat{n}; g) = \frac{\int_0^{r_s^{max}} dr_s \, g(r_s) \, \phi(\hat{n}; r_s)}{\int_0^{r_s^{max}} dr_s \, g(r_s)}. \tag{6.111}$$

The analysis of a catalogue of images leads to a map of the lensing potential $\varphi(\hat{n}; g)$, which can be expanded in spherical harmonics. It is then possible to measure the harmonic lensing power spectrum $C_l^{\varphi\varphi}$, using the same definition as for the CMB power spectrum (see Section 5.1.6), and to compare it with theoretical predictions. After equation (6.110) is expanded in spherical harmonics, it is easy to show that the theoretical lensing spectrum $C_l^{\varphi\varphi}$ is given by the convolution of $P(z, k)$ with an appropriate kernel. Typically, $C_l^{\varphi\varphi}$ depends on the average matter power spectrum $P(z, k)$ over a redshift range $0 \le z \le z_e$, where z_e is the redshift of an object located at half the distance to the source (i.e., at the Einstein radius), and over a wavenumber range corresponding to the size of objects located at the same redshifts and seen under an angle $\theta = \pi/l$. Hence, the small-scale suppression of $P(z, k)$ as a function of k due to neutrino masses is visible in the lensing spectrum $C_l^{\varphi\varphi}$ in the large l limit. However, the step in $P(k)$ (or the spoon shape of Fig. 6.11, if we take into account nonlinear corrections) is smoothed by the various integrals leading to $C_l^{\varphi\varphi}$. Still, the advantage of cosmic shear observations is that they are directly related to the density power spectrum $P(k)$: they make it possible to measure cosmological parameters without any assumption on the light-to-mass bias function $b(k)$.

If the number of source galaxies is sufficient, it is possible to split the catalogue in several redshift bins i, described by selection functions $g_i(r)$ such that $g(r) = \sum_i g_i(r)$, and to measure the lensing spectrum $C_l^{\varphi_i \varphi_i}$ in each redshift bin. Because sources in each bin are sensitive to the gravitational potential integrated up to a different comoving radius r and a different conformal times $\eta = \eta_0 - r$, this tomographic approach makes possible a three-dimensional reconstruction of the gravitational potential in our past-line cone and is sensitive to the variation of the power spectrum with redshift. As for galaxy redshift surveys, cosmic shear tomography is particularly useful for measuring neutrino masses, because it can probe the scale-dependent growth factor induced by neutrino masses, over an extended range of redshifts.

Current cosmic shear surveys make it possible to put bounds on neutrino parameters: we will refer later to results from the CFHTLS presented by Tereno et al., 2009 and denoted as WL-CFHTLS. Spectacular improvements are expected from future cosmic shear surveys like Pan-STARRS,[8] or the DES, LSST and the Euclid surveys already mentioned in Section 6.4.1.

---

[8] pan-starrs.ifa.hawaii.edu/.

6.4.4 *Cosmic microwave background lensing*

We have already described CMB lensing in Section 5.1.6. This technique leads to another lensing spectrum $C_l^{\varphi\varphi}$ which is complementary to the one inferred from tomographic weak lensing surveys. This spectrum can be seen as coming from an ultimate redshift bin corresponding to sources with a radial coordinate $r_s = r_{LS}$, lensed mainly by structures up to $z \sim 3$. The CMB lensing spectrum in a small region of the sky has been estimated by the South Pole Telescope[9] (SPT), but with too large error bars to give interesting constraints on neutrinos. In the next section, we will mention the expected sensitivity of CMB lensing observations by Planck[10] and several ground-based instruments.

6.4.5 *Lyman alpha forests*

The most luminous and distant compact objects that we can observe are quasars. Some of the photons emitted by quasars interact along the line of sight. In particular, a fraction of photons are absorbed at the Lyman alpha wavelength by hydrogen atoms located in the interstellar galactic medium (IGM). The absorbed fraction in a given point of the photon trajectory is proportional to the local density of neutral hydrogen. Because photons are continuously redshifted, absorption at a given point is seen by the observer as a depletion of the spectrum at a given frequency. Hence, inside a limited range called the *Lyman alpha forest*, the frequency dependence of quasar spectra is a tracer of the spatial fluctuations of the hydrogen density along the line of sight. Lyman alpha forests in quasar spectra offer an opportunity to reconstruct the hydrogen density fluctuation along several lines of sight in a given redshift range. After Fourier expanding each spectrum and averaging over many spectra, one gets an estimate of the so-called flux power spectrum $P_F(z, k)$.

If this spectrum were probing only linear scales, we could assume that hydrogen fluctuations are equal to total baryon fluctuations, which in turn are equal to total matter fluctuations after the baryon drag epoch. In that case, the flux power spectrum would be directly related to the linear one-dimensional matter power spectrum $P_{1D}(k)$, which is related to the usual three-dimensional $P(k)$ through a simple integral. Unfortunately, the flux power spectrum does not probe linear scales, but mildly nonlinear scales. In order to relate $P_F(z, k)$ to $P(k)$, it is necessary to perform N-body simulations with a hydrodynamic treatment of baryons, accounting for the complicated thermodynamic evolution of the IGM (which depends on star formation). Also, a limitation of this technique comes from the fact that the emitted quasar spectra already have a nontrivial frequency dependence, and that photons

---

[9] `pole.uchicago.edu/`.   [10] `www.rssd.esa.int/planck/`.

are affected by several other effects than Lyman alpha absorption along the line of sight. Nevertheless, a careful modeling of all relevant effects makes it possible to obtain interesting constraints. The fact that Lyman alpha forests probe mildly nonlinear scales rather than strongly nonlinear ones is of course crucial for keeping systematic errors under control. Lyman alpha observations typically constrain the matter power spectrum in the wavenumber range $0.3 < k < 3\,h$/Mpc and in the redshift range $2 < z < 5$.

In the next section we will mention neutrino bounds inferred from quasar spectra obtained by the SDSS and presented in Viel *et al.*, 2010, denoted as Lyα-SDSS.

6.4.6 21-cm surveys

Instead of mapping the distribution of hydrogen atoms through the absorption rate of photons traveling from quasars, it should be possible to observe directly the photons emitted by these atoms at a wavelength $\lambda = 21$ cm, when they flip from one hyperfine level to the other. While travelling towards the observer, these photons are redshifted, and they are seen with a wavelength indicating the position of the emitting atoms in redshift space.

Recent theoretical progress in this field shows that using this technique, future dedicated experiments should be able to map hydrogen and hence baryonic fluctuations at very high redshift (typically $6 < z < 12$) and to probe the matter power spectrum deep inside the matter-dominated regime on linear scales (Pritchard and Loeb, 2011). This field is still in its infancy, and the forecasts presented in the next section have to be taken with care, because of the difficulty of making a realistic estimate of systematic errors in future data sets.

6.5 Large-scale structure bounds on neutrino properties

6.5.1 Active neutrino masses

Current bounds

Using LSS observations in combination with CMB data offers an opportunity to observe (or to bound) the steplike suppression of the matter power spectrum in the presence of neutrino masses, explained in Section 6.1.4 and illustrated in Fig. 6.5. The use of CMB data is crucial to constrain parameters such as the baryon density, the primordial spectrum amplitude, the tilt and a combination of ω_M and h. Without such constraints, there would be too much freedom in the matter power spectrum fitted to LSS data for identification of a smooth steplike suppression.

Current data sets are far from reaching the sensitivity required to probe the mass splitting of the total mass $M_\nu = \sum_i m_{\nu_i}$ between different species, whose effect is

Table 6.1 *95% CL upper bounds on the total neutrino mass M_ν in eV, for various combinations of CMB, homogeneous cosmology and LSS data sets*

Data	Reference	$w = -1$	$w \neq -1$
WMAP7+Gal-LRG+H_0	(Komatsu *et al.*, 2011)	0.44	0.76
WMAP5+Gal-MegaZ+BAO+SNIa	(Thomas *et al.*, 2010)	0.325	0.491
WMAP5+Gal-MegaZ+BAO+SNIa+H_0	(Thomas *et al.*, 2010)	0.281	0.471
WMAP7+Gal-WiggleZ	(Riemer-Sørensen *et al.*, 2012)	0.60	–
WMAP7+Gal-WiggleZ+BAO+H_0	(Riemer-Sørensen *et al.*, 2012)	0.29	–
WMAP7+Gal-CFHTLS	(Xia *et al.*, 2012)	0.64(0.44)	–
WMAP7+Gal-CFHTLS+H_0	(Xia *et al.*, 2012)	0.41(0.29)	–
WMAP5+BAO+SNIa+Clus-ROSAT	(Mantz *et al.*, 2010)	0.33	0.43
WMAP5+H_0+Clus-MaxBCG	(Reid *et al.*, 2010b)	0.40	0.47
WMAP5+BAO+SNIa+WL-CFHTLS	(Tereno *et al.*, 2009)	0.53	–

Note: The first seven lines refer to galaxy power spectrum measurements, the next two lines to cluster mass function measurements and the last line to a weak lensing survey. The acronyms of CMB and homogeneous cosmology data are introduced in Section 5.4.1, those of LSS data in Sections 6.4.1, 6.4.2 and 6.4.3. Bounds in the first column assume a minimal ΛCDM model with massive neutrinos (seven parameters). In the last column, the cosmological constant was replaced with a dark energy component with arbitrary equation-of-state parameter w (eight parameters).

described in Section 6.1.4. The constraints mentioned in the following have been derived in the case of three degenerate neutrinos with mass $m_\nu = M_\nu/3$, but they roughly apply to the total mass of any scenario. Also, the bounds shown in the first column of Table 6.1 have been obtained under the assumption of a minimal ΛCDM model with massive neutrinos, featuring seven free parameters. More conservative bounds are sometimes derived for basic extensions of this model, with one or two more parameters. The constraints do not change significantly on assuming, for instance, a primordial spectrum with a running of the tilt $[d \ln n_s/d \ln k]$ or a significant contribution to the CMB from primordial gravitationnal waves (Reid *et al.*, 2010b). Parameters known to be slightly degenerate with neutrino masses and leading to weaker bounds are w, the equation-of-state parameter of a dark energy component (substituting the cosmological constant), and $N_{\rm eff}$. The degeneracy with w, explained in Hannestad, 2005a, is illustrated by the last column in Table 6.1. The degeneracy with $N_{\rm eff}$, explained in Hannestad and Raffelt, 2004 or Crotty *et al.*, 2004, will be discussed in Subsection 6.5.2.

We summarize in Table 6.1 the main constraints on M_ν at the time of writing, obtained from combinations of CMB plus homogeneous cosmology data (see Section 5.4.1), galaxy power spectrum data (see Section 6.4.1) and cluster abundance data (see Section 6.4.2). In the case of the Gal-WiggleZ, Gal-CFHTLS and WL-CFHTLS data, bounds were presented for only the seven-parameter model. For the

Gal-CFHTLS case, the first bounds correspond to a conservative analysis limited to scales for which nonlinear corrections are small and well under control, whereas the bounds in parentheses rely on including slightly smaller scales.

Including Lyman alpha data is a delicate issue. It requires extensive N-body simulations with a hydrodynamic treatment of baryons and of the thermodynamics of the IGM (see Section 6.2), as well as extra HDM particles (see Section 6.4.5). On has to keep under control systematic errors coming both from uncertainties on the flux power spectrum, and from the fact that simulations can be performed for only a limited number of models and must be extrapolated to other models following some particular scheme. A few bounds in the literature were derived without including HDM particles in simulations, or with data sets probably affected by large systematic errors (because these data indicated a global power spectrum amplitude significantly larger than the one derived from CMB data). A conservative analysis based on Ly-α-SDSS data was presented by Viel *et al.*, 2010. In this case, neutrinos were included as HDM particles in N-body simulations, and the data were marginalized over a global normalization factor. This analysis gives a bound $M_\nu < 0.9$ eV (95% CL) from Ly-α-SDSS data alone.

In conclusion, the combination of available data sets to date indicates that the total neutrino mass is below 0.3 eV at the 95%CL (0.5 eV if we allow for dark energy with arbitrary w). This means that the degenerate scenario in which all neutrinos share roughly the same mass is almost excluded. The data are about to probe the region in which masses are different from each other, and are ranked according to the NH or IH scenario (see Section 1.4.1). Other recent summaries of existing bounds have recently been presented (González-García *et al.*, 2010; Hannestad, 2010; Reid *et al.*, 2010b; Abazajian *et al.*, 2011; Komatsu *et al.*, 2011; Wong, 2011).

Future bounds

In the very near future, significant improvements in the neutrino mass bounds will be triggered by the Planck CMB satellite and the Baryon Oscillation Spectroscopic Survey (BOSS). The forecasts presented in Perotto *et al.*, 2006 predict a neutrino mass sensitivity of $\sigma(M_\nu) \sim 0.1$ eV from Planck alone, using the lensing extraction technique of Okamoto and Hu, 2003. This would be twice as good as without lensing extraction. Sekiguchi *et al.* (2010) find that the combination of Planck (with lensing extraction) with BAO-scale information from BOSS could lower the error to $\sigma(M_\nu) \sim 0.06$ eV. Gratton *et al.* (2008) find that adding Lyman alpha data from BOSS should lead to comparable sensitivities, and even better results might be expected from the addition of galaxy power spectrum data from the same survey.

With better tomographic data (for either galaxy clustering or cosmic shear), it will become possible to probe the scale dependence of the growth factor induced by neutrino masses (or in other words, the fact that the steplike suppression has an amplitude increasing with time, as illustrated in Fig. 6.10) and to reach spectacular sensitivites. We present below the typical sensitivity expected for a collection of planned surveys (not all approved). These numbers should be taken with care, because forecasts are based on an idealization of each experiment, as well as on several assumptions such as the underlying cosmological model, or even the fiducial value of the neutrino mass itself.

Lahav *et al.* (2009) find that the measurement of the galactic harmonic power spectrum in seven redshift bins by the Dark Energy Survey (DES) should lead to a sensitivity of $\sigma(M_\nu) \sim 0.06$ eV when combined with Planck data (without lensing extraction). Similar bounds were found by Namikawa *et al.*, 2010 for another combination of comparable experiments. This shows that at the horizon of 2014 or 2015, a total neutrino mass close to $M_\nu \simeq 0.1$ eV could be marginally detected at the 2σ level by cosmological observations. Because this value coincides with the lowest possible total mass in the inverted hierarchy scenario, the latter could start to be marginally ruled out in case the data still prefer $M_\nu = 0$.

Kitching *et al.* (2008) find that the sensitivity of cosmic shear data from a satellite experiment comparable to Euclid would shrink to $\sigma(M_\nu) \sim 0.03$ eV in combination with Planck (without lensing extraction). The forecast of Carbone *et al.*, 2011, based on galaxy clustering data also from Euclid (completed at small redhsift by similar data from BOSS), gives comparable numbers. Constraints based on the ground-based Large Synoptic Survey Telescope should be slightly weaker (Hannestad *et al.*, 2006). Hence, in the early 2020s, we expect that a combination of cosmological data sets could detect the total neutrino mass of the normal hierachy scenario, $M_\nu \simeq 0.05$ eV, at the 2σ level. If the total mass is instead close to $M_\nu \simeq 0.1$ eV, it will be detected at the 4σ level. However, in that case, available experiments will not have enough sensitivity to make the difference between an inverted and a normal hierarchy scenario with the same M_ν: the mass-splitting effects illustrated in Figure 6.9 are too small to be detected by such surveys (Lesgourgues *et al.*, 2004).

Even more progress could be provided by 21-cm surveys (see Section 6.4.6). It is difficult to make reliable predictions for this technique, given the number of unanswered questions needing to be addressed first. Pritchard and Pierpaoli (2008) find a sensitivity of $\sigma(M_\nu) \sim 0.075$ eV for the combination of Planck with the Square Kilometer Array (SKA) project, or $\sigma(M_\nu) \sim 0.0075$ eV with the Fast Fourier Transform Telescope (FFTT). However, they show that such impressive experiments would still fail in discriminating between the NH and IH scenarios (of course, such a question would still be around only if the detected total mass is not smaller than 0.1 eV).

An eventual post-Planck CMB satellite or post-Euclid survey would also have a great potential. Lesgourgues *et al.* (2006) find that a CMB satellite of the next generation could get $\sigma(M_\nu) \sim 0.03$ eV alone, thanks to a very precise reconstruction of the CMB lensing potential. Wang *et al.* (2005) discuss the potential of cluster surveys. Finally, Jimenez *et al.* (2010) show how far the characteristics of a hypothetical galaxy or cosmic shear survey should be pushed to discriminate between two allowed NH and IH scenarios with the same total mass.

6.5.2 Neutrino abundance and light sterile neutrinos

The relevance of LSS observations for measuring the effective neutrino number N_{eff} was stressed in Section 6.1.3. Let us recall that the matter power spectrum has a very small dependence on N_{eff} when this parameter is varied while $(\Omega_M, z_{\text{eq}}, \omega_B/\omega_C)$ are kept fixed: only the scale of BAOs is affected by such a transformation (see Fig. 6.2, solid versus dashed curve). However, both CMB and BAO data fix ω_B/ω_γ, not ω_B/ω_C. The matter power spectrum amplitude and slope do change on small scales when N_{eff} is varied while $(\Omega_M, z_{\text{eq}}, \omega_B)$ are kept fixed (see Fig. 6.2, solid versus dotted curve). Hence, a measurement of the matter power spectrum with a good resolution of the BAO scale and/or in combination with CMB data is sensitive to N_{eff}.

Bounds on N_{eff} for negligible neutrino masses

Let us first discuss the bounds on N_{eff} obtained under the assumption that all neutrinos (or other possible light relics) have negligible masses with respect to the sensitivity of current experiments (i.e., with respect to 0.1 eV). In that case, the minimal model to be fitted to the data has seven free parameters. For such a model, Reid *et al.* (2010b) find $N_{\text{eff}} = 4.16^{+0.76}_{-0.77}$ (68% CL) for WMAP7+H_0+Gal-LRG, and $N_{\text{eff}} = 3.77^{+0.67}_{-0.76}$ (68% CL) for WMAP5+H_0+Clus-MaxBCG (for the meaning of these acronyms, see Sections 5.4.1, 6.4.1 and 6.4.2). Using the most recent galaxy power spectrum data, Xia *et al.* (2012) find $N_{\text{eff}} = 3.98^{+1.04}_{-0.51}$ (68% CL) for WMAP7+H_0+Gal-CFHTLS.

By comparing these numbers with those presented in Section 5.4.2 for various combinations of CMB and homogeneous cosmology data sets, we see that current LSS experiments do not play an important role in bounding N_{eff}. The best complementary data sets to WMAP are still small-scale CMB data (from SPT and ACT), Hubble parameter data (referred as H_0 or OHD in Section 5.4.2), or BAO-scale data. Hence we refer the reader to Section 5.4.2 for a detailed discussion of current bounds on N_{eff}, on how much they depend on the underlying model, and on their possible implications.

In the future, the sensitivity of LSS experiments to N_{eff} will increase significantly. Perotto *et al.* (2006) find that using lensing extraction, Planck can lower the error bar from $\sigma(N_{\mathrm{eff}}) \sim 0.5$ to $\sigma(N_{\mathrm{eff}}) \sim 0.3$. Kitching *et al.* (2008) find that the combination of cosmic shear data from a satellite experiment comparable to Euclid would give $\sigma(N_{\mathrm{eff}}) \sim 0.1$ when combined with Planck. The forecast of Carbone *et al.* (2011) based on galaxy clustering data also from Euclid (completed at small redshift with similar data from BOSS and combined again with Planck) gives $\sigma(N_{\mathrm{eff}}) \sim 0.09$. Such a sensitivity would still not be sufficient to test the details of neutrino decoupling in the standard scenario, leading to $N_{\mathrm{eff}} = 3.046$, but it could exclude at high significance the assumption of extra light relics, unless their density is very much suppressed with respect to that of active neutrinos.

Bounds on N_{eff} in the presence of active/sterile neutrino masses

If we allow simultaneously for extra light degrees of freedom and small masses (for active neutrinos and/or for these extra relics), observational bounds become rather model-dependent. Indeed, the effect of the corresponding parameters can be degenerate, to a limited extent (Crotty *et al.*, 2004; Hannestad and Raffelt, 2004). Furthermore, mass-splitting issues can become relevant in this case. We saw in Section 6.5 that in a minimal scenario with $N_{\mathrm{eff}} \simeq 3$ and a given total mass $M_\nu \geq 0.1$ eV, the difference between NH and IH is almost impossible to detect. However, in a scenario with $N_\nu > 3$, there are a priori no oscillation data constraining the difference between active and sterile neutrino masses, so we could equally well assume that the total mass resides mainly in one sterile species, or is split equally between all species. With high enough values of N_{eff} and M_ν, these two assumptions lead to significant differences in the matter power spectrum (qualitatively similar to those illustrated in Fig. 6.8), and each of them should be studied separately.

For instance, Hamann *et al.* (2010) investigated several cases, including that of three massless active neutrinos plus N_s degenerate neutrinos of mass m_s each. For the data combination WMAP7 + ACBAR + BICEP + QuAD + H_0 + Gal-LRG, they find

$$3.046 < N_{\mathrm{eff}} < 6.15, \qquad m_s < 0.66\,\mathrm{eV} \quad (95\%\mathrm{C.L.}). \qquad (6.112)$$

When they assume instead three degenerate active neutrinos of mass m_ν each, plus N_s massless sterile neutrinos, they find for the same data

$$3.10 < N_{\mathrm{eff}} < 6.80, \qquad m_\nu < 0.42\,\mathrm{eV} \quad (95\%\mathrm{C.L.}). \qquad (6.113)$$

Finally, Giusarma *et al.* (2011) assumed a third scenario with three degenerate active neutrinos, each with mass m_ν, plus N_s degenerate neutrinos of mass m_s each. For the data combination WMAP7+H_0+Gal-LRG they quote the bound

$$3.26 < N_{\text{eff}} < 7.66, \quad m_\nu < 0.36\,\text{eV}, \quad m_s < 0.70\,\text{eV} \quad (95\%\text{C.L.}). \quad (6.114)$$

The bounds for sterile neutrinos are larger because m_s can correspond to fewer than three species, whereas m_ν is always shared among three species. These bounds are already outdated (especially by SPT), but they illustrate the difference between various cases. For each of these cases, the correlations between N_{eff} and masses (illustrated by contour plots which can be found in these references) show that bounds on N_{eff} and on masses are both relaxed with respect to simpler models with either the masses fixed to zero, or N_{eff} fixed to 3.046. The weakening of the joint bounds can reach a factor of 2 in some cases.

Giusarma *et al.* (2012) showed that the bounds on such models are rather robust, even when more general cosmological models with extra parameters are assumed. Hamann *et al.* (2011) presented an update and a more detailed discussion of some of these cases.

Concerning future joint bounds on N_{eff} and M_ν, the combination of Euclid lensing data and Planck data would give $\sigma(M_\nu) \sim 0.14$ eV and $\sigma(N_{\text{eff}}) \sim 0.12$, assuming that all species are degenerate in mass (Debono *et al.*, 2010). Giusarma *et al.* (2011) present similar forecasts in a few other cases.

Properties of possible extra relativistic relics

In all these studies, extra relics are assumed to share the same Fermi–Dirac distribution and the same temperature as active neutrinos. In the case of sterile neutrinos, this is motivated by the assumption of a large mixing angle between active and sterile neutrinos. With such assumptions, the difference $N_{\text{eff}} - 3.046$ can be directly interpreted as the number of extra species. Note that the main motivation for light sterile neutrinos comes from anomalies in LSND, MiniBoone and nuclear reactor data (see Section 1.4.1). These anomalies provide very marginal evidence for sterile neutrinos with a mass of order eV. The cases $N_{\text{eff}} \simeq 4$ or 5 and $M_\nu \sim 1$ eV are already excluded by the bounds we have just reported (see Hamann *et al.*, 2011 for more details), not even speaking of BBN bounds discussed in Section 4.4.1. Therefore, possible eV-mass sterile neutrinos should have a suppressed nonthermal distribution (and/or a lower temperature than active neutrinos). However, this is unlikely, because they are expected at the same time to have a large mixing angle.

If extra relics are assumed to have a different temperature or to be nonthermally distributed, bounds on $(N_{\text{eff}} - 3.046)$ cannot be immediately interpreted as the number of extra species. We discussed different phase-space distributions in

Section 6.1.4, and considered two cases (a) and (b). In case (a), i.e., relics with a rescaled Fermi–Dirac distribution or just a different temperature, the previous upper bounds can be translated into bounds on the rescaling factor χ or on the temperature T_s. This can be done by equating $(\Delta N_{\mathrm{eff}}, \omega_s, k_{\mathrm{nr}})$ in the thermal and nonthermal cases, because these are the three quantities really probed by observations. This situation is discussed in the analysis of (Acero and Lesgourgues, 2009). In case (b), i.e., for any other phase-space distribution function, a specific analysis is needed for each particular model.

Finally, the previous bounds assume that extra massless or light relics are collisionless, and free-stream on small scales like ordinary neutrinos. Some nonstandard interaction may change the clustering properties of these relics without affecting their background evolution. Archidiacono *et al.* (2011) address this issue by introducing more freedom in the perturbation equations of massless relics. They promote two coefficients in Eq. (5.68) to the rank of free parameters. One of these coefficients is simply the sound speed c_{eff} relating pressure perturbations to density perturbations. The other relates the shear σ to the bulk velocity θ and is called the viscosity speed. Ordinary collisionless species are described by setting $(c_{\mathrm{eff}}^2, c_{\mathrm{vis}}^2) = (1/3, 1/3)$, whereas self-interacting particles could be described by lower values. The case $(c_{\mathrm{eff}}^2, c_{\mathrm{vis}}^2) = (1/3, 0)$ is that of a relativistic perfect fluid. In this limit the Boltzmann hierarchy reduces to a simple pair of equations, the continuity and Euler equations.[11] Archidiacono *et al.* (2011) fit such a model to the data combination WMAP7+ACT-2008+SPT+H_0+Gal-LRG and find that observations prefer a standard value of the sound speed, but leave the viscosity speed unconstrained. In other words, if extra ultrarelativistic relics are present, we do not know whether they are collisionless or not.

6.5.3 Nonstandard properties of active neutrinos

Let us now assume that the universe contains the three ordinary neutrino species, possibly with nonstandard properties, and no other massless or light relics.

The simplest deviation from the standard picture would consist of a large leptonic asymmetry leading to significant chemical potentials in the Fermi–Dirac distribution of each species. If neutrino masses are considered negligible given the sensitivity of the data, the effect of chemical potentials is entirely described by an enhancement of N_{eff}, which can be computed following Eq. (4.51). The previous

---

[11] Other works have addressed this problem with different parameterizations. Some of them are problematic because the standard case of collisionless species is not one particular point in the space of free model parameters. In particular, this happens when the Boltzmann hierarchy is truncated at $l = 2$ with a given value of c_{vis}, as in the imperfect fluid approximation (Hu, 1998). In that case, taking $c_{\mathrm{vis}}^2 = 1/3$ only provides a crude approximation to the actual behavior of collisionless species.

bounds on N_{eff} can be then translated into bounds on the chemical potentials, with no information on the splitting of the leptonic asymmetry between different species. If neutrino masses are taken into account, a specific analysis is required, as in Lesgourgues and Pastor, 1999, with in principle two free parameters per species (mass and chemical potential). Extra distortions are introduced by neutrino oscillations until the BBN epoch, as explained in Section 4.4.2. We do not report results for this case, because leptonic asymmetries are still constrained more strongly by a combination of BBN and oscillation data than by CMB and LSS observations, as discussed by Castorina *et al.*, 2012. Current constraints on these asymmetries are summarized by the results of Mangano *et al.*, 2012 described in Section 4.4.2. In the future, the increasing sensitivity of CMB and LSS data to neutrino masses and abundances will provide even better limits on the leptonic asymmetry.

Besides, the phase-space distribution of active neutrinos could be distorted, e.g., by the decay of a particle into neutrinos after neutrino decoupling (Cuoco *et al.*, 2005). It could also depart from Fermi–Dirac statistics under very unusual assumptions (Dolgov *et al.*, 2005). As long as neutrino masses can be neglected, these scenarios can be probed through their impact on N_{eff}. If masses are assumed to be significant, such cases also require specific studies. The range of possible nonstandard assumptions is so wide that we do not present explicit observational bounds in this book. Particular examples of such analyses can be found, e.g., in Cuoco *et al.*, 2005 or Hannestad *et al.*, 2005.

Neutrinos could also be coupled with other species. Many nonstandard particle physics models leading to neutrino interactions have been presented in the literature (see Section 4.4.3). Cosmological bounds on these models can be obtained case by case, or through some more or less generic parameterization. Hereafter we briefly summarize some of the approaches that have been followed so far.

One can simply assume that neutrinos experience self-interactions preventing them from streaming freely. In the massless neutrino limit, this assumption would simply lead to a reduction of the neutrino anisotropic stress. Trotta and Melchiorri (2005) and De Bernardis *et al.* (2008) proposed to parameterize this effect through a viscosity speed c_{vis} accounting for the relation between the neutrino anisotropic stress σ_ν and the bulk velocity θ_ν. The limit $c_{\mathrm{vis}} = 0$ is that of strongly self-interacting massless neutrinos, behaving as a perfect relativistic fluid. The case $c_{\mathrm{vis}}^2 = 1/3$ corresponds to an imperfect fluid whose evolution closely mimics (but not perfectly) that of ordinary collisionless neutrinos, as shown in Hu, 1998. Fitting such a model to data shows that the case of a perfect relativistic fluid is strongly disfavoured with respect to ordinary free-streaming neutrinos. However, this conclusion is mainly driven by CMB data. So it is not excluded that neutrinos free-stream normally until recombination, and experience later some self-interactions driving their effective viscosity speed to zero (Basbøll *et al.*, 2009).

Several authors studied the case in which some (or all) of the neutrinos are tightly coupled to a new scalar particle. In that case the perturbations of the whole neutrino–scalar fluid can be described with one continuity and one Euler equation, just like the photon–baryon fluid in the tightly coupled limit. This would be equivalent to the previous approach in the limit $c_{vis}^2 = 0$ if the neutrino and scalar particles were ultrarelativistic. These models have a richer phenomenology when scalar and neutrino masses are taken into account. For instance, most neutrinos could decay into bosons in the early universe, so that no effects from neutrino masses could ever be observed in LSS data (Beacom *et al.*, 2004). Less extreme cases have been investigated by Hannestad, 2005b; Bell *et al.*, 2006. Current constraints on these models are still very weak, and dominated by the fact that CMB observations confirm the presence of ultrarelativistic free-streaming neutrinos around the time of photon decoupling. A positive detection of neutrino mass effects in LSS data would prove that neutrinos free-stream until today and rule out most of these models. Models where neutrinos are coupled to dark matter with mass in the MeV range have been studied in Mangano *et al.*, 2006b; Serra *et al.*, 2010. In this case, too, the bounds on the typical parameter of these scenario, the interaction cross section, are quite loose.

Finally let us mention the mass-varying neutrino (MaVaN) scenarios. In these, neutrinos are coupled to a scalar field with a dynamical vacuum expectation value, in such a way that the effective neutrino mass would depend on the field value. In this case, the neutrino evolution remains standard until the time of the nonrelativistic transition. After that time, the coupling term becomes important and triggers a nontrivial evolution both in the scalar field and in the neutrino sector, so that the average neutrino mass varies with time. The effective mass can either increase or decrease in the recent universe, depending on the model. The linear perturbation equations of neutrinos and of the scalar field can be integrated using a dedicated Boltzmann code, which yields the corresponding CMB and LSS power spectra. If the coupling is not very small (i.e., if the neutrino mass variation is significant), perturbations grow very quickly on large scales, signalling an instability in the model. Assuming that this instability is incompatible with the data, França *et al.* (2009) derived some strong bounds from current CMB and LSS observations. Instead, Mota *et al.* (2008) pointed out that unstable perturbations might back-react on the background evolution without conflicting observations. This possibility is currently being investigated with nonlinear simulations (Ayaita *et al.*, 2012).

6.5.4 Heavy sterile neutrinos (warm dark matter)

We have seen in Section 6.1.5 that WDM particles becoming nonrelativistic during radiation domination are indistinguishable from CDM particles on large scales.

Both CMB observations and LSS data limited to linear scales are perfectly compatible with a ΛCDM model. They show no evidence for a depletion of the matter power spectrum on small scales that would be imprinted below the free-streaming horizon in the presence of WDM.

So any possible evidence for WDM must be searched for in LSS data on small scales. Although small-scale galaxy clustering or weak lensing data are in principle sensitive to WDM masses, they are available today at small redshifts for which the scale of nonlinearity is relatively large. This raises several difficulties. First, we know that nonlinear corrections, redshift space distortions and nonlinear biasing are difficult to compute on nonlinear scales. Second, the nonlinear evolution tends to mask the effect of WDM. Indeed, a transfer of power from larger to smaller scales tends to move the break in the matter power spectrum to smaller scales, and to make it smoother. In other words, if we go deeper inside the nonlinear regime, we must search for WDM signatures at even smaller scales. For these reasons, given the sensitivity of current data and the status of nonlinear simulations, the most stringent cosmological bounds on WDM are those derived from Lyman alpha forest data. Indeed, this technique probes mildly nonlinear scales at very high redshift ($2 < z < 4$). On Lyman alpha scales and at $z = 3$, the WDM-induced break could be clearly visible, whereas on the same scale and at $z \simeq 1$ it could have been essentially erased. Still, Lyman alpha data must be used in combination with CMB and eventually other LSS data sets to measure all of the parameters of the ΛWDM model. To fit theoretical WDM models to Lyman alpha data, performing specific hydrodynamical simulations with WDM particles is unavoidable. Implementing WDM particles in simulations is much simpler than for HDM particles, as explained in Section 6.2.1.

We presented an explicit form for the break in the matter power spectrum in Eq. (6.97). This formula corresponds either to a thermal WDM distribution or to a rescaled Fermi–Dirac distribution. We know that the latter case is a good approximation to scenarios in which heavy sterile neutrinos are populated by nonresonant oscillations with active neutrinos, the Dodelson–Widrow scenario. A fit to Lyman alpha and other cosmological data gives a lower bound on the parameter k_{break} of Eq. (6.97), i.e., on the ratio m_x / T_x. This bound can be formulated as a lower limit for the mass of DW sterile neutrinos m_{DW} by assuming $T_x \simeq T_\nu^a$ and $m_x = m_{\text{DW}}$. It can also be translated into a bound on the mass of a thermal WDM particle (see Eq. (6.94) and the following relations). Using a combination of WMAP5, extra small-scale CMB data, SNIa, H_0, Gal-LRG and Ly-α-SDSS data sets,[12] Boyarsky *et al.* (2009d) obtained a bound $m_{\text{DW}} > 12$ keV (95%CL).

---

[12] For the meaning of these acronyms, see Sections 5.4.1 and 6.4.

ΛCWDM models in which WDM particles (either thermal or DW) would coexist with CDM particles in the universe can be described with two parameters on top of the usual six ΛCDM parameters: the WDM fraction $f_w = \omega_w/(\omega_C + \omega_w)$ and the velocity dispersion parameter m_x/T_x (or directly the mass m_{DW} if $T_x \simeq T_\nu^a$ is assumed). As mentioned in Section 6.1.5, the power spectrum of such models is steplike suppressed (at the level of linear theory). The step amplitude and location are controlled by the two WDM parameters. The possible existence of a step is much less constrained by the data than that of a break. For the same combination of data sets, Boyarsky *et al.* (2009d) derived some joint bound in (f_w, m_{DW}) parameter space. For $f_w < 0.35$, it appears that any mass is compatible with the data. For larger fractions, the bound increases gradually and reaches the previous value when $f_w = 1$.

We mentioned in Section 6.1.5 that the DW model and thermal model are just two out of many possibilities for the WDM phase-space distribution. For instance, for sterile neutrinos populated by resonant oscillations, the phase-space distribution leads roughly to the same cosmological signature as in a mixed warm plus cold model. Using this similarity, Boyarsky *et al.* (2009c) showed that for resonantly produced sterile neutrinos the bound is much weaker than in the DW model, so that a mass of 2 keV is still well compatible with the data.

Lyman alpha bounds on sterile neutrino masses are complementary to various astrophysical bounds. First, the WDM characteristics can be probed by the small-scale structure of the universe: halo profiles, number of satellite galaxies, morphology of the galactic center, etc. These indications are, however, subject to huge uncertainties, and although several arguments tend to favour WDM (see, e.g., Lovell *et al.*, 2012), they must be taken with care. Second, if sterile neutrinos are too heavy, or if their mixing angle with active neutrinos is too large, a neutrino decay line should be clearly visible in galaxy halos using X-ray observations. More precisely, it is possible for a sterile neutrino to decay into a photon and an active neutrino via a 1-loop process, with branching ratio $27\alpha/(8\pi) \approx 1/128$. The radiative decay width (Pal and Wolfenstein, 1982) is

$$\Gamma(\nu_s \to \gamma\nu_a) = \frac{9\alpha}{2048\pi^4} G_F^2 \sin^2 2\theta \, m_s^5 \simeq \frac{1}{1.5 \times 10^{32} \text{ s}} \frac{\sin^2 2\theta}{10^{-10}} \left[\frac{m_s}{\text{keV}}\right]^5 \quad (6.115)$$

in the case of Dirac neutrinos (for Majorana neutrinos the rate gets an extra factor of two). Even if sterile neutrinos are required to be cosmologically stable, a very small fraction of them will decay following this channel. Because the radiative decay is a two-body process, the signal is a monochromatic flux of X-rays with energy $E_\gamma \simeq m_s/2$. Emitted photons may be observed by X-ray instruments such as the XMM-Newton, Chandra X-ray, and Suzaku observatories, especially in the direction of dark matter halos (Abazajian *et al.*, 2001; Dolgov and Hansen, 2002;

Feng, 2010; Boyarsky *et al.*, 2006). Null results exclude the upper right region of Fig. 4.7. Future X-ray observations may extend sensitivities to the entire range of parameters plotted in Fig. 4.7 (Abazajian, 2009; den Herder *et al.*, 2009).

Hence X-ray bounds on the sterile neutrino mass depend on the mixing angle. The relic density ω_{DW} of DW sterile neutrinos can also be inferred from the mass and mixing angle. Joint constraints from cosmological and X-ray data in (m_{DW}, ω_{DW}) parameter space are incompatible with each other: the DW scenario is now excluded with a good confidence level (see, e.g., Boyarsky *et al.*, 2008). In the case of resonant production, cosmological bounds are much weaker, and a large allowed window remains open (Boyarsky *et al.*, 2009c), as illustrated in Fig. 4.7.

7

Cosmological neutrinos today

Our journey in the land of neutrino cosmology is almost over. We have seen how neutrinos have silently influenced the evolution of the universe: how they perhaps may be responsible for the production of the baryon density we observe today, and the way they leave their signature in the nuclear abundances during primordial nucleosynthesis, in the CMB anisotropies and in structure formation.

We cannot leave the patient reader, who has kindly followed us till this point, without offering some final considerations about a simple question he or she might have thought about since the very beginning: can we *directly* detect the neutrino background in a laboratory experiment as Penzias and Wilson did for the CMB radiation?

There are two major obstacles one must overcome to achieve such a goal. Neutrinos are elusive particles because they interact only weakly. Cross sections are typically small, far smaller than electromagnetic ones, which were exploited by Penzias and Wilson. Furthermore, relic neutrinos today are quite cold particles with an average momentum on the order of $c\, p_\nu \sim 3.15\, T_{\nu,0} \sim 5 \times 10^{-4}$ eV, so reaction rates are further suppressed. For the best interaction process candidate to date, neutrino capture on β-unstable nuclei such as ^{3}H, we have $\langle \sigma v/c \rangle \sim 10^{-44}$ cm^2.

The task therefore seems quite challenging. Nevertheless, the goal is so ambitious that scientists have always continued to think about possible experimental strategies, even though they are usually declaring that all attempts are frustratingly desperate. But after all, even Pauli was mistaken about the possibility of detecting neutrinos!

In this chapter we present an overview of the main possible effects and interaction processes that have been considered to measure the cosmological neutrino background. As we will see, some are in fact completely unfeasible, whereas for others there might be some hope in a not too far future.

348

7.1 The ultimate dream: detecting cosmological neutrinos

7.1.1 Scatterings: G_F^2 effects are too small

The first obvious process one might think of is neutrino and antineutrino scatterings off nucleons or capture processes

$$\nu(\bar{\nu}) + N \rightarrow \nu(\bar{\nu}) + N \tag{7.1}$$

$$\nu(\bar{\nu}) + N \rightarrow N' + l^-(l^+), \tag{7.2}$$

with l a charged lepton. Because the CNB energy is exceedingly small, capture reactions are only possible if there is some available energy to create an electron or positron. We will see that this is the case with nuclei N which spontaneously decay by β–decay processes.

The scattering cross sections are well known – see Chapter 1 – and depend on the value of neutrino mass with respect to their present mean momentum $p_\nu \sim T_{\nu,0}$. For the relativistic case $m_\nu \ll T_{\nu,0}$ one has

$$\sigma_{\nu N} \sim \frac{G_F^2}{\pi} E_\nu^2 \sim 10^{-63} \left(\frac{T_{\nu,0}}{1.9\,\mathrm{K}} \right)^2 \mathrm{cm}^2, \tag{7.3}$$

whereas for nonrelativistic neutrinos,

$$\sigma_{\nu N} \sim \frac{G_F^2}{\pi} m_\nu^2 \sim 10^{-56} \left(\frac{m_\nu}{\mathrm{eV}} \right)^2 \mathrm{cm}^2. \tag{7.4}$$

Consider a macroscopic body exposed to the CNB flux. For each scattering event we expect a transferred linear momentum of order $\Delta p \sim T_{\nu,0}$. Because the neutrino background is homogeneous and isotropic (up to small perturbations, as for the CMB), if the Earth were at rest with respect to the CNB frame the net effect would be zero. However, the motion of the solar system in the comoving frame induces a dipole effect, so that the total acceleration a of our detector due to momentum tranfer does not vanish and is proportional to the Earth's velocity, $\beta_\oplus \sim 10^{-3}$. For 1 g of material with mass number A, containing $N = N_A/A$ target nuclei, with N_A the Avogadro number, one has

$$a = n_{\nu,0} \beta_\nu \frac{N_A}{A} \sigma_{\nu N}\, \Delta p, \tag{7.5}$$

with β_ν the average neutrino velocity, and we recall that in the standard case $n_{\nu,0} \simeq 56\,\mathrm{cm}^{-3}$ per neutrino or antineutrino flavour, barring any chemical potential in their distribution. Using (7.5) and either (7.3) or (7.4), we therefore obtain

$$a \simeq n_{\nu,0} \frac{N_A}{A} \frac{G_F^2}{\pi} \beta_\oplus \begin{cases} (3.15 T_{\nu,0})^3 - \text{relativistic} \\ m_\nu^2 (3.15 T_{\nu,0}) - \text{nonrelativistic}. \end{cases} \tag{7.6}$$

In both cases, the result is very small, $a \simeq (100/A)10^{-57}$ cm s^{-2} and $a \simeq (100/A)10^{-51}$ $(m_\nu/\mathrm{eV})^2$ cm s^{-2} for relativistic and nonrelativistic neutrinos today, respectively.

From what we have seen in the previous chapters, it is quite unlikely that $n_{\nu,0}$ might be much larger than its standard value, so these estimates can change by only one order of magnitude or so. Clustering might help. For order eV neutrino mass, the local neutrino density could be a factor of 10 higher than its average value (Ringwald and Wong, 2004).

On the other hand, it was noticed (Zeldovich and Khlopov, 1981; Shvartsman *et al.*, 1982; Smith and Lewin, 1983) that because relic neutrinos have a typical wavelength $\lambda_\nu \sim 2\pi\hbar/p_\nu \simeq 0.24$ cm, they can coherently interact with a large number of nucleons. This gives a factor of A^2, because the interactions are with all nucleons in a given nucleus, and a further and much more relevant enhancement due to the total number of nuclei contained in a volume of order λ_ν^3,

$$\frac{N_A}{A} \rho \lambda_\nu^3 \simeq \frac{10^{21}}{A} \left(\frac{\rho}{\mathrm{g\ cm}^{-3}} \right) \left(\frac{1.9\,\mathrm{K}}{T_{\nu,0}} \right)^3 , \tag{7.7}$$

with ρ the density of the target. In this case accelerations can be as high as $a \sim 10^{-34}$ cm s^{-2} for relativistic neutrinos, and even higher for the nonrelativistic case.[1] Regrettably, such accelerations are many orders of magnitude below the present experimental reach. The most sensitive devices for measurement of small accelerations are Cavendish torsion–balance experiments, which currently reach $a \simeq 10^{-12}$ cm s^{-2} (Adelberger *et al.*, 1991).

7.1.2 *The order G_F interactions and the Stodolsky effect*

We have already discussed the effect of matter on neutrino propagation in Section 1.3.5 by computing the effective potential, which at first order is proportional to target number density and linear in the Fermi constant G_F; see (1.118) and (1.120). We come back to this point to see if these effects might lead to a possible mechanism for CNB detection. Unfortunately, we will argue that only the so-called Stodolsky effect can in principle be used for this purpose, but it leads to very small signals, if not vanishing entirely.

Because the CNB neutrino wavelength λ_ν today is large compared with the typical interatomic spacing, we can describe neutrino propagation in a medium such as our target detector, by introducing a neutrino (antineutrino) index of refraction

---

[1] The effect of coherent interaction also depends on the Dirac or Majorana nature of neutrinos. In the Majorana case there is a further suppression of order β_ν^2 because there are no coherent vector Z couplings in the static limits, but only axial couplings (see, e.g., Duda *et al.*, 2001).

$n_{\nu,\bar\nu}$ as we did already in Chapter 1,

$$n_{\nu,\bar\nu} = \frac{p_{\text{matter}}}{p}, \tag{7.8}$$

with p_{matter} and p the linear momentum in matter and in vacuum, respectively, for a given energy $E = \sqrt{p^2 + m_\nu^2}$. Moreover, if λ_ν is smaller than the size of the detector, one can neglect diffraction and describe neutrino propagation as a ray, in analogy with standard geometrical optics.

To make a connection with our discussion in Section 1.3.5, consider first the case of relativistic neutrinos. If V^a is the effective potential experienced by, say, a neutrino due to some species a of particles, we have $p_{\text{matter}} = p - \sum_a V^a$, so that

$$n_\nu - 1 = -\frac{1}{p} \sum_a V^a. \tag{7.9}$$

On the other hand, for small values of $n_\nu - 1$, n_ν is defined in terms of the forward scattering amplitudes $f_\nu^a(0)$ for the process $\nu + a \to \nu + a$,

$$n_\nu - 1 = \frac{2\pi}{p^2} \sum_a f_\nu^a(0) n_a, \tag{7.10}$$

where n_a is the a particle number density. We computed these amplitudes in Chapter 1 for the particular case of a nonpolarized medium, finding the charged and neutral current potentials

$$V_{\text{CC}} = \sqrt{2}\, G_{\text{F}}\, n_e \tag{7.11}$$

$$V_{\text{NC}}^f = \sqrt{2}\, G_{\text{F}}\, n_f\, g_V^f. \tag{7.12}$$

For our present purposes we have to generalize this result in two ways: introduce the effect of neutrino mass, which is not negligible for CNB neutrinos, and allow for a possible nonzero polarization $\vec\sigma$ of the target material. This is quite straightforward and, proceeding as in Section 1.3.5, one finds

$$n_{\nu,\bar\nu} - 1 = \mp \frac{\sqrt{2} G_{\text{F}} \sum_a n_a \left(g_V^a - g_A^a \vec\sigma \cdot \hat{n}_p\right)}{p} \frac{E + m_\nu}{4p} \left(1 + \frac{p}{E + m_\nu}\right)^2, \tag{7.13}$$

with $\hat{n}_p = \vec{p}/p$ and the upper (lower) sign referring to neutrinos (antineutrinos). For CNB we recall that $p_\nu \sim 5 \times 10^{-4}$ eV and taking $n_a \simeq 10^{23}$ cm^{-3}, one obtains quite small values for the index of refraction:

$$n_{\nu,\bar\nu} - 1 \simeq \begin{cases} 10^{-10} - \text{relativistic} \\ 10^{-7}\, \frac{m_\nu}{\text{eV}} - \text{nonrelativistic}. \end{cases} \tag{7.14}$$

Due to the refractive bending of a neutrino ray passing through the target, or the total external reflection on a surface when neutrino incident momentum is at an angle smaller than the critical value $\theta_{cr} = \sqrt{2(1 - n_{\nu,\bar\nu})}$, some measurable energy-momentum or angular momentum could be transferred to the target. This would be a way to directly detect CNB. Unfortunately, this is not the case. Indeed, assuming that

- the target is static and
- the time average of the incident wave is spatially homogeneous over the size of the detector, as expected for the CNB flux,

it was shown in Cabibbo and Maiani, 1982; Langacker *et al.*, 1983 that the total momentum and energy transfer to the target vanishes at order G_F. Moreover, the net angular momentum transfer also vanishes to the same order, unless the target either is polarized or carries some current, and furthermore, the CNB contains a neutrino–antineutrino asymmetry.

Let us sketch the proof of this result in the case of energy–momentum transfer, following Langacker *et al.*, 1983.

The interaction between the neutrino field and the detector can be described by a current–current Lagrangian density,

$$\mathcal{L}_{int} = \overline{\nu_L}(x)\gamma^\mu \nu_L(x)J_\mu(x), \tag{7.15}$$

where $J_\mu(\vec{x})$ is the *static* classical source representing the detector. The total energy and momentum transfer averaged over some measurement time T is given by the time-averaged stress–energy tensor $T^{\mu\nu}$ integrated over a surface enclosing the whole detector $\Sigma = \partial V$,

$$-\frac{\Delta p^\mu}{T} = < \int_\Sigma dS\, n_i T^{i\mu} >_T, \tag{7.16}$$

where n^i are the components of the outward-pointing normal to Σ and $< >_T$ is time-averaging over T.

From (7.15), one obtains

$$\partial_\nu T^{\nu\mu} = \overline{\nu_L}(x)\gamma^\rho \nu_L(x)\partial^\mu J_\rho(x). \tag{7.17}$$

Because the source is static, the total energy is conserved and the time average of $\partial_0 T^{0\mu}$ vanishes. Using this we can therefore write

$$-\frac{\Delta p^\mu}{T} = < \int_\Sigma dS\, n_i T^{i\mu} >_T = < \int_V d^3x\, \partial_i T^{i\mu} >_T$$

$$= < \int_V d^3x\, \overline{\nu_L}(x)\gamma^\rho \nu_L(x)\partial^\mu J_\rho(x) >_T\, . \tag{7.18}$$

At first order in the neutrino–detector coupling, we can substitute the (interacting) neutrino fields as the free fields, because the current $J_\mu(x)$ is already at first order in

G_F. Furthermore, the source is static, so we can take it out of the time average, and we are left with the time average of the incident neutrino current. If we finally use the second assumption, namely that the time-averaged neutrino current is spatially homogeneous, we arrive at the following result:

$$-\frac{\Delta p^\mu}{T} = < \overline{\nu_L}(x)\gamma^\rho \nu_L(x) >_T \int_V d^3x\, \partial^\mu J_\rho(x). \tag{7.19}$$

From this we immediately see that energy transfer vanishes, because $\partial^0 J_\rho(x) = 0$. Similarly, one can prove that there is no momentum exchange, using Gauss's law and the fact that the current is zero on the boundary Σ enclosing the detector.

By using the same method of reasoning, it is possible to show that the variation of the time-averaged angular momentum M^i is

$$-\frac{\Delta M^i}{T} = \epsilon^{ijk} < \overline{\nu_L}\gamma^j \nu_L >_T \int d^3x\, J^k(x), \tag{7.20}$$

which is nonvanishing only if the detector carries some current, such as a net magnetization, and moreover the CNB has a neutrino–antineutrino asymmetry; otherwise the first term on the r.h.s. vanishes, because it counts the neutrino minus antineutrino flux.

This last possibility was already considered in Stodolsky, 1974. Indeed, it was argued in this paper that, if there is an asymmetry between neutrinos and antineutrinos, then the interaction of N polarized electrons in a ferromagnet with the neutrino sea induces a torque $\tau = N\Delta E/\pi$, where ΔE is the energy difference between the two electron helicity states in the direction of the CNB velocity.

We saw already that this velocity is due to the motion of the Earth in the CNB isotropic background, so we expect that $\Delta E \propto \beta_\oplus$. In the electron rest frame the neutrino current is $\vec{J}_\nu = -\vec{\beta}_\oplus(n_\nu - n_{\bar\nu})$, whereas the electron weak current reduces to $\vec{J}_e = g_A^l \vec{\sigma}_e$, with $\vec{\sigma}_e$ the electron spin. As an order of magnitude we therefore obtain

$$\Delta E \sim -\sqrt{2}G_F g_A^l \vec{\sigma}_e \cdot \vec{\beta}_\oplus(n_\nu - n_{\bar\nu}). \tag{7.21}$$

This result can be further refined (see, e.g., Duda *et al.*, 2001), and it depends on whether neutrinos are relativistic or nonrelativistic today. In particular, the energy shift is larger in the nonrelativistic limit by a factor of $1/\langle|\beta_\nu|\rangle$, the inverse of the average modulus of neutrino velocity in the CNB frame.

In both cases the energy shift is extremely tiny. We have seen in Chapter 4 that the constraints from Big Bang nucleosynthesis on electron neutrino asymmetry and neutrino flavour oscillations conspire to give quite a strong bound on all flavour asymmetries, $n_{\nu_\alpha} - n_{\bar\nu_\alpha} = \eta_{\nu_\alpha} n_\gamma$, with $|\eta_{\nu_\alpha}| \lesssim 0.1$. If we use these bounds, we find from our simple estimate (7.21) that

$$\Delta E \lesssim 10^{-39} \text{ eV}. \tag{7.22}$$

Correspondingly, the torque exerted on macroscopic bodies would result in accelerations which are extremely small and comparable with the G_F^2 *coherent* scattering effects described in the previous section; see, e.g., the review by Duda *et al.*, 2001. Only by a dramatic change in the technology of torsion balances or exploiting entirely new experimental approaches could CNB be found by the Stodolsky effect, provided of course the neutrino asymmetry is not too small or vanishing.

7.1.3 Massive neutrinos and β-decaying nuclei

Neutrino capture processes on nuclei,

$$\nu(\bar{\nu}) + N \rightarrow N' + l^-(l^+), \tag{7.23}$$

have been extensively used to study neutrino properties since their very first discovery by Reines and Cowan. From energy conservation, the process is kinematically allowed only if the incoming neutrino has enough energy to produce the charged lepton $l^\pm$ and, if $m_{N'} > m_N$, the mass difference between the two nuclei:

$$E_\nu \geq m_{N'} - m_N + m_l. \tag{7.24}$$

This condition is far more easily satisfied for the lightest charged lepton, the electron or positron. We will consider only this case in the following.

For MeV neutrinos or antineutrinos, such as those produced in the sun or by nuclear reactors, (7.24) is typically satisfied, but here we are interested in CNB neutrinos, whose energy is very small. Only reactions with a zero energy threshold can be used in this case, $m_N - m_{N'} - m_e \geq 0$, which is to say, nuclei N which can spontaneously undergo $\beta^\pm$ decay,

$$N \rightarrow N' + e^-(e^+) + \bar{\nu}_e(\nu_e), \tag{7.25}$$

or electron capture reactions,

$$N + e^- \rightarrow N' + \nu_e. \tag{7.26}$$

The idea of using these processes as a possible way to detect CNB was already put forward 50 years ago in Weinberg, 1962. At that time there were no indications in favour of massive neutrinos, so let us consider for the moment $m_\nu = 0$. Suppose there is some large negative neutrino–antineutrino asymmetry in the neutrino background, characterized by some (negative) value of the chemical-potential-to-temperature ratio ξ_{ν_e}. The antineutrino distribution is degenerate, and energy levels up to the Fermi energy $|\xi_{\nu_e}|T_{\nu,0}$ are filled up. If $Q = m_N - m_{N'}$ is the Q-value for a specific β^- decay process, because of the Pauli principle, the emission of an antineutrino and an electron with energy close to the endpoint $E_e = Q$ is suppressed. Therefore, the Kurie function defined in (1.154) would bend before Q and vanish at $E_e \simeq Q - |\xi_{\nu_e}|T_{\nu,0}$. At the same time, the spectrum of a β^+-decaying nucleus would show a bump above the endpoint at $E_e = Q + |\xi_{\nu_e}|T_{\nu,0}$. Of course, if ξ_{ν_e} is positive and there are more neutrinos than antineutrinos, the situation is the same, with simply the role of $\beta^\pm$ processes interchanged.

The effect is a clear signature of the CNB. The problem is, however, that the neutrino temperature is very low and the neutrino degeneracy even lower. In the most optimistic case, the shift in the E_e endpoint and the location of the (small) event bump are on the order of $Q \pm 10^{-4}$ eV, even if we assume $\xi_{\nu_e} \simeq 1$, which is already excluded by primordial nucleosynthesis constraints. An energy resolution in the emitted electron spectrum on this order of magnitude is well below the present value (eV) and also too small for any realistic improvement in the near future.

However, if background neutrinos are not degenerate, they are massive particles. Consider again a decaying nucleus, such as ^{3}H. On one hand, the β spectrum has an endpoint at $Q - m_\beta$ with m_β the effective mass,

$$m_\beta = \left(c_{12}^2\, c_{13}^2\, m_1^2 + s_{12}^2\, c_{13}^2\, m_2^2 + s_{13}^2\, m_3^2\right)^{1/2}. \tag{7.27}$$

On the other, CNB neutrino capture processes produce a signal at $E_e = Q + m_\beta$, where we have neglected the narrow distribution of neutrino momenta, whose spread is on the order of $T_{\nu,0}$. If the neutrino capture rate is not too low, and the energy resolution of the apparatus is on the order of m_β, the capture peak is not completely hidden by the β-decay spectrum tail, and the observation of both these features, displaced by $2m_\beta$ around the endpoint, would be again a direct way to detect the CNB.

The idea is basically the same that we saw in the case of a degenerate neutrino background. However, the advantage is that the energy resolution scale is now related to the neutrino mass scale, rather than the neutrino chemical potential (Irvine and Humphreys, 1983; Cocco *et al.*, 2007). We have seen that m_β could be as large as fractions of eV and that the Katrin experiment will test its value soon at this level of sensitivity (Osipowicz *et al.*, 2001). In this optimistic scenario, with an order 0.1 eV neutrino mass, it is plausible that the next generation of experiments might be able to reach an energy resolution good enough to look for CNB signals. For lower values of m_β, this experimental strategy seems much more challenging, probably impossible to implement in a sufficiently near future.

Let us now elaborate in more detail the two key points of this detection method, the expected signal in neutrino capture and the role played by energy resolution, following Cocco *et al.* (2007).

Neutrino capture is an order G_F^2 process, so the first question to answer is: Why should it be better than scattering, which we saw produces an immeasurably small effect? The reason is simply phase space. Consider neutrinos with very small momentum. Because the electron is emitted with energy $E_e \simeq Q$, the capture cross section $\sigma_{\nu,c}$ can be estimated – see also Chapter 1 – as

$$\sigma_{\nu,c} \sim \frac{G_F^2 \cos^2 \theta_C}{\pi} E_e p_e \frac{1}{\beta_\nu} \simeq \frac{G_F^2 \cos^2 \theta_C}{\pi} Q \sqrt{Q^2 - m_e^2} \frac{c}{v_\nu}. \qquad (7.28)$$

Notice the dependence on the inverse neutrino velocity v_ν. This is analogous to the well-known behavior of slow neutron capture rates, and it increases the value of the cross section for small momenta. In this case, the neutrino de Broglie wavelength is much larger than the target dimension.

For a nucleus at rest, the product of cross section times relative velocity remains constant at low v_ν, reaching a constant value. For $Q \sim \mathrm{MeV}$,

$$\sigma_{\nu,c} \, v_\nu \simeq 10^{-44} \ \mathrm{cm}^2 \, c. \qquad (7.29)$$

Because the nucleus N already has sufficient energy to produce N' and the electron/positron, the neutrino is contributing with only its lepton number and the process is possible even for an extremely low velocity. Using the estimate (7.29), the CNB capture rate is then given by

$$\Gamma_{\nu,c} \sim \sigma_{\nu,c} \, v_\nu n_{\nu,0} \simeq 10^{-32} \ \mathrm{s}^{-1}. \qquad (7.30)$$

A more careful analysis makes it necessary to go further into the details of nuclear structures and β-decay processes. We have, in general,

$$\Gamma_{\nu,c} = \frac{G_F^2 \cos^2 \theta_C}{\pi} \int \frac{d^3 p_\nu}{(2\pi)^3} p_e E_e F(Z, E_e) C(E_e, p_\nu) f(p_\nu), \qquad (7.31)$$

with $f(p_\nu)$ the neutrino distribution (the Fermi–Dirac function in the standard case) and $F(Z, E_e)$ the Fermi function, due to the electromagnetic interaction between the electron and the final nucleus N' with charge $Z \pm 1$. The function $C(E_e, p_\nu)$ is called the nuclear shape factor. It depends on the various nuclear state transition amplitudes and requires the calculation of nuclear matrix elements. In Figure 7.1 are shown the values for the neutrino capture cross sections for more than 2000 $\beta^\pm$ decaying nuclei versus Q (Cocco *et al.*, 2007). For example, the case of ^3H, with $Q = 18.591$ keV and $\sigma_{\nu,c} v_\nu / c = 0.78 \times 10^{-44}$ cm^2, corresponds to the second point from the left in the upper plot.

Consider this isotope for illustration. The capture rate is exceedingly low when compared to the total β-decay rate Γ_β:

$$\Gamma_{\nu,c} \simeq 0.66 \times 10^{-23} \, \Gamma_\beta. \qquad (7.32)$$

This is always the case for all β-decaying nuclei.

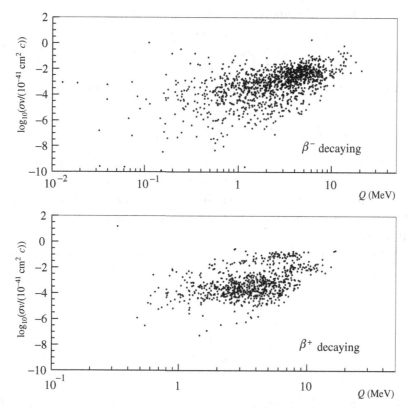

Figure 7.1 The product $\sigma_{\nu,c} v_\nu/c$ for several $\beta^\pm$ decays, versus the value of Q in the limit of small v_ν. Each point represents a specific decaying nucleus. ^{3}H ($Q = 18.591$ keV) corresponds to the second point from the left in the upper panel. (Reprinted from Cocco *et al.*, 2007, with permission from IOP.)

However, what should be compared with $\Gamma_{\nu,c}$ is rather the rate integrated only over the tail of the β spectrum near the end point, because only these events constitute the *background* for our signal. If Δ is the energy resolution of an experimental apparatus, the ratio of the event rate $\Gamma_\beta(\Delta)$ for the last electron energy bin $Q - \Delta - m_\beta < E_e < Q - m_\beta$, compared with the total CNB event rate, can easily be calculated. Using the expression $n_\nu = 3\zeta(3)T_{\nu,0}^3/(4\pi^2)$,

$$\frac{\Gamma_{\nu,c}}{\Gamma_\beta(\Delta)} = \frac{9}{2}\zeta(3)\left(\frac{T_{\nu,0}}{\Delta}\right)^3 \frac{1}{(1 + 2m_\nu/\Delta)^{3/2}}. \qquad (7.33)$$

For $\Delta \sim 0.1$ eV and $m_\nu \sim 0.2$ eV, this ratio is still quite small, of order 10^{-9}.

To get an estimate of the signal-to-background ratio, we can proceed as follows. Assuming gaussian statistics, the expected background electron events which are produced by β-decay, yet have an energy in the bin centered at $E_e = Q + m_\nu$ by a statistical fluctuation, are suppressed by the exponential factor

$$\kappa = \frac{1}{\sqrt{2\pi}} \int_{2m_\nu/\Delta-1/2}^{2m_\nu/\Delta+1/2} e^{-t^2/2} dt. \tag{7.34}$$

A signal-to-background ratio larger than unity thus means that

$$\frac{\Gamma_{\nu,c}}{\Gamma_\beta(\Delta)} = \frac{9}{2}\zeta(3)\left(\frac{T_{\nu,0}}{\Delta}\right)^3 \frac{1}{(1+2m_\nu/\Delta)^{3/2}\kappa} \geq 1. \tag{7.35}$$

This ratio is of order 3 if, for example, $\Delta = 0.1$ eV for a neutrino mass of 0.3 eV. Currently, this energy resolution seems very difficult to achieve. Nevertheless, if a large neutrino mass were to be found by ongoing β-decay experiments such as KATRIN, it is not inconceivable that a future generation of experiments might reach this goal.

Finally, given the CNB capture rate, the total number of events for a given target mass m – see also Section 7.1.1 – is

$$\Gamma_{\nu,c}\frac{N_A}{A}\frac{m}{g} \tag{7.36}$$

or

$$2.85 \times 10^{-2} \frac{\sigma_{\nu,c}v_\nu/c}{10^{-45} \text{ cm}^2}\text{yr}^{-1}\text{mol}^{-1}. \tag{7.37}$$

Consider the case of ^{3}H again. From the value of the cross section, we find that one needs a mass of order 100 g to get order 10 events per year. If neutrinos are sufficiently massive, they would cluster and their local density is larger than the standard value $n_{\nu,0}$. For order-eV neutrino mass, the increase might be by a factor of 10 (Ringwald and Wong, 2004). Even in this case, one has to deal with quite a large mass of radioactive material. In experiments such as KATRIN, where electron energy is measured with a spectrometer, the gaseous radioactive source cannot contain more than micrograms of ^{3}H to avoid rescatterings of the emitted electrons off the gas atoms. A different approach such as bolometric detectors, where the total energy emitted in β-decay is detected as a temperature increase, can be scaled up to large masses and perhaps represents a better perspective (Sangiorgio, 2006). In this case, however, a possible limitation of sensitivity is due to the so-called *pile-up*, namely the fact that more than a single decay/capture event is detected in some measurement time, so that one cannot reconstruct the energy of the electrons emitted in each interaction.

Let us finally note that there might be also interesting candidates for CNB detection among nuclei which decay by electron capture (Cocco *et al.*, 2009), in particular ^{163}Ho, with a corresponding signal-to-background neutrino capture event slightly larger than in the ^{3}H β–decay (Lusignoli and Vignati, 2011).

7.2 Beyond the ultimate dream: neutrino anisotropies in the sky

We have seen in the last section how challenging it would be to achieve direct detection of the CNB. Hence it is even more unlikely that a CNB anisotropy map could ever be measured. We will, however, conclude this chapter with a summary of what one could learn from such a map.

7.2.1 Neutrino last scattering surface

We know that neutrinos decouple at a temperature on the order of 1 MeV, corresponding to a redshift $z_{\nu LS} \sim 10^{10}$. If neutrinos were relativistic today, their last scattering surface would be part of our past light-cone. Neutrinos of whatever momentum would have decoupled at a comoving distance $\chi(z_{\nu LS})$ given by Eq. (2.67). On the other hand, nonrelativistic neutrinos propagate much more slowly at late times, with a momentum-dependent velocity given by $v \sim y/\epsilon$ in units of c. This reduces considerably the comoving radius $\chi(z_{\nu LS}, y)$ of the last scattering surface for neutrinos of given momentum $p = y/a$, now given by Eq. (2.70). As mentioned in Chapter 2, this surface is actually closer to us than the photon last scattering surface for neutrinos of mass greater than 10^{-4} eV and typical momentum $p \sim T_\nu$ (see Dodelson and Vesterinen, 2009 for more details). At the same time, the neutrino last scattering surface probes an older epoch than the photon one. There is no contradiction between these two statements because massive neutrinos do not originate from our past light-cone, but from our past neutrino-cones (which are different for each momentum). Geometrical quantities such as the comoving distance and angular diameter distance are different on each of these cones.

Scales crossing the Hubble radius at neutrino decoupling would be seen today within a tiny angle in a CNB map and would contribute to very large multipoles in its power spectrum. Let us ignore these modes and concentrate on scales crossing the Hubble radius long after neutrino decoupling (this is equivalent to studying the CMB spectrum on angular scales smaller than the angular sound horizon at decoupling, given by the Sachs–Wolfe plateau plus LISW corrections).

7.2.2 Massless neutrinos

The neutrino anisotropy spectrum was computed in the case of massless neutrinos by Hu *et al.*, 1995. At the level of primary anisotropies, the spectrum is just given by a Sachs–Wolfe plateau with the same amplitude as in the CMB spectrum, because on superhorizon scales $\delta_\gamma = \delta_\nu$ for adiabatic initial conditions. However, the spectrum is corrected by an ISW effect originating from two epochs: radiation domination and Λ domination. The latter effect is the same as for the CMB, and we know that it is small. The former effect is much larger for modes entering inside the Hubble radius during radiation domination. We know that after Hubble crossing during the radiation epoch, metric fluctuations are strongly suppressed (as can be checked by looking at Fig. 5.1). This evolution boosts CNB temperature fluctuations by roughly one order of magnitude. For all modes entering deep inside the radiation era, this ISW effect has the same amplitude because the gravitational potential decreases by approximately the same amount after Hubble crossing. Hence the spectrum of CNB anisotropies for massless neutrinos has a steplike spectrum with more power on small angular scales (large l's). The scale of the step can be inferred from the comoving scale $2\pi/k_{eq}$ projected in angular space using the angular diameter distance $d_A(z_{eq})$. It corresponds roughly to $l \sim 500$.

7.2.3 Massive neutrinos

We have seen in Section 5.2.1 that massive neutrino perturbations can be described by phase-space fluctuations $\Psi_\nu(\eta, \vec{x}, y, \hat{n})$ inducing spectral distortions. Therefore, we cannot speak of neutrino temperature anisotropies, unless temperature is defined in some effective way. For this reason, the definition of the CNB spectrum is not unique and should be adapted to the hypothetical method by which anisotropies would be measured. In Michney and Caldwell, 2007; Hannestad and Brandbyge, 2010, some generic properties of this spectrum have been studied for two different definitions, corresponding to different ways of integrating the phase-space distribution $f_\nu(\eta, \vec{x}, y, \hat{n})$ over momentum at the observer's location. Both definitions make it possible to recover the CNB temperature spectrum in the zero-mass limit. These works deliberately ignore additional complications coming from neutrino oscillations: neutrinos propagate in the universe as mass eigenstates, but a putative experiment would probably map flavour eigenstates.

Generic features in the massive CNB spectrum (compared to the massless case) are

- A mass dependence on the scale of the step described previously. We saw that this step corresponds to the comoving scale $2\pi/k_{eq}$. For a different mass, this scale would be seen by the observer under a different angle, because the neutrino angular diameter distance $d_A(z_{eq}, y)$ is momentum-dependent.
- An additional contribution accumulating along the line of sight after the time of the nonrelativistic transition. For scales above the maximum free-streaming scale $2\pi/k_{nr}$, neutrino fluctuations grow because of the infall of nonrelativistic neutrinos in the gravitational potential wells created by CDM and baryons. This neutrino clustering enhances the CNB spectrum on small l's. For large l's

corresponding to modes above the free-streaming scale, neutrino clustering is inefficient, and the spectrum is the same as for massless neutrinos.

Hannestad and Brandbyge (2010) found that this last effect boosts the CNB spectrum by several orders of magnitude. Therefore, an ideal experiment sensitive to CNB anisotropies would directly measure neutrino masses. The same authors also stress that for large masses (typically above 0.1 eV), neutrino clustering is affected by nonlinear corrections at small redshift, so that the CNB cannot be predicted by linear theory. Finally, they show that weak lensing effects on the CNB map (caused by the neighboring large-scale structure of the universe) are negligible.

References

[1] Aalseth, C. E., *et al.* 2002. *Phys. Rev. D*, **65**, 092007.

[2] Abazajian, K. N. 2003. *Astropart. Phys.*, **19**, 303.

[3] Abazajian, K. N. 2009. arXiv 0903.2040.

[4] Abazajian, K. N., Fuller, G. M., and Tucker, W. H. 2001. *Astrophys. J.*, **562**, 593.

[5] Abazajian, K. N., Beacom, J. F., and Bell, N, F. 2002. *Phys. Rev. D*, **66**, 013008.

[6] Abazajian, K. N., *et al.* 2011. *Astropart. Phys.*, **35**, 177.

[7] Abazajian, K. N., *et al.* 2012. arXiv 1204.5379.

[8] Abdurashitov, J. N., *et al.* 1999. *Phys. Rev. C*, **60**, 055801.

[9] Abe, K., *et al.* 2011. *Phys. Rev. Lett.*, **107**, 041801.

[10] Abe, Y., *et al.* 2012. *Phys. Rev. Lett.*, **108**, 131801.

[11] Acero, M. A., and Lesgourgues, J. 2009. *Phys. Rev. D*, **79**, 045026.

[12] Acquafredda, R., *et al.* 2006. *New J. Phys.*, **8**, 303.

[13] Adams, T. F. 1976. *Astron. Astrophys.*, **50**, 461.

[14] Adamson, P., *et al.* 2011. *Phys. Rev. Lett.*, **107**, 181802.

[15] Adelberger, E. G., Heckel, B. R., Stubbs, C. W., and Rogers, W. F. 1991. *Ann. Rev. Nucl. Part. Sci.*, **41**, 269.

[16] Adler, S. 1969. *Phys. Rev.*, **177**, 2426.

[17] Aguilar-Arevalo, A. A., *et al.* 2001. *Phys. Rev. D*, **64**, 112007.

[18] Aguilar-Arevalo, A. A., *et al.* 2009. *Phys. Rev. Lett.*, **102**, 101802.

[19] Aguilar-Arevalo, A. A., *et al.* 2010. *Phys. Rev. Lett.*, **105**, 181801.

[20] Ahmad, Q. R., *et al.* 2002. *Phys. Rev. Lett.*, **89**, 011301.

[21] Ahn, J. K., *et al.* 2012. *Phys. Rev. Lett.*, **108**, 191802.

[22] Akhmedov, E. K., Rubakov, V. A., and Smirnov, A. Yu. 1998. *Phys. Rev. Lett.*, **81**, 1359.

[23] Alimonti, G., *et al.* 2002. *Astropart. Phys.*, **16**, 205.

[24] Allaby, J. V., *et al.* 1987. *Z. Phys. C*, **36**, 611.

[25] Allison, W. W. M., *et al.* 1999. *Phys. Lett. B*, **449**, 137.

[26] Alpher, R. A., Bethe, H., and Gamow, G. 1948. *Phys. Rev.*, **73**, 803.

[27] Altmann, M., *et al.* 2000. *Phys. Lett. B*, **490**, 16.

[28] Ambrosio, M., *et al.* 1998. *Phys. Lett. B*, **434**, 451.

[29] An, F. P., *et al.* 2012. *Phys. Rev. Lett.*, **108**, 171803.

[30] Anisimov, A., Buchmüller, W., Drewes, M., and Mendizabal, S. 2011. *Ann. Phys.*, **326**, 1998.

[31] Apollonio, M., *et al.* 2003. *Eur. Phys. J. C*, **27**, 331.

[32] Archidiacono, M., Calabrese, E., and Melchiorri, A. 2011. *Phys. Rev. D*, **84**, 123008.

[33] Ardellier, F., *et al.* 2006. arXiv hep–ex/0606025.

[34] Argyriades, J., *et al.* 2009. *Phys. Rev. C*, **80**, 032501.

[35] Argyriades, J., *et al.* 2010. *Nucl. Phys. A*, **847**, 168.

[36] Armbruster, B., *et al.* 2002. *Phys. Rev. D*, **65**, 112001.

[37] Arnaboldi, C., *et al.* 2005. *Phys. Rev. Lett.*, **95**, 142501.

[38] Arnold, P. B., and McLerran, L. D. 1987. *Phys. Rev. D*, **36**, 581.

[39] Arnold, P. B., and McLerran, L. D. 1988. *Phys. Rev. D*, **37**, 1020.

[40] Arnold, R., *et al.* 2005. *Phys. Rev. Lett.*, **95**, 182302.

[41] Asaka, T., and Shaposhnikov, M. 2005. *Phys. Lett. B*, **620**, 17.

[42] Asaka, T., Blanchet, S., and Shaposhnikov, M. 2005. *Phys. Lett. B*, **631**, 151.

[43] Asaka, T., Laine, M., and Shaposhnikov, M. 2007. *J. High-Energy Phys.*, **0701**, 091.

[44] Astier, P., *et al.* 2001. *Nucl. Phys. B*, **611**, 3.

[45] Astier, P., *et al.* 2003. *Phys. Lett. B*, **570**, 19.

[46] Athanassopoulos, C., *et al.* 1996. *Phys. Rev. Lett.*, **77**, 3082.

[47] Auerbach, L. B., *et al.* 2001. *Phys. Rev. D*, **63**, 112001.

[48] Aver, E., Olive, K. A., and Skillman, E. D. 2010. *J. Cosmol. Astropart. Phys.*, **1005**, 003.

[49] Aver, E., Olive, K. A., and Skillman, E. D. 2011. *J. Cosmol. Astropart. Phys.*, **1103**, 043.

[50] Aver, E., Olive, K. A., and Skillman, E. D. 2012. *J. Cosmol. Astropart. Phys.*, **1204**, 004.

[51] Ayaita, Y., Weber, M., and Wetterich, C. 2012. *Phys. Rev. D*, **85**, 123010.

[52] Bandyopadhyay, A., *et al.* 2009. *Rep. Prog. Phys.*, **72**, 106201.

[53] Bania, T. M., Rood, R. T., and Balser, D. S. 2002. *Nature*, **415**, 54.

[54] Barbieri, R., and Dolgov, A. D. 1990. *Phys. Lett. B*, **237**, 440.

[55] Barbieri, R., and Dolgov, A. D. 1991. *Nucl. Phys. B*, **349**, 743.

[56] Bardeen, J. M. 1980. *Phys. Rev. D*, **22**, 1882.

[57] Bardeen, J. M., Steinhardt, P. J., and Turner, M. S. 1983. *Phys. Rev. D*, **28**, 1809.

[58] Barger, V., Langacker, P., and Lee, H. S. 2003. *Phys. Rev. D*, **67**, 075009.

[59] Barranco, J., Miranda, O. G., Moura, C. A., and Valle, J. W. F. 2006. *Phys. Rev. D*, **73**, 113001.

[60] Barranco, J., Miranda, O. G., Moura, C. A., and Valle, J. W. F. 2008. *Phys. Rev. D*, **77**, 093014.

[61] Basbøll, A., Bjaelde, O. E., Hannestad, S., and Raffelt, G. G. 2009. *Phys. Rev. D*, **79**, 043512.

[62] Bashinsky, S., and Seljak, U. 2004. *Phys. Rev. D*, **69**, 083002.

[63] Beacom, J. F., Bell, N. F., and Dodelson, S. 2004. *Phys. Rev. Lett.*, **93**, 121302.

[64] Bell, J. S., and Jackiw, R. 1969. *Nuovo Cime.*, **51**, 47.

[65] Bell, N. F., Volkas, R. R., and Wong, Y. Y. Y. 1999. *Phys. Rev. D*, **59**, 113001.

[66] Bell, N. F., Pierpaoli, E., and Sigurdson, K. 2006. *Phys. Rev. D*, **73**, 063523.

[67] Berezhiani, Z., and Rossi, A. 2002. *Phys. Lett. B*, **535**, 207.

[68] Beringer, J., *et al.* (Particle Data Group). 2012. *Phys. Rev. D*, **86**, 010001.

[69] Bernabei, R., *et al.* 2002. *Phys. Lett. B*, **546**, 23.

[70] Bernabéu, J., Papavassiliou, J., and Vidal, J. 2004. *Nucl. Phys. B*, **680**, 450.

[71] Bernstein, J. 1988. *Kinetic Theory in the Expanding Universe.* Cambridge University Press.

[72] Bilenky, S. M., Giunti, C., and Grimus, W. 1998. *Eur. Phys. J. C*, **1**, 247.

[73] Bilenky, S. M., Giunti, C., Grimus, W., and Schwetz, T. 1999. *Phys. Rev. D*, **60**, 073007.

[74] Bird, S., Viel, M., and Haehnelt, M. G. 2012. *Mon. Not. Roy. Astron. Soc.*, **420**, 2551.

[75] Blas, D., Lesgourgues, J., and Tram, T. 2011. *J. Cosmol. Astropart. Phys.*, **1107**, 034.

[76] Bode, P., Ostriker, J. P., and Turok, N. 2001. *Astrophys. J.*, **556**, 93.

[77] Bodeker, D. 1998. *Phys. Lett. B*, **426**, 351.

[78] Bolaños, A., Miranda, O. G., Palazzo, A., Tórtola, M. A., and Valle, J. W. F. 2009. *Phys. Rev. D*, **79**, 113012.

[79] Boltzmann, L. 1872. *Wiener Beri.*, **66**, 275.

[80] Boyanovsky, D., and Wu, J. 2011. *Phys. Rev. D*, **83**, 043524.

[81] Boyarsky, A., *et al.* 2006. *Phys. Rev. Lett.*, **97**, 261302.

[82] Boyarsky, A., Iakubovskyi, D., Ruchayskiy, O., and Savchenko, V. 2008. *Mon. Not. Roy. Astron. Soc.*, **387**, 1361.

[83] Boyarsky, A., Ruchayskiy, O., and Shaposhnikov, M. 2009a. *Ann. Rev. Nucl. Part. Sci.*, **59**, 191.

[84] Boyarsky, A., Ruchayskiy, O., and Iakubovskyi, D. 2009b. *J. Cosmol. Astropart. Phys.*, **0903**, 005.

[85] Boyarsky, A., Lesgourgues, J., Ruchayskiy, O., and Viel, M. 2009c. *Phys. Rev. Lett.*, **102**, 201304.

[86] Boyarsky, A., Lesgourgues, J., Ruchayskiy, O., and Viel, M. 2009d. *J. Cosmol. Astropart. Phys.*, **0905**, 012.

[87] Brandbyge, J., and Hannestad, S. 2009. *J. Cosmol. Astropart. Phys.*, **0905**, 002.

[88] Brandbyge, J., and Hannestad, S. 2010. *J. Cosmol. Astropart. Phys.*, **1001**, 021.

[89] Brandbyge, J., Hannestad, S., Haugbølle, T., and Thomsen, B. 2008. *J. Cosmol. Astropart. Phys.*, **0808**, 020.

[90] Brandbyge, J., Hannestad, S., Haugbøelle, T., and Wong, Y. Y. Y. 2010. *J. Cosmol. Astropart. Phys.*, **1009**, 014.

[91] Brandenberger, R., Kahn, R., and Press, W. H. 1983. *Phys. Rev. D*, **28**, 679.

[92] Bucher, M., Moodley, K., and Turok, N. 2000. *Phys. Rev. D*, **62**, 083508.

[93] Buchmüller, W., Di Bari, P., and Plümacher, M. 2002. *Nucl. Phys. B*, **643**, 367.

[94] Buchmüller, W., Peccei, R. D., and Yanagida, T. 2005. *Ann. Rev. Nucl. Part. Sci.*, **55**, 311.

[95] Burguet-Castell, J., *et al.* 2004. *Nucl. Phys. B*, **695**, 217.

[96] Cabibbo, N. 1963. *Phys. Rev. Lett.*, **10**, 531.

[97] Cabibbo, N., and Maiani, L. 1982. *Phys. Lett. B*, **114**, 115.

[98] Carbone, C., Verde, L., Wang, Y., and Cimatti, A. 2011. *J. Cosmol. Astropart. Phys.*, **1103**, 030.

[99] Carson, L., Li, X., McLerran, L. D., and Wang, R.T. 1990. *Phys. Rev. D*, **42**, 2127.

[100] Casas, A., Cheng, W. Y., and Gelmini, G. 1999. *Nucl. Phys. B*, **538**, 297.

[101] Castorina, E., *et al.* 2012. *Phys. Rev. D*, **86**, 023517.

[102] Castro, P. G., *et al.* 2009. *Astrophys. J.*, **701**, 857.

[103] Cervera, A., *et al.* 2000. *Nucl. Phys. B*, **579**, 17.

[104] Chang, S., and Choi, K. 1994. *Phys. Rev. D*, **49**, 12.

[105] Chen, M. C. 2007. arXiv hep–ph/0703087.

[106] Chikashige, Y., Mohapatra, R. N., and Peccei, R. D. 1981. *Phys. Lett. B*, **98**, 265.

[107] Christenson, J. H., Cronin, J. W., Fitch, V. L., and Turlay, R. 1964. *Phys. Rev. Lett.*, **13**, 138.

[108] Church, E. D., Eitel, K., Mills, G. B., and Steidl, M. 2002. *Phys. Rev. D*, **66**, 013001.

[109] Cirelli, M., Marandella, G., Strumia, A., and Vissani, F. 2005. *Nucl. Phys. B*, **708**, 215.

[110] Cleveland, B. T., *et al.* 1998. *Astrophys. J.*, **496**, 505.
[111] Cline, J. M. 1992. *Phys. Rev. Lett.*, **68**, 3137.
[112] Cocco, A. G., Mangano, G., and Messina, M. 2007. *J. Cosmol. Astropart. Phys.*, **0706**, 015.
[113] Cocco, A. G., Mangano, G., and Messina, M. 2009. *Phys. Rev. D*, **79**, 053009.
[114] Cohen, A. G., and De Rújula, A. 1997. arXiv astro–ph/9709132.
[115] Conley, A., *et al.* 2011. *Astrophys. J. Suppl.*, **192**, 1.
[116] Costantini, H., *et al.* 2009. *Rep. Prog. Phys.*, **72**, 086301.
[117] Cowsik, R., and McClelland, J. 1977. *Phys. Rev. Lett.*, **29**, 669.
[118] Crotty, P., Lesgourgues, J., and Pastor, S. 2004. *Phys. Rev. D*, **69**, 123007.
[119] Cuoco, A., Lesgourgues, J., Mangano, G., and Pastor, S. 2005. *Phys. Rev. D*, **71**, 123501.
[120] Cyburt, R. H. 2004. *Phys. Rev. D*, **70**, 023505.
[121] Cyburt, R. H., Fields, B. D., and Olive, K. A. 2002. *Astropart. Phys.*, **17**, 87.
[122] Danby, G., *et al.* 1962. *Phys. Rev. Lett.*, **9**, 36.
[123] Danevich, F. A., *et al.* 2003. *Phys. Rev. C*, **68**, 035501.
[124] Daraktchieva, Z., *et al.* 2003. *Phys. Lett. B*, **564**, 190.
[125] Davidson, S., and Ibarra, A. 2002. *Phys. Lett. B*, **535**, 25.
[126] Davidson, S., Peña Garay, C., Rius, N., and Santamaria, A. 2003. *J. High-Energy Phys.*, **0303**, 011.
[127] Davidson, S., Nardi, E., and Nir, Y. 2008. *Phys. Rep.*, **466**, 105.
[128] De Bernardis, F., *et al.* 2008. *J. Cosmol. Astropart. Phys.*, **0806**, 013.
[129] de Gouvea, A., and Jenkins, J. 2006. *Phys. Rev. D*, **74**, 033004.
[130] Debono, I., *et al.* 2010. *Mon. Not. Roy. Astron. Soc.*, **404**, 110.
[131] Dekel, A., and Ostriker, J. P. 1999. *Formation of Structure in the Universe.* Cambridge University Press.
[132] den Herder, J. W., *et al.* 2009. arXiv 0906.1788.
[133] Deniz, M., *et al.* 2010. *Phys. Rev. D*, **81**, 072001.
[134] Di Bari, P. 2002. *Phys. Rev. D*, **65**, 043509.
[135] Dicus, D. A., *et al.* 1982. *Phys. Rev. D*, **26**, 2694.
[136] Dimopoulos, S., and Susskind, L. 1978. *Phys. Rev. D*, **18**, 4500.
[137] Dodelson, S. 2003. *Modern Cosmology.* Academic Press.
[138] Dodelson, S., and Turner, M. S. 1992. *Phys. Rev. D*, **46**, 3372.
[139] Dodelson, S., and Vesterinen, M. 2009. *Phys. Rev. Lett.*, **103**, 171301.
[140] Dodelson, S., and Widrow, L. M. 1994. *Phys. Rev. Lett.*, **72**, 17.
[141] Dolgov, A. D. 1979. *Pis'ma ZhETf*, **29**, 254.
[142] Dolgov, A. D. 1981. *Sov. J. Nucl. Phys.*, **33**, 700.
[143] Dolgov, A. D. 1992. *Phys. Rep.*, **222**, 309.
[144] Dolgov, A. D. 2002. *Phys. Rep.*, **370**, 333.
[145] Dolgov, A. D., and Fukugita, M. 1992. *Phys. Rev. D*, **46**, 5378.
[146] Dolgov, A. D., and Hansen, S. H. 2002. *Astropart. Phys.*, **16**, 339.
[147] Dolgov, A. D., and Takahashi, F. 2004. *Nucl. Phys. B*, **688**, 189.
[148] Dolgov, A. D., and Villante, F. L. 2004. *Nucl. Phys. B*, **679**, 261.
[149] Dolgov, A. D., and Zeldovich, Ya. B. 1981. *Rev. Mod. Phys.*, **53**, 1.
[150] Dolgov, A. D., Hansen, S. H., and Semikoz, D. V. 1997. *Nucl. Phys. B*, **503**, 426.
[151] Dolgov, A. D., Hansen, S. H., and Semikoz, D. V. 1999. *Nucl. Phys. B*, **543**, 269.
[152] Dolgov, A. D., Hansen, S. H., Pastor, S., and Semikoz, D. V. 2000a. *Astropart. Phys.*, **14**, 79.
[153] Dolgov, A. D., Hansen, S. H., Raffelt, G., and Semikoz, D. V. 2000b. *Nucl. Phys. B*, **580**, 331.

[154] Dolgov, A. D., Hansen, S. H., Raffelt, G., and Semikoz, D. V. 2000c. *Nucl. Phys. B*, **590**, 562.

[155] Dolgov, A. D., *et al.* 2002. *Nucl. Phys. B*, **632**, 363.

[156] Dolgov, A. D., Hansen, S. H., and Smirnov, A. Yu. 2005. *J. Cosmol. Astropart. Phys.*, **0506**, 004.

[157] Doran, M. 2005. *J. Cosmol. Astropart. Phys.*, **0510**, 011. Code at http://www.cmbeasy.org.

[158] Duan, H., Fuller, G. M., and Qian, Y. Z. 2010. *Ann. Rev. Nucl. Part. Sci.*, **60**, 569.

[159] Duda, G., Gelmini, G., and Nussinov, S. 2001. *Phys. Rev. D*, **64**, 122001.

[160] Dunkley, J., *et al.* 2009. *Astrophys. J. Suppl.*, **180**, 306.

[161] Dunkley, J., *et al.* 2011. *Astrophys. J.*, **739**, 52.

[162] Dunstan, R. M., Abazajian, K. N., Polisensky, E., and Ricotti, M. 2011. arXiv 1109.6291.

[163] Dydak, F., *et al.* 1984. *Phys. Lett. B*, **134**, 281.

[164] Eisenstein, D. J., and Hu, W. 1998. *Astrophys. J.*, **496**, 605.

[165] Englert, F., and Brout, R. 1964. *Phys. Rev. Lett.*, **13**, 321.

[166] Enqvist, K., Kainulainen, K., and Maalampi, J. 1990. *Phys. Lett. B*, **244**, 186.

[167] Enqvist, K., Kainulainen, K., and Maalampi, J. 1991. *Nucl. Phys. B*, **349**, 754.

[168] Enqvist, K., Kainulainen, K., and Semikoz, V. B. 1992a. *Nucl. Phys. B*, **374**, 392.

[169] Enqvist, K., Kainulainen, K., and Thomson, M. J. 1992b. *Nucl. Phys. B*, **373**, 498.

[170] Eskut, E., *et al.* 1997. *Nucl. Instrum. Meth. A*, **401**, 7.

[171] Esposito, S., Mangano, G., Miele, G., and Pisanti, O. 1999. *Nucl. Phys. B*, **540**, 3.

[172] Esposito, S., *et al.* 2000a. *Nucl. Phys. B*, **590**, 539.

[173] Esposito, S., Mangano, G., Miele, G., and Pisanti, O. 2000b. *J. High-Energy Phys.*, **0009**, 038.

[174] Feng, J. L. 2010. *Ann. Rev. Astron. Astrophys.*, **48**, 495.

[175] Fields, B. D. 1996. *Astrophys. J.*, **456**, 478.

[176] Fields, B. D. 2011. *Ann. Rev. Nucl. Part. Sci.*, **61**, 47.

[177] Fogli, G. L., *et al.* 2011. *Phys. Rev. D*, **84**, 053007.

[178] Fogli, G. L., *et al.* 2012. *Phys. Rev. D*, **86**, 013012.

[179] Foot, R., and Volkas, R. R. 1997. *Phys. Rev. D*, **55**, 5147.

[180] Foot, R., Lew, H., He, X. G., and Joshi, G. C. 1989. *Z. Phys. C*, **44**, 441.

[181] Foot, R., Joshi, G. C., Lew, H., and Volkas, R. R. 1990. *Mod. Phys. Lett. A*, **5**, 95.

[182] Foot, R., Thomson, M. J., and Volkas, R. R. 1996. *Phys. Rev. D*, **53**, 5349.

[183] Foot, Robert, and Volkas, R. R. 1995. *Phys. Rev. Lett.*, **75**, 4350.

[184] Forero, D. V., and Guzzo, M. M. 2011. *Phys. Rev. D*, **84**, 013002.

[185] Frampton, P. H., Glashow, S. L., and Yanagida, T. 2002. *Phys. Lett. B*, **548**, 119.

[186] França, U., Lattanzi, M., Lesgourgues, J., and Pastor, S. 2009. *Phys. Rev. D*, **80**, 083506.

[187] Friedmann, A. 1922. *Z. Phys.*, **16**, 377.

[188] Friedmann, A. 1924. *Z. Phys.*, **21**, 326.

[189] Fritzsch, H., and Minkowski, P. 1975. *Ann. Phys.*, **93**, 193.

[190] Fritzsch, H., Gell-Mann, M., and Leutwyler, H. 1973. *Phys. Lett. B*, **47**, 365.

[191] Fukuda, Y., *et al.* 1994. *Phys. Lett. B*, **335**, 237.

[192] Fukuda, Y., *et al.* 1998. *Phys. Rev. Lett.*, **81**, 1562.

[193] Fukugita, M., and Yanagida, T. 1986. *Phys. Lett. B*, **174**, 45.

[194] Fukugita, M., and Yanagida, T. 2003. *Physics of Neutrinos and Applications to Astrophysics*. Springer.

[195] Geiss, J. and Gloeckler, G. 1998. *Space Sci. Rev.*, **84**, 239.

[196] Geiss, J. and Gloeckler, G. 2007. *Space Sci. Rev.*, **130**, 5.

[197] Gell-Mann, M., Ramond, P., and Slansky, R. 1979. *Supergravity*, ed. F. van Nieuwen-huizen and D. Freedman. North Holland.

[198] Gelmini, G. B., and Roncadelli, M. 1981. *Phys. Lett. B*, **99**, 411.

[199] Georgi, H., and Glashow, S. L. 1974. *Phys. Rev. Lett.*, **32**, 438.

[200] Georgi, H. M., Glashow, S. L., and Nussinov, S. 1981. *Nucl. Phys. B*, **193**, 297.

[201] Gerstein, S. S., and Zeldovich, Ya. B. 1966. *Sov. Phys. JETP Letters*, **4**, 120.

[202] Giunti, C. 2011. arXiv 1110.3914.

[203] Giunti, C., and Kim, C. W. 2007. *Fundamentals of Neutrino Physics and Astro-physics*. Oxford University Press.

[204] Giunti, C., and Laveder, M. 2011a. *Phys. Rev. D*, **84**, 073008.

[205] Giunti, C., and Laveder, M. 2011b. *Phys. Rev. C*, **83**, 065504.

[206] Giunti, C., and Laveder, M. 2011c. *Phys. Lett. B*, **706**, 200.

[207] Giusarma, E., *et al.* 2011. *Phys. Rev. D*, **83**, 115023.

[208] Giusarma, E., *et al.* 2012. *Phys. Rev. D*, **85**, 083522.

[209] Glashow, S. L. 1961. *Nucl. Phys.*, **22**, 579.

[210] Gómez-Cadenas, J. J., Martín-Albo, J., Mezzetto, M., Monrabal, F., and Sorel, M. 2012. *Riv. Nuovo Cim.*, **35**, 29.

[211] González-García, M. C., Maltoni, M., and Salvado, J. 2010. *J. High-Energy Phys.*, **1008**, 117.

[212] Grassler, H., *et al.* 1986. *Nucl. Phys. B*, **273**, 253.

[213] Gratton, S., Lewis, A., and Efstathiou, G. 2008. *Phys. Rev. D*, **77**, 083507.

[214] Grifols, J. A., and Massó, E. 1987. *Mod. Phys. Lett. A*, **2**, 205.

[215] Grifols, J. A., and Massó, E. 1989. *Phys. Rev. D*, **40**, 3819.

[216] Gross, D. J., and Wilczek, F. 1973. *Phys. Rev. Lett.*, **30**, 1343.

[217] Groth, E. J., and Peebles, P. J. E. 1975. *Astron. Astrophys.*, **41**, 143.

[218] Guralnik, G., Hagen, C. R., and Kibble, T. W. B. 1964. *Phys. Rev. Lett.*, **13**, 585.

[219] Guth, A. 1981. *Phys. Rev. D*, **23**, 347.

[220] Guth, A., and Pi, S. Y. 1982. *Phys. Rev. Lett.*, **49**, 1110.

[221] Halzen, F., and Martin, A. D. 1984. *Quarks and Leptons: An Introductory Course in Modern Particle Physics*. Wiley.

[222] Hamann, J. 2012. *J. Cosmol. Astropart. Phys.*, **1203**, 021.

[223] Hamann, J., Lesgourgues, J., and Mangano, G. 2008. *J. Cosmol. Astropart. Phys.*, **0803**, 004.

[224] Hamann, J, Hannestad, S., Raffelt, G. G., Tamborra, I., and Wong, Y. Y. Y. 2010. *Phys. Rev. Lett.*, **105**, 181301.

[225] Hamann, J., Hannestad, S., Raffelt, G. G., and Wong, Y. Y. Y. 2011. *J. Cosmol. Astropart. Phys.*, **1109**, 034.

[226] Hambye, T., *et al.* 2004. *Nucl. Phys. B*, **695**, 169.

[227] Hannestad, S. 2002. *Phys. Rev. D*, **65**, 083006.

[228] Hannestad, S. 2004. *Phys. Rev. D*, **70**, 043506.

[229] Hannestad, S. 2005a. *Phys. Rev. Lett.*, **95**, 221301.

[230] Hannestad, S. 2005b. *J. Cosmol. Astropart. Phys.*, **0502**, 011.

[231] Hannestad, S. 2010. *Prog. Part. Nucl. Phys.*, **65**, 185.

[232] Hannestad, S., and Brandbyge, J. 2010. *J. Cosmol. Astropart. Phys.*, **1003**, 020.

[233] Hannestad, S., and Madsen, J. 1995. *Phys. Rev. D*, **52**, 1764.

[234] Hannestad, S., and Raffelt, G. 2004. *J. Cosmol. Astropart. Phys.*, **0404**, 008.

[235] Hannestad, S., Ringwald, A., Tu, H., and Wong, Y. Y. Y. 2005. *J. Cosmol. Astropart. Phys.*, **0509**, 014.

[236] Hannestad, S., Tu, H., and Wong, Y-Y. Y. 2006. *J. Cosmol. Astropart. Phys.*, **0606**, 025.

[237] Hannestad, S., Tamborra, I., and Tram, T. 2012a. *J. Cosmol. Astropart. Phys.*, **1207**, 025.

[238] Hannestad, S., Haugbølle, T., and Schultz, C. 2012b. *J. Cosmol. Astropart. Phys.*, **1202**, 045.

[239] Hansen, S. H., *et al.* 2002. *Phys. Rev. D*, **65**, 023511.

[240] Harvey, J. A., and Turner, M. S. 1990. *Phys. Rev. D*, **42**, 3344.

[241] Hawking, S. 1982. *Phys. Lett. B*, **115**, 295.

[242] Heckler, A. F. 1994. *Phys. Rev. D*, **49**, 611.

[243] Herrera, M. A., and Hacyan, S. 1989. *Astrophys. J.*, **336**, 539.

[244] Hicken, M., *et al.* 2009. *Astrophys. J.*, **700**, 1097.

[245] Higgs, P. W. 1964a. *Phys. Lett.*, **12**, 132.

[246] Higgs, P. W. 1964b. *Phys. Rev. Lett.*, **13**, 508.

[247] Hirata, C. M., and Seljak, U. 2003. *Phys. Rev. D*, **68**, 083002.

[248] Hirsch, M. 2011. *Nucl. Phys. Proc. Suppl.*, **221**, 119.

[249] Hirsch, M., Nardi, E., and Restrepo, D. 2003. *Phys. Rev. D*, **67**, 033005.

[250] Holtzman, J. A. 1989. *Astrophys. J. Suppl.*, **71**, 1.

[251] Hooper, D., March-Russell, J., and West, S. M. 2009. *Phys. Lett. B*, **605**, 228.

[252] Hou, Z., *et al.* 2011. arXiv 1104.2333.

[253] Howlett, C., Lewis, A., Hall, A., and Challinor, A. 2012. *J. Cosmol. Astropart. Phys.*, **1204**, 027.

[254] Hoyle, F. 1967. *Proc. Royal Soc. (London) A*, **301**, 171.

[255] Hu, W. 1998. *Astrophys. J.*, **506**, 485.

[256] Hu, W., and Eisenstein, D. J. 1998. *Astrophys. J.*, **498**, 497.

[257] Hu, W., and Sugiyama, N. 1996. *Astrophys. J.*, **471**, 542.

[258] Hu, W., Scott, D., Sugiyama, N., and White, M. J. 1995. *Phys. Rev. D*, **52**, 5498.

[259] Hu, W., Eisenstein, D. J., and Tegmark, M. 1998. *Phys. Rev. Lett.*, **80**, 5255.

[260] Hu, W. T. 1995. Ph. D. Thesis.

[261] Huang, K. 1987. *Statistical Mechanics*. Wiley.

[262] Hubble, E. P. 1929. *Proc. Nat. Acad. Sci.*, **15**, 168.

[263] Ichikawa, K., Kawasaki, M., and Takahashi, F. 2005. *Phys. Rev. D*, **72**, 043522.

[264] Iocco, F., *et al.* 2009. *Phys. Rep.*, **472**, 1.

[265] Irvine, J. M., and Humphreys, R. 1983. *J. Phys. G*, **9**, 847.

[266] Itow, Y., *et al.* 2001. arXiv hep–ex/0106019.

[267] Itzykson, C., and Zuber, J. B. 1980. *Quantum Field Theory*. McGraw–Hill.

[268] Izotov, Yu. I., and Thuan, T. X. 2004. *Astrophys. J.*, **602**, 200.

[269] Izotov, Yu. I., and Thuan, T. X. 2010. *Astrophys. J.*, **710**, L67.

[270] Izotov, Yu. I., Thuan, T. X., and Stasinska, G. 2007. *Astrophys. J.*, **662**, 15.

[271] Jimenez, R., Kitching, T., Peña Garay, C., and Verde, L. 2010. *J. Cosmol. Astropart. Phys.*, **1005**, 035.

[272] Kainulainen, K. 1990. *Phys. Lett. B*, **244**, 191.

[273] Kang, H. S., and Steigman, G. 1992. *Nucl. Phys. B*, **372**, 494.

[274] Kaplan, D. E., Luty, M. A., and Zurek, K. M. 2009. *Phys. Rev. D*, **79**, 115016.

[275] Kawano, L. H. 1988. Preprint FERMILAB-PUB-88-034-A.

[276] Kawano, L. H. 1992. Preprint FERMILAB-PUB-92-004-A.

[277] Keisler, R., *et al.* 2011. *Astrophys. J.*, **743**, 28.

[278] Kessler, R. 2009. *Astrophys. J. Suppl.*, **185**, 32.

[279] Khlebnikov, S. Y., and Shaposhnikov, M. E. 1996. *Phys. Lett. B*, **387**, 817.

[280] Khlebnikov, S. Yu., and Shaposhnikov, M. E. 1988. *Nucl. Phys. B*, **308**, 885.

[281] Kinney, W. H., Kolb, E. W., and Turner, M. S. 1997. *Phys. Rev. Lett.*, **79**, 2620.

[282] Kirilova, D. P., and Chizhov, M. V. 1998a. *Phys. Rev. D*, **58**, 073004.

[283] Kirilova, D. P., and Chizhov, M. V. 1998b. *Nucl. Phys. B*, **534**, 447.

[284] Kirilova, D. P., and Chizhov, M. V. 2000. *Nucl. Phys. B*, **591**, 457–468.
[285] Kirilova, D. P., and Panayotova, M. P. 2006. *J. Cosmol. Astropart. Phys.*, **0612**, 014.
[286] Kitching, T. D., *et al.* 2008. *Phys. Rev. D*, **77**, 103008.
[287] Klapdor-Kleingrothaus, H. V., *et al.* 2001. *Eur. Phys. J. A*, **12**, 147.
[288] Klinkhamer, R. F., and Manton, N. S. 1984. *Phys. Rev. D*, **30**, 2212.
[289] Knop, R. A., *et al.* 2003. *Astrophys. J.*, **598**, 102.
[290] Kobayashi, M., and Maskawa, T. 1973. *Prog. Theor. Phys.*, **49**, 652.
[291] Kofman, L. A., Gnedin, N. Y., and Bahcall, N. A. 1993. *Astrophys. J.*, **413**, 1.
[292] Kolb, E. W., and Turner, M. S. 1994. *The Early Universe*. Westview Press.
[293] Komatsu, E., *et al.* 2011. *Astrophys. J. Suppl.*, **192**, 18.
[294] Kristiansen, J. R., Elgaroy, O., and Dahle, H. 2007. *Phys. Rev. D*, **75**, 083510.
[295] Kusenko, A. 2006. *Phys. Rev. Lett.*, **97**, 241301.
[296] Kuzmin, V. A., Rubakov, V. A., and Shaposhnikov, M. E. 1985. *Phys. Lett. B*, **155**, 36.
[297] Lahav, O., Kiakotou, A., Abdalla, F. B., and Blake, C. 2010. *Mon. Not. Roy. Astron. Soc.*, **405**, 168.
[298] Laine, M., and Shaposhnikov, M. 2008. *J. Cosmol. Astropart. Phys.*, **0806**, 031.
[299] Langacker, P., Leveille, J. P., and Sheiman, J. 1983. *Phys. Rev. D*, **27**, 1228.
[300] Langacker, P., Petcov, S. T., Steigman, G., and Toshev, S. 1987. *Nucl. Phys. B*, **282**, 589.
[301] Larson, D., *et al.* 2011. *Astrophys. J. Suppl.*, **192**, 16.
[302] Lazarides, G., Shafi, Q., and Wetterich, C. 1981. *Nucl. Phys. B*, **181**, 287.
[303] LEP. 2006. *Phys. Rep.*, **427**, 257.
[304] Lesgourgues, J., and Pastor, S. 1999. *Phys. Rev. D*, **60**, 103521.
[305] Lesgourgues, J., and Pastor, S. 2006. *Phys. Rep.*, **429**, 307.
[306] Lesgourgues, J., and Tram, T. 2011. *J. Cosmol. Astropart. Phys.*, **1109**, 032.
[307] Lesgourgues, J., Pastor, S., and Perotto, L. 2004. *Phys. Rev. D*, **70**, 045016.
[308] Lesgourgues, J., Perotto, L., Pastor, S., and Piat, M. 2006. *Phys. Rev. D*, **73**, 045021.
[309] Lesgourgues, J., Valkenburg, W., and Gaztañaga, E. 2008. *Phys. Rev. D*, **77**, 063505.
[310] Lesgourgues, J., Matarrese, S., Pietroni, M., and Riotto, A. 2009. *J. Cosmol. Astropart. Phys.*, **0906**, 017.
[311] Lewis, A., and Challinor, A. 2006. *Phys. Rep.*, **429**, 1.
[312] Lewis, A., Challinor, A., and Lasenby, A. 2000. *Astrophys. J.*, **538**, 473.
[313] Linde, A. 1990. *Particle Physics and Inflationary Cosmology*. Harwood Academic.
[314] Linsky, J. L., *et al.* 2006. *Astrophys. J.*, **647**, 1106.
[315] Lisi, E., Sarkar, S., and Villante, F. L. 1999. *Phys. Rev. D*, **59**, 123520.
[316] Lopez, R. E., and Turner, M. S. 1999. *Phys. Rev. D*, **59**, 103502.
[317] Lovell, M. R., *et al.* 2012. *Mon. Not. Roy. Astron. Soc.*, **420**, 2318.
[318] Lusignoli, M., and Vignati, M. 2011. *Phys. Lett. B*, **697**, 11–14.
[319] Luty, M. A. 1992. *Phys. Rev. D*, **45**, 455.
[320] Lyth, D. H., and Riotto, A. 1999. *Phys. Report.*, **314**, 1.
[321] Ma, C. P. 1996. *Astrophys. J.*, **471**, 13.
[322] Ma, C. P., and Bertschinger, E. 1995. *Astrophys. J.*, **455**, 7.
[323] Ma, E. 1998. *Phys. Rev. Lett.*, **81**, 1171.
[324] Ma, E., and Roy, D. P. 2002. *Nucl. Phys. B*, **644**, 290.
[325] Macciò, A. V., and Fontanot, F. 2009. *Mon. Not. Roy. Astron. Soc.*, **404**, L16.
[326] Macciò, A. V., Paduroiu, S., Anderhalden, D., Schneider, A., and Moore, B. 2012a. *Mon. Not. Roy. Astron. Soc.*, **424**, 1105–1112.

[327] Macciò, A. V., Ruchayskiy, O., Boyarsky, A., and Muñoz Cuartas, J. C. 2012b. arXiv 1202.2858.

[328] Magg, M., and Wetterich, C. 1980. *Phys. Lett. B*, **94**, 61.

[329] Majorana, E. 1937. *Nuovo Cim.*, **14**, 171.

[330] Maki, Z., Nakagawa, M., and Sakata, S. 1962. *Prog. Theor. Phys.*, **28**, 870.

[331] Malik, K. A., and Wands, D. 2009. *Phys. Rep.*, **475**, 1.

[332] Maltoni, M., and Schwetz, T. 2003. *Phys. Rev. D*, **68**, 033020.

[333] Maltoni, M., Schwetz, T., Tórtola, M. A., and Valle, J. W. F. 2004. *New J. Phys.*, **6**, 122.

[334] Mangano, G., and Serpico, P. D. 2011. *Phys. Lett. B*, **701**, 296.

[335] Mangano, G., Miele, G., Pastor, S., and Peloso, M. 2002. *Phys. Lett. B*, **534**, 8.

[336] Mangano, G., *et al.* 2005. *Nucl. Phys. B*, **729**, 221.

[337] Mangano, G., *et al.* 2006a. *Nucl. Phys. B*, **756**, 100.

[338] Mangano, G., *et al.* 2006b. *Phys. Rev. D*, **74**, 043517.

[339] Mangano, G., *et al.* 2011. *J. Cosmol. Astropart. Phys.*, **1103**, 035.

[340] Mangano, G., *et al.* 2012. *Phys. Lett. B*, **708**, 1.

[341] Manton, N. S. 1983. *Phys. Rev. D*, **28**, 2019.

[342] Mantz, A., Allen, S. W., Rapetti, D., and Ebeling, H. 2010. *Mon. Not. Roy. Astron. Soc.*, **406**, 1759.

[343] Mather, J. C., *et al.* 1999. *Astrophys. J.*, **512**, 511.

[344] McDonald, J. 2000. *Phys. Rev. Lett.*, **84**, 4798.

[345] McKellar, B. H. J., and Thomson, M. J. 1994. *Phys. Rev. D*, **49**, 2710.

[346] Mention, G., *et al.* 2011. *Phys. Rev. D*, **83**, 073006.

[347] Mészáros, P. 1974. *Astron. Astrophys.*, **37**, 225.

[348] Michael, D. G., *et al.* 2006. *Phys. Rev. Lett.*, **97**, 191801.

[349] Michney, R. J., and Caldwell, R. R. 2007. *J. Cosmol. Astropart. Phys.*, **0701**, 014.

[350] Mikheev, S. P., and Smirnov, A. Yu. 1985. *Sov. J. Nucl. Phys.*, **42**, 913.

[351] Mikheev, S. P., and Smirnov, A. Yu. 1986. *Nuovo Cim. C*, **9**, 17.

[352] Minkowski, P. 1977. *Phys. Lett. B*, **67**, 421.

[353] Mirizzi, A., Montanino, D., and Serpico, P. D. 2007. *Phys. Rev. D*, **76**, 053007.

[354] Mirizzi, A., Saviano, N., Miele, G., and Serpico, P. D. 2012. *Phys. Rev. D*, **86**, 053009.

[355] Mo, H., van den Bosch, F., and White, M. 2010. *Galaxy Formation and Evolution*. Cambridge University Press.

[356] Mohapatra, R. N., and Pati, J. C. 1975. *Phys. Rev. D*, **11**, 566.

[357] Mohapatra, R. N., and Senjanovic, G. 1980. *Phys. Rev. Lett.*, **44**, 912.

[358] Mohapatra, R. N., and Senjanovic, G. 1981. *Phys. Rev. D*, **23**, 165.

[359] Moresco, M., *et al.* 2012a. *J. Cosmol. Astropart. Phys.*, **1207**, 053.

[360] Moresco, M., *et al.* 2012b. *J. Cosmol. Astropart. Phys.*, **1208**, 006.

[361] Mota, D. F., Pettorino, V., Robbers, G., and Wetterich, C. 2008. *Phys. Lett. B*, **663**, 160.

[362] Mukhanov, S. V. 2005. *Physical Foundations of Cosmology*. Cambridge University Press.

[363] Mukhanov, S. V., and Chibisov, G. V. 1981. *Sov. Phys. JETP Lett.*, **33**, 532.

[364] Mukhanov, V., Feldman, H. A., and Brandenberger, R. H. 1992. *Phys. Rep.*, **215**, 203.

[365] Mukhanov, V. F. 2004. *Int. J. Theor. Phys.*, **43**, 669.

[366] Namikawa, T., Saito, S., and Taruya, A. 2010. *J. Cosmol. Astropart. Phys.*, **1012**, 027.

[367] Nir, Y. 2007. arXiv hep–ph/0702199.

[368] Nollett, K. M., and Holder, G. P. 2011. arXiv 1112.2683.
[369] Nötzold, D., and Raffelt, G. G. 1988. *Nucl. Phys. B*, **307**, 924.
[370] Novosyadlyj, B., *et al.* 2000. *Astron. Astrophys.*, **356**, 418.
[371] Okada, N., and Yasuda, O. 1997. *Int. J. Mod. Phys. A*, **12**, 3669.
[372] Okamoto, T., and Hu, W. 2003. *Phys. Rev. D*, **67**, 083002.
[373] Olive, K. A., Steigman, G., and Walker, T. P. 2000. *Phys. Rep.*, **333**, 389.
[374] Osipowicz, A., *et al.* 2001. arXiv hep–ex/0109033.
[375] Padmanabhan, T. 1993. *Structure Formation in the Universe.* Cambridge University Press.
[376] Pal, P. B., and Wolfenstein, L. 1982. *Phys. Rev. D*, **25**, 766.
[377] Pantaleone, J. T. 1992. *Phys. Lett. B*, **287**, 128.
[378] Pastor, S., Raffelt, G. G., and Semikoz, D.V. 2002. *Phys. Rev. D*, **65**, 053011.
[379] Pastor, S., Pinto, T., and Raffelt, G. G. 2009. *Phys. Rev. Lett.*, **102**, 241302.
[380] Pati, J. C., and Salam, A. 1974. *Phys. Rev. D*, **10**, 275.
[381] Peebles, P. J. E. 1968. *Astrophys. J.*, **153**, 1.
[382] Peebles, P. J. E. 1980. *The Large-Scale Structure of the Universe.* Princeton.
[383] Peimbert, M., Luridiana, V., and Peimbert, A. 2007. *Astrophys. J.*, **666**, 636.
[384] Percival, W. J., *et al.* 2010. *Mon. Not. Roy. Astron. Soc.*, **401**, 2148.
[385] Perlmutter, S., *et al.* 1999. *Astrophys. J.*, **517**, 565.
[386] Perotto, L., *et al.* 2006. *J. Cosmol. Astropart. Phys.*, **0610**, 013.
[387] Peskin, M. E., and Schroeder, D. V. 1995. *An Introduction to Quantum Field Theory.* Addison-Wesley.
[388] Petraki, K., and Kusenko, A. 2008. *Phys. Rev. D*, **77**, 065014.
[389] Pettini, M., *et al.* 2008. *Mon. Not. Roy. Astron. Soc.*, **391**, 1499.
[390] Pilaftsis, A. 1997. *Phys. Rev. D*, **56**, 5431.
[391] Pilaftsis, A., and Underwood, T. E. J. 2004. *Nucl. Phys. B*, **692**, 303.
[392] Pisanti, O., *et al.* 2008. *Comput. Phys. Commun.*, **178**, 956.
[393] Plümacher, M. 1997. *Z. Phys. C*, **74**, 549.
[394] Pogosian, D. Y., and Starobinsky, A. A. 1995. *Astrophys. J.*, **447**, 465.
[395] Polisensky, E., and Ricotti, M. 2011. *Phys. Rev. D*, **83**, 043506.
[396] Politzer, H. D. 1973. *Phys. Rev. Lett.*, **30**, 1346.
[397] Pontecorvo, B. 1957. *Sov. Phys. JETP*, **6**, 429.
[398] Pontecorvo, B. 1958. *Sov. Phys. JETP*, **7**, 172.
[399] Pontecorvo, B. 1968. *Sov. Phys. JETP*, **26**, 984–988.
[400] Porter, R. L., Ferland, G. J., and MacAdam, K. B. 2007. *Astrophys. J.*, **657**, 327.
[401] Pospelov, M., and Pradler, J. 2010. *Ann. Rev. Nucl. Part. Sci.*, **60**, 539.
[402] Prakash, M., Lattimer, J. M., Sawyer, R. F., and Volkas, R. R. 2001. *Ann. Rev. Nucl. Part. Sci.*, **51**, 295.
[403] Pritchard, J. R., and Loeb, A. 2012. *Rep. Prog. Phys.*, **75**, 086901.
[404] Pritchard, J. R., and Pierpaoli, E. 2008. *Phys. Rev. D*, **78**, 065009.
[405] Raffelt, G. G. 1999. *Phys. Rep.*, **320**, 319.
[406] Raffelt, G. G., Sigl, G., and Stodolsky, L. 1993. *Phys. Rev. Lett.*, **70**, 2363.
[407] Raha, N. C., and Mitra, B. 1991. *Phys. Rev. D*, **44**, 393.
[408] Reid, B. A., *et al.* 2010a. *Mon. Not. Roy. Astron. Soc.*, **404**, 60.
[409] Reid, B. A., Verde, L., Jimenez, R., and Mena, O. 2010b. *J. Cosmol. Astropart. Phys.*, **1001**, 003.
[410] Reines, F., Gurr, H. S., and Sobel, H. W. 1976. *Phys. Rev. Lett.*, **37**, 315.
[411] Riemer-Sørensen, S., *et al.* 2012. *Phys. Rev. D*, **85**, 081101.
[412] Riess, A. G., *et al.* 1998. *Astron. J.*, **116**, 1009.
[413] Riess, A. G., *et al.* 2009. *Astrophys. J.*, **699**, 539.

[414] Ringwald, A. 1988. *Phys. Lett. B*, **201**, 510.

[415] Ringwald, A., and Wong, Y. Y. Y. 2004. *J. Cosmol. Astropart. Phys.*, **0412**, 005.

[416] Riotto, A., and Trodden, M. 1999. *Ann. Rev. Nucl. Part. Sci.*, **49**, 35.

[417] Robertson, H. P. 1935. *Astrophys. J.*, **82**, 284.

[418] Rodejohann, W. 2011. *Int. J. Mod. Phys. E*, **20**, 1833.

[419] Ruchayskiy, O., and Ivashko, A. 2012. *J. Cosmol. Astropart. Phys.*, **1210**, 014.

[420] Sachs, R. K., and Wolfe, A. M. 1967. *Astrophys. J.*, **147**, 73.

[421] Saito, S., Takada, M., and Taruya, A. 2009. *Phys. Rev. D*, **80**, 083528.

[422] Sakharov, A. D. 1967. *Pisma Zh. Eksp. Teor. Fiz.*, **5**, 32.

[423] Salam, A. 1968. Proceedings of the 8th Nobel Symposium (Stockholm, Sweden, 1968).

[424] Salaris, M., Riello, M., Cassisi, S., and Piotto, G. 2004. *Astron. Astrophys.*, **420**, 911.

[425] Samuel, S. 1993. *Phys. Rev. D*, **48**, 1462.

[426] Sangiorgio, S. 2006. *Prog. Part. Nucl. Phys.*, **57**, 68.

[427] Sarkar, S. 1996. *Rept. Prog. Phys.*, **59**, 1493.

[428] Schechter, J., and Valle, J. W. F. 1980. *Phys. Rev. D*, **22**, 2227.

[429] Schechter, J., and Valle, J. W. F. 1981. *Phys. Rev. D*, **24**, 1883.

[430] Schechter, J., and Valle, J. W. F. 1982a. *Phys. Rev. D*, **25**, 2951.

[431] Schechter, J., and Valle, J. W. F. 1982b. *Phys. Rev. D*, **25**, 774.

[432] Schmidt, M. A. 2007. *Phys. Rev. D*, **76**, 073010.

[433] Schneider, A., Smith, R. E., Macciò, A. V., and Moore, B. 2012. *Mon. Not. Roy. Astron. Soc.*, **424**, 684.

[434] Schwarz, D. J., and Stuke, M. 2009. *J. Cosmol. Astropart. Phys.*, **0911**, 05.

[435] Schwetz, T., Tórtola, M. A., and Valle, J. W. F. 2011. *New J. Phys.*, **13**, 109401.

[436] Seager, S., Sasselov, S. S., and Scott, S. 1999. *Astrophys. J.*, **523**, L1.

[437] Sekiguchi, T., Ichikawa, K., Takahashi, T., and Greenhill, L. 2010. *J. Cosmol. Astropart. Phys.*, **1003**, 015.

[438] Seljak, U., and Zaldarriaga, M. 1996. *Astrophys. J.*, **469**, 437.

[439] Semikoz, V. B., Sokoloff, D. D., and Valle, J. W. F. 2009. *Phys. Rev. D*, **80**, 083510.

[440] Serebrov, A., *et al.* 2005. *Phys. Lett. B*, **605**, 72.

[441] Serpico, P. D., and Raffelt, G. G. 2005. *Phys. Rev. D*, **71**, 127301.

[442] Serpico, P. D., *et al.* 2004. *J. Cosmol. Astropart. Phys.*, **0412**, 010.

[443] Serra, P., *et al.* 2010. *Phys. Rev. D*, **81**, 043507.

[444] Seto, R. 1988. *AIP Conf. Proc.*, **176**, 957.

[445] Shaposhnikov, M. 2008. *J. High-Energy Phys.*, **0808**, 008.

[446] Shaposhnikov, M., and Tkachev, I. 2006. *Phys. Lett. B*, **639**, 414.

[447] Shi, X., Schramm, D. N., and Fields, B. D. 1993. *Phys. Rev. D*, **48**, 2563.

[448] Shi, X. D., and Fuller, G. M. 1999. *Phys. Rev. Lett.*, **82**, 2832.

[449] Shoji, M., and Komatsu, E. 2010. *Phys. Rev. D*, **81**, 123516.

[450] Shvartsman, V. F. 1969. *JETP Lett.*, **9**, 184.

[451] Shvartsman, V. F., *et al.* 1982. *JETP Lett.*, **36**, 277.

[452] Sigl, G., and Raffelt, G. G. 1993. *Nucl. Phys. B*, **406**, 423.

[453] Sigurdson, K., and Furlanetto, S. R. 2006. *Phys. Rev. Lett.*, **97**, 091301.

[454] Silk, J. 1968. *Astrophys. J.*, **151**, 459.

[455] Simha, V., and Steigman, G. 2008. *J. Cosmol. Astropart. Phys.*, **0808**, 011.

[456] Smith, M. S., Kawano, L. H., and Malaney, R. A. 1993. *Astrophys. J. Suppl.*, **85**, 219.

[457] Smith, P. F., and Lewin, J. D. 1983. *Phys. Lett. B*, **127**, 185.

[458] Smith, R. E., *et al.* 2003. *Mon. Not. Roy. Astron. Soc.*, **341**, 1311.

[459] Smooth, G. F., *et al.* 1992. *Astrophys. J.*, **396**, L1.
[460] Spite, F., and Spite, M. 1982. *Astron. Astrophys.*, **115**, 357.
[461] Springel, V. 2005. *Mon. Not. Roy. Astron. Soc.*, **364**, 1105.
[462] Starobinsky, A. A. 1979. *JETP Lett.*, **30**, 682.
[463] Starobinsky, A. A. 1980. *Phys. Lett. B*, **91**, 99.
[464] Starobinsky, A. A. 1982. *Phys. Lett. B*, **117**, 175.
[465] Steigman, G. 2007. *Ann. Rev. Nucl. Part. Sci.*, **57**, 463.
[466] Steigman, G., Schramm, D. N., and Gunn, J. E. 1977. *Phys. Lett. B*, **66**, 202.
[467] Steigman, G., Olive, K. A., and Schramm, D. N. 1979. *Phys. Rev. Lett.*, **43**, 239.
[468] Stodolsky, L. 1974. *Phys. Rev. Lett.*, **34**, 110.
[469] Stodolsky, L. 1987. *Phys. Rev. D*, **36**, 2273.
[470] Swanson, M. E. C., Percival, W. J., and Lahav, O. 2010. *Mon. Not. Roy. Astron. Soc.*, **409**, 1100.
[471] Szalay, A. S., and Marx, G. 1976. *Astron. Astrophys.*, **49**, 437.
[472] 't Hooft, G. 1976. *Phys. Rev. Lett.*, **37**, 8.
[473] Tegmark, M. 2005. *Phys. Scripta T*, **121**, 153.
[474] Tereno, I., *et al.* 2009. *Astron. Astrophys.*, **500**, 657.
[475] Thomas, S. A., Abdalla, F. B., and Lahav, O. 2010. *Phys. Rev. Lett.*, **105**, 031301.
[476] Toussaint, D., Treiman, S. B., Wilczek, F., and Zee, A. 1979. *Phys. Rev. D*, **19**, 1036.
[477] Tremaine, S., and Gunn, J. E. 1979. *Phys. Rev. Lett.*, **42**, 407.
[478] Trotta, R., and Hansen, S. H. 2004. *Phys. Rev. D*, **69**, 023509.
[479] Trotta, R., and Melchiorri, A. 2005. *Phys. Rev. Lett.*, **95**, 011305.
[480] Umehara, S., *et al.* 2008. *Phys. Rev. C*, **78**, 058501.
[481] Viel, M., *et al.* 2005. *Phys. Rev. D*, **71**, 063534.
[482] Viel, M., Haehnelt, M. G., and Springel, V. 2010. *J. Cosmol. Astropart. Phys.*, **1006**, 015.
[483] Viel, M., Markovic, K., Baldi, M., and Weller, J. 2012. *Mon. Not. Roy. Astron. Soc.*, **421**, 50.
[484] Vilain, P., *et al.* 1994. *Phys. Lett. B*, **335**, 246.
[485] Wagoner, R. V. 1969. *Astrophys. J. Suppl.*, **18**, 247.
[486] Wald, R. M. 1984. *General Relativity*. University of Chicago Press.
[487] Walker, A. G. 1936. *Proc. London Math. Soc.*, **42**, 90.
[488] Wang, J., and White, S. D. M. 2007. *Mon. Not. Roy. Astron. Soc.*, **380**, 93.
[489] Wang, S., *et al.* 2005. *Phys. Rev. Lett.*, **95**, 011302.
[490] Weinberg, S. 1962. *Phys. Rev.*, **128**, 1457.
[491] Weinberg, S. 1972. *Gravitation and Cosmology: Principles and Applications of the General Theory of Relativity*. Wiley.
[492] Weinberg, S. 1973. *Phys. Rev. Lett.*, **31**, 494.
[493] Weinberg, S. 1979. *Phys. Rev. Lett.*, **42**, 850.
[494] Weinberg, S. 1989. *Rev. Mod. Phys.*, **61**, 1.
[495] Weinberg, S. 1995, 1996. *The Quantum Theory of Fields*, Vols. 1, 2. Cambridge University Press.
[496] Weinberg, S. 2002. *Astrophys. J.*, **581**, 810.
[497] Weinberg, S. 2004. *Phys. Rev. D*, **69**, 023503.
[498] Weinberg, S. 2008. *Cosmology*. Oxford.
[499] Wetterich, C. 1981. *Nucl. Phys. B*, **187**, 343.
[500] Wolfenstein, L. 1978. *Phys. Rev. D*, **17**, 2369.
[501] Wong, W. Y., Moss, A., and Scott, D. 2007. *Mon. Not. Roy. Astron. Soc.*
[502] Wong, Y. Y. Y. 2002. *Phys. Rev. D*, **66**, 025015.

[503] Wong, Y. Y. Y. 2008. *J. Cosmol. Astropart. Phys.*, **0810**, 035.

[504] Wong, Y. Y. Y. 2011. *Ann. Rev. Nucl. Part. Sci.*, **61**, 69.

[505] Wood, B. E., *et al.* 2004. *Astrophys. J.*, **609**, 838.

[506] Xia, J. Q., *et al.* 2012. *J. Cosmol. Astropart. Phys.*, **1206**, 010.

[507] Yanagida, T. 1979. *Workshop on the Baryon Number of the Universe and Unified Theories, Tsukuba, Japan.*

[508] Zaldarriaga, M., and Harari, D. D. 1995. *Phys. Rev. D*, **52**, 3276.

[509] Zeldovich, Y., and Khlopov, M. 1981. *Sov. Phys. Usp.*, **24**, 755.

[510] Zeller, G. P., *et al.* 2002. *Phys. Rev. Lett.*, **88**, 091802.

Index

$M_{\rm GUT}$, 114
$SO(10)$ unified gauge theory, 114
$SU(5)$ unified gauge theory, 114
νMSM model, 25
σ_8, 331
$\Omega_{\rm C}$, 91
Ω_k, 64
$\Omega_{\rm M}$, 91
$\Omega_{\rm R}$, 64
Ω_γ, 91
Ω_Λ, 64
^{3}H β-decay, 45
   CNB detection, 356
^{4}He abundance analytic result, 167
^{4}He mass fraction Y_p, 168
^{163}Ho electron capture and CNB, 358
a_{lm} coefficients for photon temperature anisotropy,
   220
k_{nr}, 246
n/p ratio decoupling temperature, 170
^{3}He$^+$ spin flip transition, 177
PArthENoPE BBN code, 180

active–sterile neutrino oscillations, 160
adiabatic initial conditions, 210, 243
amplitude of primordial spectrum of perturbations,
   232
angular distance, 70
anisotropic stress or pressure, 203, 208, 243, 251, 266,
   343
anomaly in chiral gauge theories, 116
antimatter in the universe, 107
appearance experiments, 35
atmospheric neutrinos, 37

baryon acoustic oscillation, 268
baryon acoustic oscillations BAO, 289
baryon drag term, 285
baryon number violation in GUT theories, 115
baryon-to-photon density $\eta_{\rm B}$, 90, 108
baryon-to-photon ratio R, 206

BBN bounds on right-handed gauge bosons, 192
BBN equations, 169
BBN nuclear network, 173
BBN numerical codes, 168
Bjorken variable, 11
blue compact galaxies, 178
Boltzmann equation
   collisional, 73
   collisionless, 72
   perturbed, 205, 238, 240
bound on reheating temperature from BBN, 185
bounds on neutrino–scalar particle interactions,
   344

Cabibbo angle, 8
Cavendish torsion-balance, 350
charge conjugation, 19
charge exchange processes, 173
charged-current interaction, 5
chemical equilibrium condition, 76
Chern–Simons numbers, 117
Christoffel symbols, 57
CKM matrix, 8
cluster mass function, 331
CMB lensing, 327
CMB measurement of ^{4}He mass fraction, 179
CMB polarization modes, 234
coincidence problem, 93
cold relics, 72
collisional integral, 73
colour, 2
comoving coordinates, 55
comoving curvature perturbation, 212
comoving momentum, 58
comoving particle horizon, 66
conformal time, 56
continuity equation
   baryons, 206
   dark matter, 207
cosmic shear, 332
cosmic variance, 221

cosmological bounds on neutrino nonstandard
    interactions, 191
cosmological constant, 60
CP symmetry, 110
CP violation, 112
CP violation in the Standard Model, 9
CPT theorem, 110
critical density, 64
    today, 64
current–current weak interactions, 10
CVC hypothesis, 15

dark energy, 94
dark matter linear growing mode, 277
deceleration parameter, 65
decoherence term, 151
decoupling condition, 72
deuterium binding energy, 166
deuterium bottleneck, 167
deuterium capture, 173
diffusion length, 215, 229
Dirac mass, 16
disappearance experiments, 35
distribution function, 61
Dodelson–Widrow model, 164, 314, 345
Doppler term in CMB anisotropies, 226
drag depth, 285
dwarf irregular galaxies, 178

early integrated Sachs–Wolfe effect, 229
effective number of neutrino species N_{eff}
    bounds from BBN, 184
    bounds from CMB, 269
    bounds from LSS, 339
    definition, 103
    effect of neutrino chemical potential, 155
    effect of neutrino distribution distortion, 142
    effects on structure formation, 290
Einstein–de Sitter cosmology, 69
electroweak gauge bosons, 4
entropy relativistic degrees of freedom g_s, 82
equation of state, 60
equilibrium distributions, 76
equivalence wavenumber k_{eq}, 279, 284
Euler equation
    baryons, 206
    dark matter, 207

Fermi constant, 10
finite-temperature radiative corrections, 173
free-streaming
    horizon, 244, 246, 247
    scale, 244, 246, 247
Friedmann equation, 63
FRW metric, 55

gauge boson propagators, 9
gauge choice for perturbed metrics, 201

gauge symmetry, 2
Gell-Mann–Nishijima relation, 3
geodesic equation, 57, 202
Grand Unified Theories (GUT), 114
    baryogenesis, 115

Higgs vacuum expectation value, 6
Homestake experiment, 16
hot relics, 72
Hubble
    constant, 59
    law, 59
    parameter, 63

imperfect fluid, 251
inelasticity parameter, 11
inflaton quantum perturbations, 86
instantaneous decoupling limit, 72
    neutrinos, 100
    photons, 95
instanton, 118
integrated Sachs–Wolfe effect, 227
    early, 229, 232
    late, 230, 232
irreducible representation, 2
isocurvature initial conditions, 210, 240
    neutrino density mode, 241
    neutrino velocity mode, 241

Jarlskog invariant, 9, 112
Jeans length, 244

Kurie function, 45

last scattering surface
    neutrino, 100, 359
    photon, 95
left–right symmetric GUT models, 115
lensing potential, 327
lepton number, 18
lepton number parameters η_{ν_α}, 154
leptoquark gauge bosons, 114
light-to-mass bias parameter, 329
line-of-sight integral, 223
linear growth factor, 309
linear power spectrum, 274
Liouville operator, 72
luminosity distance, 67
Lyman alpha forest, 334

Majorana mass, 18
majoron model, 191
Mandelstam variable, 11
matter–radiation equality point, 91
MaVaN model, 344
Mészáros equation, 280
metallicity and BBN, 175
MSW effect, 34

N-body simulation
  particle mesh method, 318
  tree method, 318
N-body simulations
  massive neutrinos, 321
  warm dark matter, 323
neutral-current interaction, 5
neutrino (pseudo)scalar interactions, 191
neutrino capture processes, 354
neutrino charge radius, 190
neutrino decoupling, 135
neutrino decoupling temperature, 99
  using kinetic equation, 100
neutrino density matrix, 146
  evolution equation, 146
neutrino distribution distortion
  analytic estimate, 137
  flavour oscillation effects, 154
  numerical results, 140
neutrino drag effect, 261
neutrino effective potential
  charged current, 31
  neutral current, 31
  second-order contribution, 144
  self-interactions, 147
neutrino effective viscosity speed $c_{\rm vis}$, 252
neutrino electric dipole moment, 190
neutrino electromagnetic form factors, 189
neutrino flavour polarization vector
  effective magnetic field, 149
  equation of motion, 149
neutrino fraction f_ν, 274
neutrino magnetic dipole moment, 190
neutrino mass from β-decays, 43
neutrino mass splitting and power spectrum, 307
neutrino nonstandard interactions, 50
neutrino nonrelativistic transition redshift z_{nr}, 245
neutrino number density today, 103
neutrino oscillations, 26
neutrino refraction index, 32, 351
  in a polarized medium, 351
neutrino velocity dispersion (thermal velocity),
  246
neutrino wavelength today, 350
neutrino–dark matter interaction constraints, 344
neutrino–nucleon coherent scattering, 350
neutrinoless double β decay, 46
neutrinoless universe, 192
neutron captures, 173
neutron lifetime, 16, 171
neutron–proton mass difference, 167
Newton gauge, 201
nonlinear matter power spectrum, 317
nucleon weak rates, 12, 170
  Born limit, 16, 171
  finite-nucleon-mass corrections, 172
  finite-temperature radiative corrections, 171
  radiative corrections, 171

optical depth, 223
out-of-equilibrium decay scenario, 115, 121

PCAC hypothesis, 15
photon
  diffusion length, 216
  temperature perturbation Θ_γ, 204
physical momentum, 58
Planck mass, 60
plasmon decay into neutrinos, 190
PMNS mixing matrix, 17
Poisson equation, 207, 273
power spectrum, 211
  of matter, 273
  primordial, 213
proton and neutron stripping, 173
proton decay, 114

QAS and primordial ^{2}H measurement, 177
QED plasma corrections, 140

recombination, 95
redshift, 59
  of equality, 256
  of neutrino nonrelativistic transition, 245
redshift space distortion, 329
reduced cross section, 129
reheating stage after inflation, 87
reheating temperature $T_{\rm RH}$ after particle decay,
  125
reionization, 231
relative fluctuation of neutrino distribution, Ψ_ν, 239
relativistic degrees of freedom g_*, 87
relic abundance of massive particles, 89
repopulation function, 151
Ricci
  scalar, 60
  tensor, 60

Sachs–Wolfe
  effect, 226
  plateau, 228
Saha equation, 80
  nuclear statistical equilibrium, 166
  deuterium equilibrium abundance, 166
  recombination, 96
Sakharov conditions, 109
scalar perturbations, 202
scalar tilt index n_s, 213
scale-dependent growth factor and neutrino masses,
  309
Silk damping, 229, 285
solar neutrinos, 37
sound horizon, 214
sound speed, 209, 213, 252, 266, 342
spatial curvature parameter, 55
specific entropy, 82
  conservation of, 82

sphaleron, 118
Spite plateau, 180
standard candle, 67
Standard Model, 2
sterile neutrinos
　　as warm dark matter candidates,
　　　314
　　cosmological production mechanisms,
　　　160
　　effects on BBN, 193
　　effects on CMB, 271
　　effects on large-scale structures, 340
　　Lyman alpha bounds, 346
　　X-ray observation bounds, 347
Stodolsky effect, 353
stress–energy tensor, 60
　　in terms of distribution function, 61
　　perfect fluid, 60
　　perturbations, 203
survival probability, 28

temperature perturbation monopole $\Theta_{\gamma 0}$, 204
time reversal, 110
transfer functions, 213
triangle anomaly, 115
triple-α process, 173
type Ia supernovae, 93

unitarity of S matrix, 75, 111

visibility function, 223

warm dark matter, 312, 323, 344
weak-current form factors, 14
weak hypercharge, 2
weak lensing, 332
WIMP, 90

Yukawa couplings, 4

Zel'dovich velocity, 319

Printed in the United States
By Bookmasters